CONTEMPORARY MATHEMATICS

713

Representations of Lie Algebras, Quantum Groups and Related Topics

AMS Special Session
Representations of Lie Algebras, Quantum Groups
and Related Topics
November 12–13, 2016
North Carolina State University, Raleigh, North Carolina

Naihuan Jing
Kailash C. Misra
Editors

AMERICAN
MATHEMATICAL
SOCIETY

2010 *Mathematics Subject Classification.* Primary 05E10, 11B39, 14M15, 17A32, 17B10, 17B37, 17B67, 17B69, 20G05, 81R50.

Library of Congress Cataloging-in-Publication Data

Names: Jing, Naihuan, editor. | Misra, Kailash C., 1954- editor. | AMS Special Session on Representations of Lie Algebras, Quantum Groups, and Related Topics (2016: Raleigh, N.C.)

Title: Representations of Lie algebras, quantum groups, and related topics: AMS Special Session on Representations of Lie Algebras, Quantum Groups, and Related Topics, November 12–13, 2016, North Carolina State University, Raleigh, North Carolina / Naihuan Jing, Kailash C. Misra, editors.

Description: Providence, Rhode Island: American Mathematical Society, [2018] | Series: Contemporary mathematics; volume 713 | Includes bibliographical references.

Identifiers: LCCN 2018005031 | ISBN 9781470436964 (alk. paper)

Subjects: LCSH: Representations of Lie algebras–Congresses. | Algebra–Congresses. | AMS: Combinatorics – Algebraic combinatorics – Combinatorial aspects of representation theory. msc | Number theory – Sequences and sets – Fibonacci and Lucas numbers and polynomials and generalizations. msc | Algebraic geometry – Special varieties – Grassmannians, Schubert varieties, flag manifolds. msc | Nonassociative rings and algebras – General nonassociative rings – Leibniz algebras. msc | Nonassociative rings and algebras – Lie algebras and Lie superalgebras – Representations, algebraic theory (weights). msc | Nonassociative rings and algebras – Lie algebras and Lie superalgebras – Quantum groups (quantized enveloping algebras) and related deformations. msc | Nonassociative rings and algebras – Lie algebras and Lie superalgebras – Kac-Moody (super)algebras; extended affine Lie algebras; toroidal Lie algebras. msc | Nonassociative rings and algebras – Lie algebras and Lie superalgebras – Vertex operators; vertex operator algebras and related structures. msc | Group theory and generalizations – Linear algebraic groups and related topics – Representation theory. msc | Quantum theory – Groups and algebras in quantum theory – Quantum groups and related algebraic methods. msc

Classification: LCC QA252.3 .R47 2018 | DDC 512/.482–dc23

LC record available at https://lccn.loc.gov/2018005031

Contemporary Mathematics ISSN: 0271-4132 (print); ISSN: 1098-3627 (online)

DOI: https://doi.org/10.1090/conm/713

Color graphic policy. Any graphics created in color will be rendered in grayscale for the printed version unless color printing is authorized by the Publisher. In general, color graphics will appear in color in the online version.

∞ The paper used in this book is acid-free and falls within the guidelines
established to ensure permanence and durability.
Visit the AMS home page at `https://www.ams.org/`

10 9 8 7 6 5 4 3 2 1   23 22 21 20 19 18

Contents

Preface

Representation theory of Lie algebras and related algebraic structures such as quantum groups, vertex algebras, Lie superalgebras, Leibniz algebras, and algebraic groups is an important area of current research due to their interactions with other areas of mathematics and mathematical physics. At the Fall Southeastern Sectional Meeting of the American Mathematical Society in Raleigh, North Carolina, held from November 12–13, 2016, a special session on "Representations of Lie Algebras, Quantum Groups and Related Topics" was organized by the editors. This was a mathematically robust and engaging session in which 22 senior and junior researchers presented their work on a variety of research topics. All the speakers were invited to contribute. Eleven speakers at this special session have contributed to this volume. We believe that the diversity of topics covered in this volume constitute a valuable contribution to the literature in representation theory. This volume is a useful reference for both established mathematicians and junior researchers in representation theory.

We would like to thank all of the speakers at this special session, especially those who have contributed to this proceedings volume. We are grateful to all of the referees for their thoughtful and timely reviews. We would also like to thank Sergei Gelfand for suggesting that we put this proceedings together and Christine Thivierge for her guidance and patience through the editorial process.

The Editors

Contemporary Mathematics
Volume **713**, 2018
https://doi.org/10.1090/conm/713/14326

The two bosonizations of the CKP hierarchy: Overview and character identities

Iana I. Anguelova

Abstract. We discuss the Hirota bilinear equation for the CKP hierarchy introduced by Date, Jimbo, Kashiwara, and Miwa in [J. Phys. Soc. Jpn. 50 (1981), pp. 3813–3818], and its algebraic properties. We review in parallel the two bosonizations of the CKP hierarchy: one arising from a twisted Heisenberg algebra (see van de Leur, Orlov and Shiota in [Symmetry Integrability Geom. Methods Appl. 8 (2012), 28 pp.]), and the second from an untwisted Heisenberg algebra (see Anguelova in [J. Math. Phys. 58 (2017), 20 pp.]). In particular, we recount the respective decompositions into irreducible Heisenberg modules and the (twisted) fermionic structures of the spaces spanned by the highest weight vectors under the two Heisenberg actions. We show that the two bosonizations give rise to four different diagonalizable grading operators on the CKP Fock space, not all of them commuting among each other. We compute the various graded dimensions related to these four grading operators. We prove a sum-vs-product identity relating the bosonic vs fermionic descriptions under the untwisted Heisenberg action, utilizing the charge and the degree grading operators. As a corollary, the resulting identities relate the CKP hierarchy, the Dyson crank of a partition and the Hammond-Lewis birank of a distinct integer bipartition.

1. Introduction

There are many ties between the area of integrable hierarchies on one side, and vertex algebras and conformal field theory on the other. The fruitful relation between those two areas started with the realization that the Hirota bilinear equation associated to the Kadomtsev-Petviashvili (KP) hierarchy can be concisely written in terms of vertex operators as follows ([**DJKM81a**], [**KR87**], [**Kac98**]):

$$Res_z\Big(\psi^+(z)\otimes\psi^-(z)\Big)(\tau\otimes\tau)=0,$$

where $\psi^+(z)$ and $\psi^-(z)$ are two fermionic fields with only nontrivial Operator Product Expansion (OPE)

$$\psi^+(z)\psi^-(w)\sim\frac{1}{z-w}\sim\psi^-(z)\psi^+(w).$$

2010 *Mathematics Subject Classification.* Primary 81T40, 17B69, 17B68, 81R10.
Key words and phrases. CKP hierarchy, bosonization, vertex algebras, symplectic fermions, partitions, character identities.

The vector τ on which these vertex operators $\psi^+(z)$ and $\psi^-(z)$ act belongs to (a subspace of) the Fock space which the Clifford algebra modes of these vertex operator create from the vacuum vector $|0\rangle$. The vector τ is often referred to as "the τ function".

As is well known, the advantage of this purely algebraic form of the Hirota equation (instead of the original Hirota equation, [**Hir04**]), is twofold: first, it describes the entire hierarchy, not just the KP equation itself. And second, one can easily prove that this KP Hirota equation commutes with the action of the a_∞ Lie algebra ([**DJKM81a**], [**KR87**], [**MJD00**]), and as a result the solutions, i.e., the τ functions of the KP hierarchy, belong to the orbit

$$GL_\infty|0\rangle.$$

Bosonization is the process of writing the vertex operators (in the KP case $\psi^+(z)$ and $\psi^-(z)$) in terms of exponentiated boson fields. It is a necessary process if one is to write the purely algebraic form of the Hirota equation into a hierarchy of actual differential equations, including translating the Fock space element τ into a function in the differential equation sense, over an infinite set of variables $x_1, x_2, \ldots, x_n, \ldots$. As is well known, the bosonization process in the case of the KP hierarchy resulted in the boson-fermion correspondence, a vertex algebra isomorphism between the charged free fermions super vertex algebra and the lattice super vertex algebra of the rank one odd lattice (see e.g. [**Kac98**]).

In continuation of their work on the KP hierarchy, Date, Jimbo, Kashiwara and Miwa introduced two new hierarchies related to the KP hierarchy: the BKP and the CKP hierarchies ([**DJKM82**] and [**DJKM81b**]). These were initially defined as reductions from the KP hierarchy, by assuming conditions on the pseudo-differential operator L in the Lax form used to define those hierarchies. For both of them Date, Jimbo, Kashiwara and Miwa suggested a Hirota bilinear equation, i.e., operator approach. The bosonization of the BKP case proceeded similarly to the KP case, and from the point of view of vertex algebras the bosonization of the BKP hierarchy resulted in the boson-fermion correspondence of type B ([**DJKM82**], [**You89**]), which was later interpreted as an isomorphism of certain twisted vertex (chiral) algebras ([**Ang13a**], [**Ang13b**]).

For the CKP hierarchy, Date, Jimbo, Kashiwara and Miwa suggested in [**DJKM81b**] the following Hirota equation:

$$Res_z\big(\chi(z)\otimes\chi(-z)\big)(\tau\otimes\tau)=0, \tag{1.1}$$

where the field $\chi(z)$ has Operator Product Expansion (OPE)

$$\chi(z)\chi(w)\sim\frac{1}{z+w}.$$

If one is to rewrite this purely algebraic Hirota equation into a hierarchy of differential equations, once again one needs the process of bosonization. There are 3 stages to any bosonization:

(1) Construct a bosonic Heisenberg current from the generating fields, hence obtaining a field representation of the Heisenberg algebra on the Fock space;
(2) Decompose the Fock space into irreducible Heisenberg modules and determine the structure of the space spanned by the highest weight vectors under the Heisenberg action;

(3) Use this decomposition to express the original generating fields in terms of exponential boson fields.

The CKP hierarchy offered its first surprise among many when it was discovered that not one, but two different bosonizations are possible. The first one was suggested by the original authors of the CKP hierarchy, Date, Jimbo, Kashiwara and Miwa ([**DJKM81b**]): they suggested a twisted Heisenberg field, but did not complete the bosonization following from it. In [**vOS12**] van de Leur, Orlov and Shiota completed this twisted bosonization and derived some applications of it. The second bosonization was made possible when the CKP hierarchy was related to the $\beta - \gamma$ system of conformal field theory (also called the boson ghost system), see [**Ang15**]. As a result of this relation, a second, untwisted Heisenberg current was constructed from the generating field $\chi(z)$. This opened the process of the second bosonization of the CKP hierarchy, which was completed in [**Ang17**].

In this conference proceedings, we start by discussing two important, albeit surprising, properties of the algebraic Hirota equation (1.1) and its Fock space: that there are no finite sum solutions (Lemma 2.3), but that there are series solutions in the series completion of the Fock space (Proposition 2.5 and its corollary). Specifically, as the name CKP suggests, Proposition 2.5 shows that the Hirota equation (1.1) commutes with the action of the c_∞ Lie algebra (this was remarked upon and used in [**vOS12**], but without proof). Next, we review the two bosonizations of the CKP hierarchy in parallel, by addressing each of the three stages above. In particular, we discuss the Heisenberg decompositions, as well as the structure of the corresponding vector spaces spanned by the highest weight vectors for the two Heisenberg actions. In the untwisted case, the vector space spanned by the highest weight vectors has a structure equivalent to the symplectic fermion vertex algebra, as we showed in [**Ang17**]. In the twisted case, although in [**vOS12**] the authors don't use the language of vertex algebras, their calculations show that the vector space spanned by the highest weight vectors has a structure of a twisted fermion vertex algebra (in the sense of [**ACJ14**]), see Theorem 2.13 and Corollary 2.14. In Section II we also recount two of the gradings we will use in Section III: the charge grading, by the 0-mode $h^{\mathbb{Z}}_0$ of the untwisted Heisenberg field, (2.15), and the degree grading, by the 0-mode L_0 of one of the Virasoro fields, (2.18).

In Section III we introduce two new grading operators L^t_0 and L^h_0, derived from two of the other Virasoro fields, $L^\chi(z)$ ((3.3)) and $L^h(z^2)$ ((3.6)). We show that L^t_0 doesn't commute with the grading operators $h^{\mathbb{Z}}_0$ and L^h_0 (Lemma 3.2). L^t_0 commutes with L_0 though, and we derive the graded dimension with respect to these two gradings, see Proposition 3.4. The other three grading operators ($h^{\mathbb{Z}}_0$, L_0 and L^h_0) do commute among each other, and thus a three-parameter graded dimension can be formed. In order to calculate this three-parameter graded dimension we use the concept of Hammond-Lewis birank of a bipartition ([**HL04**], [**Gar10**]). Based on Corollary 2.14, we show that there is a one-to-one correspondence between the set $\mathfrak{P}_{tdo}$ of distinct partitions with a triangular part describing the Heisenberg decomposition in the untwisted case, Theorem 2.9, and the set of distinct integer bipartitions $\mathfrak{BP}_{DI}$, see Proposition 3.13. The Hammond-Lewis birank is in fact the charge of the highest weight vector uniquely assigned to such distinct integer bipartition. This one-to-one correspondence is not an isometry (it does not preserve the weights), but we derive a formula for the degree of the highest weight vector assigned to a given distinct integer bipartition, which in turn relates the two weights, see (3.16). This formula allows us to calculate the three-parameter graded dimension, Theorem 3.15.

An important consequence of any bosonization is that by calculating the graded dimensions on both the fermionic and the bosonic sides of the correspondence one can obtain identities relating certain product formulas to certain sum formulas. Such a sum-vs-product identity perfectly illustrates the equality between the fermionic side (the product formulas) and the bosonic side (the sum formulas). In the classical boson-fermion correspondence (of type A) the sum-vs-product identity relating the two sides is the Jacobi triple product identity, as was proved in [**Kac98**]. The sum-vs-product identity for the boson-fermion correspondence of type D-A is also the Jacobi triple product identity, in a slightly different form, as shown in [**Ang14**]. Here, as is typical for the CKP quirks, this identity is complicated by the fact that the degree operator L_0, which is the most natural grading operator to use for the CKP hierarchy, doesn't act as uniformly on the symplectic fermion side with which the highest weight vectors space F_χ^{hwv} identifies. Also, in the CKP case, the "fermionic side" is not purely fermionic, but instead one can view it as a fermion times a boson (see Corollary 2.14) Nevertheless, we derive the relevant, if more complicated, sum-vs-product formula for the CKP correspondence in Theorem 3.16. The identity that follows is somewhat surprising, see comment after Theorem 3.16, but we show in the Appendix that it can also be obtained as a specialization of the Ramanujan Psi summation formula (see e.g. [**HA40**], [**And69**], [**AA78**]). The graded dimension of the highest weight vectors space F_χ^{hwv} connects the CKP hierarchy to the Dyson's crank of a partition ([**AG88**]), via its generating function. A certain sum of the number of partitions with a given Dyson crank allows us to answer as to the number of Heisenberg highest weight vectors with given degree and charge. Finally, Corollary 3.20 gives a quadruple identity relating in one of the equalities the Dyson crank with the Hammond-Lewis birank.

2. The CKP hierarchy and its two bosonizations: overview

In [**DJKM81b**] Date, Jimbo, Kashiwara and Miwa suggested the following Hirota equation associated with the CKP hierarchy:

$$Res_z\big(\chi(z)\otimes\chi(-z)\big)(\tau\otimes\tau)=0, \tag{2.1}$$

where the twisted neutral boson field $\chi(z)$, indexed as

$$\chi(z)=\sum_{n\in\mathbb{Z}+1/2}\chi_n z^{-n-1/2}, \tag{2.2}$$

has OPE

$$\chi(z)\chi(w)\sim\frac{1}{z+w}. \tag{2.3}$$

In terms of commutation relations for the modes χ_n, $n\in\mathbb{Z}+1/2$, this OPE is equivalent to

$$[\chi_m,\chi_n]=(-1)^{m-\frac{1}{2}}\delta_{m,-n}1. \tag{2.4}$$

The modes χ_n, $n\in\mathbb{Z}+1/2$ form a Lie algebra which we denote by L_χ. The tau function τ in the above equation (2.1) is an element of (a completion of) the Fock space F_χ:

DEFINITION 2.1. Denote by F_χ the induced vacuum module of L_χ with vacuum vector $|0\rangle$, such that for $n>0$ we have $\chi_n|0\rangle=0$. The vector space F_χ has a basis

$$\{|0\rangle,\ (\chi_{-j_k})^{m_k}\dots(\chi_{-j_2})^{m_2}(\chi_{-j_1})^{m_1}|0\rangle \mid j_k>\dots>j_2>j_1>0,\quad j_i\in\mathbb{Z}+\frac{1}{2},\ m_i>0, m_i\in\mathbb{Z},\ i=1,2,\dots,k\}. \tag{2.5}$$

DEFINITION 2.2. Define the Hirota operator by

$$S^C=Res_z\ \chi(z)\otimes\chi(-z) \tag{2.6}$$

In modes S^C translates to

$$S^C=\sum_{n\in\mathbb{Z}+1/2}(-1)^{n-\frac{1}{2}}\chi_n\otimes\chi_{-n}=\chi_{\frac{1}{2}}\otimes\chi_{-\frac{1}{2}}-\chi_{-\frac{1}{2}}\otimes\chi_{\frac{1}{2}}-\chi_{\frac{3}{2}}\otimes\chi_{-\frac{3}{2}}+\chi_{-\frac{3}{2}}\otimes\chi_{\frac{3}{2}}+\dots. \tag{2.7}$$

One of the first surprises of the CKP hierarchy and its Fock space is the fact that there are no actual elements of F_χ, besides the vacuum vector $|0\rangle$, that solve the algebraic Hirota equation:

LEMMA 2.3. *If $v\in F_\chi$ solves the Hirota equation* (2.1), $Res_z\big(\chi(z)\otimes\chi(-z)\big)(v\otimes v)=0$, *then $v=|0\rangle$.*

PROOF. Let $v\in F_\chi$, and let $v\neq|0\rangle$. Then v is a sum of monomials of the form (2.5). Let $N>0$, $N\in\mathbb{Z}+1/2$ be the largest N such that there is a monomial among the summands of v which contains χ_{-N} (i.e., $-N$ is the lowest index among all the indexes of χ_j present in v). Then we can write v in the form

$$v=\chi_{-N}^mP_m\big(\chi_{-j_k},\dots\chi_{-j_2},\chi_{-j_1}\big)|0\rangle+\dots\chi_{-N}P_1\big(\chi_{-j_k},\dots\chi_{-j_2},\chi_{-j_1}\big)|0\rangle + P_0\big(\chi_{-j_k},\dots\chi_{-j_2},\chi_{-j_1}\big)|0\rangle,$$

where $P_m,\dots,P_1,P_0$ are polynomials in the variables $\chi_{-j_k},\dots\chi_{-j_2},\chi_{-j_1}$ with $j_k,\dots j_2,j_1$ all strictly lower than N, $P_m\neq0$, and $m\geq1$. Then we have

$$\begin{aligned}
S^C(v\otimes v)&=\sum_{n\in\mathbb{Z}+1/2}(-1)^{n-\frac{1}{2}}\left(\chi_n\otimes\chi_{-n}\right)v\otimes v\\
&=(-1)^{N-\frac{1}{2}}\big(\chi_N\otimes\chi_{-N}-\chi_{-N}\otimes\chi_N\big)v\otimes v+\left(\sum_{n=-N+1}^{N-1}(-1)^{n-\frac{1}{2}}\chi_n\otimes\chi_{-n}\right)v\otimes v\\
&=(-1)^{N-\frac{1}{2}}\Big(\chi_{-N}^{m-1}P_m\big(\chi_{-j_k},\dots,\chi_{-j_1}\big)|0\rangle+\dots P_1|0\rangle\Big)\\
&\quad\otimes\Big(\chi_{-N}^{m+1}P_m\big(\chi_{-j_k},\dots,\chi_{-j_1}\big)|0\rangle+\dots+\chi_{-N}P_0|0\rangle\Big)\\
&\quad-(-1)^{N-\frac{1}{2}}\Big(\chi_{-N}^{m+1}P_m\big(\chi_{-j_k},\dots,\chi_{-j_1}\big)|0\rangle+\dots+\chi_{-N}P_0|0\rangle\Big)\\
&\quad\otimes\Big(\chi_{-N}^{m-1}P_m\big(\chi_{-j_k},\dots,\chi_{-j_1}\big)|0\rangle+\dots P_1|0\rangle\Big)\\
&\quad+\left(\sum_{n=-N+1}^{N-1}(-1)^{n-\frac{1}{2}}\chi_n\otimes\chi_{-n}\right)v\otimes v.
\end{aligned}$$

Since all the $j_k,\dots,j_2,j_1$ are strictly lower than N, the sum $\left(\sum_{n=-N+1}^{N-1}(-1)^{n-\frac{1}{2}}\chi_n\otimes\chi_{-n}\right)v\otimes v$ will have terms that contain at most χ_{-N}^m in each of the tensor products. Thus there is no other similar term to cancel the nonzero $\chi_{-N}^{m+1}P_m(\chi_{-j_k},\dots\chi_{-j_2},\chi_{-j_1})|0\rangle\otimes\chi_{-N}^{m-1}P_m(\chi_{-j_k},\dots\chi_{-j_2},\chi_{-j_1})|0\rangle$. Hence the sum of tensor products above is clearly nonzero. □

This shows that there are no finite-sum solutions, in contrast to the KP case where every monomial in the charged free fermion Fock space is actually a solution to the corresponding KP Hirota equation. Thus one has to necessarily go to a completion $\widetilde{F_\chi}$ of F_χ, where one considers series of monomials instead of finite sums. Since F_χ can be viewed as isomorphic to a polynomial algebra, F_χ is dense in such a completion $\widetilde{F_\chi}$. On the other hand, in $\widetilde{F_\chi}$ we expect to have many solutions, due to the following representation theory considerations. As is known from e.g. [**KWY98**], [**Wan99**], [**ACJ14**], the field $\chi(z)$ is related to the double-infinite rank Lie algebra c_∞. Recall (see [**Kac90**]), the Lie algebra c_∞ is defined via the Lie algebra $\bar{a}_\infty$ of infinite matrices of the form

$$\bar{a}_\infty=\{(a_{ij})|\ i,j\in\mathbb{Z},\ a_{ij}=0\ \text{for}|i-j|\gg 0\}. \tag{2.8}$$

As usual denote the elementary matrices by E_{ij}.

The algebra $\bar{c}_\infty$ is the subalgebra of $\bar{a}_\infty$ consisting of the infinite matrices preserving the bilinear form $(v_i;v_j)=(-1)^i\delta_{i,1-j}$, i.e.,

$$\bar{c}_\infty=\{(a_{ij})\in\bar{a}_\infty|\ a_{ij}=(-1)^{i+j-1}a_{1-j,1-i}\}. \tag{2.9}$$

The algebra c_∞ is the central extension of $\bar{c}_\infty$ by a central element c, $c_\infty=\bar{c}_\infty\oplus\mathbb{C}c$, with cocycle C given by

$$\begin{aligned}&C(E_{ij},E_{ji})=-C(E_{ji},E_{ij})=1,\quad \text{if } i\le 0,\ j\ge 1\\ &C(E_{ij},E_{kl})=0\quad \text{in all other cases.}\end{aligned}$$

The commutation relations for the elementary matrices in the centrally extended algebra are

$$[E_{ij},E_{kl}]=\delta_{jk}E_{il}-\delta_{li}E_{kj}+C(E_{ij},E_{kl})c.$$

The elementary matrices though are not in c_∞, instead the generators for the algebra c_∞ can be written in terms of these elementary matrices as:

$$\{(-1)^jE_{i,j}-(-1)^iE_{1-j,1-i},\ i,j\in\mathbb{Z};\ \text{and}\ c\}.$$

We can arrange the non-central generators in a generating series

$$E^C(z,w)=\sum_{i,j\in\mathbb{Z}}((-1)^jE_{ij}-(-1)^iE_{1-j,1-i})z^{i-1}w^{-j}. \tag{2.10}$$

The generating series $E^C(z,w)$ obeys the relations: $E^C(z,w)=E^C(w,z)$ (and thus one can view it as bosonic), and

$$\begin{aligned}[E^C(z_1,w_1),E^C(z_2,w_2)]&=E^C(z_1,w_2)\delta(z_2+w_1)-E^C(z_2,w_1)\delta(z_1+w_2)\\&\quad-E^C(w_2,w_1)\delta(z_1+z_2)+E^C(z_1,z_2)\delta(w_2+w_1)\\&+2\iota_{z_1,w_2}\frac{1}{z_1+w_2}\iota_{w_1,z_2}\frac{1}{w_1+z_2}c-2\iota_{w_2,z_1}\frac{1}{w_2+z_1}\iota_{z_2,w_1}\frac{1}{z_2+w_1}c\\&+2\iota_{z_1,z_2}\frac{1}{z_1+z_2}\iota_{w_1,w_2}\frac{1}{w_1+w_2}c-2\iota_{z_2,z_1}\frac{1}{z_2+z_1}\iota_{w_2,w_1}\frac{1}{w_2+w_1}c.\end{aligned}$$

Hence we can show that

PROPOSITION 2.4. *The assignment* $E(z,w)\to -:\chi(z)\chi(w):,\quad c\to-\frac{1}{2}Id_{F_C}$ *gives a representation of the Lie algebra* c_∞ *on the Fock space* F_χ

Note that this is an important correction from [**ACJ14**], Proposition 6.2. In particular, the central charge is negative, which is very important, as it implies that the module F_χ is not an integrable module for the action of c_∞. This in itself implies that we will need to consider a series completion of F_χ for the purposes of exponentiating the action of c_∞ to the corresponding group; which in light of Lemma 2.3 is entirely not surprising.

PROPOSITION 2.5. *The Hirota operator $S^C = Res_z\ \chi(z) \otimes \chi(-z)$ commutes with the action of c_∞ on the space $F_\chi \otimes F_\chi$, i.e.,*

$$\left[E^C(z_1, w_1) \otimes 1 + 1 \otimes E^C(z_1, w_1),\ S^C\right] = 0,$$

where for ease of notation we identify $E(z_1, w_1)$ with its representation $-:\chi(z_1)\chi(w_1):$ on F_χ. In addition,

$$S^C\left(|0\rangle \otimes |0\rangle\right) = 0.$$

A statement similar to this proposition is mentioned in [**vOS12**], but without proof.

PROOF. The second property $S^C\left(|0\rangle \otimes |0\rangle\right) = 0$ is trivially satisfied, as the modes representation (2.7) of the Hirota operator shows, since for $n > 0$ we have $\chi_n|0\rangle = 0$, and for $n < 0$ we have $\chi_{-n}|0\rangle = 0$.

To prove the first property, we will use the formal delta function notation (the formal delta-function at $z = w$, for $\lambda = 1$, is defined by e.g. [**Kac98**], it is defined for general λ in [**ACJ14**]):

$$\delta(z; \lambda w) := \sum_{n \in \mathbb{Z}} \lambda^{-n-1} z^n w^{-n-1}.$$

By abuse of notation one sometimes writes $\delta(z - \lambda w)$, even though $\delta(z; \lambda w)$ depends on two formal variables z and w, and a parameter λ. This is especially confusing for $\lambda = -1$, as then we have the rather peculiar fact that $\delta(z_1 + z_2) = \delta(z_1; -z_2) = -\delta(z_2 + z_1) = -\delta(z_2; -z_1)$.

We calculate by Wick's Theorem

$$\begin{aligned}
&[E^C(z_1, w_1) \otimes 1 + 1 \otimes E^C(z_1, w_1),\ S^C] \\
&\quad = Res_z\Big([- : \chi(z_1)\chi(w_1) : ,\ \chi(z)] \otimes \chi(-z)\Big) \\
&\quad + Res_z\Big(\chi(z) \otimes [- : \chi(z_1)\chi(w_1) : ,\ \chi(-z)]\Big) \\
&\quad = -Res_z\Big(\delta(z_1; -z)\chi(w_1) \otimes \chi(-z)\Big) - Res_z\Big(\delta(w_1; -z)\chi(z_1) \otimes \chi(-z)\Big) \\
&\quad - Res_z\Big(\chi(z) \otimes \delta(z_1; z)\chi(w_1)\Big) - Res_z\Big(\chi(z) \otimes \delta(w_1; z)\chi(z_1)\Big) \\
&\quad = -Res_z\Big(\chi(w_1) \otimes \delta(z_1; -z)\chi(-z)\Big) - Res_z\Big(\chi(z_1) \otimes \delta(w_1; -z)\chi(-z)\Big) \\
&\quad - Res_z\Big(\delta(z_1; z)\chi(z) \otimes \chi(w_1)\Big) - Res_z\Big(\delta(w_1; z)\chi(z) \otimes \chi(z_1)\Big)
\end{aligned}$$

Now we use the properties of the delta functions established in [**ACJ14**], in particular that

$$\mathrm{Res}_z f(z)\delta(z; \lambda w) = f(\lambda w),$$

as well as the above-mentioned fact that $\delta(z_1; -z_2) = -\delta(z_2; -z_1)$, but $\delta(z_1; z_2) = \delta(z_2; z_1)$, and we get

$$[E^C(z_1, w_1) \otimes 1 + 1 \otimes E^C(z_1, w_1),\ S^C]$$
$$= \chi(w_1) \otimes \chi(z_1) + \chi(z_1) \otimes \chi(w_1) - \chi(z_1) \otimes \chi(w_1) - \chi(w_1) \otimes \chi(z_1) = 0.$$

□

This proposition ensures that when we exponentiate the action of c_∞ to a corresponding group action on the series completion $\widetilde{F_\chi}$ of F_χ, each element of the resulting group orbit of the vacuum vector $|0\rangle$ will be a solution to the Hirota equation (2.1):

COROLLARY 2.6. *Let* $\tau = \exp\left(-\sum c_{mn} : \chi_m \chi_n :\right)|0\rangle \in \widetilde{F_\chi}$. *Then* τ *is a solution of the Hirota equation* (2.1), *i.e.,*

$$Res_z\big(\chi(z) \otimes \chi(-z)\big)(\tau \otimes \tau) = 0.$$

Thus the Hirota equation (2.1) is guaranteed to have solutions in $\widetilde{F_\chi}$. In order to write such solutions as τ functions solving a hierarchy of actual differential equations, we need to bosonize the purely algebraic Hirota equation. Any bosonization has three natural stages:

(1) Construct a bosonic Heisenberg current from the generating fields, hence obtaining a field representation of the Heisenberg algebra on the Fock space;
(2) Decompose the Fock space into irreducible Heisenberg modules and determine the structure of the space spanned by the highest weight vectors under the Heisenberg action;
(3) Use this decomposition to express the original generating fields in terms of exponential boson fields.

The surprise that not one, but two different bosonizations exist in the CKP case, stems from the fact that there is not one, but two different Heisenberg fields generated by the field $\chi(z)$ and its descendant field $\chi(-z)$. The first, twisted, Heisenberg representation was suggested in the original paper [**DJKM81b**] introducing the CKP hierarchy, and the bosonization initiated by the twisted Heisenberg current was studied in [**vOS12**]. For the purposes of exposition, in this conference proceedings we will summarize their results regarding the twisted bosonization, as well as our own. The existence of a second, untwisted Heisenberg field was established in [**Ang15**], and the bosonization initiated by the untwisted Heisenberg current was completed in [**Ang17**]. The following proposition summarizes the construction of the two Heisenberg fields from the generating field $\chi(z)$:

PROPOSITION 2.7. *(*[**Ang15**]*) I. Let*

$$h_\chi^{\mathbb{Z}+1/2}(z) = \frac{1}{2} : \chi(z)\chi(-z) : .$$

We have $h_\chi^{\mathbb{Z}+1/2}(-z) = h_\chi^{\mathbb{Z}+1/2}(z)$, *and we index* $h_\chi^{\mathbb{Z}+1/2}(z)$ *as* $h_\chi^{\mathbb{Z}+1/2}(z) = \sum_{n\in\mathbb{Z}+1/2} h_n^t z^{-2n-1}$. *The field* $h_\chi^{\mathbb{Z}+1/2}(z)$ *has OPE with itself given by:*

$$h_\chi^{\mathbb{Z}+1/2}(z) h_\chi^{\mathbb{Z}+1/2}(w) \sim -\frac{z^2 + w^2}{2(z^2 - w^2)^2} \sim -\frac{1}{4}\frac{1}{(z-w)^2} - \frac{1}{4}\frac{1}{(z+w)^2}, \tag{2.11}$$

and its modes, h^t_n, $n\in\mathbb{Z}+1/2$, generate a ***twisted*** *Heisenberg algebra $\mathcal{H}_{\mathbb{Z}+1/2}$ with relations*
$[h^t_m, h^t_n] = -m\delta_{m+n,0}1,\ \ m,n\in\mathbb{Z}+1/2.$
II. Let

$$h^{\mathbb{Z}}_{\chi}(z)=\frac{1}{4z}\left(:\chi(z)\chi(z):-:\chi(-z)\chi(-z):\right).$$

We have $h^{\mathbb{Z}}_{\chi}(-z)=h^{\mathbb{Z}}_{\chi}(z)$, and we index $h^{\mathbb{Z}}_{\chi}(z)$ as $h^{\mathbb{Z}}_{\chi}(z)=\sum_{n\in\mathbb{Z}}h^{\mathbb{Z}}_n z^{-2n-2}$. The field $h^{\mathbb{Z}}_{\chi}(z)$ has OPE with itself given by:

$$h^{\mathbb{Z}}_{\chi}(z)h^{\mathbb{Z}}_{\chi}(w)\sim-\frac{1}{(z^2-w^2)^2}, \tag{2.12}$$

and its modes, $h^{\mathbb{Z}}_n$, $n\in\mathbb{Z}$, generate an ***untwisted*** *Heisenberg algebra $\mathcal{H}_{\mathbb{Z}}$ with relations $[h^{\mathbb{Z}}_m, h^{\mathbb{Z}}_n]=-m\delta_{m+n,0}1,\ \ m,n\in\mathbb{Z}$.*

Next we follow with the decomposition in terms of irreducible Heisenberg modules. The reason a bosonization procedure of the algebraic Hirota equation is essential in order to produce actual differential equations lies in the well known fact (see e.g. [**KR87**], [**FLM88**]) that any irreducible highest weight module of the untwisted Heisenberg algebra $\mathcal{H}_{\mathbb{Z}}$ is isomorphic to the polynomial algebra with infinitely many variables $B_{\lambda}\cong\mathbb{C}[x_1,x_2,\dots,x_n,\dots]$ where $v\mapsto 1$ and we choose the scaling:

$$h^{\mathbb{Z}}_n\mapsto i\partial_{x_n},\quad h^{\mathbb{Z}}_{-n}\mapsto inx_n\cdot,\quad \text{for any }\ n\in\mathbb{N},\quad h^{\mathbb{Z}}_0\mapsto\lambda\cdot,\ \lambda\in\mathbb{C}. \tag{2.13}$$

Similarly for the twisted Heisenberg algebra: $\mathcal{H}_{\mathbb{Z}+1/2}$ has (up-to isomorphism) only one irreducible highest weight module $B_{1/2}\cong\mathbb{C}[t_1,t_3,\dots,t_{2n-1},\dots]$, via (we choose the same scaling as in [**vOS12**]):

$$h^t_{\frac{2n-1}{2}}\mapsto\partial_{t_{2n-1}},\quad h^t_{-\frac{2n-1}{2}}\mapsto-\frac{2n-1}{2}t_{2n-1}\cdot,\quad \text{for any }\ n\in\mathbb{N}. \tag{2.14}$$

Before we proceed with the decomposition in terms of irreducible Heisenberg modules we first need to recall the grading operators that act on F_{χ}. In [**Ang17**] we introduced two important gradings on F_{χ}, one derived from the action of $h^{\mathbb{Z}}_0$ of the untwisted Heisenberg field (the charge grading), and the second from one of the families of Virasoro fields that we discussed in [**Ang15**] (the degree grading). There are also two additional grading operators which we will introduce in the next section.

The *charge* grading *chg* is derived from the action of the "charge" operator $h^{\mathbb{Z}}_0$:

$$h^{\mathbb{Z}}_0=\sum_{k\in\mathbb{Z}_{\geq 0}+1/2}:\chi_{-k}\chi_k:=\chi_{-\frac{1}{2}}\chi_{\frac{1}{2}}+\chi_{-\frac{3}{2}}\chi_{\frac{3}{2}}+\dots, \tag{2.15}$$

where : : denotes the normal ordered product defined in the usual way (see e.g. [**Ang17**]).

$h^{\mathbb{Z}}_0$ is diagonalizable on F_{χ} and thus it gives F_{χ} a $\mathbb{Z}$ grading, which we called *charge* and denote *chg* (as it is similar to the charge grading in the usual boson-fermion correspondence, of type A), by
(2.16)

$$chg(|0\rangle)=0;\ chg\Big((\chi_{-j_k})^{m_k}\dots(\chi_{-j_2})^{m_2}(\chi_{-j_1})^{m_1}|0\rangle\Big)=\sum_{j_i\in 2\mathbb{Z}+1/2}m_i-\sum_{j_i\in 2\mathbb{Z}-1/2}m_i.$$

Example: $chg(\chi_{-\frac{1}{2}}|0\rangle)=1;\ \ chg(\chi_{-\frac{3}{2}}|0\rangle)=-1;\ chg(\chi_{-\frac{3}{2}}\chi_{-\frac{1}{2}}|0\rangle)=0.$

Denote the linear span of monomials of charge n by $F_\chi^{(n)}$. The Fock space F_χ has a charge decomposition

$$F_\chi = \oplus_{n\in\mathbb{Z}} F_\chi^{(n)}, \quad \text{with} \quad h_0^{\mathbb{Z}} v = chg(v)\cdot v = nv, \quad \text{for any } v\in F_\chi^{(n)}.$$

The second grading we used in [**Ang17**] is the *degree* grading, resulting from the action of one of the Virasoro fields. Namely, we considered the Virasoro field with central charge $c=-\frac{1}{4}$ given by

(2.17)
$$L^{-\frac{1}{4}}(z^2) = \sum_{n\in\mathbb{Z}} L_n (z^2)^{-n-2} = \sum_{n\in\mathbb{Z}} \Big(\sum_{k+l=n} \frac{k+1-3l}{4} :\chi_{2k+\frac{1}{2}}\chi_{2l-\frac{1}{2}}: \Big)(z^2)^{-n-2},$$

in particular

$$L_0 = \frac{1}{2}\Big(\frac{1}{2} :\chi_{-\frac{1}{2}}\chi_{\frac{1}{2}}: - \frac{3}{2} :\chi_{-\frac{3}{2}}\chi_{\frac{3}{2}}: + \frac{5}{2} :\chi_{-\frac{5}{2}}\chi_{\frac{5}{2}}: - \dots\Big). \tag{2.18}$$

Hence

$$\begin{aligned} &L_0\Big((\chi_{-j_k})^{m_k} \dots (\chi_{-j_2})^{m_2} (\chi_{-j_1})^{m_1} |0\rangle \Big) \\ &= \frac{1}{2}(m_k\cdot j_k + \dots m_2\cdot j_2 + m_1\cdot j_1)\Big((\chi_{-j_k})^{m_k} \dots (\chi_{-j_2})^{m_2} (\chi_{-j_1})^{m_1} |0\rangle \Big), \end{aligned} \tag{2.19}$$

$j_k > \dots > j_2 > j_1 > 0$, $j_i \in \mathbb{Z}+\frac{1}{2}$, $m_i>0, m_i\in\mathbb{Z}$, $i=1,2,\dots,k$. The degree grading is a $\frac{1}{2}\mathbb{Z}$ grading, given by the action of $2L_0$, i.e.,

$$2L_0 v = deg(v)\cdot v,$$

where v is any monomial $(\chi_{-j_k})^{m_k} \dots (\chi_{-j_2})^{m_2} (\chi_{-j_1})^{m_1} |0\rangle$. The degree grading is also used in [**vOS12**], although without its connection to the Virasoro field.

Next, we will need some notations for the corresponding indexing sets in the decompositions.

NOTATION 2.8. Let $\mathcal{ODP}$ denote the set of distinct partitions of the type

(2.20)
$$\mathcal{ODP} = \{\mathfrak{p} = (\lambda_1,\lambda_2,\dots,\lambda_k) \mid \lambda_1>\lambda_2>\dots>\lambda_k, \lambda_i\in\frac{1}{2}+\mathbb{Z}_{\geq 0},\ i=1,\dots,k\}.$$

Denote by T_m the m-th triangular number— $T_m := 1+2+\dots+m = \frac{m(m+1)}{2}$, with $T_0=0$. Let $\mathfrak{P}_{tdo}$ denote the set of distinct partitions of the type

$$\begin{aligned} \mathfrak{P}_{tdo} = \{\mathfrak{p} = (T_m, \lambda_1,\lambda_2,\dots,\lambda_k) \mid T_m &- \text{triangular number}, \\ \lambda_1>\lambda_2>\dots>\lambda_k, \lambda_i &\in \frac{1}{2}+\mathbb{Z}_{\geq 0},\ i=1,\dots,k\}. \end{aligned} \tag{2.21}$$

As usual, the weight $|\mathfrak{p}|$ of a partition $\mathfrak{p}$ is the sum of its parts, $|\mathfrak{p}| := T_m+\lambda_1+\lambda_2+\dots+\lambda_k$.

We can now formulate the decomposition in terms of Heisenberg modules.

THEOREM 2.9. *I. For the action of the twisted Heisenberg algebra $\mathcal{H}_{\mathbb{Z}+1/2}$ on F_χ, the number of highest weight vectors of degree $n\in\frac{1}{2}\mathbb{Z}$ equals the number of partitions $\mathfrak{p}\in\mathcal{ODP}$ of weight n. Thus as twisted Heisenberg modules*

$$F_\chi \cong \oplus_{\mathfrak{p}\in\mathcal{ODP}} \mathbb{C}[t_1,t_3,\dots,t_{2n-1},\dots]. \tag{2.22}$$

*II. ([**Ang17**]) For the action of the untwisted Heisenberg algebra $\mathcal{H}_{\mathbb{Z}}$ on F_χ, the number of highest weight vectors of degree $n \in \frac{1}{2}\mathbb{Z}$ equals the number of partitions $\mathfrak{p} \in \mathfrak{P}_{tdo}$ of weight n. Thus as untwisted Heisenberg modules*

$$F_\chi \cong \oplus_{\mathfrak{p}\in\mathfrak{P}_{tdo}} \mathbb{C}[x_1, x_2, \dots, x_n, \dots]. \tag{2.23}$$

The Heisenberg decomposition in the twisted case (the first part of the theorem above) is not stated in [**vOS12**], but can be implied from the calculations there. The second part of this theorem is established in [**Ang17**]. The proof of the first part is similar to the proof of the second part, and so we omit it.

EXAMPLE 2.10. The first few highest weight vectors for $\mathcal{H}_{\mathbb{Z}}$ arranged by degree are

$$\begin{array}{r|l}
1 & |0\rangle \\
q^{\frac12} & \chi_{-\frac12}|0\rangle \\
q & \chi^2_{-\frac12}|0\rangle \\
2q^{\frac32} & \chi^3_{-\frac12}|0\rangle; \quad \chi_{-\frac32}|0\rangle \\
q^2 & \chi^4_{-\frac12}|0\rangle \\
2q^{\frac52} & \chi^5_{-\frac12}|0\rangle; \quad \chi_{-\frac32}\chi^2_{-\frac12}|0\rangle + 2\chi_{-\frac52}|0\rangle \\
3q^3 & \chi^6_{-\frac12}|0\rangle; \quad \chi_{-\frac32}\chi^3_{-\frac12}|0\rangle + 3\chi_{-\frac52}\chi_{-\frac12}|0\rangle; \quad \chi^2_{-\frac32}|0\rangle
\end{array}$$

Two important families of examples of highest weight vectors for $\mathcal{H}_{\mathbb{Z}}$ are the families $\chi^n_{-\frac12}|0\rangle$ and $\chi^n_{-\frac32}|0\rangle$. They are the "minimal degree" highest weight vectors: at each negative fixed charge $-n < 0$, the vector $\chi^n_{-3/2}|0\rangle$ is the vector with minimal degree with charge $-n$; and at each fixed charge $n > 0$, the vector $\chi^n_{-3/2}|0\rangle$ is the vector with minimal degree of that charge n.

There are no similar families of highest weight vectors for $\mathcal{H}_{\mathbb{Z}+1/2}$, as there is no equivalent charge operator in $\mathcal{H}_{\mathbb{Z}+1/2}$. The first few highest weight vectors by degree are

$$\begin{array}{r|l}
1 & |0\rangle \\
q^{\frac12} & \chi_{-\frac12}|0\rangle \\
q & \text{none} \\
q^{\frac32} & \chi_{-\frac32}|0\rangle - \frac13\chi^3_{-\frac12}|0\rangle \\
q^2 & \chi_{-\frac32}\chi_{-\frac12}|0\rangle - \frac16\chi^4_{-\frac12}|0\rangle \\
q^{\frac52} & \chi_{-\frac52}|0\rangle - \chi_{-\frac32}\chi^2_{-\frac12}|0\rangle + \frac{1}{10}\chi^5_{-\frac12}|0\rangle
\end{array}$$

Next, we proceed with the structure of the spaces spanned by the corresponding highest weight vectors.

NOTATION 2.11. Denote by F_χ^{hwv} the vector space spanned by the highest weight vectors for the untwisted Heisenberg algebra representation on F_χ, and by F_χ^{t-hwv} the vector space spanned by the highest weight vectors for the twisted Heisenberg algebra representation on F_χ.

From the Proposition above, we have the following isomorphisms as vectors spaces:

$$F_\chi \cong F_\chi^{hwv} \otimes \mathbb{C}[x_1, x_2, \dots, x_n, \dots], \tag{2.24}$$

$$F_\chi \cong F_\chi^{t-hwv} \otimes \mathbb{C}[t_1, t_3, \dots, t_{2n-1}, \dots]. \tag{2.25}$$

Each of the spaces F_χ^{hwv} and F_χ^{t-hwv} has additional structure, which we need in order to complete the corresponding bosonizations.

DEFINITION 2.12. Let

$$V^-(z) = \exp\Big(-\sum_{n>0}\frac{1}{n}h_n^{\mathbb{Z}}z^{-2n}\Big); \qquad V^-(z)^{-1} = \exp\Big(\sum_{n>0}\frac{1}{n}h_n^{\mathbb{Z}}z^{-2n}\Big); \tag{2.26}$$

and

$$V^+(z) = \exp\Big(\sum_{n>0}\frac{1}{n}h_{-n}^{\mathbb{Z}}z^{2n}\Big); \qquad V^+(z)^{-1} = \exp\Big(-\sum_{n>0}\frac{1}{n}h_{-n}^{\mathbb{Z}}z^{2n}\Big) \tag{2.27}$$

Define

$$\beta_\chi(z^2) = \frac{\chi(z)-\chi(-z)}{2z}; \quad \gamma_\chi(z^2) = \frac{\chi(z)+\chi(-z)}{2}. \tag{2.28}$$

and

$$H^\beta(z^2) = V^+(z)^{-1}\beta_\chi(z^2)z^{-2h_0^{\mathbb{Z}}}V^-(z)^{-1}, \qquad H^\gamma(z^2) = V^+(z)\gamma_\chi(z^2)z^{2h_0^{\mathbb{Z}}}V^-(z). \tag{2.29}$$

($V^-(z)$ and $V^+(z)$ are actually functions of z^2, so the notation is unambiguous).

Similarly denote

$$V_t^-(z)=\exp\Big(\sum_{n>0}\frac{2}{2n-1}h^t_{\frac{2n-1}{2}}z^{-2n+1}\Big); V_t^-(z)^{-1}=\exp\Big(-\sum_{n>0}\frac{2}{2n-1}h^t_{\frac{2n-1}{2}}z^{-2n+1}\Big); \tag{2.30}$$

$$V_t^+(z)=\exp\Big(-\sum_{n>0}\frac{2}{2n-1}h^t_{-\frac{2n-1}{2}}z^{2n-1}\Big); V_t^+(z)^{-1}=\exp\Big(\sum_{n>0}\frac{2}{2n-1}h^t_{-\frac{2n-1}{2}}z^{2n-1}\Big); \tag{2.31}$$

and

$$H^\chi(z) = V_t^+(z)^{-1}\chi(z)V_t^-(z)^{-1}. \tag{2.32}$$

THEOREM 2.13. *I. (*[**Ang17**]*) The vector space F_χ^{hwv} spanned by the highest weight vectors for the untwisted Heisenberg algebra $\mathcal{H}_\mathbb{Z}$ has a structure of a super vertex algebra, strongly generated by the fields $H^\beta(z)$ and $H^\gamma(z)$, with vacuum vector $|0\rangle$, translation operator $T = L_{-1}^{hwv}$, and vertex operator map induced by*

$$Y(\chi_{-1/2}|0\rangle, z) = H^\gamma(z), \quad Y(\chi_{-3/2}|0\rangle, z) = H^\beta(z). \tag{2.33}$$

This vertex algebra structure is a realization of the symplectic fermion vertex algebra, indicated by the OPEs:

$$H^\beta(z)H^\gamma(w) \sim \frac{1}{(z-w)^2}, \quad H^\gamma(z)H^\beta(w) \sim -\frac{1}{(z-w)^2}; \tag{2.34}$$

$$H^\beta(z)H^\beta(w) \sim 0; \quad H^\gamma(z)H^\gamma(w) \sim 0. \tag{2.35}$$

II. The vector space F_χ^{t-hwv} spanned by the highest weight vectors for the twisted Heisenberg algebra $\mathcal{H}_{\mathbb{Z}+1/2}$ has a structure of an $N = 2$ twisted vertex algebra, generated by the field $H^\chi(z)$, with vacuum vector $|0\rangle$, and vertex operator map induced by

$$Y(\chi_{-1/2}|0\rangle, z) = H^\chi(z). \tag{2.36}$$

This twisted vertex algebra structure is twisted fermionic, indicated by the OPEs:

$$H^\chi(z)H^\chi(w) \sim \frac{z-w}{(z+w)^2}. \tag{2.37}$$

The first part of this theorem was proved in [**Ang17**]. The second part uses the notion of a twisted vertex algebra developed in [**Ang13b**] and [**ACJ14**]. Its proof is similar to the proof of the first part, and quite technical, so we omit it. The basic OPE indicated in (2.37) can be read from the calculations in [**vOS12**].

COROLLARY 2.14. *I. Define ([**Abe07**])*

$$SF := \{H^{\beta}_{(m_k)}\dots H^{\beta}_{(m_2)}H^{\beta}_{(m_1)}H^{\gamma}_{(n_s)}\dots H^{\gamma}_{(n_2)}H^{\gamma}_{(n_1)}|0\rangle \mid \\ \mid m_k<\dots m_2<m_1,\ n_s<\dots n_2<n_1; \\ m_i,n_j\in\mathbb{Z}_{<0},\ i=1,2,\dots,k; j=1,2,\dots s\}.$$

We have as vertex algebras

$$F_{\chi}^{hwv}\cong SF; \tag{2.38}$$

and as vector spaces

$$F_{\chi}\cong F_{\chi}^{hwv}\otimes\mathbb{C}[x_1,x_2,\dots,x_n,\dots]\cong SF\otimes\mathbb{C}[x_1,x_2,\dots,x_n,\dots]. \tag{2.39}$$

II. Define

$$SF^t := \{H^{\chi}_{(n_s)}\dots H^{\chi}_{(n_2)}H^{\chi}_{(n_1)}|0\rangle \mid n_s<\dots n_2<n_1;\ n_j\in\mathbb{Z}+1/2, n_j<0,\ j=1,2,\dots s\}. \tag{2.40}$$

We have as twisted vertex algebras

$$F_{\chi}^{t-hwv}\cong SF^t; \tag{2.41}$$

and as vector spaces

$$F_{\chi}\cong F_{\chi}^{t-hwv}\otimes\mathbb{C}[t_1,t_3,\dots,t_{2n-1},\dots]\cong SF^t\otimes\mathbb{C}[t_1,t_3,\dots,t_{2n-1},\dots]. \tag{2.42}$$

The Theorem above allows us to express the generating field in terms of the (correspondingly twisted or untwisted) Heisenberg fields, plus the (twisted or untwisted) symplectic fermion fields, thereby completing the 3rd stage of the bosonization process. From [**vOS12**], we have for the twisted bosonization

$$\chi(z)=V_t^{+}(z)H^{\chi}(z)V_t^{-}(z). \tag{2.43}$$

Unfortunately, we do not know if the field $H^{\chi}(z)$ can be bosonized further. In [**vOS12**] super, or fermionic, variables were introduced to describe the space F_{χ}^{t-hwv}. I.e., introduce the Grassmann variables $t_{\frac{2n-1}{2}}$, $n\in\mathbb{Z}$, so that

$$t_{\frac{2m-1}{2}}t_{\frac{2n-1}{2}}=-t_{\frac{2n-1}{2}}t_{\frac{2m-1}{2}},\ \text{for } m\neq n;\quad \left(t_{\frac{2n-1}{2}}\right)^2=0. \tag{2.44}$$

Consequently, we consider the Grassmann algebra generated by the Grassmann variables $t_{\frac{2n-1}{2}}$, $n\in\mathbb{Z}$, which we will denote as $\Lambda[t_{\frac{1}{2}},t_{\frac{3}{2}},\dots,t_{\frac{2n-1}{2}},\dots]$. The Grassman variables $t_{\frac{2n-1}{2}}$, $n\in\mathbb{Z}$, commute with the commutative variables t_{2n-1}, $n\in\mathbb{Z}$. We of course have $\Lambda[t_{\frac{1}{2}},t_{\frac{3}{2}},\dots,t_{\frac{2n-1}{2}},\dots]\cong SF^t$.

THEOREM 2.15. [**vOS12**] *The twisted bosonization of the CKP hierarchy is the isomorphism σ^t between the twisted vertex algebra generated by the field $\chi(z)$ and its descendant $\chi(-z)$ on the Fock space F_{χ} and the twisted vertex algebra generated by the field $\sigma^t\chi(z)(\sigma^t)^{-1}:=V_{\sigma t}^{+}(z)H^{\sigma\chi}(z)V_{\sigma t}^{-}(z)$, where the invertible map σ^t is defined as*

$$\sigma^t: F_{\chi}\to\Lambda[t_{\frac{1}{2}},t_{\frac{3}{2}},\dots,t_{\frac{2n-1}{2}},\dots]\otimes\mathbb{C}[t_1,t_3,\dots,t_{2n-1},\dots],\quad |0\rangle\mapsto 1 \tag{2.45}$$

by

(2.46)

$$\sigma^t h^t_{\frac{2n-1}{2}}(\sigma^t)^{-1} := \partial_{t_{2n-1}}, \quad \sigma^t h^t_{-\frac{2n-1}{2}}(\sigma^t)^{-1} := -\frac{2n-1}{2} t_{2n-1}\cdot, \quad \textit{for any } n\in\mathbb{N};$$

(2.47)

$$\sigma^t H^{\chi}_{(n)}(\sigma^t)^{-1} = H^{\sigma\chi}_{(n)} := \partial_{t_{\frac{2n-1}{2}}}, \sigma^t H^{\chi}_{(-n)}(\sigma^t)^{-1} := -\frac{2n-1}{2} t_{\frac{2n-1}{2}}\cdot, \textit{for any } n\in\mathbb{N}.$$

We have

$$(2.48)\qquad \begin{aligned} V^+_{\sigma t}(z) &= \sigma^t V^+_t(z)(\sigma^t)^{-1} := \exp\Big(\sum_{n>0} t_{2n-1} z^{2n-1}\Big),\\ V^-_{\sigma t}(z) &= \sigma^t V^-_t(z)(\sigma^t)^{-1} := \exp\Big(\sum_{n>0} \frac{2}{2n-1}\partial_{t_{2n-1}} z^{-2n+1}\Big); \end{aligned}$$

and

$$(2.49)\qquad H^{\sigma\chi}(z) = \sigma^t H^{\chi}(z)(\sigma^t)^{-1} := \sum_{n>0} \partial_{t_{\frac{2n-1}{2}}} z^{-n} - \sum_{n>0} \frac{2n-1}{2} t_{\frac{2n-1}{2}} z^n.$$

Now we move to the second, untwisted bosonization. We have

$$\chi(z) = \gamma_\chi(z^2) + z\beta_\chi(z^2),$$

and the fields $\beta_\chi(z^2)$ and $\gamma_\chi(z^2)$ can be written as

(2.50)

$$\beta_\chi(z^2) = V^+(z)H^\beta(z^2)V^-(z)z^{2h_0}, \qquad \gamma_\chi(z^2) = V^+(z)^{-1}H^\gamma(z^2)V^-(z)^{-1}z^{-2h_0}.$$

In this case of the untwisted bosonization, even though they are again fermionic, as in the twisted case, the fields $H^\beta(z)$ and $H^\gamma(z)$ can be bosonized further, via the identification with the symplectic fermions (for more details see [**Ang17**]), namely

$$(2.51)\qquad H^\beta(z) \to e_y^{-\alpha}(z) = \exp(-\sum_{n\ge 1} y_n z^n)\exp(\sum_{n\ge 1}\frac{\partial}{n\partial y_n} z^{-n})e^{-\alpha} z^{-h_0^y}$$

$$(2.52)\qquad H^\gamma(z) \to \partial_z e_y^\alpha(z) =: h^y(z)\exp(\sum_{n\ge 1} y_n z^n)\exp(-\sum_{n\ge 1}\frac{\partial}{n\partial y_n} z^{-n})e^{\alpha} z^{h_0^y} :.$$

where the lattice fields $e_y^\alpha(z)$, $e_y^{-\alpha}(z)$ act on the bosonic vector space $\mathbb{C}[e^\alpha, e^{-\alpha}]\otimes\mathbb{C}[y_1, y_2, \dots, y_n \dots]$ by

$$\begin{aligned} e_y^\alpha(z) &= \exp(\sum_{n\ge 1} y_n z^n)\exp(-\sum_{n\ge 1}\frac{\partial}{n\partial y_n} z^{-n})e^{\alpha} z^{\partial_\alpha},\\ e_y^{-\alpha}(z) &= \exp(-\sum_{n\ge 1} y_n z^n)\exp(\sum_{n\ge 1}\frac{\partial}{n\partial y_n} z^{-n})e^{-\alpha} z^{-\partial_\alpha}, \end{aligned}$$

We use the index y in $e_y^\alpha(z)$, $e_y^{-\alpha}(z)$ to indicate these are the exponentiated boson fields acting on the variables $y_1, y_2, \dots, y_n \dots$. We introduce similarly the Heisenberg field $h^y(z)$,

$$(2.53)\qquad h^y(z) = \sum_{n\ge 1}\frac{\partial}{\partial y_n} z^{-n-1} + h_0^y z^{-1} + \sum_{n\ge 1} n y_n z^{n-1},$$

where h_0^y acts on $\mathbb{C}[e^\alpha, e^{-\alpha}]\otimes\mathbb{C}[y_1, y_2, \dots, y_n \dots]$ by $h_0^y e^{m\alpha}P(y_1, y_2, \dots, y_n \dots) = m e^{m\alpha}P(y_1, y_2, \dots, y_n \dots)$.

Finally, the complete second bosonization of the CKP hierarchy is summarized in the following:

THEOREM 2.16. *([**Ang17**]) The generating field $\chi(z)$ of the CKP hierarchy can be written as*

$$\chi(z) = \gamma_\chi(z^2) + z\beta_\chi(z^2),$$

where the fields $\beta_\chi(z)$ and $\gamma_\chi(z)$ can be bosonized as follows:

(2.54)

$$\sigma: \ \beta_\chi(z) \mapsto \ \exp\Big(i\sum_{n>0}(x_n + iy_n)z^n\Big)\exp\Big(-i\sum_{n>0}\frac{1}{n}\left(\frac{\partial}{\partial x_n} + i\frac{\partial}{\partial y_n}\right)z^{-n}\Big)e^{-\alpha},$$

(2.55)

$$\sigma: \ \gamma_\chi(z) \mapsto :\exp\Big(-i\sum_{n>0}(x_n + iy_n)z^n\Big)h^y(z)\exp\Big(i\sum_{n>0}\frac{1}{n}\left(\frac{\partial}{\partial x_n} + i\frac{\partial}{\partial y_n}\right)z^{-n}\Big)e^{\alpha}: .$$

The Fock space F_χ is mapped via the bosonization map σ to a subspace of the bosonic space $\mathbb{C}[e^\alpha, e^{-\alpha}]\otimes\mathbb{C}[x_1, x_2, \ldots, x_n, \ldots; y_1, y_2, \ldots, y_n \ldots]$, with $|0\rangle \mapsto 1$. The Hirota equation (2.1) *is equivalent to*

$$Res_z\Big(\beta_\chi(z)\otimes\gamma_\chi(z) - \gamma_\chi(z)\otimes\beta_\chi(z)\Big) = 0. \tag{2.56}$$

This theorem allows us to write the purely algebraic Hirota equation as an infinite hierarchy of actual differential equations. One proceeds similarly to the exposition in [**KR87**], Chapter 7, by employing the Hirota derivatives technique. These calculations and the types of solutions produced will be detailed in a separate article, as they require some length. In the next section, we will instead focus on one of the applications of these bosonizations, namely the various characters (graded dimensions) and the identities one can derive from the comparison between the purely bosonic vs the boson+symplectic fermion sides.

3. Graded dimensions and character identities

As an application to any bosonization, one can typically obtain identities relating certain product formulas on the fermionic side to certain sum formulas on the bosonic side. Recall for instance that the Jacobi triple product identity can be obtained from such a bosonization — it was derived in [**Kac98**] for the classical boson-fermion correspondence of type A, and in [**Ang14**] for the bosonization of type D-A. Such a sum-vs-product identity perfectly illustrates the equality between the fermionic side (the product formulas) and the bosonic side (the sum formulas). Here, as is typical for the CKP quirks, we will show that the sum-vs-product identity representing the untwisted bosonization is much more complicated due to the fact that the degree operator L_0 that we had to use for the Heisenberg decomposition doesn't act uniformly on the symplectic fermion side with which the highest weight vectors space F_χ^{hwv} identifies. But first, we will explore the fact that due to the two bosonizations, in the CKP case we have additional grading operators, besides the charge and the degree gradings.

The degree grading on the Fock space F_χ is given by the operator L_0, but in fact, we can consider a one-parameter λ-degree grading on F_χ. Recall, the degree grading is induced by the action of the 0-mode L_0 of the Virasoro field $L^{-\frac{1}{4}}(z^2)$.

This Virasoro field is an element of the more general family of Virasoro fields $L^{\lambda}(z)$, translated from the $\beta-\gamma$ system ([**Ang15**], for simplicity parameter μ is set to 0):

$$L^{\lambda}(z)=\lambda :(\partial_z\beta(z))\,\gamma(z): +(\lambda+1):\beta(z)\,(\partial_z\gamma(z)): . \tag{3.1}$$

We have

$$L^{\lambda}_0=-\sum_{n\in\mathbb{Z}}(\lambda+k):\chi_{-2k+\frac{1}{2}}\chi_{2k-\frac{1}{2}}: . \tag{3.2}$$

We want to note that for both the operators L^{λ}_0 and $h^{\mathbb{Z}}_0$ the monomials from (2.5) form an eigenspace basis, as we showed in the previous section.

We made the particular choice to use $\lambda=-\frac{1}{4}$ for L_0 not only because it would simplify the notation — for general λ the behavior of the operator L^{λ}_0 is very similar to that of L_0. But also this is the most natural choice of grading operator for F_{χ}, in particular, as we will see it is the only choice from the family L^{λ}_0 that commutes with the operator L^t which we will introduce below.

The third grading on F_{χ}, corresponding most closely to the twisted bosonization, is obtained from the Virasoro field $L^{\chi}(z)$ with central charge $c=1$ ([**FMS86**], [**Ang15**], for simplicity we set $\kappa=0$):

$$L^{\chi}(z)=\left(-\frac{1}{2z^2}:h^{\mathbb{Z}+1/2}(z)^2: +\frac{1}{16z^4}\right). \tag{3.3}$$

We have

$$L^{\chi}_0=-\frac{1}{2}\sum_{n\in\mathbb{Z}+1/2}:h^t_{-n}h^t_n: +\frac{1}{16}. \tag{3.4}$$

For simplicity we will consider the modified grading operator L^t_0, where

$$L^t_0=L^{\chi}_0-\frac{1}{16}=-\sum_{\substack{n\in\mathbb{Z}+1/2\\ n>0}}:h^t_{-n}h^t_n: . \tag{3.5}$$

L^t_0 is diagonalizable, due to Proposition 2.9, but we indeed are always forced to use the Heisenberg decomposition of Theorem 2.9 to calculate the action of L^t_0 on the elements of F_{χ} (example of such calculation is given below, in the proof of Lemma 3.2). This is because, unlike for the operators L^{λ}_0 and $h^{\mathbb{Z}}_0$, the monomials from (2.5) do not form an eigenspace basis (an example is the monomial $\chi_{-\frac{3}{2}}|0\rangle$, which is not an eigenvector for L^t_0).

There is a fourth grading on F_{χ}, induced by the yet another family of Virasoro fields on F_{χ}: for any $a,b\in\mathbb{C}$ the field

$$L^h(z^2)=-\frac{1}{2}:h^{\mathbb{Z}}(z^2)^2: +a\partial_{z^2}h^{\mathbb{Z}}(z^2)+\frac{b}{z^2}h^{\mathbb{Z}}(z^2)+\frac{2ab-b^2}{2z^4}. \tag{3.6}$$

is a Virasoro field with central charge $1+12a^2$, the central charge is independent of b ([**Ang15**]). We have

$$L^h_0=-\frac{1}{2}(h^{\mathbb{Z}}_0)^2-\sum_{n\in\mathbb{Z}_+}h^{\mathbb{Z}}_{-n}h^{\mathbb{Z}}_n+(b-a-an)\,h^{\mathbb{Z}}_n. \tag{3.7}$$

For simplicity here we will only consider the case $a=b=0$. L^h_0 is also diagonalizable, again due to Theorem 2.9, its second part. The monomials from (2.5) do not form an eigenspace basis here either: an example is the monomial $\chi_{-\frac{5}{2}}|0\rangle$, which is not an eigenvector for L^h_0.

REMARK 3.1. The three Virasoro fields are different, neither is part of the family of any of the other two, as we explained in [**Ang15**].

Since we have multiple gradings, we can consider multiparameter characters (graded dimensions). The potential obstruction to such a multiparameter character is the lack of commuting of the respective grading operators. In our case, the grading operators L_0 and $h_0^{\mathbb{Z}}$ commute among themselves. Consequently, in [**Ang17**] we already introduced the following character with respect to the L_0 and $h_0^{\mathbb{Z}}$ grading operators:

$$dim_{q,z}^{L_0,h_0^{\mathbb{Z}}} F_\chi := tr_{F_\chi} q^{2L_0} z^{h_0^{\mathbb{Z}}}. \tag{3.8}$$

But, the third grading operator L_0^t does not commute with the charge grading operator $h_0^{\mathbb{Z}}$, and consequently, with the 4th grading by L_0^h either:

LEMMA 3.2. *The operators L_0^t and $h_0^{\mathbb{Z}}$ do not commute. Therefore L_0^t and $h_0^{\mathbb{Z}}$ cannot be simultaneously diagonalized. The same is true for the operators L_0^t and L_0^h.*

PROOF. Consider the highest weight vector $v_t = \chi_{-\frac{3}{2}}|0\rangle - \frac{1}{3}\chi^3_{-\frac{1}{2}}|0\rangle$ for the twisted Heisenberg algebra $\mathcal{H}_{\mathbb{Z}+1/2}$. We have from the definition of L_0^t that on highest weight vectors it acts as 0, and so

$$L_0^t\left(\chi_{-\frac{3}{2}}|0\rangle - \frac{1}{3}\chi^3_{-\frac{1}{2}}|0\rangle\right) = L_0^t v_t = 0.$$

Therefore $h_0^{\mathbb{Z}} L_0^t v_t = 0$. But

$$h_0^{\mathbb{Z}} v_t = h_0^{\mathbb{Z}}\left(\chi_{-\frac{3}{2}}|0\rangle - \frac{1}{3}\chi^3_{-\frac{1}{2}}|0\rangle\right) = -1\cdot\chi_{-\frac{3}{2}}|0\rangle - 3\cdot\frac{1}{3}\chi^3_{-\frac{1}{2}}|0\rangle = -\chi_{-\frac{3}{2}}|0\rangle - \chi^3_{-\frac{1}{2}}|0\rangle.$$

As we mentioned above, L_0^t is diagonalizable, but we have to use the Heisenberg decomposition of Theorem 2.9 to calculate the action of L_0^t on non-highest weight vectors, such as the vector $\left(-\chi_{-\frac{3}{2}}|0\rangle - \chi^3_{-\frac{1}{2}}|0\rangle\right)$. In particular, we have two basis vectors of degree 3, as there are 2 odd partitions of degree 3, namely (3) and $(1,1,1)$. Thus we have to use as basis eigenvectors of degree 3 the highest weight vector v_t of degree 3, and the non-highest-weight vector

$$h_{-\frac{1}{2}}\chi_{-\frac{1}{2}}|0\rangle = \left(\frac{1}{2}\chi^2_{-\frac{1}{2}} - \chi_{-\frac{3}{2}}\chi_{\frac{1}{2}}\right)\chi_{-\frac{1}{2}}|0\rangle = \frac{1}{2}\chi^3_{-\frac{1}{2}}|0\rangle - \chi_{-\frac{3}{2}}|0\rangle.$$

Note we used that $\chi_{-\frac{1}{2}}|0\rangle$ is the single highest weight vector of degree 1, and we have from Proposition 2.7 that

$$h_{-\frac{1}{2}} = \frac{1}{2}\sum_{n\in\mathbb{Z}+1/2}(-1)^{-n-1/2}\chi_{-n-1}\chi_n = \frac{1}{2}\chi^2_{-\frac{1}{2}} - \chi_{-\frac{3}{2}}\chi_{\frac{1}{2}} + \chi_{-\frac{5}{2}}\chi_{\frac{3}{2}} - \dots.$$

We can write

$$-\chi_{-\frac{3}{2}}|0\rangle - \chi^3_{-\frac{1}{2}}|0\rangle = 7v_t - 8h_{-\frac{1}{2}}\chi_{-\frac{1}{2}}|0\rangle.$$

And thus since $[h_{\frac{1}{2}}, h_{-\frac{1}{2}}] = -\frac{1}{2}$ we get

$$L_0^t\left(-\chi_{-\frac{3}{2}}|0\rangle - \chi^3_{-\frac{1}{2}}|0\rangle\right) = L_0^t\left(7v_t - 8h_{-\frac{1}{2}}\chi_{-\frac{1}{2}}|0\rangle\right) = 0 - 4h_{-\frac{1}{2}}\chi_{-\frac{1}{2}}|0\rangle.$$

In any case, $L_0^t h_0^{\mathbb{Z}} v_t \neq 0$, therefore $L_0^t h_0^{\mathbb{Z}} \neq h_0^{\mathbb{Z}} L_0^t$ on F_χ. The proof that the operators L_0^t and L_0^h do not commute is similar, and so we omit it. □

Thus we cannot consider a character $dim_{t,z}^{L_0^t,h_0^{\mathbb{Z}}} F_\chi$, as such a trace $tr_{F_\chi} t^{L_0^t} z^{h_0^{\mathbb{Z}}}$ would require the simultaneous diagonalization of L_0^t and $h_0^{\mathbb{Z}}$, and so does not exist. Consequently, a simultaneous diagonalization of all 4 grading operators, including the L_0^t grading, is not possible; and as a result a 4-parameter graded dimension with these 4 grading operators does not exist either.

But we have the following Lemma, which will ensure that the operators L_0 and L_0^t can be simultaneously diagonalized:

LEMMA 3.3. *([**vOS12**]) For any $v \in F_\chi$ a homogeneous element of given degree we have*

$$deg(h^t_{-\frac{2n-1}{2}} v) = (2n-1) + deg(v), \quad \forall\ n \in \mathbb{Z}_+. \tag{3.9}$$

An eigenvector basis for the operator L_0 consists of all the highest weight vectors for the twisted Heisenberg algebra $\mathcal{H}_{\mathbb{Z}+1/2}$, together with the twisted Heisenberg monomials induced from them:

$$h^t_{-\frac{2n_s-1}{2}} \dots h^t_{-\frac{2n_2-1}{2}} h^t_{-\frac{2n_1-1}{2}} v_t,$$

where v_t is any highest weight vector for the twisted Heisenberg algebra $\mathcal{H}_{\mathbb{Z}+1/2}$, $n_1, n_2, \dots n_s \in \mathbb{Z}_+$ are not necessarily different.

PROOF. By direct observation of the form of $h^t_{-\frac{2n_s-1}{2}}$ by Proposition 2.7, namely

$$h^t_{-\frac{2n-1}{2}} = \frac{1}{2} \sum_{\substack{k,l \in \mathbb{Z}+1/2 \\ k+l=-2n+1}} (-1)^{-l-1/2} \chi_k \chi_l.$$

□

Therefore we can consider the 2 parameter character induced by the grading operators L_0 and L_0^t

$$dim_{q,t}^{L_0,L_0^t} F_\chi := tr_{F_\chi} q^{2L_0} t^{L_0^t}. \tag{3.10}$$

By applying the Lemma above, and Theorem 2.9 (recall $\mathcal{ODP}$ stands for the set of all distinct partitions with parts in $\mathbb{Z}+1/2$), we get

PROPOSITION 3.4.

$$dim_{q,t}^{L_0,L_0^t} F_\chi = \sum_{\mathfrak{p} \in \mathcal{ODP}} \frac{q^{|\mathfrak{p}|}}{\prod_{n\geq 1}(1 - q^{2n-1} t^{\frac{2n-1}{2}})} \tag{3.11}$$

PROOF. From the decomposition of Theorem 2.9, F_χ is a a direct sum of irreducible highest weight Heisenberg $\mathcal{H}_{\mathbb{Z}+1/2}$ modules, each isomorphic to

$$\mathbb{C}[h^t_{-\frac{1}{2}}, h^t_{-\frac{3}{2}}, \dots, h^t_{-\frac{2n-1}{2}}, \dots] \cdot v_t,$$

where v_t is a highest weight vector for $\mathcal{H}_{\mathbb{Z}+1/2}$. On any element such as

$$h^t_{-\frac{2n_s-1}{2}} \dots h^t_{-\frac{2n_2-1}{2}} h^t_{-\frac{2n_1-1}{2}} v_t$$

here $n_1, n_2, \dots n_s \in \mathbb{Z}_+$ are not necessarily different, we have

$$\begin{aligned} &2L_0 h^t_{-\frac{2n_s-1}{2}} \dots h^t_{-\frac{2n_2-1}{2}} h^t_{-\frac{2n_1-1}{2}} v_t \\ &\quad = \big((2n_s-1) + \dots (2n_2-1) + (2n_1-1)\big) h^t_{-\frac{2n_s-1}{2}} \dots h^t_{-\frac{2n_2-1}{2}} h^t_{-\frac{2n_1-1}{2}} v_t. \end{aligned}$$

To calculate the action of L_0^t on such an element, we need the following Lemma:

LEMMA 3.5. *The following relations hold:*

$$[L_0^t, h^t_{-\frac{2n-1}{2}}] = \frac{2n-1}{2}\cdot h^t_{-\frac{2n-1}{2}}, \quad \forall\, n\in\mathbb{Z}, \tag{3.12}$$

PROOF. We can use Wick's Theorem here (the version generalized to twisted vertex algebras, see [**ACJ14**]), and taking into account that the field $h_\chi^{\mathbb{Z}+1/2}(z)$ really depends only on z^2, see Proposition 2.7, we calculate:

$$\begin{aligned}
L^\chi(z^2)h_\chi^{\mathbb{Z}+1/2}(w) &\sim -\frac{1}{2z^2} : h^{\mathbb{Z}+1/2}(z)^2 : h_\chi^{\mathbb{Z}+1/2}(w) \sim \frac{1}{z^2}\cdot\frac{z^2+w^2}{2(z^2-w^2)^2}h_\chi^{\mathbb{Z}+1/2}(z)\\
&\sim \left(\frac{1}{w^2}-\frac{z^2-w^2}{w^4}+\dots\right)\left(\frac{1}{2(z^2-w^2)}+\frac{w^2}{(z^2-w^2)^2}\right)h_\chi^{\mathbb{Z}+1/2}(z)\\
&\sim \left(-\frac{1}{2w^2(z^2-w^2)}+\frac{1}{(z^2-w^2)^2}\right)h_\chi^{\mathbb{Z}+1/2}(z)\\
&\sim -\frac{1}{2w^2(z^2-w^2)}h_\chi^{\mathbb{Z}+1/2}(w)+\frac{1}{(z^2-w^2)^2}h_\chi^{\mathbb{Z}+1/2}(w)+\frac{1}{z^2-w^2}\partial_{w^2}h_\chi^{\mathbb{Z}+1/2}(w).
\end{aligned}$$

And so collecting the term at z^{-4} we get

$$[L_0^t,\ h_\chi^{\mathbb{Z}+1/2}(w)] = -\frac{1}{2}h_\chi^{\mathbb{Z}+1/2}(w)+h_\chi^{\mathbb{Z}+1/2}(w)+w^2\partial_{w^2}h_\chi^{\mathbb{Z}+1/2}(w).$$

In terms of modes we have

$$\sum_{n\in\mathbb{Z}+1/2}[L_0^t, h_n^t w^{-2n-1}] = \frac{1}{2}\sum_{n\in\mathbb{Z}+1/2}h_n^t w^{-2n-1}+\frac{-2n-1}{2}\sum_{n\in\mathbb{Z}+1/2}h_n^t(w^2)^{\frac{-2n-1}{2}};$$

thus

$$[L_0^t, h_n^t] = -nh_n^t, \quad \text{for any} \quad n\in\mathbb{Z}+1/2.$$

□

Now this allows us to go back to the proof of the proposition, as this shows that

$$\begin{aligned}
&L_0^t h^t_{-\frac{2n_s-1}{2}}\dots h^t_{-\frac{2n_2-1}{2}}h^t_{-\frac{2n_1-1}{2}}v_t\\
&\quad=\frac{1}{2}\big((2n_s-1)+\dots(2n_2-1)+(2n_1-1)\big)h^t_{-\frac{2n_s-1}{2}}\dots h^t_{-\frac{2n_2-1}{2}}h^t_{-\frac{2n_1-1}{2}}v_t.
\end{aligned}$$

Thus the graded dimension of the irreducible module $\mathbb{C}[h^t_{-\frac{1}{2}}, h^t_{-\frac{3}{2}},\dots,h^t_{-\frac{2n-1}{2}},\dots]\cdot v_t$ is

$$dim_{q,t}^{L_0,L_0^t}\mathbb{C}[h^t_{-\frac{1}{2}}, h^t_{-\frac{3}{2}},\dots,h^t_{-\frac{2n-1}{2}},\dots]\cdot v_t = \frac{q^{deg(v_t)}}{\prod_{n\geq 1}(1-q^{2n-1}t^{\frac{2n-1}{2}})}.$$

Hence the proof is completed by Theorem 2.9, as the highest weight vectors v_t in the decomposition are indexed by $\mathcal{ODP}$. □

REMARK 3.6. Setting $t=1$ we obtain the not terribly interesting result that

$$dim_q^{L_0}F_\chi = \sum_{\mathfrak{p}\in\mathcal{ODP}}\frac{q^{|\mathfrak{p}|}}{\prod_{n\geq 1}(1-q^{2n-1})}. \tag{3.13}$$

This result is not surprising at all, as of course we have

$$\sum_{\mathfrak{p}\in\mathcal{ODP}}q^{|\mathfrak{p}|} = \prod_{n\geq 1}(1+q^{\frac{2n-1}{2}}),$$

and thus

$$dim_q^{L_0} F_\chi = \sum_{\mathfrak{p}\in\mathcal{ODP}} \frac{q^{|\mathfrak{p}|}}{\prod_{n\geq 1}(1-q^{2n-1})} = \frac{\prod_{n\geq 1}(1+q^{\frac{2n-1}{2}})}{\prod_{n\geq 1}(1-q^{2n-1})} = \frac{1}{\prod_{n\geq 1}(1-q^{\frac{2n-1}{2}})},$$

which we knew already.

Since L_0^t doesn't commute with any of the other two grading operators, we now exclude L_0^t and look at the combinations between the other three grading operators. Ultimately, we want to derive the tri-parameter graded dimension

$$dim_{q,z,r}^{L_0, h_0^{\mathbb{Z}}, L_0^h} F_\chi := tr_{F_\chi} q^{2L_0} z^{h_0^{\mathbb{Z}}} r^{L_0^h}. \tag{3.14}$$

For this graded dimension we will need to introduce some additional partition notation. In particular, we want to make the connection between this character and the birank of a bipartition (two-colored partition), introduced in [**HL04**], see also [**Gar10**]. In fact, there are two important sets of bipartitions we need to consider. The first set of bipartitions is the set $\mathfrak{BP}_{HI}$, consisting of bipartitions $(\pi_1 \mid \pi_2)$, such that π_1 and π_2 are partitions with parts in $\mathbb{Z}_+ + 1/2$.

To each each monomial $(\chi_{-j_k})^{m_k} \dots (\chi_{-j_2})^{m_2} (\chi_{-j_1})^{m_1} |0\rangle$ from the basis (2.5) of F_χ we can assign a bipartition $(\pi_1 \mid \pi_2) \in \mathfrak{BP}_{HI}$ as follows: the partition π_1 would consists of the $j_s \in 2\mathbb{Z} - 1/2$, and π_2 would consists of the $j_s \in 2\mathbb{Z} + 1/2$.

EXAMPLE 3.7. To the monomial $\chi^3_{-\frac{9}{2}} \chi^2_{-\frac{5}{2}} \chi_{-\frac{3}{2}} \chi_{-\frac{1}{2}} |0\rangle$ we associate the bipartition $(\pi_1 \mid \pi_2)$ with

$$\pi_1 = \left(\frac{3}{2}\right), \quad \pi_2 = \left(\frac{9}{2}, \frac{9}{2}, \frac{9}{2}, \frac{5}{2}, \frac{5}{2}, \frac{1}{2}\right).$$

We can also write the partition π_2 in the notation $\pi_2 = \left((\frac{9}{2})^3, (\frac{5}{2})^2, (\frac{1}{2})^1 \right)$.

NOTATION 3.8. Denote by $\#\pi$ the number of nonzero parts in the partition π. $|(\pi_1 \mid \pi_2)|$ will denote the weight of the bipartition $(\pi_1 \mid \pi_2)$, i.e., $|(\pi_1 \mid \pi_2)| := |\pi_1| + |\pi_2|$.

In the example above $\#\pi_1 = 1$, $\#\pi_2 = 6$, $|(\pi_1 \mid \pi_2)| = 3 \cdot \frac{9}{2} + 3 \cdot \frac{5}{2} + \frac{1}{2} + \frac{3}{2}$.

DEFINITION 3.9. ([**HL04**], [**Gar10**]) The Hammond-Lewis birank of a bipartition $(\pi_1 \mid \pi_2)$ is defined to be

$$birank = \#\pi_2 - \#\pi_1. \tag{3.15}$$

REMARK 3.10. Observe that the Hammond-Lewis birank of a bipartition $(\pi_1|\pi_2)$ corresponding to a monomial in the basis (2.5) is precisely the charge of the monomial.

REMARK 3.11. Originally, the Hammond-Lewis birank was defined for bipartitions with integer parts, with a (immaterial) minus sign difference . We extended the definition as above because of the following consideration: In [**Ang15**] we proved that the field $\chi(z)$, and its descendant field $\chi(-z)$, generate a twisted vertex algebra on the space F_χ, which is twisted-vertex-algebra-isomorphic to the $\beta-\gamma$ system and its Fock space (i.e., with singularities both at $z = w$ and $z = -w$ formally allowed). If we consider then the isomorphic $\beta - \gamma$ system, the monomials corresponding to the basis (2.5) are indexed there with an integer indexing set. And so if we use the image of the untwisted Heisenberg field in the $\beta - \gamma$ system, and the image of the monomial basis (2.5), we would get a usual type of bipartition assigned to each monomial– a bipartition with integer parts, and the charge of the monomial will correspond exactly to the originally defined Hammond-Lewis birank as in [**HL04**], [**Gar10**].

Consider now the subspace of the highest weight vectors F_{χ}^{hwv} of F_{χ}. We know that it has a basis indexed by $\mathfrak{P}_{tdo}$, where $\mathfrak{P}_{tdo}$ denoted the set of distinct partitions with the first part a triangular number, and the other parts being distinct half-integers from $\mathbb{Z}+1/2$. Namely, to each partition from $\mathfrak{P}_{tdo}$ we can assign a highest weight vector, due to the Heisenberg decomposition Theorem 2.9. But the basis of F_{χ}^{hwv} can be indexed also in another way:

DEFINITION 3.12. Denote by $\mathfrak{BP}_{DI}$ the set of distinct integer bipartitions, namely bipartitions $(\pi_1 \mid \pi_2)$ such that each of π_1 and π_2 is a partition with distinct integer parts: $\pi_1=(m_k,\dots,m_2,m_1)$, $\pi_2=(n_s,\dots,n_2,n_1)$, where $m_k>\dots m_2>m_1$; $n_s>\dots n_2>n_1$, $m_i,n_j\in\mathbb{Z}_{>0}$, for any $i=1,2,\dots,k; j=1,2,\dots s$. Note that the same integer is allowed to occur in both π_1 and π_2.

According to Corollary 2.14, since the highest weight vectors can be considered to be elements of SF, we can assign a highest weight vector to each distinct integer bipartition $(\pi_1 \mid \pi_2)\in\mathfrak{BP}_{DI}$, as follows: if

$$(\pi_1 \mid \pi_2)=\big((m_k,\dots,m_2,m_1) \mid (n_s,\dots,n_2,n_1)\big),$$

where $m_k>\dots m_2>m_1$; $n_s>\dots n_2>n_1$; $m_i,n_j\in\mathbb{Z}_{>0}$, $i=1,2,\dots,k; j=1,2,\dots s$, then we assign to $(\pi_1 \mid \pi_2)$ the highest weight vector

$$H^{\beta}_{(-m_k)}\cdots H^{\beta}_{(-m_2)}H^{\beta}_{(-m_1)}H^{\gamma}_{(-n_s)}\cdots H^{\gamma}_{(-n_2)}H^{\gamma}_{(-n_1)}|0\rangle.$$

PROPOSITION 3.13. *There is a one-to-one correspondence between the set $\mathfrak{P}_{tdo}$ and the set of the distinct integer bipartitions $\mathfrak{BP}_{DI}$, via the basis of F_{χ}^{hwv}:*

$$\mathfrak{P}_{tdo}\overset{\sim}{\longleftrightarrow}\quad \textit{basis of } F_{\chi}^{hwv}\quad\overset{\sim}{\longleftrightarrow}\quad\mathfrak{BP}_{DI}.$$

Further, the Hammond-Lewis birank of a distinct integer bipartition $(\pi_1 \mid \pi_2)\in\mathfrak{BP}_{DI}$ is precisely the charge of the highest weight vector corresponding uniquely to the bipartition $(\pi_1 \mid \pi_2)\in\mathfrak{BP}_{DI}$.

This one-to-one correspondence is not an isometry, instead for $\mathfrak{P}_{tdo}\ni\mathfrak{p}\longleftrightarrow(\pi_1 \mid \pi_2)\in\mathfrak{BP}_{DI}$, the weights $|\mathfrak{p}|$ and $|(\pi_1 \mid \pi_2)|$ are connected via the following function $\mathcal{W}:\mathfrak{BP}_{DI}\to\mathbb{Z}_{\geq 0}+\frac{1}{2}$:

$$|\mathfrak{p}|=\mathcal{W}((\pi_1 \mid \pi_2)):=2|(\pi_1 \mid \pi_2)|+2ns-\frac{n(2n-1)}{2}-\frac{s(2s+1)}{2}, \tag{3.16}$$

Here $n=\#\pi_1$, the number of nonzero parts in the partition π_1, $s=\#\pi_2$, the number of nonzero parts in the partition π_2, T_n (correspondingly T_s) is the n-th (correspondingly s-th) triangular number.

The first part of the proposition follows directly as a corollary to Theorem 2.9 and Corollary 2.14. Formula (3.16) will hinge on the following Lemma that derives the degree of the highest weight vector corresponding to a given distinct integer bipartition.

LEMMA 3.14. *I. Let $v \in F_{\chi}^{hwv}$ corresponds to a bipartition $(\pi_1 \mid \emptyset)$, where π_2 is the empty partition, and $\pi_1 = (m_n, \dots, m_2, m_1)$, $m_n > \dots m_2 > m_1$, $m_i \in \mathbb{Z}_{>0}$, for any $i = 1, 2, \dots, n$. Then*

$$deg(v) = \frac{3n}{2} + 2|\pi_1| - 2T_n = 2|\pi_1| - \frac{n(2n-1)}{2}. \tag{3.17}$$

Here $n = \#\pi_1$, the number of nonzero parts in the partition π_1, T_n is the n-th triangular number.
II. Let $v \in F_{\chi}^{hwv}$ corresponds to a bipartition $(\emptyset \mid \pi_2)$, where π_1 is the empty partition, and $\pi_2 = (m_n, \dots, m_2, m_1)$, $m_n > \dots m_2 > m_1$, $m_i \in \mathbb{Z}_{>0}$, for any $i = 1, 2, \dots, n$. Then

$$deg(v) = \frac{n}{2} + 2|\pi_1| - 2T_n = 2|\pi_1| - \frac{n(2n+1)}{2}. \tag{3.18}$$

Here $n = \#\pi_2$, T_n is the n-th triangular number.
III. Let $v \in F_{\chi}^{hwv}$ corresponds to a bipartition $(\pi_1 \mid \pi_2)$, where $\pi_1 = (m_n, \dots, m_2, m_1)$, $m_n > \dots m_2 > m_1$, $m_i \in \mathbb{Z}_{>0}$, for any $i = 1, 2, \dots, n$; and $\pi_2 = (l_s, \dots, l_2, l_1)$, $l_s > \dots l_2 > l_1$, $l_i \in \mathbb{Z}_{>0}$, for any $i = 1, 2, \dots, s$. Then

$$\begin{aligned} deg(v) &= \frac{3n}{2} + 2|\pi_1| - 2T_n + 2ns + \frac{s}{2} + 2|\pi_2| - 2T_s \\ &= 2|\pi_1| + 2|\pi_2| + 2ns - \frac{n(2n-1)}{2} - \frac{s(2s+1)}{2}, \end{aligned} \tag{3.19}$$

Here $n = \#\pi_1$, the number of nonzero parts in the partition π_1, $s = \#\pi_2$, the number of nonzero parts in the partition π_2, T_n (correspondingly T_s) is the n-th (correspondingly s-th) triangular number.

The cases I and II above are actually special cases of III (when s =0, or correspondingly n=0), but we chose to state the formulas on their own for clarity.

PROOF. We will look at three types of representative examples which will illustrate two important principles when counting the degrees, without more of the already excessive indexing. We start with the basic example of action on the vacuum vector $|0\rangle$. By using directly the formulas (2.29) for the fields $H^{\beta}(z^2)$ and $H^{\gamma}(z^2)$ we can calculate

$$\begin{aligned} H^{\beta}(z^2)|0\rangle &= H^{\beta}_{(-1)}|0\rangle + H^{\beta}_{(-2)}|0\rangle z^2 + \dots + H^{\beta}_{(-k)}|0\rangle z^{2k-2} + \dots \\ &= V^+(z)^{-1}\beta_{\chi}(z^2)|0\rangle = \left(1 - h^{\mathbb{Z}}_{-1}z^2 + \dots\right)\left(\chi_{-3/2}|0\rangle + \chi_{-7/2}|0\rangle z^2 + \dots\right) \\ &= \chi_{-3/2}|0\rangle + \left(2\chi_{-7/2}|0\rangle - \chi^2_{-3/2}\chi_{-1/2}|0\rangle\right) z^2 + \dots \end{aligned}$$

Thus we see that acting on $|0\rangle$, $H^{\beta}_{(-1)}$ raises the degree by $\frac{3}{2}$, $H^{\beta}_{(-2)}$ raises the degree by $\frac{7}{2}$, and in general, acting on $|0\rangle$ $H^{\beta}_{(-k)}$ raises the degree by $2k - \frac{1}{2}$. We have

$$deg\big(H^{\beta}_{(-1)}|0\rangle\big) = \frac{3}{2}, \quad deg\big(H^{\beta}_{(-2)}|0\rangle\big) = \frac{7}{2} = \frac{3}{2} + 2, \quad \dots,$$
$$deg\big(H^{\beta}_{(-k)}|0\rangle\big) = 2k - \frac{1}{2} = \frac{3}{2} + 2(k-1).$$

Similarly, acting on $|0\rangle$, $H^{\gamma}_{(-n)}$ raises the degree by $2n - \frac{3}{2}$.

But, on elements of other charges this is not the case, and so next we consider the example of charge -1:

$$\begin{aligned}H^\beta(z^2)\big(\chi_{-3/2}|0\rangle\big) &= H^\beta(z^2)\big(H^\beta_{(-1)}|0\rangle\big)\\ &= H^\beta_{(-2)}\big(H^\beta_{(-1)}|0\rangle\big)z^2 + H^\beta_{(-3)}\big(H^\beta_{(-1)}|0\rangle\big)z^4 + \ldots + H^\beta_{(-k)}H^\beta_{(-1)}|0\rangle z^{2k-2} + \ldots\\ &= \chi^2_{-3/2}|0\rangle z^2 + \big(\chi_{-7/2}\chi_{-3/2}|0\rangle - h^{\mathbb{Z}}_{-1}\chi^2_{-3/2}|0\rangle\big)z^4 + \ldots.\end{aligned}$$

So we see that acting on $\chi_{-3/2}|0\rangle = H^\beta_{(-1)}|0\rangle$, $H^\beta_{(-1)}$ acts as 0 (as it is fermionic), but instead $H^\beta_{(-2)}$ raises the degree by $\frac{3}{2}$ (and not as before by $\frac{7}{2}$), and in general, acting on $\chi_{-3/2}|0\rangle$, $H^\beta_{(-k)}$, $k \geq 2$, raises the degree by $2k - \frac{5}{2}$. Thus we have a "resetting of the counter" principle, where each $H^\beta_{(-k)}$ adds a degree contribution depending on the charge of the vector it is acting on (or equivalently, the number of nonzero partition parts in the bipartition describing that highest weight vector). Specifically, we have

$$\begin{aligned}&deg\big(H^\beta_{(-2)}H^\beta_{(-1)}|0\rangle\big) = \frac{2\cdot 3}{2}, \quad deg\big(H^\beta_{(-3)}H^\beta_{(-1)}|0\rangle\big) = \frac{7}{2} + \frac{3}{2} = \frac{2\cdot 3}{2} + 2,\\ &deg\big(H^\beta_{(-k)}H^\beta_{(-1)}|0\rangle\big) = \frac{3}{2} + 2k - \frac{5}{2} = \frac{2\cdot 3}{2} + 2(k-2).\end{aligned}$$

"Resetting of the counter" refers to the fact that when we considering adding new contributions to the degree, we always start at the lowest possible addition (in the example above, we started at $\frac{3}{2}$, even though it was $H^\beta_{(-2)}$ that acted).

At each fixed charge $-n < 0$ (here $n > 0$), the vector $\chi^n_{-3/2}|0\rangle$ is the vector with minimal degree , $\frac{3n}{2}$, with that charge $-n$. This vector is given by

$$\chi^n_{-\frac{3}{2}}|0\rangle = H^\beta_{(-n)}\cdots H^\beta_{(-2)}H^\beta_{(-1)}|0\rangle, \tag{3.20}$$

and corresponds to the standard distinct partition with n parts $(n, n-1, n-2, \ldots 1)$. Acting on $\chi^n_{-3/2}|0\rangle$, the first nonzero possible action is by $H^\beta_{(-n-1)}$, and that will add $\frac{3}{2}$ to the degree, by the "resetting of the counter" principle.

Let us consider the highest weight vector

$$v = H^\beta_{(-m_n)}\cdots H^\beta_{(-m_2)}\cdots H^\beta_{(-m_1)}|0\rangle,$$

corresponding to the bipartition $(\emptyset \mid \pi_2)$, where π_1 is the empty partition, and $\pi_2 = (m_k, \ldots, m_2, m_1)$. The same "resetting of the counter" principle as above applies. And so to calculate its degree we will have to see "how much" this vector differs from the vector with minimal degree of that same charge (the charge of that vector equals $-n$, with n the number of nonzero parts in the partition π_2). Thus we see that the degree is

$$deg(v) = \frac{3n}{2} + 2(m_n - n) + \cdots + 2(m_2 - 2) + 2(m_1 - 1),$$

which gives the formula (3.17). Formula (3.18) is proved the same way.

To prove the formula (3.19), we need to discuss the second important principle when counting the degrees of the highest weight vectors. Consider the following example, it is the easiest example of a "mixture" between the β and γ fields, i.e., neither the π_1, nor the π_2 are the empty partitions:

$$\begin{aligned}H^{\beta}(z^2)\big(\chi_{-1/2}|0\rangle\big) &= H^{\beta}(z^2)H^{\gamma}_{(-1)}|0\rangle\big)\\ &= H^{\beta}_{(1)}H^{\gamma}_{(-1)}|0\rangle\cdot\frac{1}{z^4}+H^{\beta}_{(0)}H^{\gamma}_{(-1)}|0\rangle\cdot\frac{1}{z^2}+H^{\beta}_{(-1)}H^{\gamma}_{(-1)}|0\rangle\\ &\quad+\dots+H^{\beta}_{(-k)}H^{\gamma}_{(-1)}|0\rangle z^{2k-2}+\dots\\ &= \frac{|0\rangle}{z^4}+0\cdot\frac{1}{z^2}+v_{4;0}+v_{6;0}z^2+\dots+v_{2k+4;0}z^{2k}+\dots,\end{aligned}$$

By direct comparison with the formula (2.29):

$$\begin{aligned}H^{\beta}(z^2)\big(\chi_{-1/2}|0\rangle\big) &= V^{+}(z)^{-1}\beta_{\chi}(z^2)\cdot\frac{1}{z^2}\big(\chi_{-1/2}|0\rangle\big)\\ &= \frac{1}{z^2}\big(1-h^{\mathbb{Z}}_{-1}z^2+\dots\big)\Big(\frac{|0\rangle}{z^2}+\chi_{-3/2}\chi_{-1/2}|0\rangle+\chi_{-7/2}\chi_{-1/2}|0\rangle z^2+\dots\Big)\\ &= \frac{|0\rangle}{z^4}+\big(-h^{\mathbb{Z}}_{-1}|0\rangle+\chi_{-3/2}\chi_{-1/2}|0\rangle\big)\cdot\frac{1}{z^2}+\\ &\quad+\Big(\frac{1}{2}\big((h^{\mathbb{Z}}_{-1})^2-h^{\mathbb{Z}}_{-2}\big)|0\rangle-h^{\mathbb{Z}}_{-1}\chi_{-3/2}\chi_{-1/2}|0\rangle+\chi_{-7/2}\chi_{-1/2}|0\rangle\Big)+\dots,\end{aligned}$$

we see that the vector $v_{4;0}$ is of charge 0 and degree 4. One can calculate that it is given by

$$v_{4;0}=\chi^2_{-\frac{3}{2}}\chi^2_{-\frac{1}{2}}|0\rangle-2\chi_{-\frac{7}{2}}\chi_{-\frac{1}{2}}|0\rangle+2\chi_{-\frac{5}{2}}\chi_{-\frac{5}{2}}|0\rangle.$$

More generally, in the above formula $v_{2k+4;0}$ is of charge 0 and degree $2k+4$. Thus we see that action of $H^{\beta}_{(1)}$ on $H^{\gamma}_{(-1)}|0\rangle=\chi_{-1/2}|0\rangle$ lowers the degree by $\frac{1}{2}$, H^{β}_{0} acts as 0, $H^{\beta}_{(-1)}$ raises the degree by $\frac{7}{2}$, $H^{\beta}_{(-2)}$ raises the degree by $\frac{11}{2}$, and in general, acting on $\chi_{-1/2}|0\rangle$ $H^{\beta}_{(-k)}$ raises the degree by $2k+\frac{3}{2}$. Now if the partition π_2 was empty, the $H^{\beta}_{(-1)}$ would have raised the degree by $\frac{3}{2}$, but instead, it now raised the degree by $\frac{7}{2}$. This is due to the positively indexed $H^{\beta}_{(1)}$ acting nontrivially, and thus the presence of the H^{β}_{0}, which always acts as 0, but nominally should have raised the degree by $\frac{3}{2}$. That forces the $H^{\beta}_{(-1)}$ to "skip" $\frac{3}{2}$, but instead go to the next available $\frac{7}{2}=\frac{3}{2}+2\cdot 1$.

Similarly, consider one more example of the same kind:

$$\begin{aligned}H^{\beta}(z^2)\big(\chi^2_{-1/2}|0\rangle\big) &= H^{\beta}(z^2)\big(H^{\gamma}_{(-2)}H^{\gamma}_{(-1)}|0\rangle\big)\\ &= H^{\beta}_{(2)}\big(H^{\gamma}_{(-2)}H^{\gamma}_{(-1)}|0\rangle\big)\frac{1}{z^6}+H^{\beta}_{(1)}\big(H^{\gamma}_{(-2)}H^{\gamma}_{(-1)}|0\rangle\big)\frac{1}{z^4}\\ &\quad+H^{\beta}_{(0)}\big(\chi^2_{-1/2}|0\rangle\big)\frac{1}{z^2}+H^{\beta}_{(-1)}\big(\chi^2_{-1/2}|0\rangle\big)+\dots\\ &= \frac{H^{\gamma}_{(-1)}|0\rangle}{z^6}+\frac{-H^{\gamma}_{(-2)}|0\rangle}{z^4}+0\cdot\frac{1}{z^2}+v_{\frac{13}{2};0}+v_{\frac{17}{2};0}z^2+\dots+\\ &= \frac{\chi_{-1/2}|0\rangle}{z^6}-\frac{2\chi_{-5/2}|0\rangle+\chi_{-3/2}\chi^2_{-1/2}|0\rangle}{z^4}+0\cdot\frac{1}{z^2}+v_{\frac{13}{2};0}+\dots\end{aligned}$$

Hence action of $H^{\beta}_{(2)}$ on $\chi^2_{-1/2}|0\rangle$ lowers the degree by $\frac{1}{2}$, $H^{\beta}_{(1)}$ raises the degree by $\frac{3}{2}$, H^{β}_{0} acts as 0 (but nominally should have raised the degree by $\frac{7}{2}$), $H^{\beta}_{(-1)}$ raises the degree by $\frac{11}{2}$, $H^{\beta}_{(-2)}$ raises the degree by $\frac{15}{2}$, and in general, acting on $\chi^2_{-1/2}|0\rangle$,

$H^{\beta}_{(-k)}$ raises the degree by $2k+\frac{7}{2}$. Hence, we again see that $H^{\beta}_{(-1)}$ which would have raised the degree by $\frac{3}{2}$, now raised the degree instead by $\frac{11}{2}$. This is due to the fact that since the charge of $\chi^2_{-1/2}|0\rangle$ is 2, $H^{\beta}_{(-1)}$ now has to "skip" two steps, from $\frac{3}{2}$ to $\frac{11}{2}=\frac{3}{2}+2\cdot 2$. One can easily see that the "skipping s steps" principle will hold when $H^{\beta}_{(-1)}$ acts on any vector of charge s.

Consider then the bipartition $(\pi_1 \mid \pi_2) \in \mathfrak{BP}_{DI}$

$$(\pi_1 \mid \pi_2) = \big((m_k,\dots,m_2,m_1) \mid (l_s,\dots,l_2,l_1)\big),$$

where $m_k > \dots m_2 > m_1$; $l_s > \dots l_2 > l_1$; $m_i, l_j \in \mathbb{Z}_{>0}$, $i=1,2,\dots,k; j = 1,2,\dots s$, and the highest weight vector v assigned to it:

$$v = H^{\beta}_{(-m_k)}\cdots H^{\beta}_{(-m_2)}H^{\beta}_{(-m_1)}H^{\gamma}_{(-l_s)}\cdots H^{\gamma}_{(-l_2)}H^{\gamma}_{(-l_1)}|0\rangle.$$

Applying both the principle of "skipping" s steps (because the charge of the vector $H^{\gamma}_{(-l_s)}\cdots H^{\gamma}_{(-l_2)}H^{\gamma}_{(-l_1)}|0\rangle$ is s), and then the "resetting the counter" principle, we see that

$$deg(v) = n\left(\frac{3}{2}+2s\right)+2(m_n-n)+\dots+2(m_2-2)+2(m_1-1)+\frac{s}{2}$$
$$+2(l_s-s)+2\dots(l_2-2)+2(l_1-1),$$

which gives the formula (3.19). □

The Proposition 3.13 then follows directly. This leads to the following:

THEOREM 3.15.

(3.21)
$$dim_{q,z,r}^{L_0,h_0^{\mathbb{Z}},L_0^h} F_{\chi} = \sum_{(\pi_1|\pi_2)\in\mathfrak{BP}_{DI}} \frac{q^{\mathcal{W}((\pi_1|\pi_2))}z^{birank((\pi_1|\pi_2))}r^{-birank((\pi_1|\pi_2))^2/2}}{\prod_{n\geq 1}(1-q^{2n}r^n)}.$$

Here $\mathcal{W}((\pi_1|\pi_2))$ *is the weight function from* (3.16).

PROOF. The proof of this theorem is similar to the proof of Proposition 3.4, and hinges on the fact that

(3.22) $$[L_0^h, h_n^{\mathbb{Z}}] = -nh_n^{\mathbb{Z}}, \quad \text{for any } n\in\mathbb{Z},$$

as well as (3.7), and the Lemma above. □

An important consequence of any bosonization is that by calculating the character (graded dimension) of both the fermionic and the bosonic side of the correspondence one can obtain identities relating certain product formulas to certain sum formulas. Such a sum-vs-product identity perfectly illustrates the equality between the fermionic side (the product formulas) and the bosonic side (the sum formulas). For example, from the classical boson-fermion correspondence (of type A) one can directly obtain the Jacobi triple product identity — as was done in [**Kac98**], and in [**Ang14**] for the bosonization of type D-A. To derive the relevant sum-vs-product formula for the CKP, we continue by considering just the first two grading operators, $h_0^{\mathbb{Z}}$ and L_0, as they are in some sense the most natural grading operators for the Fock space F_{χ}. In the CKP case, as we noticed, nothing is quite as straightforward as in the usual boson-fermion correspondence case. Here the "fermionic side" is not purely fermionic, but instead one can view it as a fermion times a boson (see Corollary 2.14) . In [**Ang17**] we showed the bosonic side of the character:

$$dim_{q,z}^{L_0,h_0^{\mathbb{Z}}} F_\chi = \frac{1}{\prod_{j\in\mathbb{Z}_+}\left(1-zq^{2j-\frac{3}{2}}\right)\left(1-z^{-1}q^{2j-\frac{1}{2}}\right)}$$

Here we obtain the new (and rather complicated formulas):

THEOREM 3.16.

$$(3.23)\quad dim_{q,z}^{L_0,h_0^{\mathbb{Z}}} F_\chi^{hwv}$$
$$= \frac{\prod_{l=1}^{\infty}\left(1+zq^{2l-\frac{3}{2}}\right)\prod_{l=1}^{\infty}\left(1+z^{-1}q^{2l-\frac{1}{2}}\right)}{2\prod_{i=1}^{\infty}(1-q^{4i})(1+q^{2i})}\cdot\left(\sum_{k=0}^{\infty}\frac{z^k q^{\frac{k}{2}}}{1+z^{-1}q^{2k+\frac{3}{2}}}+\sum_{k=0}^{\infty}\frac{z^{-k}q^{\frac{3k}{2}}}{1+zq^{2k+\frac{1}{2}}}\right).$$

$$(3.24)\quad dim_{q,z}^{L_0,h_0^{\mathbb{Z}}} F_\chi$$
$$= \frac{\prod_{l=1}^{\infty}\left(1+zq^{2l-\frac{3}{2}}\right)\prod_{l=1}^{\infty}\left(1+z^{-1}q^{2l-\frac{1}{2}}\right)}{2\prod_{l=1}^{\infty}\left(1-q^{4l}\right)^2}\cdot\left(\sum_{k=0}^{\infty}\frac{z^k q^{\frac{k}{2}}}{1+z^{-1}q^{2k+\frac{3}{2}}}+\sum_{k=0}^{\infty}\frac{z^{-k}q^{\frac{3k}{2}}}{1+zq^{2k+\frac{1}{2}}}\right)$$

Comparing the two formulas for $dim_{q,z}^{L_0,h_0^{\mathbb{Z}}} F_\chi$ *we obtain the identity*

$$(3.25)\quad \sum_{k=0}^{\infty}\frac{z^k q^{\frac{k}{2}}}{1+z^{-1}q^{2k+\frac{3}{2}}}+\sum_{k=0}^{\infty}\frac{z^{-k}q^{\frac{3k}{2}}}{1+zq^{2k+\frac{1}{2}}} = \frac{2\prod_{l=1}^{\infty}(1-q^{4l})^2}{\prod_{l=1}^{\infty}\left(1-z^2q^{4l-3}\right)\prod_{l=1}^{\infty}\left(1-z^{-2}q^{4l-1}\right)}$$

REMARK 3.17. The character (3.23), besides being necessary for deriving (3.24), is interesting on its own, because the vector space F_χ^{hwv} can be identified as the vector space of what we are told physicists would regard as the coset symmetry algebra of the $\beta-\gamma$ system by the Heisenberg field $h(z)$, see [**Rid10**] and [**CR13**] (as opposed to the coset vertex subalgebra as considered by mathematicians, see e.g. [**Kac98**], [**LL04**], which would be the charge 0 subspace of F_χ^{hwv} here).

REMARK 3.18. The last identity (3.25) is somewhat surprising, as the right-hand side manifestly only includes integral powers of q and even powers of z, which implies that the left-hand side must do too (not obvious from its form). We show in the Appendix that this identity can be obtained also as a specialization of the Ramanujan Psi summation formula (see e.g. [**HA40**], [**And69**], [**AA78**])

$$\sum_{k=-\infty}^{\infty}\frac{(a;q)_k}{(b;q)_k}x^k = \frac{(q;q)_\infty\left(\frac{b}{a};q\right)_\infty\left(\frac{q}{ax};q\right)_\infty(ax;q)_\infty}{(b;q)_\infty\left(\frac{b}{ax};q\right)_\infty\left(\frac{q}{a};q\right)_\infty(x;q)_\infty}.$$

Here as usual we denote

$$(b;q)_\infty := \prod_{i=0}^{\infty}(1-bq^i),\quad (b;q)_n := \frac{(b;q)_\infty}{(bq^n;q)_\infty}.$$

PROOF. Since all elements of each fixed Heisenberg irreducible module have the same charge, the following holds for any Heisenberg irreducible module V in the decomposition into irreducible modules:

$$dim_{q,z}^{L_0,h_0^{\mathbb{Z}}} V = \frac{z^\lambda q^{deg(v)}}{\prod_{l=1}^{\infty}(1-q^{2l})}.$$

From (2.24) we have

$$dim_{q,z}^{L_0,h_0^{\mathbb{Z}}} F_\chi = \frac{dim_{q,z}^{L_0,h_0^{\mathbb{Z}}} F_\chi^{hwv}}{\prod_{l=1}^{\infty}(1-q^{2l})}. \tag{3.26}$$

The character formula (3.24) will then result from (3.23). Hence we now proceed to prove (3.23), by using the fermionic description of the vector space F_χ^{hwv}. From Corollary 2.14 we have that F_χ^{hwv} is spanned by vectors of the form

$$H^\beta_{(n_1)} H^\beta_{(n_2)} \cdots H^\beta_{(n_k)} H^\gamma_{(m_1)} H^\gamma_{(m_2)} \cdots H^\gamma_{(m_l)} |0\rangle,$$

where $n_s < 0, s = 1, 2, \dots k$ and $m_s < 0, s = 1, 2, \dots l$. The problem is that although an action by a $H^\gamma_{(m_s)}$ always raises the charge by 1, and action by a $H^\beta_{(n_s)}$ always lowers the charge by 1, those actions are not so uniform with respect to the degrees, as we saw in Lemma 3.14 and its proof. The $H^\beta_{(n)}$ (and similarly $H^\gamma_{(n)}$) raise the degree differently depending on the charge of the element on which they act, but in a predictable manner, via the 'resetting the counter" and the "skipping steps" principles. The formula we need has to take this into account. Thus we have to start by considering how the product type contribution will depend on the charge of the vector it is generated from. We start from the vector $\chi^n_{-3/2}|0\rangle$, which is the minimal degree vector of charge $-n$. We can see from Lemma 3.14 that acting by all possible $H^\beta_{(-m)}$, $-m < -n$, on $\chi^n_{-3/2}|0\rangle$ will produce all the various factors of type $z^{-1}q^{2m-\frac{1}{2}}$, $m \geq 1$. Thus we see that for the various vectors of the type

$$H^\beta_{(-m_1)} H^\beta_{(-m_2)} \cdots H^\beta_{(-m_k)} \chi^n_{-3/2}|0\rangle,$$

where $-m_s < -n, s = 1, 2, \dots k$, the following product is a "first iteration" to this part of the contribution to the graded dimension:

$$z^{-n} q^{\frac{3n}{2}} \prod_{m=1}^{\infty} \left(1 + z^{-1} q^{2m-\frac{1}{2}}\right). \tag{3.27}$$

This product is factually wrong, but "morally correct". It is factually wrong due to the fact that in this product the contribution coming from $H^\beta_{(-n-1)}\chi^n_{-3/2}|0\rangle$ is correct ($z^{-n}q^{\frac{3n}{2}} \cdot z^{-1}q^{\frac{3}{2}}$), but the contribution corresponding to $H^\beta_{(-n-2)} H^\beta_{(-n-1)} \chi^n_{-3/2}|0\rangle$ is off by a factor of q^2: due to the "resetting the counter" principle of Lemma 3.14 it should be

$$z^{-n} q^{\frac{3n}{2}} \cdot z^{-1} q^{\frac{3}{2}} \cdot z^{-1} q^{\frac{3}{2}},$$

but in the product above it is

$$z^{-n} q^{\frac{3n}{2}} \cdot z^{-1} q^{\frac{3}{2}} \cdot z^{-1} q^{\frac{7}{2}}.$$

Nevertheless, as we will see, these discrepancies will be to our advantage, as they will allow us, when counting the infinite number of repetitions, to count each repetition with a different factor, and thereby avoid infinite coefficients when summing up infinitely repeated elements. The reason is that we will have to consider all the varying products generated from the minimal degree vectors, not just a single product. Thus we will have the same vector $\chi^n_{-3/2}|0\rangle$ first appear directly when generating from itself, as above, then appear again when generating from $\chi^{n-1}_{-3/2}|0\rangle$ as $H^\beta_{(-n)}\chi^{n-1}_{-3/2}|0\rangle$, and again from $\chi^{n-2}_{-3/2}|0\rangle$ as $H^\beta_{(-n)} H^\beta_{(-n+1)} \chi^{n-2}_{-3/2}|0\rangle$, etc... The products (3.27), when considering all $n \in \mathbb{N}$, count each element $\chi^n_{-3/2}|0\rangle$ twice with a coefficient 1, then once with a coefficient q^2, and in general, once with a coefficient q^{2T_m}, where T_m is the mth triangular number.

Similarly, at each fixed charge $n>0$, the vector $\chi^n_{-1/2}|0\rangle$ is the vector of minimal degree with that charge n. Action by $H^\gamma_{(-m)}$, $-m<-n$, on $\chi^n_{-1/2}|0\rangle$ will produce all the various factors of type $zq^{2m-\frac{3}{2}}$. The various vectors of the type

$$H^\gamma_{(-m_1)}H^\gamma_{(-m_2)}\cdots H^\gamma_{(-m_k)}\chi^n_{-1/2}|0\rangle,$$

where $-m_s<-n, s=1,2,\dots k$, will result in the following product contributions to the graded dimension:

$$z^nq^{\frac{n}{2}}\prod_{m=1}^{\infty}\left(1+zq^{2m-\frac{3}{2}}\right), \tag{3.28}$$

of course again taking into account that in these products the "repetitions" are counted with different factors, not with their exact contributions (i..e, the products as above are again factually wrong, but "morally correct").

The two types of products above, (3.27) and (3.28) represent the actions by the $H^\beta_{(n)}$ on highest weight vectors of negative charge, as well as $H^\gamma_{(n)}$ on highest weight vectors of poisitve charge (not in an exact, but in a "morally correct", manner). Now, let us consider the actions by the $H^\beta_{(n)}$ on highest weight vectors of positive charge, as well as $H^\gamma_{(n)}$ on highest weight vectors of negative charge. Due to the already heavy indexing, we will consider representative examples, instead of working with generally indexed highest weight vectors. The first example of a "mixture" $H^\beta(z^2)(\chi_{-1/2}|0\rangle)$ shows, see Lemma 3.14, that according to the "skipping steps" principle, the factor $(1+z^{-1}q^{\frac{3}{2}})$ is missing in the would-be product contribution, due to the fact that H^β_0 always acts as 0 on F^{hwv}_χ. We see that in this "mixture" we would have a product of the type (always keeping in mind that this product is "morally correct", but factually wrong):

$$\frac{zq^{\frac{1}{2}}\prod_{m=0}^{\infty}\left(1+z^{-1}q^{2m-\frac{1}{2}}\right)}{1+z^{-1}q^{\frac{3}{2}}}. \tag{3.29}$$

A final example:

$$\begin{aligned}
H^\gamma(z^2)\big(\chi^2_{-3/2}|0\rangle\big) &= H^\gamma(z^2)\big(H^\beta_{(-2)}H^\beta_{(-1)}|0\rangle\big)\\
&= H^\gamma_{(2)}\big(H^\beta_{(-2)}H^\beta_{(-1)}|0\rangle\big)\frac{1}{z^6}+H^\gamma_{(1)}\big(H^\beta_{(-2)}H^\beta_{(-1)}|0\rangle\big)\frac{1}{z^4}\\
&\quad+H^\gamma_{(0)}\big(\chi^2_{-3/2}|0\rangle\big)\frac{1}{z^2}+H^\gamma_{(-1)}\big(\chi^2_{-3/2}|0\rangle\big)+\dots\\
&= \frac{-H^\beta_{(-1)}|0\rangle}{z^6}+\frac{H^\beta_{(-2)}|0\rangle}{z^4}+0\cdot\frac{1}{z^2}+v_{\frac{15}{2};0}+v_{\frac{19}{2};0}z^2+\dots+\\
&= \frac{-\chi_{-3/2}|0\rangle}{z^6}+\frac{2\chi_{-7/2}|0\rangle-\chi^2_{-3/2}\chi_{-1/2}|0\rangle}{z^4}+0\cdot\frac{1}{z^2}+v_{\frac{15}{2};0}+\dots
\end{aligned}$$

Hence action of $H^\gamma_{(2)}$ on $\chi^2_{-3/2}|0\rangle$ lowers the degree by $\frac{3}{2}$, $H^\gamma_{(1)}$ raises the degree by $\frac{1}{2}$, H^γ_0 acts as 0, $H^\gamma_{(-1)}$ raises the degree by $\frac{9}{2}$, $H^\gamma_{(-2)}$ raises the degree by $\frac{13}{2}$, and in general, acting on $\chi^2_{-3/2}|0\rangle$, $H^\gamma_{(-n)}$ raises the degree by $2n+\frac{5}{2}$. Most importantly, the factor $(1+zq^{\frac{5}{2}})$ is missing, as per the "skipping steps" principle, due to H_0^γ always acting as 0 on F_χ^{hwv}.

Similar calculation can be done in general for the actions of $H_{(m)}^\beta$ on $\chi_{-1/2}^n|0\rangle$; as well as for the actions of $H_{(m)}^\gamma$ on $\chi_{-3/2}^n|0\rangle$; always remembering that the factors corresponding to the actions of H_0^β and H_0^γ, are missing.

The next crucial consideration is that any element of F_χ^{hwv} can be generated from any of the "minimal degree" vectors of any fixed charge (the $\chi_{-3/2}^n|0\rangle$ if the charge $-n$ is negative, or $\chi_{-3/2}^2|0\rangle$ if the charge n is positive), by action of $H_{(m)}^\gamma$ and $H_{(m)}^\beta$, but with $m\in\mathbb{Z}$ (positive and negative indexes allowed).

Thus, before we even try to calculate the exact sums of (factually correct) products resulting from the corresponding actions on each of the "minimal degree" vectors $\chi_{-1/2}^n|0\rangle$ and $\chi_{-3/2}^n|0\rangle$, we need to consider that if we do just that we would have counted each basis vector infinitely many times, and thus such a sum would be pointless to calculate. Counting the degree contributions "by charge" (and therefore infinite multiple counting) is unavoidable, because of the way the degree contributions change depends on the charge. Thus we do not have a single infinite product, but instead we have to sum infinitely products generated from each of the minimal degree vectors, which will entail encountering each basis vector infinitely many times. And so instead of trying to calculate the exact products contributions (we are not sure it is possible), we will consider instead adding the "morally correct" (but factually wrong) product contributions. Not only will this actually uniformize the products, but using the not-factually exact, but "morally"-correct products will allow us to count each basis vector with a different coefficient each time it is encountered. Namely, we consider the products, for each $n\in\mathbb{Z}_{\geq 0}$:

$$\frac{z^n q^{\frac{n}{2}}\prod_{l=1}^{\infty}\left(1+zq^{2l-\frac{3}{2}}\right)\prod_{l=1}^{\infty}\left(1+z^{-1}q^{2l-\frac{1}{2}}\right)}{1+z^{-1}q^{2n+\frac{3}{2}}},\quad\text{and}\qquad \frac{z^{-n} q^{\frac{3n}{2}}\prod_{l=1}^{\infty}\left(1+zq^{2l-\frac{3}{2}}\right)\prod_{l=1}^{\infty}\left(1+z^{-1}q^{2l-\frac{1}{2}}\right)}{1+zq^{2n+\frac{1}{2}}} \tag{3.30}$$

Observe in the above that we "shifted", for instance compare the first product above, for $n=1$, vs the product of (3.29)— we shifted the "missing factor" from $(1+z^{-1}q^{\frac{3}{2}})$ in (3.29) to $(1+z^{-1}q^{\frac{7}{2}})$ in (3.30), as well as the starting factor from $(1+z^{-1}q^{-\frac{1}{2}})$ to $(1+z^{-1}q^{\frac{3}{2}})$. Similarly, we know a factor would be missing from $\prod_{l=1}^{\infty}\left(1+z^{-1}q^{2l-\frac{1}{2}}\right)$ due to $H_{(m)}^\beta$ always acting by 0, but we shifted all the factors from $1+z^{-1}q^{2m-\frac{1}{2}}$ in (3.29) to $1+z^{-1}q^{2m+\frac{3}{2}}$ in the product in (3.30). For example, this will entail that the minimal degree vector $\chi_{-3/2}|0\rangle = H_{(-1)}^\beta|0\rangle$, which can be obtained by the consecutive action $H_{(-1)}^\beta H_{(1)}^\beta\chi_{-3/2}|0\rangle$, is counted with a different factor of $q^{2+4}=q^{2T_2}$ in the first product in (3.30). Thus the shifts are introduced to allow for the "skipping the steps" principle, but also to make sure that we don't sum the same exact term (corresponding to the same exact basis vector) infinitely many times, instead we sum it with a different degree coefficient each time.

Thus, generalizing the observation from above, we see that if we sum the products (3.30) we would have counted each basis vector infinitely many times, but with a different degree coefficient each time. That coefficient is q^{2T_m}, where T_m is the m-th triangular number— in fact we encounter the coefficient q^{2T_m} exactly twice among the infinite number of times we encounter each basis vector. Thus the sum of all the products from (3.30) equals

$$\left(2\sum_{m\in\mathbb{Z}_{\geq 0}} q^{2T_m}\right)\cdot dim_{q,z}^{L_0,h_0^{\mathbb{Z}}} F_{\chi}^{hwv}, \tag{3.31}$$

and so

$$\left(2\sum_{m\in\mathbb{Z}_{\geq 0}} q^{2T_m}\right)\cdot dim_{q,z}^{L_0,h_0^{\mathbb{Z}}} F_{\chi}^{hwv} \tag{3.32}$$
$$=\prod_{l=1}^{\infty}(1+zq^{2l-\frac{3}{2}})\prod_{l=1}^{\infty}(1+z^{-1}q^{2l-\frac{1}{2}})\cdot\left(\sum_{n=0}^{\infty}\frac{z^nq^{\frac{n}{2}}}{1+z^{-1}q^{2n+\frac{3}{2}}}+\sum_{n=0}^{\infty}\frac{z^{-n}q^{\frac{3n}{2}}}{1+zq^{2n+\frac{1}{2}}}\right).$$

Now we use the following Jacobi formula for the triangular numbers [1]

$$\sum_{m\in\mathbb{Z}_{\geq 0}} q^{2T_m} = 1+q^2+q^4+q^{12}+q^{20}+\cdots+q^{2T_m}+\dots$$
$$=\frac{\prod_{i=1}^{\infty}(1-q^{4i})}{\prod_{i=1}^{\infty}(1-q^{4i-2})}=\prod_{i=1}^{\infty}(1-q^{4i})(1+q^{2i}).$$

Hence we get

$$dim_{q,z}^{L_0,h_0^{\mathbb{Z}}} F_{\chi}^{hwv} \tag{3.33}$$
$$=\frac{\prod_{l=1}^{\infty}(1+zq^{2l-\frac{3}{2}})\prod_{l=1}^{\infty}(1+z^{-1}q^{2l-\frac{1}{2}})}{2\prod_{i=1}^{\infty}(1-q^{4i})(1+q^{2i})}\cdot\left(\sum_{n=0}^{\infty}\frac{z^nq^{\frac{n}{2}}}{1+z^{-1}q^{2n+\frac{3}{2}}}+\sum_{n=0}^{\infty}\frac{z^{-n}q^{\frac{3n}{2}}}{1+zq^{2n+\frac{1}{2}}}\right),$$

which is precisely (3.23). Thus using (3.26) we derive (3.24). □

COROLLARY 3.19. *The following equalities hold:*

$$dim_{q,z}^{L_0,h_0^{\mathbb{Z}}} F_{\chi}^{hwv} = \sum_{(\pi_1|\pi_2)\in\mathfrak{BP}_{DI}} q^{\mathcal{W}((\pi_1|\pi_2))}z^{birank((\pi_1|\pi_2))}$$
$$=\prod_{n\geq 1}\frac{(1-q^{2n})}{(1-zq^{2n-\frac{3}{2}})(1-z^{-1}q^{2n-\frac{1}{2}})}$$
$$=\prod_{l=1}^{\infty}\frac{(1+zq^{2l-\frac{3}{2}})(1+z^{-1}q^{2l-\frac{1}{2}})}{2(1-q^{4l})(1+q^{2l})}\cdot\left(\sum_{k=0}^{\infty}\frac{z^kq^{\frac{k}{2}}}{1+z^{-1}q^{2k+\frac{3}{2}}}+\sum_{k=0}^{\infty}\frac{z^{-k}q^{\frac{3k}{2}}}{1+zq^{2k+\frac{1}{2}}}\right).$$

Finally, we would like to underline a connection between the CKP hierarchy, its bosonization (and thus the $\beta-\gamma$ system), and Dyson's crank of a partition. Recall the Dyson crank of a partition λ is defined as follows ([**AG88**]): Let $l(\lambda)$ denote the largest part of λ, $\omega(\lambda)$ denote the number of 1's in λ, and $\mu(\lambda)$ denote the number of parts of λ larger than $\omega(\lambda)$. The crank $c(\lambda)$ is given by

$$c(\lambda)=\begin{cases} l(\lambda) & \text{if } \omega(\lambda)=0\\ \mu(\lambda)-\omega(\lambda) & \text{if } \omega(\lambda)>0.\end{cases}$$

[1]This identity can be found on page 185 of the original manuscript by Jacobi, [**Jac29**].

Denote by $N'(m,n)$ the number of partitions of n with crank equal to m, with the exception that for $n=1$ we set $N'(-1,1)=-N'(0,1)=N'(1,1)=1$. For convenience we also set $N'(m,n)=0$ whenever m or n is not an integer. Specifically, the generating function for $N'(m,n)$ is ([**AG88**]):

$$\sum_{n=0}^{\infty}\sum_{m=-\infty}^{\infty} N'(m,n)z^m q^n = \prod_{n=1}^{\infty}\frac{(1-q^n)}{(1-zq^n)(1-z^{-1}q^n)}. \tag{3.34}$$

Comparing this generating function to the identities of Corollary 3.19, we see that:

COROLLARY 3.20.

(3.35)

$$\begin{aligned} dim_{q,z}^{L_0,h_0^{\mathbb{Z}}} F_{\chi}^{hwv} &= \sum_{n=0}^{\infty}\sum_{m=-\infty}^{\infty}\sum_{l=0}^{\infty} N'(m-l,n)z^m q^{2n+\frac{m}{2}} \\ &= \sum_{(\pi_1|\pi_2)\in\mathfrak{BP}_{DI}} q^{\mathcal{W}((\pi_1|\pi_2))} z^{birank((\pi_1|\pi_2))} \end{aligned}$$

(3.36)

$$\begin{aligned} &= \prod_{n\geq 1}\frac{(1-q^{2n})}{\left(1-zq^{2n-\frac{3}{2}}\right)\left(1-z^{-1}q^{2n-\frac{1}{2}}\right)} \\ &= \prod_{l=1}^{\infty}\frac{\left(1+zq^{2l-\frac{3}{2}}\right)\left(1+z^{-1}q^{2l-\frac{1}{2}}\right)}{2(1-q^{4l})(1+q^{2l})}\cdot\left(\sum_{k=0}^{\infty}\frac{z^k q^{\frac{k}{2}}}{1+z^{-1}q^{2k+\frac{3}{2}}}+\sum_{k=0}^{\infty}\frac{z^{-k}q^{\frac{3k}{2}}}{1+zq^{2k+\frac{1}{2}}}\right). \end{aligned}$$

PROOF. Note that although there are several apparently infinite sums in $\sum_{n=0}^{\infty}\sum_{m=-\infty}^{\infty}\sum_{l=0}^{\infty} N'(m-l,n)z^m q^{2n+\frac{m}{2}}$, such as from $m=-\infty$ to $m=\infty$, in fact at each n only finitely many of the $N'(m,n)$ are nonzero. We obtain the first equality by setting $q\to q^2$ and $z\to zq^{\frac{1}{2}}$ in (3.34):

$$\sum_{n=0}^{\infty}\sum_{m=-\infty}^{\infty} N'(m,n)z^m q^{2n+\frac{m}{2}} = \prod_{n=1}^{\infty}\frac{(1-q^{2n})}{(1-zq^{2n+\frac{1}{2}})(1-z^{-1}q^{2n-\frac{1}{2}})}.$$

Thus we have

$$\frac{1}{1-zq^{\frac{1}{2}}}\left(\sum_{n=0}^{\infty}\sum_{m=-\infty}^{\infty} N'(m,n)z^m q^{2n+\frac{m}{2}}\right) = \prod_{n=1}^{\infty}\frac{(1-q^{2n})}{\left(1-zq^{2n-\frac{3}{2}}\right)\left(1-z^{-1}q^{2n-\frac{1}{2}}\right)}.$$

We now expand and re-sum (assuming $|zq^{\frac{1}{2}}|<1$):

$$\begin{aligned} &\frac{1}{1-zq^{\frac{1}{2}}}\left(\sum_{n=0}^{\infty}\sum_{m=-\infty}^{\infty} N'(m,n)z^m q^{2n+\frac{m}{2}}\right) \\ &= \left(\sum_{l=0}^{\infty} z^l q^{\frac{l}{2}}\right)\left(\sum_{n=0}^{\infty}\sum_{m=-\infty}^{\infty} N'(m,n)z^m q^{2n+\frac{m}{2}}\right) \\ &= \sum_{l=0}^{\infty}\sum_{n=0}^{\infty}\sum_{m=-\infty}^{\infty} N'(m,n)z^{m+l}q^{2n+\frac{m+l}{2}} = \sum_{n=0}^{\infty}\sum_{m=-\infty}^{\infty}\sum_{l=0}^{\infty} N'(m-l,n)z^m q^{2n+\frac{m}{2}}. \end{aligned}$$

□

In [**Ang17**] we asked the following questions: It would be interesting to derive a formula giving a correspondence between a partition $\mathfrak{p}\in\mathfrak{P}_{tdo}$ of weight n and the highest weight vector corresponding to that partition, or even the charge of that highest weight vector. As the weights of the partitions grow, the charges are less straightforward to calculate. For example at weight $\frac{13}{2}$ there are 7 partitions from $\mathfrak{P}_{tdo}$ and one can calculate by brute force that there is a highest weight vector of charge 13, a highest weight vector of charge 9, two highest weight vectors of charge 5, two highest weight vectors of charge 1 and a highest weight vector of charge -3.

We cannot yet give a direct formula between a $\mathfrak{p} \in \mathfrak{P}_{tdo}$ of weight n and the highest weight vector corresponding to that partition $\mathfrak{p}$, although we have an indirect correspondence in Proposition 3.13. But we can answer the question about the charges of the highest weight vectors with degree n:

COROLLARY 3.21. *The number of highest weight vectors with degree n and charge m equals $\sum_{l=0}^{\infty} N'(m-l, \frac{2n-m}{4})$.*

REMARK 3.22. We would also note that the above formula confirms that $2deg(v) \equiv chg(v) \pmod 4$, where $deg(v)$ is the degree of the highest weight vector v, and $chg(v)$ is its charge.

We would like to thank Thomas Creutzig for the helpful discussion, in particular he confirmed that such a connection with the Dyson crank should hold, and explained the physicist concept of a coset space. We are also grateful to Kailash Misra and Naihuan Jing, for organizing this AMS Special Session, and for their invitation.

4. Appendix

In this Appendix we give another derivation of the identity (3.25) from the Ramanujan summation formula (see e.g. [**HA40**], [**And69**], [**AA78**]) by appropriate changes of variables. It serves as a confirmation of the validity of this rather curious identity.

We start with the Ramanujan Psi summation formula

$$ {}_1\psi_1(a,b;q,x) = \sum_{k=-\infty}^{\infty} \frac{(a;q)_k}{(b;q)_k} x^k = \frac{(q;q)_\infty \left(\frac{b}{a};q\right)_\infty \left(\frac{q}{ax};q\right)_\infty (ax;q)_\infty}{(b;q)_\infty \left(\frac{b}{ax};q\right)_\infty \left(\frac{q}{a};q\right)_\infty (x;q)_\infty}. $$

The specialization we need is $b = aq$, which reduces the Ramanujan summation formula to

$$ (1-a)\sum_{k=-\infty}^{\infty} \frac{x^k}{1-aq^k} = \frac{(q;q)_\infty^2 \left(\frac{q}{ax};q\right)_\infty (ax;q)_\infty}{(aq;q)_\infty \left(\frac{q}{x};q\right)_\infty \left(\frac{q}{a};q\right)_\infty (x;q)_\infty}. $$

We now replace $x = qy$ and we get

$$ \frac{1}{1-a}+\sum_{k=1}^{\infty} \frac{y^k q^k}{1-aq^k} - \frac{1}{a}\sum_{k=1}^{\infty} \frac{y^{-k}}{1-a^{-1}q^k} = \frac{1}{(1-a)} \frac{(q;q)_\infty^2 \left((ay)^{-1};q\right)_\infty (qay;q)_\infty}{(aq;q)_\infty (y^{-1};q)_\infty \left(\frac{q}{a};q\right)_\infty (qy;q)_\infty}. $$

Now set $q \mapsto q^2$:

$$ \frac{1}{1-a}+\sum_{k=1}^{\infty} \frac{y^k q^{2k}}{1-aq^{2k}} - \frac{1}{a}\sum_{k=1}^{\infty} \frac{y^{-k}}{1-a^{-1}q^{2k}} $$

$$ = \frac{1}{(1-a)} \frac{(q^2;q^2)_\infty^2 \left((ay)^{-1};q^2\right)_\infty (q^2ay;q^2)_\infty}{(aq^2;q^2)_\infty (y^{-1};q^2)_\infty \left(\frac{q^2}{a};q^2\right)_\infty (q^2y;q^2)_\infty}, $$

followed by $y = zq^{\frac{-3}{2}}$:

$$\frac{1}{1-a}+\sum_{k=1}^{\infty}\frac{z^k q^{\frac{k}{2}}}{1-aq^{2k}}-\frac{1}{a}\sum_{k=1}^{\infty}\frac{z^{-k}q^{\frac{3k}{2}}}{1-a^{-1}q^{2k}}$$
$$=\frac{1}{(1-a)}\frac{\left(q^2;q^2\right)_\infty^2\left(a^{-1}z^{-1}q^{\frac{3}{2}};q^2\right)_\infty\left(azq^{\frac{1}{2}};q^2\right)_\infty}{(aq^2;q^2)_\infty\left(z^{-1}q^{\frac{3}{2}};q^2\right)_\infty\left(\frac{q^2}{a};q^2\right)_\infty\left(zq^{\frac{1}{2}};q^2\right)_\infty}.$$

Finally, letting $a = -z^{-1}q^{\frac{3}{2}}$ we get

$$\frac{1}{1+z^{-1}q^{\frac{3}{2}}}+\sum_{k=1}^{\infty}\frac{z^k q^{\frac{k}{2}}}{1+z^{-1}q^{2k+\frac{3}{2}}}+\sum_{k=1}^{\infty}\frac{z^{-k+1}q^{\frac{3k-3}{2}}}{1+zq^{2k-\frac{3}{2}}}$$
$$=\frac{1}{(1+z^{-1}q^{\frac{3}{2}})}\frac{\left(q^2;q^2\right)_\infty^2\left(-1;q^2\right)_\infty\left(q^2;q^2\right)_\infty}{\left(-z^{-1}q^{\frac{7}{2}};q^2\right)_\infty\left(z^{-1}q^{\frac{3}{2}};q^2\right)_\infty\left(-zq^{\frac{1}{2}};q^2\right)_\infty\left(zq^{\frac{1}{2}};q^2\right)_\infty}.$$

Which simplifies to

$$\sum_{k=0}^{\infty}\frac{z^k q^{\frac{k}{2}}}{1+z^{-1}q^{2k+\frac{3}{2}}}+\sum_{k=0}^{\infty}\frac{z^{-k}q^{\frac{3k}{2}}}{1+zq^{2k+\frac{1}{2}}}$$
$$=\frac{2\prod_{n=1}^{\infty}\left(1-q^{2n}\right)^2\prod_{n=1}^{\infty}\left(1+q^{2n}\right)^2}{\prod_{n=1}^{\infty}\left(1+z^{-1}q^{2n-\frac{1}{2}}\right)\prod_{n=1}^{\infty}\left(1-z^{-1}q^{2n-\frac{1}{2}}\right)\prod_{n=1}^{\infty}\left(1-zq^{2n-\frac{3}{2}}\right)\prod_{n=1}^{\infty}\left(1+zq^{2n-\frac{3}{2}}\right)}.$$

References

[AA78] George E. Andrews and Richard Askey, *A simple proof of Ramanujan's summation of the ${}_1\psi_1$*, Aequationes Math. **18** (1978), no. 3, 333–337. MR522519

[Abe07] Toshiyuki Abe, *A $\mathbb{Z}_2$-orbifold model of the symplectic fermionic vertex operator superalgebra*, Math. Z. **255** (2007), no. 4, 755–792. MR2274534

[ACJ14] Iana I. Anguelova, Ben Cox, and Elizabeth Jurisich, *N-point locality for vertex operators: normal ordered products, operator product expansions, twisted vertex algebras*, J. Pure Appl. Algebra **218** (2014), no. 12, 2165–2203. MR3227299

[AG88] George E. Andrews and F. G. Garvan, *Dyson's crank of a partition*, Bull. Amer. Math. Soc. (N.S.) **18** (1988), no. 2, 167–171. MR929094

[And69] George E. Andrews, *On Ramanujan's summation of ${}_1\psi_1(a;\, b;\, z)$*, Proc. Amer. Math. Soc. **22** (1969), 552–553. MR0241703

[Ang13a] Iana I. Anguelova, *Boson-fermion correspondence of type B and twisted vertex algebras*, Lie theory and its applications in physics, Springer Proc. Math. Stat., vol. 36, Springer, Tokyo, 2013, pp. 399–410. MR3070667

[Ang13b] Iana I. Anguelova, *Twisted vertex algebras, bicharacter construction and boson-fermion correspondences*, J. Math. Phys. **54** (2013), no. 12, 121702, 37. MR3156092

[Ang14] Iana I. Anguelova, *Boson-fermion correspondence of type D-A and multi-local Virasoro representations on the Fock space $F^{\otimes\frac{1}{2}}$*, J. Math. Phys. **55** (2014), no. 11, 111704, 23. MR3390486

[Ang15] Iana I. Anguelova, *Multilocal bosonization*, J. Math. Phys. **56** (2015), no. 12, 121702, 13. MR3432321

[Ang17] Iana I. Anguelova, *The second bosonization of the CKP hierarchy*, J. Math. Phys. **58** (2017), no. 7, 071707, 20. MR3679003

[CR13] Thomas Creutzig and David Ridout, *Relating the archetypes of logarithmic conformal field theory*, Nuclear Phys. B **872** (2013), no. 3, 348–391. MR3048507

[DJKM81a] Etsurō Date, Michio Jimbo, Masaki Kashiwara, and Tetsuji Miwa, *Transformation groups for soliton equations. III. Operator approach to the Kadomtsev-Petviashvili equation*, J. Phys. Soc. Japan **50** (1981), no. 11, 3806–3812. MR638807

[DJKM81b] Etsurō Date, Michio Jimbo, Masaki Kashiwara, and Tetsuji Miwa, *Transformation groups for soliton equations. VI. KP hierarchies of orthogonal and symplectic type*, J. Phys. Soc. Japan **50** (1981), no. 11, 3813–3818. MR638808

[DJKM82] Etsurō Date, Michio Jimbo, Masaki Kashiwara, and Tetsuji Miwa, *Transformation groups for soliton equations. IV. A new hierarchy of soliton equations of KP-type*, Phys. D **4** (1981/82), no. 3, 343–365. MR657739

[FLM88] Igor Frenkel, James Lepowsky, and Arne Meurman, *Vertex operator algebras and the Monster*, Pure and Applied Mathematics, vol. 134, Academic Press, Inc., Boston, MA, 1988. MR996026

[FMS86] Daniel Friedan, Emil Martinec, and Stephen Shenker, *Conformal invariance, supersymmetry and string theory*, Nuclear Phys. B **271** (1986), no. 1, 93–165. MR845945

[Gar10] F. G. Garvan, *Biranks for partitions into 2 colors*, Ramanujan rediscovered, Ramanujan Math. Soc. Lect. Notes Ser., vol. 14, Ramanujan Math. Soc., Mysore, 2010, pp. 87–111. MR2856959

[HA40] G. H. Hardy, *Ramanujan. Twelve lectures on subjects suggested by his life and work*, Cambridge University Press, Cambridge, England; Macmillan Company, New York, 1940. MR0004860

[Hir04] Ryogo Hirota, *The direct method in soliton theory*, Cambridge Tracts in Mathematics, vol. 155, Cambridge University Press, Cambridge, 2004. Translated from the 1992 Japanese original and edited by Atsushi Nagai, Jon Nimmo and Claire Gilson; With a foreword by Jarmo Hietarinta and Nimmo. MR2085332

[HL04] Paul Hammond and Richard Lewis, *Congruences in ordered pairs of partitions*, Int. J. Math. Math. Sci. **45-48** (2004), 2509–2512. MR2102870

[Jac29] C.G.J. Jacobi. *Fundamenta nova theoriae functionum ellipticarum.* Sumtibus fratrum, 1829.

[Kac90] Victor G. Kac, *Infinite-dimensional Lie algebras*, 3rd ed., Cambridge University Press, Cambridge, 1990. MR1104219

[Kac98] Victor Kac, *Vertex algebras for beginners*, 2nd ed., University Lecture Series, vol. 10, American Mathematical Society, Providence, RI, 1998. MR1651389

[KR87] V. G. Kac and A. K. Raina, *Bombay lectures on highest weight representations of infinite-dimensional Lie algebras*, Advanced Series in Mathematical Physics, vol. 2, World Scientific Publishing Co., Inc., Teaneck, NJ, 1987. MR1021978

[KWY98] Victor G. Kac, Weiqiang Wang, and Catherine H. Yan, *Quasifinite representations of classical Lie subalgebras of $\mathcal{W}_{1+\infty}$*, Adv. Math. **139** (1998), no. 1, 56–140. MR1652526

[LL04] James Lepowsky and Haisheng Li, *Introduction to vertex operator algebras and their representations*, Progress in Mathematics, vol. 227, Birkhäuser Boston, Inc., Boston, MA, 2004. MR2023933

[MJD00] T. Miwa, M. Jimbo, and E. Date, *Solitons*, Cambridge Tracts in Mathematics, vol. 135, Cambridge University Press, Cambridge, 2000. Differential equations, symmetries and infinite-dimensional algebras; Translated from the 1993 Japanese original by Miles Reid. MR1736222

[Rid10] David Ridout, *$\widehat{\mathfrak{sl}}(2)_{-1/2}$ and the triplet model*, Nuclear Phys. B **835** (2010), no. 3, 314–342. MR2640231

[vOS12] Johan W. van de Leur, Alexander Yu. Orlov, and Takahiro Shiota, *CKP hierarchy, bosonic tau function and bosonization formulae*, SIGMA Symmetry Integrability Geom. Methods Appl. **8** (2012), Paper 036, 28. MR2942803

[Wan99] Weiqiang Wang, *Duality in infinite-dimensional Fock representations*, Commun. Contemp. Math. **1** (1999), no. 2, 155–199. MR1696098

[You89] Yuching You, *Polynomial solutions of the BKP hierarchy and projective representations of symmetric groups*, Infinite-dimensional Lie algebras and groups (Luminy-Marseille, 1988), Adv. Ser. Math. Phys., vol. 7, World Sci. Publ., Teaneck, NJ, 1989, pp. 449–464. MR1026962

Department of Mathematics, College of Charleston, Charleston, South Carolina 29424

Email address: `anguelovai@cofc.edu`

Contemporary Mathematics
Volume **713**, 2018
https://doi.org/10.1090/conm/713/14311

Inhomogeneous supersymmetric bilinear forms

Bojko Bakalov and McKay Sullivan

ABSTRACT. We consider inhomogeneous supersymmetric bilinear forms, i.e., forms that are neither even nor odd. We classify such forms up to dimension seven in the case when the restrictions of the form to the even and odd parts of the superspace are nondegenerate. As an application, we introduce a new type of oscillator Lie superalgebra.

1. Introduction

Many important Lie superalgebras can be constructed as subalgebras of oscillator superalgebras (see, e.g., [**3**]). For other good references on Lie superalgebras see [**4**], [**1**], and [**6**]. To obtain an oscillator Lie superalgebra one may start with a superspace $V = V_{\bar{0}} \oplus V_{\bar{1}}$ over $\mathbb{C}$ and a *skew-supersymmetric* bilinear form $(\cdot|\cdot)\colon V \times V \to \mathbb{C}$, i.e., such that

$$(a|b) = -(-1)^{p(a)p(b)}(b|a)$$

for all homogeneous $a, b \in V$ of parities $p(a)$ and $p(b)$, respectively. One usually takes this form to be nondegenerate and *even*, i.e., $(V_{\bar{0}}|V_{\bar{1}}) = 0$. These two conditions imply that the restrictions of the form to $V_{\bar{0}}$ and $V_{\bar{1}}$ are also nondegenerate. Then a Lie superalgebra structure is given to the extension $A = V \oplus \mathbb{C}K$ by declaring K to be an even central element and letting

$$[a, b] = (a|b)K, \qquad a, b \in V.$$

The universal enveloping algebra $\mathcal{U}(A)$ is known as the *oscillator algebra*. It admits highest weight representations, which can be restricted to give representations of its subalgebras.

We will call a bilinear form $(\cdot|\cdot)$ on a vector superspace V *inhomogeneous* if $(V_{\bar{0}}|V_{\bar{1}}) \neq 0$ and $(V_i|V_i) \neq 0$ for some $i \in \{\bar{0}, \bar{1}\}$. Our original motivation for studying inhomogeneous forms was to investigate representations of subalgebras of the oscillator algebras obtained by relaxing the assumption that the bracket of an even (bosonic) oscillator and an odd (fermionic) oscillator must be zero. Though skew-supersymmetric forms are used to construct oscillator algebras, there is a natural one-to-one correspondence between supersymmetric forms on V and skew-supersymmetric forms on V^{Π}, where V^{Π} is the superspace obtained by reversing the parity of the elements of V:

$$V^{\Pi}_{\bar{0}} = V_{\bar{1}}, \qquad V^{\Pi}_{\bar{1}} = V_{\bar{0}}.$$

2010 *Mathematics Subject Classification.* Primary 15A21, 17B60.

The first author was supported in part by a Simons Foundation grant 279074.

The inhomogeneous supersymmetric bilinear forms whose restrictions to $V_{\bar{0}}$ and $V_{\bar{1}}$ are nondegenerate will be called *pre-oscillator forms*, and the algebras obtained as universal enveloping algebras of central extensions of V^{Π} *inhomogeneous oscillator algebras*.

Throughout the rest of the paper, every bilinear form $(\cdot|\cdot)\colon V \times V \to \mathbb{C}$ will be assumed to be a pre-oscillator form. Then we can choose a homogeneous basis for V so that the Gram matrix of the form is a block matrix of type

$$G = \left[\begin{array}{c|c} I_k & B \\ \hline B^T & J_{2\ell} \end{array}\right], \tag{1.1}$$

where $\dim V_{\bar{0}} = k$, $\dim V_{\bar{1}} = 2\ell$, and

$$J_{2\ell} = \operatorname{diag}(J, \dots, J), \qquad J = \begin{bmatrix} 0 & 1 \\ -1 & 0 \end{bmatrix}.$$

REMARK 1.1. Further relaxing the assumption that the restrictions of the form to even and odd parts are nondegenerate also leads to oscillator-like algebras. An example is the superspace spanned by an even vector v_1 and an odd vector v_2, with the form given by $(v_1|v_1) = (v_1|v_2) = (v_2|v_1) = 1$, $(v_2|v_2) = 0$.

A bilinear form $(\cdot|\cdot)$ on a superspace V will be called *reducible* if $V = U \oplus W$ is an orthogonal direct sum of subsuperspaces, i.e., if $(U|W) = 0$. Otherwise $(\cdot|\cdot)$ is *irreducible*. A natural first step in investigating inhomogeneous oscillator algebras is to classify all irreducible pre-oscillator forms on a given superspace V. We call two bilinear forms $(\cdot|\cdot)_1$ and $(\cdot|\cdot)_2$ *equivalent* if there exists an even automorphism $\varphi : V \to V$ satisfying

$$(\varphi u|\varphi v)_1 = (u|v)_2 \tag{1.2}$$

for all $u, v \in V$.

In Section 2 we introduce invariants that help us distinguish between equivalence classes of forms. Then in Section 3 we find representatives of equivalence classes of irreducible forms and use the invariants to prove that they are irreducible and distinct. This allows us to obtain a classification of pre-oscillator forms on superspaces of dimension up to 7. Finally, in Section 4 we introduce oscillator Lie superalgebras obtained from inhomogeneous bilinear forms and discuss a 3-dimensional example.

2. Invariants

We aim to find invariants that distinguish the equivalence classes of forms on a given superspace V. We assume $(\cdot|\cdot)_1$ and $(\cdot|\cdot)_2$ are equivalent forms on V, and choose bases for V so that the Gram matrices G_i $(i = 1, 2)$ take the form (1.1) with $B = B_i$. If M is the matrix of an even automorphism φ satisfying (1.2), then (1.2) can be written as a matrix equation

$$M^T G_1 M = G_2. \tag{2.1}$$

Since φ is even, M is a block matrix of the form

$$M = \left[\begin{array}{c|c} X & 0 \\ \hline 0 & Y \end{array}\right].$$

Then (2.1) holds if and only if we have:

$$\begin{aligned} X &\in \mathrm{O}(k) = \{X \in \mathrm{GL}(k)\colon X^T X = I_k\}, \\ Y &\in \mathrm{Sp}(2\ell) = \{Y \in \mathrm{GL}(2\ell)\colon Y^T J_{2\ell} Y = J_{2\ell}\}, \\ B_2 &= X^T B_1 Y. \end{aligned}$$

Hence finding the equivalence class of a bilinear form with Gram matrix (1.1) is equivalent to finding the orbit of B under the right action of $\mathrm{O}(k)\times\mathrm{Sp}(2\ell)$ on $\mathbb{C}^{k\times 2\ell}$ defined by

$$B\cdot(X,Y) = X^T BY. \tag{2.2}$$

Given $X \in \mathrm{O}(k)$ and $Y \in \mathrm{Sp}(2\ell)$, let $A = X^T BY$. Then $A^T A = Y^T B^T BY$ is independent of X. Thus any invariant of the right action of $\mathrm{Sp}(2\ell)$ on $\mathbb{C}^{2\ell\times 2\ell}$ given by

$$C\cdot Y = Y^T CY$$

is an invariant of $B^T B$ under the action (2.2). Similarly $AJ_{2\ell}A^T = X^T BJ_{2\ell}B^T X$ is independent of Y. Thus any invariant of the right action of $\mathrm{O}(k)$ on $\mathbb{C}^{k\times k}$ given by

$$C\cdot X = X^T CX \tag{2.3}$$

is an invariant of $BJ_{2\ell}B^T$ under the joint action (2.2).

Let us consider the action (2.3). Since $\det(X) = \pm 1$, it is apparent that $\det(X^T CX) = \det(C)$. Thus the determinant of $BJ_{2\ell}B^T$ is an invariant of the action (2.2). More generally, we can write

$$X^T(BJ_{2\ell}B^T - \lambda I_k)X = (X^T BY)J_{2\ell}(X^T BY)^T - \lambda I_k. \tag{2.4}$$

This implies that the characteristic polynomials of the two matrices $BJ_{2\ell}B^T$ and $(X^T BY)J_{2\ell}(X^T BY)^T$ are equal. Therefore, the characteristic polynomial

$$P_B(\lambda) = \det(BJ_{2\ell}B^T - \lambda I_k)$$

is invariant under the joint action (2.2). We have obtained the following theorem.

THEOREM 2.1. *Let $V = V_{\bar 0}\oplus V_{\bar 1}$ be a superspace with $\dim V_{\bar 0} = k$ and $\dim V_{\bar 1} = 2\ell$, and let $(\cdot|\cdot)\colon V\times V\to\mathbb{C}$ be an inhomogeneous nondegenerate supersymmetric bilinear form on V with Gram matrix (1.1) in a given homogeneous basis. Then each coefficent of the characteristic polynomial $P_B(\lambda)$ is an invariant for the joint action (2.2) of $\mathrm{O}(k)\times\mathrm{Sp}(2\ell)$ on $B\in\mathbb{C}^{k\times 2\ell}$.*

By a similar argument, the polynomial

$$Q_B(\lambda) = \det(B^T B - \lambda J_{2\ell})$$

is also invariant under the action (2.2). However, it is essentially the same as $P_B(\lambda)$.

LEMMA 2.2. *With the above notation, we have*

$$Q_B(\lambda) = (-1)^k\lambda^{2\ell-k}P_B(-\lambda). \tag{2.5}$$

PROOF. Sylvester's determinant identity (see, e.g., [**7**]) states that if U and W are matrices of size $m\times n$ and $n\times m$ respectively, then

$$\det(I_m + UW) = \det(I_n + WU).$$

We replace U with $\lambda^{-1}A^{-1}U$, where A is an invertible $m\times m$ matrix, thus obtaining

$$\det(\lambda A + UW) = \lambda^{m-n}\det(\lambda I_n + WA^{-1}U)\det(A).$$

Letting $m = 2\ell$, $n = k$, $A = -J_{2\ell}$, and $U^T = W = B$, we get (2.5). □

Along with the constant terms

$$p_0(B) = P_B(0) = \det(BJ_{2\ell}B^T),$$
$$q_0(B) = Q_B(0) = \det(B^T B),$$

the coefficient of λ^{k-2} in $P_B(\lambda)$ will also be useful in the following section. Let $R_1, \dots, R_k$ be the rows of B. Then up to a sign, this coefficient is given by

$$(2.6)\qquad \begin{aligned} p_{k-2}(B) &= \sum_{1\le s_1<s_2\le k} \big(R_{s_i} J_{2\ell} R_{s_j}^T\big)^2 \\ &= \sum_{1\le s_1<s_2\le k} \left(\sum_{t=1}^{\ell} \det\begin{bmatrix} b_{s_1,2t-1} & b_{s_1,2t} \\ b_{s_2,2t-1} & b_{s_2,2t}\end{bmatrix}\right)^2. \end{aligned}$$

REMARK 2.3. The following observation will be useful in the classification given in the next section. Assume the matrix B satisfies $C^TC = 0$ for every linear combination C of columns of B. Then this same property holds for every matrix in the orbit of B under the action (2.2).

3. Classification up to dimension 7

In this section, for each superspace V with $\dim V \le 7$, we will find a representative of each equivalence class of irreducible pre-oscillator forms on V. First, let $b = (b_1, \dots, b_k)^T \in \mathbb{C}^k$ be a column vector and $q(b) = b^T b$. We aim to find a canonical form for a representative of the orbit of b under the left action Xb of $X \in \mathrm{O}(k)$.

PROPOSITION 3.1. *Let $b \in \mathbb{C}^k$ be a column vector. If $q(b) \ne 0$, then under the left action of $\mathrm{O}(k)$, b is in the orbit of*

$$(3.1)\qquad (\sqrt{q(b)}, 0, \dots, 0)^T,$$

where $\sqrt{a}$ is defined to be the unique element γ of

$$\mathbb{C}^+ = \{\gamma \in \mathbb{C} : \operatorname{Re}\gamma > 0 \text{ or } \operatorname{Re}\gamma = 0 \text{ and } \operatorname{Im}\gamma > 0\}$$

satisfying $\gamma^2 = a$. On the other hand, if $q(b) = 0$, then either $b = 0$ or b is in the orbit of

$$(3.2)\qquad (1, \mathrm{i}, 0, \dots, 0)^T.$$

PROOF. If $q(b) \ne 0$, it is enough to give the proof in the case when $k = 2$, because the general case can be reduced to that. Let $b = (b_1, b_2)^T$ be such that $q(b) \ne 0$. Then the matrix

$$X = \frac{1}{\sqrt{q(b)}}\begin{bmatrix} b_1 & b_2 \\ -b_2 & b_1 \end{bmatrix} \in \mathrm{O}(2)$$

satisfies $Xb = (\sqrt{q(b)}, 0)^T$.

Now assume $q(b) = 0$. If $k = 2$, then $q(b) = 0$ implies $b = (b_1, \pm \mathrm{i} b_1)^T$. These two possible forms for b are in the same orbit, so without loss of generality $b = (b_1, \mathrm{i} b_1)^T$. Then $Xb = (1, \mathrm{i})^T$ for

$$(3.3)\qquad X = \frac{1}{2b_1}\begin{bmatrix} b_1^2 + 1 & \mathrm{i}(b_1^2 - 1) \\ -\mathrm{i}(b_1^2 - 1) & b_1^2 + 1 \end{bmatrix} \in \mathrm{O}(2).$$

Now suppose $k \geq 3$ and b is nonzero. Then possibly after reordering, we may assume $b_1 \neq 0$. Then

$$b_2^2 + \cdots + b_k^2 = -b_1^2 \neq 0.$$

There exists an orthogonal transformation X that leaves b_1 invariant and replaces $(b_2, \ldots, b_k)^T$ with a vector of the form (3.1). Thus we obtain

$$Xb = (b_1, \mathrm{i}b_1, 0, \ldots, 0)^T.$$

Then using an orthogonal transformation that acts as (3.3) on rows 1 and 2 and as the identity on the remaining rows, we obtain (3.2). □

Now we consider the right action of the symplectic group: $B \mapsto BY$ for $B \in \mathbb{C}^{k\times 2\ell}$ and $Y \in \mathrm{Sp}(2\ell)$. Let $C_1, \ldots, C_{2\ell}$ be the columns of B. For $1 \leq i \leq \ell$, we will say columns C_{2i-1}, C_{2i} are *paired*. Using suitable $Y \in \mathrm{Sp}(2\ell)$, we can perform the following elementary operations on paired columns of B.

i. Rescaling by $\lambda \neq 0$:

$$(\ldots, C_{2i-1}, C_{2i}, \ldots) \mapsto (\ldots, \lambda C_{2i-1}, \lambda^{-1} C_{2i}, \ldots).$$

ii. Adding any multiple of a column to its paired column:

$$(\ldots, C_{2i-1}, C_{2i}, \ldots) \mapsto (\ldots, C_{2i-1}, C_{2i} + \lambda C_{2i-1}, \ldots).$$

iii. Switching columns:

$$(\ldots, C_{2i-1}, C_{2i}, \ldots) \mapsto (\ldots, C_{2i}, -C_{2i-1}, \ldots).$$

The following are elementary operations outside of pairs.

iv. Adding a multiple of a column to a column other than its pair:

$$\begin{aligned}&(\ldots, C_{2i-1}, C_{2i}, \ldots, C_{2j-1}, C_{2j}, \ldots)\\ &\qquad \mapsto (\ldots, C_{2i-1} - \lambda C_{2j-1}, C_{2i}, \ldots, C_{2j-1}, C_{2j} + \lambda C_{2i}, \ldots).\end{aligned}$$

v. Switching pairs of columns:

$$\begin{aligned}&(\ldots, C_{2i-1}, C_{2i}, \ldots, C_{2j-1}, C_{2j}, \ldots)\\ &\qquad \mapsto (\ldots, C_{2j-1}, C_{2j}, \ldots, C_{2i-1}, C_{2i}, \ldots).\end{aligned}$$

From the above discussion, we see that using the orthogonal action, we can always reduce to at most 4ℓ nonzero rows. Also, using the symplectic action we can always reduce so that at most the first $2k$ columns are nonzero. Thus if V has an irreducible bilinear form, its even and odd dimensions must satisfy $\ell \leq k \leq 4\ell$. It follows that to obtain a complete classification of irreducible pre-oscillator forms up to dimension 7, we only need to consider the following cases.

Case $k = \ell = 1$. It is easy to see that B is in the orbit of

$$\mathrm{B}_1 = \begin{bmatrix} 1 & 0 \end{bmatrix}.$$

Case $k = \ell = 2$. If there exists a linear combination $C = \lambda_1 C_1 + \cdots + \lambda_{2\ell} C_{2\ell}$ of columns satisfying $q(C) \neq 0$, then we can use the symplectic action to put C in the first column of B. Then using the orthogonal action and rescaling we obtain

$$B = \begin{bmatrix} 1 & b_{12} & b_{13} & b_{14} \\ 0 & b_{22} & b_{23} & b_{24} \end{bmatrix}.$$

Using the symplectic action we eliminate b_{13} and b_{14} followed by b_{12} obtaining a matrix of the form

$$B = \begin{bmatrix} 1 & 0 & 0 & 0 \\ 0 & b_{22} & b_{23} & b_{24} \end{bmatrix}.$$

Then again via the symplectic action we get

$$B = \begin{bmatrix} 1 & 0 & 0 & 0 \\ 0 & \alpha & 0 & 0 \end{bmatrix}.$$

If $\alpha = 0$, this reduces to B_1. If $\alpha \neq 0$, then this reduces to the case

$$\mathrm{B}_{2,\alpha} = \begin{bmatrix} 1 & 0 \\ 0 & \alpha \end{bmatrix} \qquad (\alpha \in \mathbb{C}^+).$$

Finally, assume every linear combination C of the columns of B satisfies $q(C) = 0$. Then using the orthogonal action we obtain

$$B = \begin{bmatrix} 1 & b_{12} & b_{13} & b_{14} \\ \mathrm{i} & \mathrm{i}b_{12} & \mathrm{i}b_{13} & \mathrm{i}b_{14} \end{bmatrix}.$$

Using the symplectic action we eliminate columns 2, 3, and 4. Thus this reduces to the case

$$\mathrm{B}_3 = \begin{bmatrix} 1 & 0 \\ \mathrm{i} & 0 \end{bmatrix}.$$

Notice that this argument also shows that every irreducible matrix B of size $k = 2, \ell = 1$ is in the orbit of $\mathrm{B}_{2,\alpha}$ or B_3. The details of the remaining cases are similar to those already shown, so we omit them and provide the results.

Case $k = 3, \ell = 1$. The matrix B is in the orbit of

$$\mathrm{B}_4 = \begin{bmatrix} 1 & 0 \\ 0 & 1 \\ 0 & \mathrm{i} \end{bmatrix}, \tag{3.4}$$

or it can be reduced to B_1, $\mathrm{B}_{2,\alpha}$, or B_3.

Case $k = 3, \ell = 2$. Either the matrix B is in the orbit of

$$\mathrm{B}_5 = \begin{bmatrix} 1 & 0 & 0 & 0 \\ 0 & 1 & 1 & 0 \\ 0 & 0 & 0 & \mathrm{i} \end{bmatrix},$$

or it can be reduced to an orthogonal direct sum involving the previous four irreducible cases.

Case $k = 4, \ell = 1$. Either B is in the orbit of

$$\mathrm{B}_6 = \begin{bmatrix} 1 & 0 \\ \mathrm{i} & 0 \\ 0 & 1 \\ 0 & \mathrm{i} \end{bmatrix},$$

or it reduces to one of B_1, $\mathrm{B}_{2,\alpha}$, B_3, or B_4.

THEOREM 3.2. *Let $V = V_{\bar{0}} \oplus V_{\bar{1}}$ be a superspace of dimension ≤ 7. The following is a complete list up to equivalence of inhomogeneous irreducible supersymmetric bilinear forms on V whose restrictions to $V_{\bar{0}}$ and $V_{\bar{1}}$ are nondegenerate* ($\alpha \in \mathbb{C}^+$):

$\dim V_{\bar{0}}$	$\dim V_{\bar{1}}$	B
1	2	B_1
2	2	$\mathrm{B}_{2,\alpha}$, B_3
3	2	B_4
3	4	B_5
4	2	B_6

.

PROOF. We have shown that every matrix B corresponding to a bilinear form on a superspace of dimension up to seven is either reducible or in the orbit of one of the representatives listed in the table. In order to complete the proof of the theorem we need to check that the entries in the last column of the table are irreducible and distinct, i.e., in different orbits.

The matrix B_1 clearly corresponds to an irreducible form. We note that $\det(\mathrm{B}_3)^2 = 0$ and $\det(\mathrm{B}_{2,\alpha})^2 = \alpha^2$. Thus these matrices are all in distinct orbits. If any of these matrices reduces, it must be in the orbit of

$$\begin{bmatrix} 1 & 0 \\ 0 & 0 \end{bmatrix}, \tag{3.5}$$

but this matrix has determinant 0 and therefore cannot share an orbit with $\mathrm{B}_{2,\alpha}$. So $\mathrm{B}_{2,\alpha}$ is irreducible for $\alpha \in \mathbb{C}^+$. That B_3 is not in the orbit of (3.5) follows from Remark 2.3.

The matrix B_4 has rank 2 and so reduces only if it is in the orbit of the matrix

$$\begin{bmatrix} 1 & 0 \\ 0 & \alpha \\ 0 & 0 \end{bmatrix} \tag{3.6}$$

for some $\alpha \in \mathbb{C}^+$. But the invariant p_1 (cf. (2.6)) evaluated on the matrix (3.6) is $\alpha^2 \neq 0$, whereas $p_1(\mathrm{B}_4) = 0$. So B_4 is irreducible. The matrix B_6 has rank 2 and is therefore reducible only if it is in the orbit of B_4 or the matrix

$$\begin{bmatrix} 1 & 0 \\ 0 & \alpha \\ 0 & 0 \\ 0 & 0 \end{bmatrix} \tag{3.7}$$

for some $\alpha \in \mathbb{C}^+$. By Remark 2.3, B_6 is not in the orbit of B_4 or (3.7). So B_6 is irreducible.

B_5 has rank three and so is reducible only if it is in the orbit of the matrix

$$\begin{bmatrix} 1 & 0 & 0 & 0 \\ 0 & \alpha & 0 & 0 \\ 0 & 0 & 1 & 0 \end{bmatrix} \tag{3.8}$$

for some $\alpha \in \mathbb{C}^+$. But the invariant p_1 shows that B_5 is not in the same orbit as (3.8). So B_5 is irreducible. □

Applying our methods to higher dimensions, we have found that the classification seems to become increasingly complicated as $\dim V$ increases. Though it should be possible to extend the classification to dimension 8 or 9 with the methods of this paper, a more general approach will be needed for a complete classification in any dimension. We plan to address this question in the future by using the theory of *θ-groups* (see [**2**, **5**]).

4. Oscillator Lie superalgebras

As we explained in the introduction, oscillator Lie superalgebras can be obtained from skew-supersymmetric bilinear forms. Now we show how one obtains an inhomogoneous oscillator superalgebra from an inhomogeneous pre-oscillator form. Then we use an example of such a superalgebra to construct a trivial abelian extension of $\mathfrak{osp}(1|2)$.

Let $V = V_{\bar{0}} \oplus V_{\bar{1}}$ be a superspace and $(\cdot|\cdot)\colon V \times V \to \mathbb{C}$ be a skew-supersymmetric pre-oscillator form. Consider the extension

$$A = V \oplus \mathbb{C}K \oplus \mathbb{C}\kappa. \tag{4.1}$$

We give this extension the structure of a Lie superalgebra by declaring K and κ to be even and odd central elements respectively and letting

$$[a,b] = \begin{cases} (a|b)K, & p(a) = p(b) \\ (a|b)\kappa, & p(a) \neq p(b) \end{cases} \tag{4.2}$$

for all homogeneous vectors $a, b \in V$.

Before considering an example of such a form, let us recall a construction of $\mathfrak{osp}(1|2)$ as a subalgebra of a homogeneous oscillator algebra. Assume $k = \ell = 1$ and $(\cdot|\cdot)$ is a nondegenerate, even, skew-supersymmetric bilinear form. Then there exist bases $\{b_1, b_2\}$ of $V_{\bar{0}}$ and $\{a\}$ of $V_{\bar{1}}$ such that the Gram matrix of the bilinear form is

$$G = \left[\begin{array}{cc|c} 0 & 1 & 0 \\ -1 & 0 & 0 \\ \hline 0 & 0 & 1 \end{array}\right].$$

The central extension $A = V \oplus \mathbb{C}K$ has nonzero brackets

$$[b_1, b_2] = K, \qquad \{a, a\} = K,$$

where we use $\{\cdot,\cdot\}$ to denote the superbracket of odd vectors. Then $\mathfrak{osp}(1|2)$ is realized as a subalgebra of the oscillator algebra $\mathcal{U}(A)$ as follows (see, e.g., [**3**]):

$$\begin{aligned} H &= \frac{1}{4}b_2b_1 + \frac{1}{4}b_1b_2, & F^+ &= \frac{1}{4}\,ab_2 + \frac{1}{4}b_2a, \\ E^+ &= \frac{1}{2}b_2^2, & F^- &= \frac{1}{4}\,ab_1 + \frac{1}{4}b_1a, \\ E^- &= -\frac{1}{2}b_1^2, \end{aligned} \tag{4.3}$$

with brackets

$$\begin{aligned} [H, E^{\pm}] &= \pm E^{\pm}, & [E^+, E^-] &= 2H, \\ [H, F^{\pm}] &= \pm\frac{1}{2}F^{\pm}, & \{F^+, F^-\} &= \frac{1}{2}H, \\ [E^{\pm}, F^{\mp}] &= -F^{\pm}, & \{F^{\pm}, F^{\pm}\} &= \pm\frac{1}{2}E^{\pm}. \end{aligned}$$

We obtain a highest weight representation of $\mathcal{U}(A)$ on a Fock space $\mathcal{F} = \mathbb{C}[x] \otimes \mathbb{C}\xi$ where ξ is an odd indeterminate satisfying $\xi^2 = \frac{1}{2}$. The action of A on $\mathcal{F}$ is given by

$$b_1 \mapsto \partial_x, \qquad b_2 \mapsto x, \qquad a \mapsto \xi, \qquad K \mapsto \mathrm{Id}.$$

Now consider the case when the Gram matrix is given instead by

$$G = \left[\begin{array}{cc|c} 0 & 1 & 1 \\ -1 & 0 & 0 \\ \hline -1 & 0 & 1 \end{array}\right].$$

This inhomogeneous form gives rise to the central extension (4.1) with nonzero brackets

$$[b_1, b_2] = K, \qquad [b_1, a] = \kappa, \qquad \{a, a\} = K.$$

We consider the same elements (4.3) of $\mathcal{U}(A)$ we used to construct $\mathfrak{osp}(1|2)$ in the previous example. Then the brackets become

$$\begin{aligned} [H, E^{\pm}] &= \pm E^{\pm}, & [E^-, F^-] &= \kappa E^-, \\ [H, F^+] &= \frac{1}{2}F^+ + \frac{\kappa}{2}E^+, & [E^+, E^-] &= 2H, \\ [H, F^-] &= -\frac{1}{2}F^- + \frac{\kappa}{2}H, & \{F^+, F^-\} &= \frac{1}{2}H - \frac{\kappa}{2}F^+, \\ [E^+, F^-] &= -F^+, & \{F^+, F^+\} &= \frac{1}{2}E^+, \\ [E^-, F^+] &= -F^- - \kappa H, & \{F^-, F^-\} &= -\frac{1}{2}E^- - \kappa F^-. \end{aligned} \tag{4.4}$$

Observe that $\kappa^2 = 0$ in $\mathcal{U}(A)$. Thus the brackets of the $\mathfrak{osp}(1|2)$ subalgebra have been modified by elements of the abelian ideal

$$M = \operatorname{span}\{\kappa H, \kappa E^{\pm}, \kappa F^{\pm}\}.$$

We note that $\operatorname{span}\{H, E^{\pm}, F^{\pm}\}$ acts on M as the adjoint representation of $\mathfrak{osp}(1|2)$, and thus we have a Lie superalgebra structure on the vector space $L = \mathfrak{osp}(1|2) \oplus M$ such that the projection $\pi\colon L \to \mathfrak{osp}(1|2)$ is a surjective homomorphism, and the restriction of the adjoint representation of L to M yields the original action of $\mathfrak{osp}(1|2)$ on M. Thus L is an abelian extension of $\mathfrak{osp}(1|2)$ by its adjoint representation viewed as an abelian Lie superalgebra with parities reversed.

Denote by $[\cdot,\cdot]_L$ the bracket on L given by (4.4) and let

$$\gamma\colon \mathfrak{osp}(1|2) \times \mathfrak{osp}(1|2) \to \mathfrak{osp}(1|2)$$

satisfy

$$[a, b]_L = [a, b] + \kappa\gamma(a, b).$$

This extension is trivial if there exists an odd linear map $f\colon \mathfrak{osp}(1|2) \to \mathfrak{osp}(1|2)$ such that

$$\gamma(a, b) = (-1)^{p(a)}[a, f(b)] - (-1)^{(p(a)+1)p(b)}[b, f(a)] - f([a, b]) \tag{4.5}$$

for all $a, b \in \mathfrak{osp}(1|2)$. It is straightforward to check that the following choice of f satisfies (4.5):

$$f(H) = f(E^{\pm}) = 0, \qquad f(F^{+}) = E^{+}, \qquad f(F^{-}) = H.$$

As before, we can represent A on the Fock space $\mathcal{F} = \mathbb{C}[x] \otimes \bigwedge(\xi, \kappa)$ where ξ is odd with $\xi^2 = \frac{1}{2}$ and κ acts as an odd indeterminate that we denote again by κ. Then the action of A on $\mathcal{F}$ is given by

$$b_1 \mapsto \partial_x + \kappa\partial_\xi, \qquad b_2 \mapsto x, \qquad a \mapsto \xi, \qquad K \mapsto \mathrm{Id}, \qquad \kappa \mapsto \kappa.$$

Acknowledgments

We are grateful to Dimitar Grantcharov and Victor Kac for many valuable discussions. We would like to thank the referee for carefully reading the manuscript and suggesting improvements.

References

[1] S.-J. Cheng and W. Wang, *Dualities and representations of Lie superalgebras*, Graduate Studies in Mathematics, vol. 144, American Mathematical Society, Providence, RI, 2012. MR3012224

[2] J. Dadok and V. Kac, *Polar representations*, J. Algebra **92** (1985), no. 2, 504–524. MR778464

[3] L. Frappat, A. Sciarrino, and P. Sorba, *Dictionary on Lie algebras and superalgebras*, Academic Press, Inc., San Diego, CA, 2000. With 1 CD-ROM (Windows, Macintosh and UNIX). MR1773773

[4] V. G. Kac, *Lie superalgebras*, Advances in Math. **26** (1977), no. 1, 8–96. MR0486011

[5] V. G. Kac, *Some remarks on nilpotent orbits*, J. Algebra **64** (1980), no. 1, 190–213. MR575790

[6] I. M. Musson, *Lie superalgebras and enveloping algebras*, Graduate Studies in Mathematics, vol. 131, American Mathematical Society, Providence, RI, 2012. MR2906817

[7] C. Pozirikis, *An introduction to grids, graphs, and networks*. Oxford Univ. Press, New York, 2014.

Department of Mathematics, North Carolina State University, Raleigh, North Carolina 27695
Email address: bojko_bakalov@ncsu.edu

Department of Mathematics, Dixie State University, Saint George, Utah 84770
Email address: mckay.sullivan@dixie.edu

Contemporary Mathematics
Volume **713**, 2018
https://doi.org/10.1090/conm/713/14319

Imaginary crystal bases for $U_q(\widehat{\mathfrak{sl}(2)})$-modules in category $\mathcal{O}^q_{\mathbf{red,im}}$

Ben Cox, Vyacheslav Futorny, and Kailash C. Misra

ABSTRACT. Recently we defined imaginary crystal bases for $U_q(\widehat{\mathfrak{sl}(2)})$- modules in category $\mathcal{O}^q_{\text{red,im}}$ and showed the existence of such bases for reduced quantized imaginary Verma modules for $U_q(\widehat{\mathfrak{sl}(2)})$. In this paper we show the existence of imaginary crystal basis for any object in the category $\mathcal{O}^q_{\text{red,im}}$.

1. Introduction

Consider the affine Lie algebra $\widehat{\mathfrak{g}} = \widehat{\mathfrak{sl}(2)}$ with Cartan subalgebra $\widehat{\mathfrak{h}}$. Let $\{\alpha_0, \alpha_1\}$ be the simple roots, $\delta = \alpha_0 + \alpha_1$ the null root and Δ the set of roots for $\widehat{\mathfrak{g}}$ with respect to $\widehat{\mathfrak{h}}$. Then we have a natural (standard) partition of $\Delta = \Delta_+ \cup \Delta_-$ into set of positive and negative roots which is closed (i.e. $\alpha, \beta \in \Delta_+$ and $\alpha + \beta \in \Delta$ implies $\alpha + \beta \in \Delta_+$). Corresponding to this standard partition we have a standard Borel subalgebra from which we induce the standard Verma module. Let $S = \{\alpha_1 + k\delta \mid k \in \mathbb{Z}\} \cup \{l\delta \mid l \in \mathbb{Z}_{>0}\}$. Then $\Delta = S \cup -S$ is another closed partition of the root system Δ which is not Weyl group conjugate to the standard partition. The classification of closed partitions of the root system for affine Lie algebras was obtained by Jakobsen and Kac [**JK85, JK89**], and independently by Futorny [**Fut90, Fut92**]. For the affine Lie algebra $\widehat{\mathfrak{g}}$ the partition $\Delta = S \cup -S$ is the only nonstandard closed partition which gives rise to a nonstandard Borel subalgebra. The Verma module $M(\lambda)$ with highest weight λ induced by this nonstandard Borel subalgebra is called the imaginary Verma module for $\widehat{\mathfrak{sl}(2)}$. Unlike the standard Verma module, the imaginary Verma module $M(\lambda)$ contains both finite and infinite dimensional weight spaces.

For generic q, consider the associated quantum affine algebra $U_q(\widehat{\mathfrak{sl}(2)})$ ([**Dri85**], [**Jim85**]). Lusztig [**Lus88**] proved that the integrable highest weight modules of $\widehat{\mathfrak{sl}(2)}$ can be deformed to those over $U_q(\widehat{\mathfrak{sl}(2)})$ in such a way that the dimensions of the weight spaces are invariant under the deformation. Following the framework of [**Lus88**] and [**Kan95**], it was shown in ([**CFKM97**], [**FGM98**]) that the

2010 *Mathematics Subject Classification.* Primary 17B37, 17B10; Secondary 17B67, 17B65.
Key words and phrases. Quantum affine algebras, imaginary Verma modules, Kashiwara algebras, imaginary crystal bases.
The first author was partially support by a Simons Foundation Grant #319261. The second author was supported in part by the CNPq grant #301320/2013-6 and by the FAPESP grant #2014/09310-5. The third author was partially support by the Simons Foundation Grant #307555.

imaginary Verma modules $M(\lambda)$ can also be q-deformed to the quantum imaginary Verma modules $M_q(\lambda)$ in such a way that the weight multiplicities, both finite and infinite-dimensional, are preserved.

Lusztig [**Lus90**] from a geometric view point and Kashiwara [**Kas91**] from an algebraic view point introduced the notion of canonical bases (equivalently, global crystal bases) for standard Verma modules $V_q(\lambda)$ and integrable highest weight modules $L_q(\lambda)$. The crystal basis ([**Kas90**]) can be thought of as the $q = 0$ limit of the canonical basis. An important ingredient in the construction of crystal basis by Kashiwara in [**Kas91**], is a subalgebra $\mathcal{B}_q$ of the quantum group which acts on the negative part of the quantum group by left multiplication. This subalgebra $\mathcal{B}_q$, which is called the Kashiwara algebra, played an important role in the definition of the Kashiwara operators which defines the crystal basis. The algebra $\mathcal{B}_q$ has been defined in greater generality for integrable cases in [**Mas09**]. In [**CFM10**] we constructed an analog of Kashiwara algebra, denoted by $\mathcal{K}_q$ for the imaginary Verma module $M_q(\lambda)$ for the quantum affine algebra $U_q(\widehat{\mathfrak{sl}(2)})$ by introducing certain Kashiwara-type operators. Then we proved that a certain quotient $\mathcal{N}_q^-$ of $U_q(\widehat{\mathfrak{g}})$ is a simple $\mathcal{K}_q$-module and gave a necessary and sufficient condition for a particular quotient $\bar{M}_q(\lambda)$ (called reduced imaginary Verma module) of $M_q(\lambda)$ to be simple. These results were generalized to all untwisted affine Lie algebras in [**CFM14**].

In [**CFM17**] we considered a category $\mathcal{O}^q_{\mathrm{red,im}}$ of $U_q(\widehat{\mathfrak{sl}(2)})$-modules and defined a crystal-like basis which we called imaginary crystal basis for modules in this category. We showed that the reduced imaginary Verma modules $\bar{M}_q(\lambda)$ are simple objects in $\mathcal{O}^q_{\mathrm{red,im}}$ and any module in $\mathcal{O}^q_{\mathrm{red,im}}$ is a direct sum of reduced imaginary Verma modules for $U_q(\widehat{\mathfrak{sl}(2)})$. Then we proved the existence of imaginary crystal basis for a reduced imaginary Verma module $\bar{M}_q(\lambda)$. In this paper we prove the existence of imaginary crystal basis for any object in the category $\mathcal{O}^q_{\mathrm{red,im}}$.

2. Quantum affine algebra $U_q(\widehat{\mathfrak{sl}(2)})$

Let $\mathbb{F}$ denote a field of characteristic zero and q be an indeterminant (not a root of unity). The *quantum affine algebra* $U_q(\widehat{\mathfrak{sl}(2)})$ is the associative $\mathbb{F}(q^{1/2})$-algebra with 1 generated by

$$E_0, E_1, F_0, F_1, K_0^{\pm 1}, K_1^{\pm 1}, D^{\pm 1}$$

with defining relations:

$$\begin{aligned}
&DD^{-1} = D^{-1}D = K_iK_i^{-1} = K_i^{-1}K_i = 1,\\
&E_iF_j - F_jE_i = \delta_{ij}\frac{K_i - K_i^{-1}}{q - q^{-1}},\\
&K_iE_iK_i^{-1} = q^2E_i, \quad K_iF_iK_i^{-1} = q^{-2}F_i,\\
&K_iE_jK_i^{-1} = q^{-2}E_j, \quad K_iF_jK_i^{-1} = q^2F_j, \quad i \neq j,\\
&K_iK_j - K_jK_i = 0, \quad K_iD - DK_i = 0,\\
&DE_iD^{-1} = q^{\delta_{i,0}}E_i, \quad DF_iD^{-1} = q^{-\delta_{i,0}}F_i,\\
&E_i^3E_j - [3]E_i^2E_jE_i + [3]E_iE_jE_i^2 - E_jE_i^3 = 0, \quad i \neq j,\\
&F_i^3F_j - [3]F_i^2F_jF_i + [3]F_iF_jF_i^2 - F_jF_i^3 = 0, \quad i \neq j,
\end{aligned}$$

where, $[n] = \frac{q^n - q^{-n}}{q - q^{-1}}$ and $i, j \in \{0, 1\}$.

There is an alternative realization for $U_q(\widehat{\mathfrak{sl}(2)})$, due to Drinfeld [**Dri85**], which we need. Let U_q be the associative algebra with 1 over $\mathbb{F}(q^{1/2})$ generated by the elements $x_k^{\pm}$ ($k \in \mathbb{Z}$), h_l ($l \in \mathbb{Z} \setminus \{0\}$), $K^{\pm 1}$, $D^{\pm 1}$, and $\gamma^{\pm\frac{1}{2}}$ with the following defining relations:

$$DD^{-1} = D^{-1}D = KK^{-1} = K^{-1}K = 1, \tag{2.1}$$

$$[\gamma^{\pm\frac{1}{2}}, u] = 0 \quad \forall u \in U, \tag{2.2}$$

$$[h_k, h_l] = \delta_{k+l,0}\frac{[2k]}{k}\frac{\gamma^k - \gamma^{-k}}{q-q^{-1}}, \tag{2.3}$$

$$[h_k, K] = 0, \quad [D, K] = 0, \tag{2.4}$$

$$Dh_kD^{-1} = q^k h_k, \tag{2.5}$$

$$Dx_k^{\pm}D^{-1} = q^k x_k^{\pm}, \tag{2.6}$$

$$Kx_k^{\pm}K^{-1} = q^{\pm 2}x_k^{\pm}, \tag{2.7}$$

$$[h_k, x_l^{\pm}] = \pm\frac{[2k]}{k}\gamma^{\mp\frac{|k|}{2}}x_{k+l}^{\pm}, \tag{2.8}$$

$$\begin{aligned} x_{k+1}^{\pm}x_l^{\pm} &- q^{\pm 2}x_l^{\pm}x_{k+1}^{\pm} \\ &= q^{\pm 2}x_k^{\pm}x_{l+1}^{\pm} - x_{l+1}^{\pm}x_k^{\pm}, \end{aligned} \tag{2.9}$$

$$[x_k^+, x_l^-] = \frac{1}{q-q^{-1}}\left(\gamma^{\frac{k-l}{2}}\psi(k+l) - \gamma^{\frac{l-k}{2}}\phi(k+l)\right), \tag{2.10}$$

$$\text{where } \sum_{k=0}^{\infty}\psi(k)z^{-k} = K\exp\left((q-q^{-1})\sum_{k=1}^{\infty}h_k z^{-k}\right), \tag{2.11}$$

$$\sum_{k=0}^{\infty}\phi(-k)z^{k} = K^{-1}\exp\left(-(q-q^{-1})\sum_{k=1}^{\infty}h_{-k} z^{k}\right). \tag{2.12}$$

The isomorphism between $U_q(\widehat{\mathfrak{sl}(2)})$ and U_q is given by:

$$\begin{aligned} &E_0 \mapsto x_1^- K^{-1}, \quad F_0 \mapsto Kx_{-1}^+, \\ &E_1 \mapsto x_0^+, \quad F_1 \mapsto x_0^-, \\ &K_0 \mapsto \gamma K^{-1}, \quad K_1 \mapsto K, \quad D \mapsto D. \end{aligned}$$

If one uses the formal sums

$$\phi(u) = \sum_{p\in\mathbb{Z}}\phi(p)u^{-p}, \;\; \psi(u) = \sum_{p\in\mathbb{Z}}\psi(p)u^{-p}, \;\; x^{\pm}(u) = \sum_{p\in\mathbb{Z}}x_p^{\pm}u^{-p} \tag{2.13}$$

Drinfeld's relations (2.3), (2.8)-(2.10) can be written as

$$[\phi(u), \phi(v)] = 0 = [\psi(u), \psi(v)] \tag{2.14}$$

$$\phi(u)x^{\pm}(v)\phi(u)^{-1} = g(uv^{-1}\gamma^{\mp 1/2})^{\pm 1}x^{\pm}(v) \tag{2.15}$$

$$\psi(u)x^{\pm}(v)\psi(u)^{-1} = g(vu^{-1}\gamma^{\mp 1/2})^{\mp 1}x^{\pm}(v) \tag{2.16}$$

$$(u - q^{\pm 2}v)x^{\pm}(u)x^{\pm}(v) = (q^{\pm 2}u - v)x^{\pm}(v)x^{\pm}(u) \tag{2.17}$$

$$[x^+(u), x^-(v)] = (q-q^{-1})^{-1}(\delta(u/v\gamma)\psi(v\gamma^{1/2}) - \delta(u\gamma/v)\phi(u\gamma^{1/2})) \tag{2.18}$$

where $g(t) = g_q(t) = \sum_{k\geq 0} g(r)t^k$ is the Taylor series at $t = 0$ of the function $(q^2t-1)/(t-q^2)$ and $\delta(z) = \sum_{k\in\mathbb{Z}} z^k$ is the formal Dirac delta function.

REMARK 2.1. *Writing $g(t) = g_q(t) = \sum_{r\geq 0} g(r)t^r$ we have*

$$g(r) = g_q(r) = g_{q^{-1}}(r) = \begin{cases} q^2 & \text{if } r = 0 \\ (1-q^{-4})q^{2(r+1)} = (q^4-1)q^{2(r-1)}, & \text{if } r > 0. \end{cases} \tag{2.19}$$

Considering Serre's relation with $k = l$, we get

$$x_k^- x_{k+1}^- = q^2 x_{k+1}^- x_k^- \tag{2.20}$$

The product on the right side is in the correct order for a basis element. If $k+1 > l$ and $k \neq l$ in (2.9), then $k+1 > l+1$ so that $k \geq l+1$, and thus we can write

$$x_l^- x_{k+1}^- = q^2 x_{k+1}^- x_l^- - x_k^- x_{l+1}^- + q^2 x_{l+1}^- x_k^- \tag{2.21}$$

and then after repeating the above identity, we will eventually arrive at sums of terms that are in the correct order. This is the opposite ordering of monomials as we had previously.

3. Ω-operators and the Kashiwara algebra $\mathcal{K}_q$

Let $\mathbb{N}^{\mathbb{N}^*}$ denote the set of all functions from $\{k\delta \mid k \in \mathbb{N}^*\}$ to $\mathbb{N}$ with finite support. Then we write

$$h^+ = h_+^{(s_k)} := h_{r_1}^{s_1} \cdots h_{r_k}^{s_l}, \quad h^- := h_-^{(s_k)} = h_{-r_1}^{s_1} \cdots h_{-r_k}^{s_l}$$

for $f = (s_m) \in \mathbb{N}^{\mathbb{N}^*}$ where $f(r_m) = s_m$ and $f(t) = 0$ for $t \neq r_i, m \leq i \leq l$.

Let $\mathcal{N}_q^-$, be the subalgebra of $U_q(\widehat{\mathfrak{sl}(2)})$ generated by $\gamma^{\pm 1/2}$, and x_n^-, $n \in \mathbb{Z}$. Consider $x^-(v) = \sum_n x_n^- v^{-n}$ as a formal power series of left multiplication operators $x_n^- : \mathcal{N}_q^- \to \mathcal{N}_q^-$.

As in [**CFM10**], for fixed k we set

$$\bar{P} = x^-(v_1) \cdots x^-(v_k)$$
$$\bar{P}_l = x^-(v_1) \cdots x^-(v_{l-1}) x^-(v_{l+1}) \cdots x^-(v_k),$$

and

$$G_l = G_l^{1/q} := \prod_{j=1}^{l-1} g_{q^{-1}}(v_j/v_l), \quad G_l^q = \prod_{j=1}^{l-1} g(v_l/v_j)$$

where $G_1 := 1$. Then we define a collection of operators $\Omega_\psi(k), \Omega_\phi(k) : \mathcal{N}_q^- \to \mathcal{N}_q^-$, $k \in \mathbb{Z}$, in terms of the generating functions as follows.

$$\Omega_\psi(u) = \sum_{l\in\mathbb{Z}} \Omega_\psi(l) u^{-l}, \quad \Omega_\phi(u) = \sum_{l\in\mathbb{Z}} \Omega_\phi(l) u^{-l}$$

where

$$\Omega_\psi(u)(\bar{P}) := \gamma^m \sum_{l=1}^{k} G_l \bar{P}_l \delta(u/v_l\gamma) \tag{3.1}$$

$$\Omega_\phi(u)(\bar{P}) := \gamma^m \sum_{l=1}^{k} G_l^q \bar{P}_l \delta(u\gamma/v_l). \tag{3.2}$$

Note that $\Omega_\psi(u)(1) = \Omega_\phi(u)(1) = 0$. More generally, let us write

$$\bar{P} = x^-(v_1) \cdots x^-(v_k) = \sum_{n\in\mathbb{Z}} \sum_{\substack{n_1, n_2, \ldots, n_k \in\mathbb{Z} \\ n_1+\cdots+n_k=n}} x_{n_1}^- \cdots x_{n_k}^- v_1^{-n_1} \cdots v_k^{-n_k}$$

Then

$$\begin{aligned}&\psi(u\gamma^{-1/2})\Omega_\psi(u)(\bar{P})\\&=\sum_{k\geq 0}\sum_{p\in\mathbb{Z}}\sum_{n_i\in\mathbb{Z}}\gamma^{k/2}\psi(k)\Omega_\psi(p)(x^-_{n_1}\cdots x^-_{n_k})v_1^{-n_1}\cdots v_k^{-n_k}u^{-k-p}\\&=\sum_{n_i\in\mathbb{Z}}\sum_{m\in\mathbb{Z}}\sum_{k\geq 0}\gamma^{k/2}\psi(k)\Omega_\psi(m-k)(x^-_{n_1}\cdots x^-_{n_k})v_1^{-n_1}\cdots v_k^{-n_k}u^{-m}\end{aligned}$$

while

$$[x^+(u),\bar{P}]=\sum_{m\in\mathbb{Z}}\sum_{n_1,n_2,\dots,n_k\in\mathbb{Z}}[x^+_m,x^-_{n_1}\cdots x^-_{n_k}]v_1^{-n_1}\cdots v_k^{-n_k}u^{-m}.$$

Thus for a fixed m and k-tuple $(n_1,\dots,n_k)$ the sum

$$\sum_{k\geq 0}\gamma^{k/2}\psi(k)\Omega_\psi(m-k)(x^-_{n_1}\cdots x^-_{n_k})$$

must be finite. Hence

$$\Omega_\psi(m-k)(x^-_{n_1}\cdots x^-_{n_k})=0, \tag{3.3}$$

for k sufficiently large.

PROPOSITION 3.1 ([**CFM10**], Prop. 4.0.3). *We have*

$$\Omega_\psi(u)x^-(v)=\delta(v\gamma/u)+g_{q^{-1}}(v\gamma/u)x^-(v)\Omega_\psi(u), \tag{3.4}$$

$$\Omega_\phi(u)x^-(v)=\delta(u\gamma/v)+g(u\gamma/v)x^-(v)\Omega_\phi(u) \tag{3.5}$$

$$(q^2u_1-u_2)\Omega_\psi(u_1)\Omega_\psi(u_2)=(u_1-q^2u_2)\Omega_\psi(u_2)\Omega_\psi(u_1) \tag{3.6}$$

$$(q^2u_1-u_2)\Omega_\phi(u_1)\Omega_\phi(u_2)=(u_1-q^2u_2)\Omega_\phi(u_2)\Omega_\phi(u_1) \tag{3.7}$$

$$(q^2\gamma^2u_1-u_2)\Omega_\phi(u_1)\Omega_\psi(u_2)=(\gamma^2u_1-q^2u_2)\Omega_\psi(u_2)\Omega_\phi(u_1). \tag{3.8}$$

The identities in Proposition 3.1 can be rewritten as

$$(q^2v\gamma-u)\Omega_\psi(u)x^-(v)=(q^2v\gamma-u)\delta(v\gamma/u)+(q^2v\gamma-u)x^-(v)\Omega_\psi(u), \tag{3.9}$$

$$(q^2v-u\gamma)\Omega_\phi(u)x^-(v)=(q^2v-u\gamma)\delta(v/u\gamma)+(v-q^2u\gamma)x^-(v)\Omega_\phi(u) \tag{3.10}$$

which may be written out in terms of components as

$$\begin{aligned}&q^2\gamma\Omega_\psi(m)x^-(n+1)-\Omega_\psi(m+1)x^-_n\\&\quad=(q^2\gamma-1)\delta_{m,-n-1}+\gamma x^-_{n+1}\Omega_\psi(m)-q^2x^-_n\Omega_\psi(m+1),\end{aligned} \tag{3.11}$$

$$\begin{aligned}&q^2\Omega_\phi(m)x^-(n+1)-\gamma\Omega_\phi(m+1)x^-(n)\\&\quad=(q^2-\gamma)\delta_{m,-n-1}+x^-(n+1)\Omega_\psi(m)-q^2\gamma x^-(n)\Omega_\psi(m+1).\end{aligned} \tag{3.12}$$

We also have by (3.8)

$$\Omega_\psi(k)\Omega_\phi(m)=\sum_{r\geq 0}g(r)\gamma^{2r}\Omega_\phi(r+m)\Omega_\psi(k-r), \tag{3.13}$$

as operators on $\mathcal{N}_q^-$.

We can also write (3.4) in terms of components and as operators on $\mathcal{N}_q^-$

$$\Omega_\psi(k)x^-(m)=\delta_{k,-m}\gamma^k+\sum_{r\geq 0}g_{q^{-1}}(r)x^-(m+r)\Omega_\psi(k-r)\gamma^r. \tag{3.14}$$

The sum on the right hand side turns into a finite sum when applied to an element in $\mathcal{N}_q^-$, due to (3.3).

PROPOSITION 3.2. [**CFM10**] *There is a unique nondegenerate symmetric bilinear form* (,) *defined on* $\mathcal{N}_q^-$ *satisfying*

$$(x_m^- a, b) = (a, \Omega_\psi(-m)b), \quad (1,1) = 1.$$

For $\mathbf{m} = (m_1, \dots, m_n)$ set

$$x_{\mathbf{m}} = x_{m_1}^- \cdots x_{m_n}^-$$

and define the length of such a Poincare-Birkhoff-Witt basis element to be $|\mathbf{m}| = n$.

PROPOSITION 3.3. [**CFM17**] *For* $\mathbf{m}=(m_1, \dots, m_n) \in \mathbb{Z}^n$, *and* $\mathbf{k}=(k_1, \dots, k_l) \in \mathbb{Z}^l$, *if* $n > l$, *then*

$$(x_{\mathbf{m}}, x_{\mathbf{k}}) = 0. \tag{3.15}$$

On the other hand if $n = l$ *with*

$$m_1 \geq m_2 \geq \cdots \geq m_n, \quad k_1 \geq k_2 \geq \cdots \geq k_n,$$
$$\sum_{i=1}^n m_i = \sum_{i=1}^n k_i$$

we have

$$(x_{\mathbf{m}}, x_{\mathbf{k}}) \equiv \delta_{\mathbf{m},\mathbf{k}} \mod q^2\mathbb{Z}[\![q]\!]. \tag{3.16}$$

The Kashiwara algebra $\mathcal{K}_q$ is defined to be the $\mathbb{F}(q^{1/2})$-subalgebra of End $(\mathcal{N}_q^-)$ generated by $\Omega_\psi(m), x_n^-, \gamma^{\pm 1/2}$, $m, n \in \mathbb{Z}$, $\gamma^{\pm 1/2}$. Then the $\gamma^{\pm 1/2}$ are central and the following relations (which are implied by (3.14)) are satisfied

$$\begin{aligned} (3.17)\quad & q^2\gamma\Omega_\psi(m)x_{n+1}^- - \Omega_\psi(m+1)x_n^- \\ &= (q^2\gamma - 1)\delta_{m,-n-1} + \gamma x_{n+1}^-\Omega_\psi(m) - q^2 x_n^-\Omega_\psi(m+1) \end{aligned}$$

(3.18)

$$q^2\Omega_\psi(k+1)\Omega_\psi(l) - \Omega_\psi(l)\Omega_\psi(k+1) = \Omega_\psi(k)\Omega_\psi(l+1) - q^2\Omega_\psi(l+1)\Omega_\psi(k)$$

$$x_l^- x_{k+1}^- - q^2 x_{k+1}^- x_l^- = q^2 x_{l+1}^- x_k^- - x_k^- x_{l+1}^- \tag{3.19}$$

together with

$$\gamma^{1/2}\gamma^{-1/2} = 1 = \gamma^{-1/2}\gamma^{1/2}.$$

Then $\mathcal{N}_q^-$ is a simple $\mathcal{K}_q$-module ([**CFM10**], Theorem 7.0.14).

4. Quantized imaginary Verma modules and category $\mathcal{O}^q_{\mathbf{red,im}}$

We begin by recalling some basic facts for the affine Kac-Moody algebra $\widehat{\mathfrak{sl}(2)}$ over field $\mathbb{F}$ with generalized Cartan matrix $A = (a_{ij})_{0\leq i,j\leq 1} = \begin{pmatrix} 2 & -2 \\ -2 & 2 \end{pmatrix}$ and its imaginary Verma modules. We use notations in [**Kac90**].

$$\mathfrak{g} = \widehat{\mathfrak{sl}(2)} = \mathfrak{sl}(2) \otimes \mathbb{F}[t, t^{-1}] \oplus \mathbb{F}c \oplus \mathbb{F}d$$

with Lie bracket relations

$$[x \otimes t^n, y \otimes t^m] = [x, y] \otimes t^{n+m} + n\delta_{n+m,0}(x, y)c,$$
$$[x \otimes t^n, c] = 0 = [d, c], \qquad [d, x \otimes t^n] = nx \otimes t^n,$$

for $x, y \in \mathfrak{s}l(2)$, $n, m \in \mathbb{Z}$, where $(\ , \ )$ denotes the trace form on $\mathfrak{s}l(2)$. For $x \in \mathfrak{s}l(2)$ and $n \in \mathbb{Z}$, we write $x(n)$ for $x \otimes t^n$. We consider the usual basis for $\mathfrak{s}l(2)$:

$$\left\{ e = \begin{pmatrix} 0 & 1 \\ 0 & 0 \end{pmatrix}, \quad f = \begin{pmatrix} 0 & 0 \\ 1 & 0 \end{pmatrix}, \quad h = \begin{pmatrix} 1 & 0 \\ 0 & -1 \end{pmatrix} \right\}.$$

Then $\{e_0 = f(1), e_1 = e(0), f_0 = e(-1), f_1 = f(0), h_0 = -h(0) + c, h_1 = h(0), d\}$ generate $\mathfrak{g}$ and $\mathfrak{h} = \mathrm{span}\{h_0, h_1, d\}$ is the Cartan subalgebra.

Let Δ denote the root system of $\widehat{\mathfrak{sl}(2)}$, $\{\alpha_0, \alpha_1\}$ the simple roots, $\delta = \alpha_0 + \alpha_1$ the minimal imaginary root , $\{\Lambda_0, \Lambda_1\}$ fundamental weights and $P = \mathbb{Z}\Lambda_0 \oplus \mathbb{Z}\Lambda_1 \oplus \mathbb{Z}\delta$ the weight lattice. Then

$$\Delta = \{\pm\alpha_1 + n\delta \mid n \in \mathbb{Z}\} \cup \{k\delta \mid k \in \mathbb{Z} \setminus \{0\}\}.$$

A subset S of the root system Δ is called *closed* if $\alpha, \beta \in S$ and $\alpha + \beta \in \Delta$ implies $\alpha + \beta \in S$. The subset S is called a *closed partition* of the roots if S is closed, $S \cap (-S) = \emptyset$, and $S \cup -S = \Delta$ [**JK85**],[**JK89**],[**Fut90**],[**Fut92**]. The set

$$S = \{\alpha_1 + k\delta \mid k \in \mathbb{Z}\} \cup \{l\delta \mid l \in \mathbb{Z}_{>0}\}$$

is a closed partition of Δ and is $W \times \{\pm 1\}$-inequivalent to the standard partition of the root system into positive and negative roots [**Fut94**].

Let $\mathfrak{g}^{(S)}_{\pm} = \sum_{\alpha \in S} \hat{\mathfrak{g}}_{\pm\alpha}$. We have that $\mathfrak{g}^{(S)}_{+}$ is the subalgebra generated by $e(k)$ $(k \in \mathbb{Z})$ and $h(l)$ $(l \in \mathbb{Z}_{>0})$ and $\mathfrak{g}^{(S)}_{-}$ is the subalgebra generated by $f(k)$ $(k \in \mathbb{Z})$ and $h(-l)$ $(l \in \mathbb{Z}_{>0})$. Since S is a partition of the root system, the algebra has a direct sum decomposition:

$$\mathfrak{g} = \mathfrak{g}^{(S)}_{-} \oplus \mathfrak{h} \oplus \mathfrak{g}^{(S)}_{+}.$$

Let $U(\mathfrak{g}^{(S)}_{\pm})$ be the universal enveloping algebra of $\mathfrak{g}^{(S)}_{\pm}$. Then, by the PBW theorem, we have

$$U(\mathfrak{g}) \cong U(\mathfrak{g}^{(S)}_{-}) \otimes U(\mathfrak{h}) \otimes U(\mathfrak{g}^{(S)}_{+}),$$

where $U(\mathfrak{g}^{(S)}_{+})$ is generated by $e(k)$ $(k \in \mathbb{Z})$, $h(l)$ $(l \in \mathbb{Z}_{>0})$, $U(\mathfrak{g}^{(S)}_{-})$ is generated by $f(k)$ $(k \in \mathbb{Z})$, $h(-l)$ $(l \in \mathbb{Z}_{>0})$ and $U(\mathfrak{h})$, the universal enveloping algebra of $\mathfrak{h}$. A $U(\mathfrak{g})$-module V is called a *weight* module if $V = \oplus_{\mu \in P} V_\mu$, where

$$V_\mu = \{v \in V \mid h \cdot v = \mu(h)v, c \cdot v = \mu(c)v, d \cdot v = \mu(d)v\}.$$

Any submodule of a weight module is a weight module. A $U(\mathfrak{g})$-module V is called an *S-highest weight module* with highest weight $\lambda \in P$ if there is a non-zero $v_\lambda \in V$ such that (i) $u^+ \cdot v_\lambda = 0$ for all $u^+ \in U(\mathfrak{g}^{(S)}_{+}) \backslash \mathbb{F}^*$, (ii) $h \cdot v_\lambda = \lambda(h)v_\lambda$, $c \cdot v_\lambda = \lambda(c)v_\lambda$, $d \cdot v_\lambda = \lambda(d)v_\lambda$, (iii) $V = U(\hat{\mathfrak{g}}) \cdot v_\lambda = U(\mathfrak{g}^{(S)}_{-}) \cdot v_\lambda$. An S-highest weight module is a weight module.

For $\lambda \in P$, let $I_S(\lambda)$ denote the ideal of $U(\mathfrak{g})$ generated by $e(k)$ $(k \in \mathbb{Z})$, $h(l)$ $(l > 0)$, $h - \lambda(h)1$, $c - \lambda(c)1$, $d - \lambda(d)1$. Then $M(\lambda) = U(\mathfrak{g})/I_S(\lambda)$ is the *imaginary Verma module* of $\mathfrak{g}$ with highest weight λ. Imaginary Verma modules have many structural features similar to those of standard Verma modules, with the exception of the infinite-dimensional weight spaces [**Fut94**]. It was shown in [**CFKM97**] that the imaginary Verma modules $M(\lambda)$ for $\mathfrak{g}$ can be q-deformed to the quantum imaginary Verma module $M_q(\lambda)$ for $U_q(\mathfrak{g})$ in such a way that the weight multiplicities, both finite and infinite-dimensional, are preserved. Indeed for $\lambda \in P$ the quantum imaginary Verma module $M_q(\lambda)$ can be described as follows. Denote by $I^q(\lambda)$ the ideal of U_q generated by $x^+(k)$, $k \in \mathbb{Z}$, $a(l), l > 0$, $K^{\pm 1} - q^{\lambda(h)}1$, $\gamma^{\pm\frac{1}{2}} - q^{\pm\frac{1}{2}\lambda(c)}1$ and $D^{\pm 1} - q^{\pm\lambda(d)}1$. Then

$$M_q(\lambda) = U_q/I^q(\lambda).$$

THEOREM 4.1 ([**CFKM97**], Theorem 3.6). *The imaginary Verma module $M_q(\lambda)$ is simple if and only if $\lambda(c) \neq 0$.*

Suppose now that $\lambda(c) = 0$. Then $\gamma^{\pm\frac{1}{2}}$ acts on $M_q(\lambda)$ by 1. Consider the ideal $J^q(\lambda)$ of U_q generated by $I^q(\lambda)$ and $a(l)$ for all l. Denote

$$\bar{M}_q(\lambda) = U_q/J^q(\lambda).$$

Then $\bar{M}_q(\lambda)$ is a homomorphic image of $M_q(\lambda)$ which is called the *reduced quantized imaginary Verma module.* The module $\bar{M}_q(\lambda)$ has a P-gradation:

$$\bar{M}_q(\lambda) = \sum_{\xi \in P} \bar{M}_q(\lambda)_\xi.$$

THEOREM 4.2 ([**CFM10**]). *Let $\lambda \in P$ be such that $\lambda(c) = 0$. Then the module $\bar{M}_q(\lambda)$ is simple if and only if $\lambda(h) \neq 0$.*

Consider the set $\mathfrak{h}^*_{red} := \{\lambda \in \mathfrak{h}^* \mid \lambda(c) = 0, \lambda(h) \neq 0\}$. Let G_q be the quantized Heisenberg subalgebra generated by $h_k, k \in \mathbb{Z} \setminus \{0\}$ and γ. We say that a nonzero $U_q(\mathfrak{g})$-module V is G_q-compatible if V has a decomposition $V = TF(V) \oplus T(V)$ into a sum of nonzero G_q-submodules such that G_q is bijective on $TF(V)$ (that any nonzero element $g \in G_q$ is a bijection on $TF(V)$) and $TF(V)$ has no nonzero $U_q(\mathfrak{g})$-submodule, and $G_q \cdot T(V) = 0$.

The category $\mathcal{O}^q_{\text{red,im}}$ has as objects $U_q(\mathfrak{g})$-modules M such that

(1)
$$M = \bigoplus_{\nu \in \mathfrak{h}^*_{red}} M_\nu, \quad \text{where} \quad M_\nu = \{m \in M \mid Km = K^{\nu(h)}m,\ Dm = q^{\nu(d)}m\},$$

(2) x_n^+, $n \in \mathbb{Z}$ act locally nilpotently,
(3) M is G_q-compatible.

The morphisms in this category are the $U_q(\mathfrak{g})$-module homomorphisms. Then for $M \in \mathcal{O}^q_{\text{red,im}}$ there exists $\lambda_i \in \mathfrak{h}^*_{red}$, $i \in I$, with $M \cong \bigoplus_{i \in I} \bar{M}_q(\lambda_i)$ ([**CFM17**], Theorem 6.0.4).

For $M \in \mathcal{O}^q_{\text{red,im}}$, we can write $M = \oplus_i \bar{M}_q(\lambda_i)$ with

$$\bar{M}_q(\lambda_i) = \oplus \mathbb{F}(q^{1/2}) x^-_{n_1} \cdots x^-_{n_k} v_{\lambda_i}.$$

We define $\tilde{\Omega}_\psi(m)$ and $\tilde{x}^-_m$ on each $\bar{M}_q(\lambda_i)$ as in (3.1):

$$\tilde{\Omega}_\psi(m)(x^-_{n_1} \cdots x^-_{n_k} v_{\lambda_i}) := \Omega_\psi(m)(x^-_{n_1} \cdots x^-_{n_k}) v_{\lambda_i} \tag{4.1}$$

$$\tilde{x}^-_m(x^-_{n_1} \cdots x^-_{n_k} v_{\lambda_i}) := x^-_m x^-_{n_1} \cdots x^-_{n_k} v_{\lambda_i}. \tag{4.2}$$

Hence the following result follows.

THEOREM 4.3. [**CFM17**] *The operators $\tilde{\Omega}_\psi(m)$ and $\tilde{x}^-_m$ are well defined on objects in the category $\mathcal{O}^q_{red,im}$. Moreover on each summand $\bar{M}_q(\lambda_i) \cong \mathcal{N}^-_q$ they agree with the $\Omega_\psi(m)$ respectively left multiplication by x^-_m defined as in (3.1).*

THEOREM 4.4. *The operators $\tilde{\Omega}_\psi(m)$ and $\tilde{x}_m^-$ commute with $U_q(\mathfrak{g})$-module homomorphisms.*

PROOF. It is enough to prove the statement for a $U_q(\mathfrak{g})$-module homomorphism $\nu : \bar{M}_q(\lambda) \to M$ from a highest weight module $\bar{M}_q(\lambda)$ to any module $M \in \mathcal{O}^q_{\text{ref,im}}$ for some $\lambda \in \mathfrak{h}^*_{\text{red}}$. Now since $\bar{M}_q(\lambda)$ is simple, its image under ν is isomorphic to $\bar{M}_q(\lambda)$.

Consider now a basis element of $\bar{M}_q(\lambda)$ of the form $x_{n_1}^- \cdots x_{n_k}^- v_\lambda$. Then

$$\begin{aligned}
\nu\left(\tilde{\Omega}_\psi(m)(x_{n_1}^- \cdots x_{n_k}^- v_\lambda)\right) &= \nu\left(\Omega_\psi(m)(x_{n_1}^- \cdots x_{n_k}^-) v_\lambda\right) \\
&= \Omega_\psi(m)(x_{n_1}^- \cdots x_{n_k}^-)\nu\left(v_\lambda\right) \\
&= \tilde{\Omega}_\psi(m)\left(x_{n_1}^- \cdots x_{n_k}^- \nu(v_\lambda)\right) \\
&= \tilde{\Omega}_\psi(m)\nu\left(x_{n_1}^- \cdots x_{n_k}^- v_\lambda\right)
\end{aligned}$$

and

$$\begin{aligned}
\nu\left(\tilde{x}_m^-(x_{n_1}^- \cdots x_{n_k}^-) v_\lambda\right) &= \nu\left(x_m^-(x_{n_1}^- \cdots x_{n_k}^-) v_\lambda\right) \\
&= x_m^-(x_{n_1}^- \cdots x_{n_k}^-)\nu\left(v_\lambda\right) \\
&= \tilde{x}_m^-\left(x_{n_1}^- \cdots x_{n_k}^- \nu(v_\lambda)\right) \\
&= \tilde{x}_m^-\nu\left(x_{n_1}^- \cdots x_{n_k}^- v_\lambda\right).
\end{aligned}$$

□

5. Imaginary crystal lattice and imaginary crystal basis

We first recall the definition of imaginary crystal lattice and imaginary crystal basis for modules in category $\mathcal{O}^q_{\text{red,im}}$. Let $\mathbb{A}_0$ to be the ring of rational functions in $q^{1/2}$ with coefficients in a field $\mathbb{F}$ of characteristic zero, regular at 0. Let $\mathbb{A} = \mathbb{F}[q^{1/2}, q^{-1/2}, \frac{1}{[n]_q}, n > 1]$, and $\pi = \{-k\alpha_1 + m\delta \mid k > 0, m \in \mathbb{Z}\} \cup \{0\}$. Let $M \in \mathcal{O}^q_{\text{red,im}}$. We call a free $\mathbb{A}_0$-submodule $\mathcal{L}$ of M an *imaginary crystal $\mathbb{A}_0$-lattice* of M if the following hold

(i) $\mathbb{F}(q^{1/2}) \otimes_{\mathbb{A}_0} \mathcal{L} \cong M$,
(ii) $\mathcal{L} = \oplus_{\lambda\in\pi}\mathcal{L}_\lambda$ and $\mathcal{L}_\lambda = \mathcal{L} \cap M_\lambda$,
(iii) $\tilde{\Omega}_\psi(m)\mathcal{L} \subseteq \mathcal{L}$ and $\tilde{x}_m^-\mathcal{L} \subseteq \mathcal{L}$ for all $m \in \mathbb{Z}$.

An *imaginary crystal basis* of a $U_q(\mathfrak{g})$-module $M \in \mathcal{O}^q_{\text{red,im}}$ is a pair $(\mathcal{L}, \mathcal{B})$ satisfying

(1) $\mathcal{L}$ is an imaginary crystal lattice of M,
(2) $\mathcal{B}$ is an $\mathbb{F}$-basis of $\mathcal{L}/q\mathcal{L} \cong \mathbb{F} \otimes_{\mathcal{A}_0} \mathcal{L}$,
(3) $\mathcal{B} = \sqcup_{\mu\in\pi}\mathcal{B}_\mu$, where $\mathcal{B}_\mu = \mathcal{B} \cap (\mathcal{L}_\mu/q\mathcal{L}_\mu)$,
(4) $\tilde{x}_m^-\mathcal{B} \subset \pm\mathcal{B} \cup \{0\}$ and $\tilde{\Omega}_\psi(m)\mathcal{B} \subset \pm\mathcal{B} \cup \{0\}$,
(5) For $m \in \mathbb{Z}$, if $\tilde{\Omega}_\psi(-m)b \neq 0$ and $\tilde{x}_m^- b \neq 0$ for $b \in \mathcal{B}$, then $\tilde{x}_m^-\tilde{\Omega}_\psi(-m)b = \tilde{\Omega}_\psi(-m)\tilde{x}_m^- b$. .

For $\lambda \in \mathfrak{h}^*_{red}$, $\bar{M}_q(\lambda) \in \mathcal{O}^q_{\text{red,im}}$, define

$$\mathcal{L}(\lambda) := \sum_{\substack{k\geq 0 \\ i_1\geq\cdots\geq i_k,\, i_j\in\mathbb{Z}}} \mathbb{A}_0 x_{i_1}^- \cdots x_{i_k}^- v_\lambda \subset \mathcal{N}_q^- v_\lambda = \bar{M}_q(\lambda).$$

Then $\mathcal{L}(\lambda)$ is a imaginary crystal $\mathbb{A}_0$-lattice [**CFM17**] since

(i) $\mathbb{F}(q^{1/2}) \otimes_{\mathbb{A}_0} \mathcal{L}(\lambda) \cong \bar{M}_q(\lambda)$,
(ii) $\mathcal{L}(\lambda) = \oplus_{\mu\in\pi} \mathcal{L}(\lambda)_\mu$ where $\mathcal{L}(\lambda)_\mu = \mathcal{L}(\lambda) \cap \bar{M}_q(\lambda)_\mu$,
(iii) $\tilde{\Omega}_\psi(m)\mathcal{L}(\lambda) \subseteq \mathcal{L}(\lambda)$ and $\tilde{x}_m^- \mathcal{L}(\lambda) \subseteq \mathcal{L}(\lambda)$ for all $m \in \mathbb{Z}$.

PROPOSITION 5.1. [**CFM17**] *For $\lambda \in \mathfrak{h}^*_{red}$ we have*

$$\mathcal{L}(\lambda) = \left\{ u \in \bar{M}_q(\lambda) \,|\, (u, \bar{M}_q(\lambda)) \subset \mathbb{A}_0 \right\}.$$

For $\lambda \in \mathfrak{h}^*$ define

$$\mathcal{B}(\lambda) := \left\{ \tilde{x}_{i_1}^- \cdots \tilde{x}_{i_k}^- v_\lambda + q\mathcal{L}(\lambda) \in \mathcal{L}(\lambda)/q\mathcal{L}(\lambda) \,|\, i_1 \geq \cdots \geq i_k \right\}.$$

THEOREM 5.2. [**CFM17**] *For $\lambda \in \mathfrak{h}^*_{red}$, the pair $(\mathcal{L}(\lambda), \mathcal{B}(\lambda))$ is an imaginary crystal basis of the reduced imaginary Verma module $\bar{M}_q(\lambda)$.*

Now suppose $M \in \mathcal{O}^q_{\text{red,im}}$. Then there exists $\lambda_i \in \mathfrak{h}^*_{red}$, $i \in I$, with $M \cong \bigoplus_{i\in I} \bar{M}_q(\lambda_i)$. Let $(\mathcal{L}(\lambda_i), \mathcal{B}(\lambda_i))$ be the imaginary crystal basis for $\bar{M}_q(\lambda_i)$ for $i \in I$. Set $\mathcal{L} = \bigoplus_{i\in I} \mathcal{L}(\lambda_i)$ and $\mathcal{B} = \bigsqcup_{i\in I} \mathcal{B}(\lambda_i)$.

THEOREM 5.3. *The pair $(\mathcal{L}, \mathcal{B})$ is an imaginary crystal basis for $M \in \mathcal{O}^q_{red,im}$.*

PROOF. First let us see that $\mathcal{L}$ is an imaginary crystal lattice:

(i) $\mathbb{F}(q^{1/2}) \otimes_{\mathbb{A}_0} \mathcal{L} \cong \oplus_{i\in I} \mathbb{F}(q^{1/2}) \otimes_{\mathbb{A}_0} \mathcal{L}(\lambda_i) \cong \oplus_{i\in I} \bar{M}_q(\lambda_i) = M$,
(ii) First we show that $\mathcal{L}_\mu = (\oplus_{i\in I} \mathcal{L}(\lambda_i))_\mu = \oplus_{i\in I} \mathcal{L}(\lambda_i)_\mu$ where $\mathcal{L}(\lambda_i)_\mu = \mathcal{L}(\lambda_i) \cap M_\mu$. This follows because if $u \in (\oplus_{i\in I} \mathcal{L}(\lambda_i))_\mu$, then

$$u = \sum_{i\in I} u_i$$

with $u_i \in \mathcal{L}(\lambda_i)$ and $Ku = q^{\mu(h)}u$. Moreover since $\mathcal{L}(\lambda_i) = \oplus_{\mu_i \in\pi} \mathcal{L}(\lambda_i)_{\mu_i}$,

$$u_i = \sum_{\mu_i\in\pi} u_{i,\mu_i}$$

with $u_{i,\mu_i} \in \mathcal{L}(\lambda_i)$ and

$$\sum_{i\in I} \sum_{\mu_i\in\pi} q^{\mu(h)} u_{i,\mu_i} = \sum_{i\in I} q^{\mu(h)} u_i = q^{\mu(h)} u = Ku = \sum_{i\in I} K u_i$$

$$= \sum_{i\in I} \sum_{\mu_i\in\pi} K u_{i,\mu_i} = \sum_{i\in I} \sum_{\mu_i\in\pi} q^{\mu_i(h)} u_{i,\mu_i}.$$

Now since the sum $\oplus_{i\in I} \mathcal{L}(\lambda_i)$ is direct, the above gives us

$$\sum_{\mu_i\in\pi} \left(q^{\mu(h)} - q^{\mu_i(h)} \right) u_{i,\mu_i} = 0$$

for each i. Since $\mathcal{L}(\lambda_i) = \oplus_{\mu_i\in\pi} \mathcal{L}(\lambda_i)_{\mu_i}$ is a direct sum we have $\mu = \mu_i$ for all i. Thus $u_{i,\mu_i} \in \mathcal{L}(\lambda_i) \cap M_\mu = \mathcal{L}(\lambda_i)_\mu$. Finally we have $\mathcal{L} = \oplus_{i\in I} \mathcal{L}(\lambda_i) = \oplus_{i\in I} \oplus_{\mu\in\pi} \mathcal{L}(\lambda_i)_\mu = \oplus_{\mu\in\pi} \oplus_{i\in I} \mathcal{L}(\lambda_i)_\mu = \oplus_{\mu\in\pi} \mathcal{L}_\mu$ and $\mathcal{L}_\mu = \mathcal{L} \cap M_\mu$.
(iii) $\tilde{\Omega}_\psi(m)\mathcal{L} = \tilde{\Omega}_\psi(m) (\oplus_{i\in I} \mathcal{L}(\lambda_i)) = \left(\oplus_{i\in I} \tilde{\Omega}_\psi(m) \mathcal{L}(\lambda_i) \right) \subseteq \oplus_{i\in I} \mathcal{L}(\lambda_i) = \mathcal{L}$ and $\tilde{x}_m^- \mathcal{L} = x_m^- (\oplus_{i\in I} \mathcal{L}(\lambda_i)) = (\oplus_{i\in I} \tilde{x}_m^- \mathcal{L}(\lambda_i)) \subseteq \oplus_{i\in I} \mathcal{L}(\lambda_i) = \mathcal{L}$ for all $m \in \mathbb{Z}$.

This proves that $\mathcal{L}$ is an imaginary crystal lattice.

We know that $\mathcal{B}(\lambda_i)$ is a $\mathbb{F}$-basis of $\mathcal{L}(\lambda_i)/q\mathcal{L}(\lambda_i) = \mathbb{F}\otimes_{\mathcal{A}_0}\mathcal{L}(\lambda_i)$ for each $i\in I$. Now

$$\begin{aligned}\mathcal{L}/q\mathcal{L} &= \left(\oplus_{i\in I}\mathcal{L}(\lambda_i)\right)/q\left(\oplus_{i\in I}\mathcal{L}(\lambda_i)\right)\\ &\cong \oplus_{i\in I}\left(\mathcal{L}(\lambda_i)/q\mathcal{L}(\lambda_i)\right)\cong\oplus_{i\in I}\left(\mathbb{F}\otimes_{\mathcal{A}_0}\mathcal{L}(\lambda_i)\right)\ \cong\mathbb{F}\otimes_{\mathcal{A}_0}\mathcal{L}\end{aligned}$$

since $\mathcal{L}=\oplus_{i\in I}\mathcal{L}(\lambda_i)$. Hence $\mathcal{L}/q\mathcal{L}$ has the $\mathbb{F}$-basis $\sqcup_{i\in I}\mathcal{B}(\lambda_i)=\mathcal{B}$.

For (3) we have

$$\mathcal{B}=\sqcup_{i\in I}\mathcal{B}(\lambda_i)=\sqcup_{i\in I}\sqcup_{\mu\in\pi}\mathcal{B}(\lambda_i)_\mu=\sqcup_{\mu\in\pi}\sqcup_{i\in I}\mathcal{B}(\lambda_i)_\mu=\sqcup_{\mu\in\pi}\mathcal{B}_\mu$$

where $\mathcal{B}_\mu:=\sqcup_{i\in I}\mathcal{B}(\lambda_i)_\mu$ and

$$\begin{aligned}\mathcal{B}_\mu &= \sqcup_{i\in I}\left(\mathcal{B}(\lambda_i)\cap(\mathcal{L}(\lambda_i)_\mu/q\mathcal{L}(\lambda_i)_\mu)\right)\\ &=\sqcup_{i\in I}\mathcal{B}(\lambda_i)\cap\left(\oplus_{j\in I}(\mathcal{L}(\lambda_j)_\mu/q\mathcal{L}(\lambda_j)_\mu)\right)\\ &=(\sqcup_{j\in I}\mathcal{B}(\lambda_i))\cap\mathcal{L}_\mu/q\mathcal{L}_\mu\\ &=\mathcal{B}\cap\mathcal{L}_\mu/q\mathcal{L}_\mu.\end{aligned}$$

For (4) one notes that $\tilde{x}^-_m\mathcal{B}=\sqcup_{i\in I}\tilde{x}^-_m\mathcal{B}(\lambda_i)\subseteq\pm\sqcup_{i\in I}\mathcal{B}(\lambda_i)\cup\{0\}\subset\pm\mathcal{B}\cup\{0\}$ and similarly $\tilde{\Omega}_\psi(m)\mathcal{B}\subset\pm\mathcal{B}\cup\{0\}$,

For (5) we take $b\in\mathcal{B}$ and suppose $\tilde{\Omega}_\psi(-m)b\neq 0$ and $\tilde{x}^-_m b\neq 0$. Now $b\in\mathcal{B}(\lambda_i)$ for some $i\in I$. Since $(\mathcal{L}(\lambda_i),\mathcal{B}(\lambda_i))$ is an imaginary crystal basis, we have $\tilde{x}^-_m\tilde{\Omega}_\psi(-m)b=\tilde{\Omega}_\psi(-m)\tilde{x}^-_m b$.

□

We also have the following partial converse.

THEOREM 5.4. *Let $M=M_1\oplus M_2$ where M_1 and M_2 are modules in the category $\mathcal{O}^q_{red,im}$ and suppose $(\mathcal{L},\mathcal{B})$ is an imaginary crystal basis for M. Furthermore, suppose that there exists $\mathcal{A}_0$-submodules $\mathcal{L}_j\subset M_j$, and subsets $\mathcal{B}_j\subset\mathcal{L}_j/q\mathcal{L}_j$, $j=1,2$ such that $\mathcal{L}=\mathcal{L}_1\oplus\mathcal{L}_2$ and $\mathcal{B}=\mathcal{B}_1\sqcup\mathcal{B}_2$. Then $(\mathcal{L}_j,\mathcal{B}_j)$ is an imaginary crystal basis of M_j, $j=1,2$.*

PROOF. It is straightforward to see that $\mathbb{F}(q^{1/2})\otimes_{\mathcal{A}_0}(\mathcal{L}_j)_\mu\cong(M_j)_\mu$ $(\mu\in\pi)$, $\mathcal{L}_j=\mathcal{L}\cap M_j$ and $\mathcal{B}_j=\mathcal{B}\cap(\mathcal{L}_j/q\mathcal{L}_j)$ for $j=1,2$ (see for instance [**HK02**, Theorem 4.2.10]).

Let $u\in\mathcal{L}_\mu$, for some $\mu\in\pi$. Then $u=u_1+u_2\in\mathcal{L}_1\oplus\mathcal{L}_2$ and

$$Ku_1+Ku_2=Ku=q^{\mu(h)}=q^{\mu(h)}u_1+q^{\mu(h)}u_2$$

Then $Ku_1-q^{\mu(h)}u_1=-Ku_2+q^{\mu(h)}u_2$ which must be zero since $\mathcal{L}\cap\mathcal{L}_2=\{0\}$. Thus $u_j\in\mathcal{L}_{j,\mu}$, $j=1,2$. Hence $\mathcal{L}_\mu=\mathcal{L}_{1,\mu}\oplus\mathcal{L}_{2,\mu}$.

For $u\in\mathcal{L}_j\subset\mathcal{L}$, $u\in\oplus_{\mu\in\pi}\mathcal{L}_\mu$, so as a consequence $u=\sum_{\mu\in\pi}u_\mu$ with $u_\mu\in\mathcal{L}_\mu$ and we can write $u_\mu=u_{1,\mu}+u_{2,\mu}$ with $u_{j,\mu}\in(\mathcal{L}_j)_\mu:=\mathcal{L}_j\cap M_\mu$. Consequently $u-\sum_{\mu\in\pi}u_{j,\mu}=\sum_{\lambda\in\pi}u_{k,\mu}\in\mathcal{L}_j\cap\mathcal{L}_k$ with $k\neq j$. Hence $u-\sum_{\mu\in\pi}u_{j,\mu}=0$ and we have $u\in\oplus_{\mu\in\pi}\mathcal{L}_{j,\mu}$.

Let $p_j:M\to M_j$ denote the natural projections which are $U_q(\mathfrak{g})$-module homomorphisms . Consider $u_j\in\mathcal{L}_j$. Since $\Omega_\psi(m)\mathcal{L}\subset\mathcal{L}$ and $\bar{x}^-_m\mathcal{L}\subset\mathcal{L}$ we write $\Omega_\psi(m)u_1=\bar{u}_1+\bar{u}_2$ and $\bar{x}^-_m u_1=\check{u}_1+\check{u}_2$ with $\check{u}_j,\bar{u}_j\in\mathcal{L}_j$. By Theorem 4.4 we have

$$\Omega_\psi(m)(u_1)=\Omega_\psi(m)p_1(u_1)=p_1\Omega_\psi(m)(u_1)=\bar{u}_1$$

and

$$\bar{x}^-_m(u_1)=\bar{x}^-_m p_1(u_1)=p_1\bar{x}^-_m(u_1)=\check{u}_1$$

Hence $\Omega_\psi(m)(u_1) \in \mathcal{L}_1$ and $\bar{x}_m^-(u_1) \in \mathcal{L}_1$. Similarly $\Omega_\psi(m)(\mathcal{L}_2) \subset \mathcal{L}_2$ and $\bar{x}_m^-(\mathcal{L}_2) \subset \mathcal{L}_2$. This concludes the proof that $\mathcal{L}_j$ are imaginary crystal lattices.

We can write

$$\mathcal{L}_1/q\mathcal{L}_1 \oplus \mathcal{L}_2/q\mathcal{L}_2 \cong \mathcal{L}/q\mathcal{L} \cong \mathbb{F} \otimes_{\mathcal{A}_0} \mathcal{L} \cong \mathbb{F} \otimes_{\mathcal{A}_0} \mathcal{L}_1 \oplus \mathbb{F} \otimes_{\mathcal{A}_0} \mathcal{L}_2$$

Using this isomorphism we have $\mathcal{B}_j = \mathcal{B} \cap (\mathcal{L}_j/q\mathcal{L}_j)$ is an $\mathbb{F}$-basis of $\mathcal{L}_j/q\mathcal{L}_j \cong \mathbb{F} \otimes_{\mathcal{A}_0} \mathcal{L}_j$. Next we have

$$\mathcal{B} = \mathcal{B}_1 \sqcup \mathcal{B}_2$$

and thus $\mathcal{B}_j = \sqcup_{\mu \in \pi} (\mathcal{B}_j)_\mu$ where $(\mathcal{B}_j)_\mu = \mathcal{B}_j \cap ((\mathcal{L}_j)_\mu / q(\mathcal{L}_j)_\mu)$.

Since the operators $\Omega_\psi(m)$ and $\bar{x}_m^-$ commute with $U_q(\mathfrak{g})$-module homomorphisms in the category $\mathcal{O}^q_{\mathrm{red,im}}$ (Theorem 4.4), we have that $\Omega_\psi(m)\mathcal{B}_j \subset \mathcal{B}_j \cup \{0\}$ and $\bar{x}_m^- \mathcal{B}_j \subset \mathcal{B}_j \cup \{0\}$ for all m and $j = 1, 2$.

For $m \in \mathbb{Z}$, if $\tilde{\Omega}_\psi(-m)b \neq 0$ and $\tilde{x}_m^- b \neq 0$ for $b \in \mathcal{B}_j$ (and hence $b \in \mathcal{B}$), then $\tilde{x}_m^- \tilde{\Omega}_\psi(-m)b = \tilde{\Omega}_\psi(-m)\tilde{x}_m^- b$.

This completes the proof that $(\mathcal{L}_j, \mathcal{B}_j)$ is an imaginary crystal basis for $j = 1, 2$. □

References

[CFKM97] B. Cox, V. Futorny, S.-J. Kang, and D. Melville, *Quantum deformations of imaginary Verma modules*, Proc. London Math. Soc. (3) **74** (1997), no. 1, 52–80. MR1416726

[CFM96] B. Cox, V. Futorny, and D. Melville, *Categories of nonstandard highest weight modules for affine Lie algebras*, Math. Z. **221** (1996), no. 2, 193–209. MR1376292

[CFM10] B. Cox, V. Futorny, and K. C. Misra, *Imaginary Verma modules and Kashiwara algebras for $U_q(\widehat{\mathfrak{sl}(2)})$*, Quantum affine algebras, extended affine Lie algebras, and their applications, Contemp. Math., vol. 506, Amer. Math. Soc., Providence, RI, 2010, pp. 105–126. MR2642563

[CFM14] B. Cox, V. Futorny, and K. C. Misra, *Imaginary Verma modules and Kashiwara algebras for $U_q(\hat{\mathfrak{g}})$*, J. Algebra **424** (2015), 390–415. MR3293226

[CFM17] B. Cox, V. Futorny, and K. C. Misra, *Imaginary Verma modules for $U_q(\widehat{\mathfrak{sl}(2)})$ and crystal-like bases*, J. Algebra **481** (2017), 12–35. MR3639466

[Cox94] B. Cox, *Structure of the nonstandard category of highest weight modules*, Modern trends in Lie algebra representation theory (Kingston, ON, 1993), Queen's Papers in Pure and Appl. Math., vol. 94, Queen's Univ., Kingston, ON, 1994, pp. 35–47. MR1281180

[Dri85] V. G. Drinfel′d, *Hopf algebras and the quantum Yang-Baxter equation* (Russian), Dokl. Akad. Nauk SSSR **283** (1985), no. 5, 1060–1064. MR802128

[FGM98] V. M. Futorny, A. N. Grishkov, and D. J. Melville, *Quantum imaginary Verma modules for affine Lie algebras* (English, with English and French summaries), C. R. Math. Acad. Sci. Soc. R. Can. **20** (1998), no. 4, 119–123. MR1662112

[Fut90] V. M. Futorny, *Parabolic partitions of root systems and corresponding representations of the affine Lie algebras* (English, with Russian summary), Akad. Nauk Ukrain. SSR Inst. Mat. Preprint **8** (1990), 30–39. MR1078876

[Fut92] V. M. Futorny, *The parabolic subsets of root system and corresponding representations of affine Lie algebras*, Proceedings of the International Conference on Algebra, Part 2 (Novosibirsk, 1989), Contemp. Math., vol. 131, Amer. Math. Soc., Providence, RI, 1992, pp. 45–52. MR1175820

[Fut94] V. M. Futorny, *Imaginary Verma modules for affine Lie algebras*, Canad. Math. Bull. **37** (1994), no. 2, 213–218. MR1275706

[HK02] J. Hong and S.-J. Kang, *Introduction to quantum groups and crystal bases*, Graduate Studies in Mathematics, vol. 42, American Mathematical Society, Providence, RI, 2002. MR1881971

[Jim85] M. Jimbo, *A q-difference analogue of $U(\mathfrak{g})$ and the Yang-Baxter equation*, Lett. Math. Phys. **10** (1985), no. 1, 63–69. MR797001

[JK85] H. P. Jakobsen and V. G. Kac, *A new class of unitarizable highest weight representations of infinite-dimensional Lie algebras*, Nonlinear equations in classical and quantum field theory (Meudon/Paris, 1983/1984), Lecture Notes in Phys., vol. 226, Springer, Berlin, 1985, pp. 1–20. MR802097

[JK89] H. P. Jakobsen and V. Kac, *A new class of unitarizable highest weight representations of infinite-dimensional Lie algebras. II*, J. Funct. Anal. **82** (1989), no. 1, 69–90. MR976313

[Kac90] V. G. Kac, *Infinite-dimensional Lie algebras*, 3rd ed., Cambridge University Press, Cambridge, 1990. MR1104219

[Kan95] S.-J. Kang, *Quantum deformations of generalized Kac-Moody algebras and their modules*, J. Algebra **175** (1995), no. 3, 1041–1066. MR1341758

[Kas90] M. Kashiwara, *Crystalizing the q-analogue of universal enveloping algebras*, Comm. Math. Phys. **133** (1990), no. 2, 249–260. MR1090425

[Kas91] M. Kashiwara, *On crystal bases of the Q-analogue of universal enveloping algebras*, Duke Math. J. **63** (1991), no. 2, 465–516. MR1115118

[Lus88] G. Lusztig, *Quantum deformations of certain simple modules over enveloping algebras*, Adv. in Math. **70** (1988), no. 2, 237–249. MR954661

[Lus90] G. Lusztig, *Canonical bases arising from quantized enveloping algebras*, J. Amer. Math. Soc. **3** (1990), no. 2, 447–498. MR1035415

[Mas09] A. Masuoka, *Generalized q-boson algebras and their integrable modules*, J. Algebra **322** (2009), no. 6, 2199–2219. MR2542839

Department of Mathematics, University of Charleston, Charleston, South Carolina 27695

Email address: coxbl@cofc.edu

Department of Mathematics, University of São Paulo, São Paulo, Brazil

Email address: futorny@ime.usp.br

Department of Mathematics, North Carolina State University, Raleigh, North Carolina 27695

Email address: misra@ncsu.edu

Contemporary Mathematics
Volume **713**, 2018
https://doi.org/10.1090/conm/713/14312

On the module structure of the center of hyperelliptic Krichever-Novikov algebras

Ben Cox and Mee Seong Im

ABSTRACT. We consider the coordinate ring of a hyperelliptic curve and let $\mathfrak{g} \otimes R$ be the corresponding current Lie algebra where $\mathfrak{g}$ is a finite dimensional simple Lie algebra defined over $\mathbb{C}$. We give a generator and relations description of the universal central extension of $\mathfrak{g} \otimes R$ in terms of certain families of polynomials $P_{k,i}$ and $Q_{k,i}$ and describe how the center Ω_R/dR decomposes into a direct sum of irreducible representations when the automorphism group is C_{2k} or D_{2k}.

1. Introduction

In [**Cox16a**], the author describes the action of the automorphism group of the ring $R = \mathbb{C}[t, (t-a_1)^{-1}, \ldots, (t-a_n)^{-1}]$ on the center of the current Krichever-Novikov algebra whose coordinate ring is R, where $a_1, \ldots, a_n$ are pairwise distinct complex numbers (see also [**CGLZ14**]). In that setting, the five Kleinian groups C_n, D_n, A_4, S_4 and A_5 appear as automorphism groups of R for particular choices of $a_1, \ldots, a_n$. These five groups naturally appear in the McKay correspondence, which ties together the representation theory of finite subgroups G of $SL_n(\mathbb{C})$ to the resolution of singularities of quotient orbifolds $\mathbb{C}^n/G$.

It is known that ℓ-adic cohomology groups tend to be acted on by Galois groups, and the way in which these cohomology groups decompose can give interesting and important number theoretic information (see for example R. Taylor's review of Tate's conjecture [**Tay04**]). Moreover it is an interesting and very difficult problem to describe the group $\mathrm{Aut}(R)$ where R is the space of meromorphic functions on a compact Riemann surface X and to determine the module structure of its induced action on the module of holomorphic differentials $\mathcal{H}^1(X)$ (see [**Bre00**]). Now if one realizes the fact that the cyclic homology group $HC_1(R) = \Omega_R^1/dR$ can be identified with the $H_2(\mathfrak{sl}(R), \mathbb{C})$ which gives the space of 2-cocycles (see [**Blo81**]), it is natural to ask how Ω_R^1/dR decomposes into a direct sum of irreducible modules under the action of the $\mathrm{Aut}(R)$.

2010 *Mathematics Subject Classification.* Primary 22E60, 22E66, 22E99, 16W25, 16W20.

Key words and phrases. Hyperelliptic Krichever-Novikov algebras, universal central extensions, Kähler differentials, current algebras, Riemann surfaces with punctures.

The first author was partially supported by a collaboration grant from the Simons Foundation #319261.

The second author was partially supported by B. Cox's Simons Collaboration Grant #319261 and the Department of Mathematical Sciences at the United States Military Academy.

One of our main results includes Theorem 5.1, where we describe the universal central extension of the hyperelliptic Lie algebra as a $\mathbb{Z}_2$-graded Lie algebra. In this theorem we give a description of the bracket of two basis elements in the universal central extension of $\mathfrak{g}\otimes R$ in terms of polynomials $P_{k,i}$ and $Q_{k,i}$ defined recursively

$$(2k+r+3)P_{k,i} = -\sum_{j=1}^{r}(3j+2k-2r)a_jP_{k-r+j-1,i} \tag{1}$$

for $k\geq 0$ with the initial condition $P_{l,i}=\delta_{l,i}$, $-r\leq i,l\leq -1$ and

$$(2m-3)a_1Q_{m,i} = \sum_{j=2}^{r+1}(3j-2m)a_jQ_{m-j+1,i} \tag{2}$$

with initial condition $Q_{m,i}=\delta_{m,-i}$ for $1\leq m\leq r$ and $-r\leq i\leq -1$. In this paper $\mathfrak{g}$ is assumed throughout to be a finite dimensional simple Lie algebra defined over the complex numbers. The generating series for these polynomials can be written in terms of hyperelliptic integrals (28) and (34) using Bell polynomials and Faá de Bruno's formula (see §4). One can compare this result to that given in [**Cox16b**] and also in [**CZ17**].

We also describe in this paper (see Theorem 7.2) how Kähler differentials modulo exact forms Ω_R/dR decompose under the action of the automorphism group of the coordinate ring $R:=R_2(p)=\mathbb{C}[t^{\pm1},u:u^2=p(t)]$, where $p(t)=t(t-\alpha_1)\cdots(t-\alpha_{2n})=\sum_{i=1}^{2n+1}a_it^i$, with the α_i being pairwise distinct roots. In this setting, we first observe that we have the following result due to M. Bremner (see [**Bre94**])

$$\Omega_R/dR=\bigoplus_{i=0}^{2n}\mathbb{C}\omega_i, \tag{3}$$

where $\omega_0=\overline{t^{-1}\,dt}$, $\omega_i=\overline{t^{-i}u\,dt}$ for $i=1,\dots,2n$.

The possible automorphism groups for the hyperelliptic curve

$$R=\mathbb{C}[t^{\pm1},u:u^2=t(t-\alpha_1)\cdots(t-\alpha_{2n})]$$

are the groups C_{2k}, D_{2k} or one of the groups

$$\begin{aligned}\mathbb{V}_{2k}&:=\langle x,y:x^4,y^{2k},(xy)^2,(x^{-1}y)^2\rangle,\\ \mathbb{U}_k&:=\langle x,y:x^2,y^{2n},xyxy^{k+1}\rangle\\ Dic_k&:=\langle a,x:a^{2k}=1,\ x^2=a^k,\ x^{-1}ax=a^{-1}\rangle\end{aligned}$$

(see Theorem 6.2 below, [**CGLZ17**, Corollary 15], [**BGG93**] and [**Sha03**]).

The above polynomials $P_{k,i}$ help us to describe how the center decomposes under the group of automorphisms of R. The automorphism group of R has a canonical action on Ω_R/dR and so it is natural to ask how this representation decomposes into a direct sum of irreducible representations. When the automorphism group is C_{2k} we can rewrite (3) as a direct sum of 1-dimensional irreducible C_{2k}-representations. More precisely the center decomposes as:

$$\Omega_R/dR\cong U_0\oplus\ldots\oplus U_{k-1}, \tag{4}$$

where $U_r = \bigoplus_{i\equiv r \mod k, 1\leq i\leq 2n} \mathbb{C}\omega_i$ for $r = 1, \ldots, k-1$ is a sum of one-dimensional irreducible representation of C_{2k} with character $\chi_r(s) = \exp(2\pi\imath rs/2k)$, each occurring with multiplicity l and

$$U_0 = \mathbb{C}\omega_0 \oplus \bigoplus_{i=1}^{l} \mathbb{C}\omega_{ki}.$$

If the automorphism group is D_{2k} (with a certain parameter $c^{2n} = a_1$ and $k|2n$) the center decomposes under the action of D_{2k} as

$$\Omega_R/dR \cong \mathbb{C}\omega_0 \oplus \bigoplus_{i=3}^{4} U_i^{\oplus \Upsilon_i(\epsilon_i,\nu_i)} \oplus \bigoplus_{h=1}^{k-1} V_h^{\oplus \frac{(1-(-1)^h)n}{k}} \tag{5}$$

where

$$\Upsilon_i(\epsilon_i, \nu_i) = \frac{(1-(-1)^k)n}{2k}(\delta_{i,3} + \delta_{i,4}) + (-1)^i \frac{1-(-1)^n}{4} + \frac{1}{2}(-1)^i \sum_{i=n+3}^{2n} c^{n+3-2i} P_{i-n-3,-i}.$$

and U_i, $i = 1, 2, 3, 4$, are the irreducible one dimensional representations for D_{2k} with character ρ_i and V_h are the irreducible 2-dimensional representations for D_{2k} with character χ_h, $1 \leq h \leq k-1$ (see Theorem 7.2 below). Note $\mathbb{C}\omega_0$ and U_1 are the trivial representations.

We use classical representation theory techniques found for example in [**Ser77**] by Serre and [**FH91**] by Fulton and Harris to prove our results.

The remaining cases where the automorphism group is D_{2k} when $c^{2n} = -a_1$, $\mathbb{V}_{2k}$, $\mathbb{U}_k$ or Dic_k will be studied in a future publication.

The authors would like to thank Xiangqian Guo and Kaiming Zhao for useful discussions and pointing some corrections.

2. Background

2.1. Universal central extensions. An *extension* of a Lie algebra $\mathfrak{g}$ is a short exact sequence of Lie algebras

$$0 \longrightarrow \mathfrak{k} \xrightarrow{f} \mathfrak{g}' \xrightarrow{g} \mathfrak{g} \longrightarrow 0. \tag{6}$$

A homomorphism from one extension $\mathfrak{g}' \xrightarrow{g} \mathfrak{g}$ to another extension $\mathfrak{g}'' \xrightarrow{g'} \mathfrak{g}$ is a Lie algebra homomorphism $\mathfrak{g}' \xrightarrow{h} \mathfrak{g}''$ such that $g' \circ h = g$. A central extension $\widehat{\mathfrak{g}} \xrightarrow{u} \mathfrak{g}$ is a universal central extension if there is a unique homomorphism from $\widehat{\mathfrak{g}} \xrightarrow{u} \mathfrak{g}$ to any other central extension $\mathfrak{g}' \xrightarrow{g} \mathfrak{g}$.

Now let R be a commutative ring over $\mathbb{C}$ and let $\mathfrak{g}$ be a finite-dimensional simple Lie algebra over $\mathbb{C}$. Let $F = R \otimes R$ be the left R-module with the action $a(b \otimes c) = ab \otimes c$, where $a, b, c \in R$. Let K be the submodule of F generated by elements of the form $1 \otimes ab - a \otimes b - b \otimes a$. Then $\Omega_R^1 = F/K$ is the module of Kähler differentials. The canonical map $d : R \to \Omega_R$ sends $da = 1 \otimes a + K$, so we will write $c\,da := c \otimes a + K$. Exact differentials consist of elements in the subspace dR and we write $\overline{c\,da}$ as the coset of $c\,da$ modulo dR. It is a classical result by C.

Kassel (1984) that the universal central extension of the current algebra $\mathfrak{g} \otimes R$ is the vector space $\widehat{\mathfrak{g}} = (\mathfrak{g} \otimes R) \oplus \Omega_R/dR$, with the Lie bracket:

$$[x \otimes a, y \otimes b] = [x, y] \otimes ab + (x, y)\overline{a\,db}, \quad [x \otimes a, \omega] = 0, \quad [\omega, \omega'] = 0, \tag{7}$$

where $x, y \in \mathfrak{g}$, $a, b \in R$, $\omega, \omega' \in \Omega_R/dR$, and $(\cdot,\cdot)$ is the Killing form on $\mathfrak{g}$. Since the center of the universal central extension is defined to be $Z(\widehat{\mathfrak{g}}) \subseteq \ker(\widehat{\mathfrak{g}} \to \mathfrak{g} \otimes R)$, Kassel showed that $Z(\widehat{\mathfrak{g}})$ is precisely Ω_R/dR. In this paper, we will fix $R = R_2(p) := \mathbb{C}[t^{\pm 1}, u : u^2 = p(t)]$, where $p(t) = t(t-\alpha_1)\cdots(t-\alpha_{2n}) \in \mathbb{C}[t]$ and α_i's are pairwise distinct roots.

2.2. Lie algebra 2-cocycles. Given a Lie algebra $\mathfrak{g}$ over $\mathbb{C}$, a *Lie algebra 2-cocycle* for $\mathfrak{g}$ is a bilinear map $\psi : \mathfrak{g} \times \mathfrak{g} \to \mathbb{C}$ satisfying:

(1) $\psi(x, y) = -\psi(y, x)$ for $x, y \in \mathfrak{g}$, and
(2) $\psi([x, y], z) + \psi([y, z], x) + \psi([z, x], y) = 0$ for $x, y, z \in \mathfrak{g}$.

In particular, $\psi : (\mathfrak{g} \otimes R) \times (\mathfrak{g} \otimes R) \to \mathbb{C}$ is given by

$$\psi(x \otimes a, y \otimes b) = (x, y)\overline{a\,db}, \tag{8}$$

which is a 2-cocycle on $\mathfrak{g} \otimes R$.

Since we do not need the degree of the polynomial $p(t)$ to be odd, we will first let $\deg p(t) = r+1$, up until Section 6. The reason we first work in the more general setting is that it allows us to fill in the remaining case which was not covered in [**Cox16b**] (in this manuscript, the author required that the constant term a_0 of p to be $a_0 \neq 0$). In Sections 6 and 7, we restrict to the case of $r = 2n$, which allows us to use the results in [**CGLZ17**] on automorphism groups of such algebras. So let

$$p(t) = t(t-\alpha_1)\cdots(t-\alpha_r) = \sum_{i=1}^{r+1} a_i t^i,$$

where the α_i are pairwise distinct nonzero complex numbers with $a_1 = (-1)^r \prod_{i=1}^{r} \alpha_i \neq 0$ and $a_{r+1} = 1$.

Note that $R = \mathbb{C}[t^{\pm 1}, u : u^2 = p(t)]$ is a regular ring when α_i are distinct complex numbers, and $\operatorname{Der}(R)$ is a simple infinite dimensional Lie algebra (see [**CGLZ17**], [**Jor86**], [**Skr88**] and [**Skr04**]).

We recall:

LEMMA 2.1 ([**CF11**], Lemma 2.0.2). *If $u^m = p(t)$ and $R = \mathbb{C}[t^{\pm 1}, u : u^m = p(t)]$, then one has in Ω^1_R/dR the congruence*

$$((m+1)(r+1) + im)t^{r+i}u\,dt \equiv -\sum_{j=0}^{r}((m+1)j + mi)a_j t^{i+j-1}u\,dt \mod dR. \tag{9}$$

Motivated by Lemma 2.1 with $m = 2$ and $a_0 = 0$, we let $P_{k,i} := P_{k,i}(a_1, \ldots, a_r)$, $k \geq -r$, $-r \leq i \leq -1$ be the polynomials in the a_i satisfying the recursion relations:

$$(2k + r + 3)P_{k,i} = -\sum_{j=1}^{r}(3j + 2k - 2r)a_j P_{k-r+j-1,i} \tag{10}$$

for $k \geq 0$ with the initial condition $P_{l,i} = \delta_{l,i}$, $-r \leq i, l \leq -1$.

3. Cocycles

Let $p(t) = t^{r+1} + a_r t^r + \ldots + a_1 t$, where $a_i \in \mathbb{C}$. Fundamental to the description of $\widehat{\mathfrak{g}}$ is the following:

THEOREM 3.1 ([**Bre94**], Theorem 3.4). *Let $R = \mathbb{C}[t^{\pm 1}, u : u^2 = p(t)]$. The set*

$$\{\overline{t^{-1}\,dt}, \overline{t^{-1}u\,dt}, \ldots, \overline{t^{-r}u\,dt}\} \tag{11}$$

forms a basis of Ω^1_R/dR.

Let

$$\omega_0 := \overline{t^{-1}\,dt} \quad \text{and} \quad \omega_k := \overline{t^{-k}u\,dt} \qquad \text{for } 1 \le k \le r. \tag{12}$$

We will first describe the cocyles contributing to the *even* part $\mathbb{C}\omega_0$ of the center of the universal central extension of the hyperelliptic current algebra:

LEMMA 3.2 ([**Bre94**], Proposition 4.2). *For $i, j \in \mathbb{Z}$ one has*

$$\overline{t^i\,d(t^j)} = j\delta_{i+j,0}\omega_0 \tag{13}$$

and

$$\overline{t^i u\,d(t^j u)} = \sum_{k=1}^{r+1}\left(j + \frac{1}{2}k\right) a_k \delta_{i+j,-k}\omega_0. \tag{14}$$

For the *odd* part $\mathbb{C}\omega_1 \oplus \ldots \oplus \mathbb{C}\omega_r$ of the center, we generalize Proposition 4.2 in [**Bre94**] via the following result:

PROPOSITION 3.3. *For $i, j \in \mathbb{Z}$, one has*

$$\overline{t^i u\,d(t^j)} = j\begin{cases} \displaystyle\sum_{k=1}^{r} P_{i+j-1,-k}\omega_k & \text{if } i+j \ge -r+1, \\ \displaystyle\sum_{k=1}^{r} Q_{-i-j+1,-k}\omega_k & \text{if } i+j < -r+1, \end{cases} \tag{15}$$

where $P_{m,i}$ is the recursion relation in Equation (10) and $Q_{m,i}$ satisfies

$$(2m-3)a_1 Q_{m,i} = \left(\sum_{j=2}^{r+1}(3j-2m)a_j Q_{m-j+1,i}\right) \tag{16}$$

with initial condition $Q_{m,i} = \delta_{m,-i}$ for $1 \le m \le r$ and $-r \le i \le -1$.

PROOF. We set $m = 2$ and replace j in the summation in Equation (9) by k, and then replace i with $-r+i+j-1$ to obtain:

$$(2(i+j)+r+1)t^{i+j-1}u\,dt \equiv -\sum_{k=1}^{r}(3k+2(i+j)-2(r+1))a_k t^{i+j-(r+1)+k-1}u\,dt \mod dR,$$

and similarly

$$(2(i+j)+r+1)P_{i+j-1,\iota} = -\sum_{k=1}^{r}(3k+2(i+j)-2(r+1))a_k P_{i+j-1+k-(r+1),\iota}. \tag{17}$$

So now assume for $\iota \geq -r$,

$$\overline{t^\iota u\,dt} = \sum_{k=1}^{r} P_{\iota,k-(r+1)}\omega_{r+1-k}. \tag{18}$$

It is clear that Equation (18) holds when $\iota = -r, \dots, -1$ as $P_{l,i} = \delta_{l,i}$ for $-r \leq i, l \leq -1$. Then the induction step is:

$$\begin{aligned}\overline{t^{\iota+1}u\,dt} &= -\sum_{k=1}^{r}\left(\frac{3k+2\iota-2r+2}{2\iota+r+5}\right)a_k\overline{t^{\iota+k-r}u\,dt}\\ &= -\sum_{l=1}^{r}\sum_{k=1}^{r}\left(\frac{3k+2\iota-2r+2}{2\iota+r+5}\right)a_kP_{\iota+k-r,l-(r+1)}\omega_{r+1-l}\\ &= \sum_{l=1}^{r}P_{\iota+1,l-(r+1)}\omega_{r+1-l}.\end{aligned}$$

Now, for $i+j \geq -r+1$, we have

$$\overline{t^iu\,d(t^j)} = j\overline{t^{i+j-1}u\,dt} = j\sum_{l=1}^{r}P_{i+j-1,l-(r+1)}\omega_{r+1-l} = j\sum_{k=1}^{r}P_{i+j-1,-k}\omega_k. \tag{19}$$

Again consider (9) and set $r+i = k-1$ or $i = k-(r+1)$:

$$(2k+r+1)\overline{t^{k-1}u\,dt} = -\sum_{j=1}^{r}(3j+2k-2(r+1))a_j\overline{t^{k+j-1-(r+1)}u\,dt}, \tag{20}$$

and write it as

$$(-2(m-1)+r+1)\overline{t^{-m}u\,dt} = -\sum_{j=1}^{r}(3j-2m+2-2(r+1))a_j\overline{t^{-(m-j+r+1)}u\,dt}. \tag{21}$$

Then

$$\begin{aligned}0 &= -\sum_{j=1}^{r+1}(3j+2k-2(r+1))a_j\overline{t^{k+j-1-(r+1)}u\,dt}\\ &= -(2k-2r+1)a_1\overline{t^{k-(r+1)}u\,dt} - \dots - (2k+r-2))a_r\overline{t^{k-2}u\,dt} - (2k+r+1)\overline{t^{k-1}u\,dt},\end{aligned}$$

as $a_{r+1} = 1$. We rewrite this as

$$\begin{aligned}\overline{t^{k-(r+1)}u\,dt} &= \frac{-1}{(2k-2r+1)a_1}\Big((2k-2r+4)a_2\overline{t^{k-r}u\,dt} + \dots\\ &\qquad + (2k+r-2)a_r\overline{t^{k-2}u\,dt} + (2k+r+1)\overline{t^{k-1}u\,dt}\Big)\\ &= \frac{-1}{(2k-2r+1)a_1}\left(\sum_{j=2}^{r+1}(3j+2k-2(r+1))a_j\overline{t^{k+j-1-(r+1)}u\,dt}\right).\end{aligned}$$

For $k = 0, -1, -2$ we have for instance

$$\overline{t^{-(r+1)}u\,dt} = \frac{1}{(-2r+1)a_1}\left(-(-2r+4)a_2\overline{t^{-r}u\,dt} - \ldots - (r-2)a_r\overline{t^{-2}u\,dt} - (r+1)\overline{t^{-1}u\,dt}\right)$$

$$\overline{t^{-r-2}u\,dt} = \frac{1}{(-2r-1)a_1}\left(-(-2r+2)a_2\overline{t^{-r-1}u\,dt} - \ldots - (r-4)a_r\overline{t^{-3}u\,dt} - (r-1)\overline{t^{-2}u\,dt}\right)$$

$$\overline{t^{-r-3}u\,dt} = \frac{1}{(-2r-3)a_1}\left(-(-2r)a_2\overline{t^{-r-2}u\,dt} - \ldots - (r-6)a_r\overline{t^{-4}u\,dt} - (r-3)\overline{t^{-3}u\,dt}\right).$$

Setting $-m = k - r - 1$, we get $k = -m + r + 1$ and

$$\overline{t^{-m}u\,dt} = \frac{1}{(2m-3)a_1}\left(\sum_{j=2}^{r+1}(3j-2m)a_j\overline{t^{-m+j-1}u\,dt}\right) \tag{22}$$

for $m \geq r+1$. This leads us to the recursion relation:

$$Q_{m,i} = \frac{1}{(2m-3)a_1}\left(\sum_{j=2}^{r+1}(3j-2m)a_jQ_{m-j+1,i}\right) \tag{23}$$

for $m \geq r+1$ with the initial condition $Q_{m,i} = \delta_{m,-i}$, $1 \leq m \leq r$ and $-r \leq i \leq -1$.

So now assume for $\iota \geq 1$,

$$\overline{t^{-\iota}u\,dt} = \sum_{k=0}^{r-1} Q_{\iota,k-r}\omega_{r-k} = \sum_{k=1}^{r} Q_{\iota,-k}\omega_k. \tag{24}$$

It is clear that Equation (24) holds for $\iota = 1, \ldots, r$ as $Q_{m,i} = \delta_{m,-i}$, $1 \leq m \leq r$ and $-r \leq i \leq -1$.

For $\iota \geq r$, we have by (22), (23) and the induction hypothesis:

$$\begin{aligned}\overline{t^{-(\iota+1)}u\,dt} &= \sum_{j=2}^{r+1}\frac{(3j-2\iota-2)a_j}{(2\iota-1)a_1}\overline{t^{-\iota+j-2}u\,dt}\\ &= \sum_{k=0}^{r-1}\sum_{j=2}^{r+1}\frac{(3j-2i-2)a_j}{(2i-1)a_1}Q_{\iota-j+2,k-r}\omega_{r-k}\\ &= \sum_{k=0}^{r-1}Q_{\iota+1,k-r}\omega_{r-k},\end{aligned}$$

which proves (22) for $m = \iota + 1$.

We conclude for $i + j - 1 < -r$, we have

(25)

$$\overline{t^iu\,d(t^j)} = j\overline{t^{i+j-1}u\,dt} = j\sum_{k=1}^{r}Q_{-i-j+1,k-(r+1)}\omega_{r+1-k} = j\sum_{k=1}^{r}Q_{-i-j+1,-k}\omega_k.$$

□

4. Faá de Bruno's formula and Bell polynomials

Now consider the formal power series

$$P_i(z) := P_i(a_1, \ldots, a_r, z) := \sum_{k \geq -r} P_{k,i} z^{k+r} = \sum_{k \geq 0} P_{k-r,i} z^k \tag{26}$$

for $-r \leq i \leq -1$. We will find an integral formula for $P_i(z)$ below. One can show that $P_i(z)$ must satisfy the first order differential equation

$$\frac{d}{dz} P_i(z) - \frac{Q(z)}{2zT(z)} P_i(z) = \frac{R_i(z)}{2zT(z)}, \tag{27}$$

where

$$T(z) := \sum_{j=1}^{r+1} a_j z^{r+1-j}, \quad Q(z) := zT'(z) + (r-3)T(z),$$

and

$$R_i(z) := \sum_{j=1}^{r+1} \left(\sum_{1-j \leq k < 0} (3j + 2k - 2r) a_j \delta_{k+j-r-1,i} z^{k+r} \right)$$

since indeed, we have

$$\begin{aligned}
&2zT(z)\frac{d}{dz}P_i(z) - Q(z)P_i(z) \\
&= \sum_{k \geq 0} \left(\sum_{j=1}^{r+1} 2k a_j P_{k-r,i} z^{r+k+1-j} - \sum_{j=1}^{r+1} (2r - j - 2) a_j P_{k-r,i} z^{r+k+1-j} \right) \\
&= \sum_{k \geq 0} \left(\sum_{j=1}^{r+1} (2k - 2r + j + 2) a_j P_{k-r,i} z^{r+1+k-j} \right) \\
&= \sum_{k \geq 0} \left(\sum_{j=1}^{r+1} (2k + 3j - 2r) a_j P_{k+j-r-1,i} z^{r+k} \right) \\
&\quad + \sum_{j=1}^{r+1} \left(\sum_{1-j \leq k < 0} (2k + 3j - 2r) a_j P_{k+j-r-1,i} z^{r+k} \right) \\
&= R_i(z),
\end{aligned}$$

where the first summation in the second to last equality is zero due to (10).

An integrating factor is

$$\mu(z) = \exp \int -\frac{Q(z)}{2zT(z)}\, dz = \frac{1}{z^{(r-3)/2}\sqrt{T(z)}},$$

and so

$$P_i(z) := z^{(r-3)/2}\sqrt{T(z)} \int \frac{R_i(z)}{2z^{(r-1)/2}T(z)^{3/2}}\, dz. \tag{28}$$

The way we interpret the right hand hyperelliptic integral ($T(0) = a_{r+1} = 1 \neq 0$) is to expand $R_i(z)/T(z)^{3/2}$ in terms of a Taylor series about $z = 0$ and then formally integrate term by term. We then multiply the result by series for $z^{(r-3)/2}\sqrt{T(z)}$. Let us explain this more precisely.

One can expand both $\sqrt{T(z)}$ and $1/T(z)^{3/2}$ using Bell polynomials and Faà di Bruno's formula as follows. Bell polynomials in the variables $z_1, z_2, z_3, \ldots, z_{m-k+1}$ are defined to be

$$B_{m,k}(z_1, \ldots, z_{m-k+1}) := \sum \frac{m!}{l_1! l_2! \cdots l_{m-k+1}!} \left(\frac{z_1}{1!}\right)^{l_1} \cdots \left(\frac{z_{m-k+1}}{(m-k+1)!!}\right)^{l_{m-k+1}},$$

where the sum is over $l_1 + l_2 + \ldots + l_{m-k+1} = k$ and $l_1 + 2l_2 + 3l_3 + \ldots + (m-k+1)l_{m-k+1} = m$ (see [**Bel28**]).

Now Faà di Bruno's formula ([**FdB55**] and [**FdB57**]; discovered earlier by Arbogast [**Arb00**]) for the m-derivative of $f(g(x))$ is

$$\frac{d^m}{dx^m} f(g(x)) = \sum_{l=0}^{m} f^{(l)}(g(x)) B_{m,l}(g'(x), g''(x), \ldots, g^{(m-l+1)}(x)).$$

Here $f(x) = x^{-3/2}$, $g(x) = T(x)$, so we get

$$f^{(m)}(x) = \frac{(-1)^m (2m+1)!!}{2^m x^{(2m+3)/2}} \tag{29}$$

where

$$(2k-1)!! = \Gamma(k + (1/2)) 2^k / \sqrt{\pi}.$$

Then $(-1)!! = 1$ and $T^{(k)}(0) = k! a_{r+1-k}$ so that

$$\left.\frac{d^m}{dx^m} f(g(x))\right|_{x=0} = \sum_{l=0}^{m} \frac{(-1)^l (2l+1)!!}{2^l} B_{m,l}(a_r, 2a_{r-1}, \ldots, (m-l+1)! a_{r-m+l}).$$

As a consequence

$$\begin{aligned} &\frac{1}{T(z)^{3/2}} \\ &\quad = \sum_{m=0}^{\infty} \frac{1}{m!} \left.\frac{d^m}{dz^m} f(g(z))\right|_{z=0} z^m \\ &\quad = \sum_{m=0}^{\infty} \frac{1}{m!} \left(\sum_{l=0}^{m} \frac{(-1)^l (2l+1)!!}{2^l} B_{m,l}(a_{m-1}, 2a_{m-2}, \ldots, (m-l+1)! a_{r-m+l}) \right) z^m, \end{aligned}$$

and hence

$$\begin{aligned} &T_m(a_1, \ldots, a_{m-1}) \\ &\quad = \frac{1}{m!} \sum_{l=0}^{m} \frac{(-1)^l (2l+1)!!}{2^l} B_{m,l}(a_{m-1}, 2a_{m-2}, \ldots, (m-l+1)! a_{r-m+l}), \end{aligned} \tag{30}$$

where $T_m(a_1, \ldots, a_{m-1})$ are defined through the equation

$$\frac{1}{T(z)^{3/2}} = \sum_{m=0}^{\infty} T_m(a_1, \ldots, a_{m-1}) z^m.$$

Similarly for $\sqrt{T(z)}$, we set $f(z) = \sqrt{z}$ so that

$$f^{(k)}(z) = \frac{(-1)^{k+1} (2k-3)!!}{2^k z^{(2k-1)/2}}$$

for $m \geq 0$ and thus

$$\sqrt{T(z)} = \sum_{m=0}^{\infty} \frac{1}{m!} \left(\sum_{l=0}^{m} \frac{(-1)^{l+1}(2l-3)!!}{2^l} B_{m,l}(a_{m-1}, 2a_{m-2}, \dots, (m-l+1)!a_{l-1}) \right) z^m.$$

We then form the formal power series

$$Q_i(z) := Q_i(a_1, \dots, a_r, z) = \sum_{k \geq r+2} Q_{k-(r+1),i} z^k = \sum_{k \geq 1} Q_{k,i} z^{k+r+1} \tag{31}$$

for $1 \leq i \leq r+1$. Similar to above, we see that this formal series must satisfy

$$\frac{d}{dz} Q_i(z) - \frac{Q(z)}{2zP(z)} Q_i(z) = \frac{S_i(z)}{2zP(z)}, \tag{32}$$

where

$$P(z) := \sum_{j=1}^{r+1} a_j z^j, \quad Q(z) := zP'(z) + 2(r+2)P(z), \tag{33}$$

and

$$S_i(z) := -\sum_{m=1}^{r+1} \left(\sum_{j=1}^{m-1} (3j - 2m + 2) a_j Q_{m-j,i} \right) z^{m+r+1}.$$

Indeed

$$\begin{aligned}
&2zP(z)\frac{d}{dz}Q_i(z) - Q(z)Q_i(z) \\
&= \sum_{k \geq 1} \sum_{j=1}^{r+1} (2(k+r+1) - j - 2(r+1) - 2) a_j Q_{k,i} z^{j+k+r+1} \\
&= -\sum_{k \geq 1} \sum_{j=1}^{r+1} (j - 2k + 2) a_j Q_{k,i} z^{j+k+r+1} \\
&= -\sum_{m \geq r+2} \left(\sum_{j=1}^{r+1} (3j - 2m + 2) a_j Q_{m-j,i} \right) z^{m+r+1} \\
&\quad - \sum_{m=1}^{r+1} \left(\sum_{j=1}^{m-1} (3j - 2m + 2) a_j Q_{m-j,i} \right) z^{m+r+1} \\
&= S_i(z).
\end{aligned}$$

An integrating factor is

$$\mu(z) = \exp \int -\frac{Q(z)}{2zP(z)}\, dz = \frac{1}{z^{r+2}\sqrt{P(z)}},$$

and so

$$Q_i(z) := z^{r+2}\sqrt{P(z)} \int \frac{S_i(z)}{2z^{r+3}P(z)^{3/2}}\, dz. \tag{34}$$

4.1. Example. Let $p(t) = t^{2n+1} - t = t(t-\zeta)(t-\zeta^2)\cdots(t-\zeta^{2n})$, where $\zeta = \exp(\pi\imath/n)$ is a primitive $2n$-th root of unity. Then $a_1 = -1$ and $a_j = 0$ for $2 \leq j \leq 2n$, and hence the recursion relation (10) becomes

$$P_{k,i} = \frac{-4n+2k+3}{2n+2k+3} P_{k-2n,i}, \tag{35}$$

where $k \geq 0$. Since $P_{\ell,i} = \delta_{\ell,i}$ for all $-2n \leq \ell, i \leq -1$, the closed form is:

$$P_{k,i} = \prod_{j=1}^{s} \frac{-4(jn)+2k+3}{(2-4(j-1))n+2k+3} P_{k-s(2n),i} \quad \text{where } k \geq 0. \tag{36}$$

This implies when $k = s(2n) + i$ or $s = \frac{k-i}{2n}$, we have

$$P_{k,i} = \prod_{j=1}^{\frac{k-i}{2n}} \frac{-4(jn)+2k+3}{(2-4(j-1))n+2k+3} \quad \text{where } k \geq 0.$$

Similarly, the recursion relation (16) becomes

$$Q_{k,i} = \frac{6n-2k+3}{3-2k} Q_{k-2n,i}, \tag{37}$$

with initial conditions $Q_{m,i} = \delta_{m,-i}$, where $1 \leq m \leq 2n$ and $-2n \leq i \leq -1$. So the closed form is

$$Q_{k,i} = \prod_{j=1}^{v} \frac{(6+4(j-1))n-2k+3}{(4(j-1))n-2k+3} Q_{k-v(2n),i} \quad \text{where } k \geq 0. \tag{38}$$

So when $k = v(2n) - i$ or $v = \frac{k+i}{2n}$, then

$$Q_{k,i} = \prod_{j=1}^{\frac{k+i}{2n}} \frac{(6+4(j-1))n-2k+3}{(4(j-1))n-2k+3}.$$

Thus,

$$P_i(z) = P_i(-1,0,\ldots,0,z) = \sum_{k\geq 0} \prod_{j=1}^{\frac{k-i}{2n}-1} \frac{-4(j+1)n+2k+3}{(2-4j)n+2k+3} \delta_{\bar{k},\bar{i}} z^k, \tag{39}$$

where $-2n \leq i \leq -1$, $\bar{a}$ is the congruence class of a mod $2n$, and

$$Q_i(z) = Q_i(-1,0,\ldots,0,z) = \sum_{k\geq 1} \prod_{j=1}^{\frac{k+i}{2n}} \frac{(6+4(j-1))n-2k+3}{(4(j-1))n-2k+3} \delta_{\overline{k+i},\bar{0}} z^{k+2n+1}, \tag{40}$$

where $1 \leq i \leq 2n+1$.

4.2. Example. We consider now the particular example $p(t) = t^5 - 2ct^3 + t$. Here $r = 4$, $a_0 = 0 = a_2 = a_4$, $a_1 = 1 = a_5$ and $a_3 = -2c$. We have

$$T(z) = z^4 - 2cz^2 + 1,$$

and

$$R_{-1}(z) = \sum_{j=1}^{5}\left(\sum_{1-j\le k<0}(3j+2k-8)a_j\delta_{k+j,4}z^{k+4}\right) = 5z^3,$$

$$R_{-2}(z) = \sum_{j=1}^{5}\left(\sum_{1-j\le k<0}(3j+2k-8)a_j\delta_{k+j,3}z^{k+4}\right) = 3z^2,$$

$$R_{-3}(z) = \sum_{j=1}^{5}\left(\sum_{1-j\le k<0}(3j+2k-8)a_j\delta_{k+j,2}z^{k+4}\right) = 2cz^3 + z,$$

$$R_{-4}(z) = \sum_{j=1}^{5}\left(\sum_{1-j\le k<0}(3j+2k-8)a_j\delta_{k+j,1}z^{k+4}\right) = 6cz^2 - 1.$$

Thus

$$\begin{aligned} P_{-1}(z) &= 5z^{1/2}\sqrt{z^4-2cz^2+1}\int \frac{z^3}{2z^{3/2}(z^4-2cz^2+1)^{3/2}}\,dz \\ &= z^3 + \frac{2cz^5}{3} + \left(\frac{28c^2}{39} - \frac{1}{13}\right)z^7 + \left(\frac{616c^3}{663} - \frac{196c}{663}\right)z^9 \\ &\quad + \frac{\left(6160c^4 - 3388c^2 + 153\right)z^{11}}{4641} + O\left(z^{13}\right), \end{aligned}$$

$$\begin{aligned} P_{-2}(z) &= 3z^{1/2}\sqrt{z^4-2cz^2+1}\int \frac{z^2}{2z^{3/2}(z^4-2cz^2+1)^{3/2}}\,dz \\ &= z^2 + \frac{2cz^4}{7} + \left(\frac{20c^2}{77} + \frac{1}{11}\right)z^6 + \left(\frac{24c^3}{77} + \frac{4c}{77}\right)z^8 \\ &\quad + \left(\frac{624c^4}{1463} - \frac{36c^2}{1463} - \frac{7}{209}\right)z^{10} + O\left(z^{12}\right), \end{aligned}$$

$$\begin{aligned} P_{-3}(z) &= z^{1/2}\sqrt{z^4-2cz^2+1}\int \frac{2cz^3+z}{2z^{3/2}(z^4-2cz^2+1)^{3/2}}\,dz \\ &= z + \frac{z^5}{3} + \frac{14cz^7}{39} + \left(\frac{308c^2}{663} - \frac{5}{51}\right)z^9 + \left(\frac{440c^3}{663} - \frac{1364c}{4641}\right)z^{11} + O\left(z^{13}\right), \end{aligned}$$

$$\begin{aligned} P_{-4}(z) &= z^{1/2}\sqrt{z^4-2cz^2+1}\int \frac{6cz^2-1}{2z^{3/2}(z^4-2cz^2+1)^{3/2}}\,dz \\ &= 1 + \frac{5z^4}{7} + \frac{50cz^6}{77} + \left(\frac{60c^2}{77} - \frac{1}{7}\right)z^8 + \frac{12c\left(130c^2 - 53\right)z^{10}}{1463} + O\left(z^{13}\right). \end{aligned}$$

Here the integrals are from 0 to z.

The polynomials $P_{k,i} = P_{k,i}(c)$ satisfy the recursion:

$$(2k+7)P_{k,i} = -(2k-5)P_{k-4,i} + 2c(2k+1)P_{k-2,i} \tag{41}$$

for $k \ge 0$ with initial conditions $P_{l,i} = \delta_{l,i}$, $-r \le i, l \le -1$. We see that $P_{k,i}$ agree with the coefficients given above in the generating series.

To get explicit generating formulae for the $Q_{k,i}$ (see (23)), we have

$$S_{-1}(z) = -\sum_{m=1}^{5}\left(\sum_{j=1}^{m-1}(3j-2m+2)a_jQ_{m-j,-1}\right)z^{m+5} = -z^7+6cz^9,$$
$$S_{-2}(z) = -\sum_{m=1}^{5}\left(\sum_{j=1}^{m-1}(3j-2m+2)a_jQ_{m-j,-2}\right)z^{m+5} = z^8+2cz^{10},$$
$$S_{-3}(z) = -\sum_{m=1}^{5}\left(\sum_{j=1}^{m-1}(3j-2m+2)a_jQ_{m-j,-3}\right)z^{m+5} = 3z^9,$$
$$S_{-4}(z) = -\sum_{m=1}^{5}\left(\sum_{j=1}^{m-1}(3j-2m+2)a_jQ_{m-j,-4}\right)z^{m+5} = 5z^{10},$$

and thus

$$\begin{aligned}Q_{-1}(z) &= z^6\sqrt{z^5-2cz^3+z}\int\frac{-z^7+6cz^9}{2z^7(z^5-2cz^3+z)^{3/2}}\,dz\\ &= z^6+\frac{5z^{10}}{7}+\frac{50cz^{12}}{77}+\left(\frac{60c^2}{77}-\frac{1}{7}\right)z^{14}+\frac{12c\left(130c^2-53\right)z^{16}}{1463}+O\left(z^{18}\right),\end{aligned}$$

$$\begin{aligned}Q_{-2}(z) &= z^6\sqrt{z^5-2cz^3+z}\int\frac{z^8+2cz^{10}}{2z^7(z^5-2cz^3+z)^{3/2}}\,dz\\ &= z^7+\frac{z^{11}}{3}+\frac{14cz^{13}}{39}+\frac{1}{663}\left(308c^2-65\right)z^{15}+\frac{44c\left(70c^2-31\right)z^{17}}{4641}+O\left(z^{19}\right),\end{aligned}$$

$$\begin{aligned}Q_{-3}(z) &= z^6\sqrt{z^5-2cz^3+z}\int\frac{3z^9}{2z^7(z^5-2cz^3+z)^{3/2}}\,dz\\ &= z^8+\frac{2cz^{10}}{7}+\frac{1}{77}\left(20c^2+7\right)z^{12}+\frac{4}{77}\left(6c^3+c\right)z^{14}\\ &\quad+\frac{\left(624c^4-36c^2-49\right)z^{16}}{1463}+O\left(z^{17}\right),\end{aligned}$$

$$\begin{aligned}Q_{-4}(z) &= z^6\sqrt{z^5-2cz^3+z}\int\frac{5z^{10}}{2z^7(z^5-2cz^3+z)^{3/2}}\,dz\\ &= z^9+\frac{2cz^{11}}{3}+\frac{1}{39}\left(28c^2-3\right)z^{13}+\frac{28}{663}c\left(22c^2-7\right)z^{15}\\ &\quad+\frac{\left(6160c^4-3388c^2+153\right)z^{17}}{4641}+O\left(z^{19}\right).\end{aligned}$$

The recurrence relation for $Q_{m,i}$ is (23):

$$Q_{m,i} = \frac{1}{(2m-3)a_1}\left(\sum_{j=2}^{5}(3j-2m)a_jQ_{m-j+1,i}\right) = \frac{2c(2m-9)Q_{m-2,i}+(15-2m)Q_{m-4,i}}{2m-3} \tag{42}$$

for $m \geq 5$ and $Q_{m,i} = \delta_{m,-i}$, $1 \leq m \leq 4$. This agrees with the coefficients of the generating series given above for $Q_i(z)$.

4.3. Example. Let us take $p(t) = t^7 - 2bt^4 + t$. For this example, we limit ourselves to writing down just the first few terms of the generating series $P_{-1}(z)$. The recursion relation for the $P_{k,i}$'s using (10) is

$$(2k+9)P_{k,i} = -\sum_{j=1}^{6}(3j+2k-12)a_jP_{k-r+j-1,i} = 4bkP_{k-3,i} - (2k-9)P_{k-6,i} \tag{43}$$

for $k \geq 0$ with the initial condition $P_{l,i} = \delta_{l,i}$, $-6 \leq i, l \leq -1$. One can calculate by hand for example the first three nonzero nonconstant polynomials for $i = -1$, which are

$$P_{2,-1} = \frac{8b}{13}, \quad P_{5,-1} = \frac{(-13+160b^2)}{247}, \quad P_{8,-1}\frac{8b(-37+128b^2)}{1235}.$$

In this setting of $p(t)$, we have $R_{-1}(z) = 7z^5$, $T(z) = z^6 - 2bz^3 + 1$ and as an example using Faà di Bruno's formula and Bell polynomials, we get

$$\begin{aligned} P_{-1}(z) &= z^{3/2}\sqrt{z^6 - 2bz^3 + 1}\int \frac{7z^5}{2z^{5/2}(z^6-2bz^3+1)^{3/2}}\,dz \\ &= z^5 + \frac{8bz^8}{13} + \frac{1}{247}\left(160b^2 - 13\right)z^{11} + \frac{8b\left(128b^2-37\right)z^{14}}{1235} + O\left(z^{17}\right). \end{aligned}$$

Note in the integral we take the constant of integration to be 0.

5. Lie algebra generators and relations for $\widehat{\mathfrak{g}\otimes R}$.

Theorem 5.1 is a generalization of the main theorem in [**Cox08**].

THEOREM 5.1. *Let $a_1 \neq 0$. Let $\mathfrak{g}$ be a simple finite dimensional Lie algebra over the complex numbers with Killing form $(\cdot|\cdot)$ and for $\mathbf{a} = (a_1, \ldots, a_r)$ define $\psi_{ij}(\mathbf{a}) \in \Omega^1_R/dR$ by*

$$\psi_{ij}(\mathbf{a}) = \begin{cases} \displaystyle\sum_{k=1}^{r} P_{i+j-1,-k}\omega_k & \text{if } i+j \geq -r+1, \\ \displaystyle\sum_{k=1}^{r} Q_{-i-j+1,-k}\omega_k & \text{if } i+j < -r+1. \end{cases} \tag{44}$$

The universal central extension of the hyperelliptic Lie algebra $\mathfrak{g}\otimes R$ is the $\mathbb{Z}_2$-graded Lie algebra

$$\widehat{\mathfrak{g}} = \widehat{\mathfrak{g}}^0 \oplus \widehat{\mathfrak{g}}^1,$$

where

$$\widehat{\mathfrak{g}}^0 = (\mathfrak{g}\otimes\mathbb{C}[t,t^{-1}]) \oplus \mathbb{C}\omega_0, \qquad \widehat{\mathfrak{g}}^1 = (\mathfrak{g}\otimes\mathbb{C}[t,t^{-1}]u) \oplus \bigoplus_{k=1}^{r}(\mathbb{C}\omega_k)$$

with bracket

$$[x \otimes t^i, y \otimes t^j] = [x,y] \otimes t^{i+j} + \delta_{i+j,0} j(x,y)\omega_0, \tag{45}$$

$$[x \otimes t^i u, y \otimes t^j u] = [x,y] \otimes t^{i+j} p(t) + \sum_{k=1}^{r+1} \left(j + \frac{1}{2}k\right) a_k \delta_{i+j,-k}\omega_0, \tag{46}$$

$$[x \otimes t^i u, y \otimes t^j] = [x,y]u \otimes t^{i+j} u + j(x,y)\psi_{ij}(\mathbf{a}). \tag{47}$$

PROOF. The identities (45) and (46) follow from Lemma 3.2 whereas (47) follows from Proposition 3.3. □

6. Automorphism group for $R = \mathbb{C}[t, t^{-1}, u : u^2 = p(t) = t(t-\alpha_1)\cdots(t-\alpha_{2n})]$.

In this section, we restrict to the case of $r = 2n$ which allows us to use the results in [**CGLZ17**], [**BGG93**] and [**Sha03**] on automorphism groups of such algebras.

6.1. Automorphisms of $Z(\widehat{\mathfrak{g}})$ of the current algebra. Let S_{2n} be the symmetry group on the finite set $\{1, 2, \ldots, 2n\}$.

First we recall some background material.

THEOREM 6.1 ([**BGG93**] and [**Sha03**]). *The automorphism group of a hyperelliptic curve $A = \mathbb{C}[X, Y : Y^2 = P(X)]$ is isomorphic to one of the following groups:*

$$D_n, \mathbb{Z}_n, \mathbb{V}_n, \mathbb{H}_n, \mathbb{G}_n, \mathbb{U}_n, GL_2(3), W_2, W_3$$

where

$$\begin{aligned}
\mathbb{V}_n &:= \langle x, y : x^4, y^n, (xy)^2, (x^{-1}y)^2 \rangle, \\
\mathbb{H}_n &:= \langle x, y : x^4, y^2x^2, (xy)^n \rangle, \\
\mathbb{G}_n &:= \langle x, y : x^2y^n, y^{2n}, x^{-1}yxy \rangle, \\
\mathbb{U}_n &:= \langle x, y : x^2, y^{2n}, xyxy^{n+1} \rangle, \\
W_2 &:= \langle x, y : x^4, y^3, yx^2y^{-1}x^2, (xy)^4 \rangle, \\
W_3 &:= \langle x, y : x^4, y^3, x^2(xy)^4, (xy)^8 \rangle.
\end{aligned}$$

In [**Sha03**] a description of the reduced automorphism group is described for a given polynomial $P(X)$. In our paper we don't work with the reduced automorphism group and our coordinate ring is the localization $\mathbb{C}[t, t^{-1}, u : u^2 = p(t) = t(t-\alpha_1)\cdots(t-\alpha_{2n})]$ of A.

The result below describes the action of automorphisms of the algebra of the hyperelliptic curve $u^2 = p(t)$. The Theorem below corrects some errors that occur in [**CGLZ17**], Corollary 15.

THEOREM 6.2 (Corollary 15, [**CGLZ17**]). *Let $p(t) = t(t-\alpha_1)\cdots(t-\alpha_{2n})$, where α_i are distinct nonzero roots. Two possible types of automorphisms $\phi \in \mathrm{Aut}(R_2(p))$ of the algebra $R_2(p)$ are the following:*

(1) *If $\alpha_{\gamma(i)} = \zeta\alpha_i$ for some $2n$-th root of unity ζ and $\gamma \in S_{2n}$, then*

$$\phi(t) = \zeta t = \xi^2 t, \quad \phi(u) = \pm\xi^{2n+1}u = \pm\xi u \tag{48}$$

where $\xi = \exp(2\pi r\imath/2k)$, $\xi^2 = \zeta$ has order k with $k|2n$ and r and $2k$ are relatively prime. Denote these automorphisms by $\phi_\xi^\pm$ which satisfy $(\phi_\xi^\pm)^{2k} = id$, $(\phi_\xi^+)^k = \phi_1^-$, and $(\phi_\xi^+)^j = \phi_{\xi^j}^+$ for all j. Consequently $C_{2k} \cong \langle\phi_\xi^+\rangle$.

(2) *If there exists $\gamma \in S_{2n}$ and $c \in \mathbb{C}$ such that $\alpha_i\alpha_{\gamma(i)} = c^2$ for all i, then $\phi(t) = \zeta t = \xi^2 t$ and $\phi(u) = \pm\xi u$ (ξ as above), and $\psi(t) = c^2t^{-1}$ and*

$$\psi(u) = \pm t^{-n-1}c^{n+1}u \quad \text{if } a_1 = \prod_{i=1}^{2n}\alpha_i = c^{2n}, \tag{a}$$

or

$$\psi(u) = \pm t^{-n-1}\imath c^{n+1}u \quad \text{if } a_1 = \prod_{i=1}^{2n}\alpha_i = -c^{2n}. \tag{b}$$

Denote these automorphisms by $\psi_c^\pm$, respectively which satisfy $(\psi_c^\pm)^2 = id$ if $a_1 = c^{2n}$ and $(\psi_c^-)^4 = id$ but $(\psi_c^\pm)^2 = \phi_1^-$, if $a_1 = -c^{2n}$.

For case (a) we have if $l = (2n)/k$ is even, then $Aut(R_2(p)) = \langle\phi_\xi^+, \psi_c^+\rangle$ is isomorphic to $D_{2k} = \langle r, s : r^2 = s^{2k} = (rs)^2 = 1\rangle$. If $l = (2n)/k$ is odd, then $Aut(R_2(p)) = \langle\phi_\xi^+, \psi_c^+\rangle$ is isomorphic to $\mathbb{U}_k$.

For (b) if $l = (2n)/k$ is odd, $Aut(R_2(p)) = \langle\phi_\xi^+, \psi_c^+\rangle$ is isomorphic to $\mathbb{V}_{2k}$. If l is even, then $Aut(R_2(p)) = \langle\phi_\xi^+, \psi_c^+\rangle$ is isomorphic to $\mathbb{G}_k = Dic_k$.

PROOF. Part (1). Let ϕ be an automorphism of $R_2(p)$. Then since the group of units of $R_2(p)$ is $\mathbb{C}^*\{t^a : a \in \mathbb{Z}\}$, we know either $\phi(t) = \zeta t$ for some $\zeta \in \mathbb{C}^*$ or $\phi(t) = c^2/t$ for some $c \in \mathbb{C}^*$. In the first case we have

$$\phi(p(t)) = \zeta t(\zeta t - \alpha_1)\cdots(\zeta t - \alpha_{2n}) = \zeta^{2n+1}t(t - \zeta^{-1}\alpha_1)\cdots(t - \zeta^{-1}\alpha_{2n}) = f^2 p(t)$$

as one can show $\phi(u) = fu$ for some $f \in \mathbb{C}^*\{t^k : k \in \mathbb{Z}\}$.

Since the α_i/ζ are distinct we must have that there exists $\gamma \in S_{2n}$ such that $\alpha_{\gamma(i)} = \zeta\alpha_i$ for all $1 \le i \le 2n$. Then $\alpha_{\gamma^a(i)} = \zeta\alpha_{\gamma^{a-1}(i)} = \zeta^a\alpha_i$. Suppose γ is a product of disjoint cycles $(c_{i_1}, \dots, c_{i_k})$ with i in $\{c_{i_1}, \dots, c_{i_k}\}$. Then $\alpha_i = \alpha_{\gamma^k(i)} = \zeta^k\alpha_i$ for some $k \ge 2$ and $\zeta^k = 1$ with k minimal and $\zeta = \exp(2\pi\imath r/k)$ where r is relatively prime to k. Suppose $(d_{i_1}, \dots, d_{i_q})$ is an q-cycle appearing in γ. Then $\zeta^l = 1$ as well, so $k|q$. Now

$$\alpha_{d_{i_1}}, \alpha_{d_{i_2}} = \alpha_{\gamma(d_{i_1})} = \zeta\alpha_{d_{i_1}}, \dots, \alpha_{d_{i_q}} = \zeta^q\alpha_{d_{i_1}}$$

are supposed to be distinct. So ζ is also a q-th primitive root of unity, which implies that $q|k$. Thus $q = k$. So γ is a product of k-cycles, and $k|2n$ since $p(t)/t$ has only $2n$ distinct roots. After reordering the indices we may assume $r = 1$.

In addition $f^2 = \zeta^{2n+1} = \zeta = \xi^2$ and hence $f = \pm\xi$. Now $\xi^2 = \zeta = \exp(2\pi\imath/k) = (\exp(2\pi\imath/2k))^2$ so that $\xi = \pm\exp(2\pi\imath/2k)$. We can replace ξ by $-\xi$ in (48) if necessary so as to assume $\xi = \exp(2\pi\imath/2k)$. Keep in mind below the fact that $\xi^k = \exp(\pi\imath) = -1$.

It is also easy to check $(\phi_\xi^+)^j = \phi_{\xi^j}^+$. Let us point out in particular

$$(\phi_\xi^+)^{-1} = \phi_{\xi^{2k-1}}^+ = (\phi_\xi^+)^{2k-1}. \tag{49}$$

Note also the following

$$\phi_{\xi^k}^+(t) = (\phi_\xi^+)^k(t) = \xi^{2k}t = t = \phi_1^-(t), \quad \phi_{\xi^k}^+(u) = (\phi_\xi^+)^k(u) = \xi^k u = -u = \phi_1^-(u).$$

We can thus write

$$\phi_{\xi^a}^-(t) = \xi^{2a}t = \phi_1^-(\phi_\xi^+)^a(t) = (\phi_\xi^+)^{k+a}(t),$$
$$\phi_{\xi^a}^-(u) = -\xi^a u = \phi_1^-(\phi_\xi^+)^a(u) = (\phi_\xi^+)^{k+a}(u).$$

Consequently all of the automorphisms of the first type are in the subgroup generated by ϕ_ξ^+ and this subgroup of $\mathrm{Aut}(R)$ in turn generates a group isomorphic to C_{2k}.

We know $kl = 2n$ for some positive integer l.

Part (2). In the case (a) $c^{2n} = \prod_{i=1}^l \prod_{j=0}^k \zeta^j \alpha_i$, we have

$$(\psi_c^\pm)^2(t) = c^2\psi_c^\pm(t^{-1}) = t,$$
$$(\psi_c^\pm)^2(u) = \pm c^{n+1}\psi_c^\pm(t^{-n-1}u) = c^{n+1}c^{-2n-2}t^{n+1}c^{n+1}t^{-n-1}u = u.$$

Then $(\psi_c^\pm)^2 = \mathrm{id}$.

Moreover we have

$$\begin{aligned}
\psi_c^+\phi_\xi^+(\psi_c^+)^{-1}(u) &= \psi_c^+\phi_\xi^+\psi_c^+(u) \\
&= \psi_c^+\phi_\xi^+(t^{-n-1}c^{n+1}u) \\
&= \psi_c^+(\xi^{-2n-2}t^{-n-1}c^{n+1}\xi u) \\
&= (-1)^l\xi^{-1}c^{-2n-2}t^{n+1}c^{n+1}t^{-n-1}c^{n+1}u \\
&= (-1)^l\xi^{-1}u.
\end{aligned}$$

If l is even (for instance if $k|n$) we conclude

$$\psi_c^+\phi_\xi^+(\psi_c^+)^{-1}(u) = (\phi_\xi^+)^{-1}(u).$$

Furthermore,

$$\begin{aligned}
\psi_c^+\phi_\xi^+(\psi_c^+)^{-1}(t) &= \psi_c^+\phi_\xi^+\psi_c^+(t) = \psi_c^+\phi_\xi^+(c^2t^{-1}) \\
&= \psi_c^+(c^2\xi^{-2}t^{-1}) = c^2\xi^{-2}c^{-2}t = \xi^{-2}t \\
&= (\phi_\xi^+)^{-1}(t).
\end{aligned}$$

Finally note

$$\psi_c^-(u) = -c^{n+1}t^{-n-1}u = \phi_1^-\psi_c^+(u), \quad \psi_c^-(t) = c^2/t = \phi_1^-\psi_c^+(t)$$

so that for $r = \psi_c^+$, and $s = \phi_\xi^+$ we have $\psi_c^- \in \langle r, s\rangle$. In conclusion we have for case (a) with l even, $r = \psi_c^+$ has order 2 and $s = \phi_\xi^+$ has order $2k$, so they generate the dihedral group $D_{2k} = \langle r, s : r^2 = s^{2k} = (rs)^2 = 1\rangle$.

If l is odd, then

$$\psi_c^+\phi_\xi^+(\psi_c^+)^{-1}(u) = -\xi^{-1}u = \phi_\xi^{k-1}(u),$$
$$\psi_c^+\phi_\xi^+(\psi_c^+)^{-1}(t) = \psi_c^+\phi_\xi^+(c^{-2}t^{-1}) = \xi^{-2}t = \phi_\xi^{k-1}(t).$$

Thus $\psi_c^+\phi_\xi^+(\psi_c^+)^{-1} = (\phi_\xi^+)^{k-1}$ and hence $\psi_c^+\phi_\xi^+\psi_c^+(\phi_\xi^+)^{k+1} = \mathrm{id}$ so $\langle\phi_\xi^+\rangle$ is a normal subgroup of $\mathrm{Aut}(R_2(p))$ and $\mathrm{Aut}(R_2(p)) = \langle\phi_\xi^+, \psi_c^+\rangle \cong \mathbb{U}_k$.

In the case (b) $-c^{2n} = \prod_{i=1}^{l}\prod_{j=1}^{k}\zeta^j\alpha_i^k$, we have

$$(\psi_c^{\pm})^2(t) = c^2\psi_c^{\pm}(t^{-1}) = t,$$
$$(\psi_c^{\pm})^2(u) = \pm\imath c^{n+1}\psi_c^{\pm}(t^{-n-1}u) = \imath c^{n+1}c^{-2n-2}t^{n+1}\imath c^{n+1}t^{-n-1}u = -u.$$

Then $(\psi_c^{\pm})^2 = \phi_1^-$.

Moreover we have

$$\begin{aligned}\psi_c^+\phi_\xi^+(\psi_c^+)^{-1}(u) &= -\psi_c^+\phi_\xi^+\psi_c^+(u)\\ &= -\psi_c^+\phi_\xi^+(t^{-n-1}\imath c^{n+1}u)\\ &= -\psi_c^+(\xi^{-2n-2}t^{-n-1}\imath c^{n+1}\xi u)\\ &= (-1)^{l+1}\imath\xi^{-1}c^{-2n-2}t^{n+1}c^{n+1}t^{-n-1}\imath c^{n+1}u\\ &= (-1)^l\xi^{-1}u,\end{aligned}$$

as $\xi^{2n} = \xi^{kl} = (-1)^l$. Now if l is odd we conclude

$$\psi_c^+\phi_\xi^+(\psi_c^+)^{-1}(u) = (\phi_\xi^+)^{k-1}(u).$$

In addition

$$\begin{aligned}\psi_c^+\phi_\xi^+(\psi_c^+)^{-1}(t) &= \psi_c^+\phi_\xi^+\psi_c^+(t)\\ &= \psi_c^+\phi_\xi^+(c^2/t)\\ &= \psi_c^+(c^2\xi^{-2}t^{-1})\\ &= c^2\xi^{-2}c^{-2}t\\ &= \xi^{-2}t = (\phi_\xi^+)^{k-1}(t).\end{aligned}$$

Thus

$$\psi_c^+\phi_\xi^+(\psi_c^+)^{-1} = (\phi_\xi^+)^{k-1}.$$

Note that this means

$$\psi_c^+\phi_\xi^+\psi_c^+ = \psi_c^+\phi_\xi^+(\psi_c^+)^{-3} = \psi_c^+\phi_\xi^+(\psi_c^+)^{-1}\phi_1^- = (\phi_\xi^+)^{k-1}\phi_1^- = \phi_\xi^{-1}$$

so that $(\psi_c^+\phi_\xi^+)^2 = 1$. Similarly

$$(\psi_c^+)^{-1}\phi_\xi^+(\psi_c^+)^{-1}\phi_\xi^+ = (\psi_c^+)^3\phi_\xi^+(\psi_c^+)^3\phi_\xi^+ = 1$$

as $(\psi_c^+)^2 = \phi_1^-$.

We conclude in the case that l is odd that

$$(\phi_\xi^+)^2 = (\phi_\xi^+)^k, \quad (\phi_\xi^+)^{2k} = \mathrm{id}, \quad (\psi_c^+\phi_\xi^+)^2 = 1 = ((\psi_c^+)^{-1}\phi_\xi^+)^2$$

and

$$\langle\phi_\xi^+, \psi_c^+\rangle \cong \mathbb{V}_{2k}.$$

When l is even we get

$$(\phi_\xi^+)^2 = (\phi_\xi^+)^k, \quad (\phi_\xi^+)^{2k} = \mathrm{id}, \quad \psi_c^+\phi_\xi^+(\psi_c^+)^{-1} = (\phi_\xi^+)^{-1}$$

and

$$\langle\phi_\xi^+, \psi_c^+\rangle \cong \mathbb{G}_k = \mathrm{Dic}_k.$$

□

REMARK 6.3. *In the above cited paper we wrote* $(\psi_c^{\pm})^2 = id$ *but this was in error in case (b) as* $\psi_c^{\pm}$ *has order* 4. *Observe also* $\phi_\xi^- = \phi_1^-\phi_\xi^+$.

We add to this another

COROLLARY 6.4. *Let $p(t) = t(t-\alpha_1)\cdots(t-\alpha_{2n})$, where α_i are distinct roots. Two possible types of automorphisms $\phi \in \operatorname{Aut}(R_2(p))$ of the algebra $R_2(p)$ are the following:*

(1) *If $\alpha_{\gamma(i)} = \zeta\alpha_i$ for some $2n$-th root of unity ζ and $\gamma \in S_{2n}$, then*

$$\phi(t) = \zeta t, \quad \phi(u) = \pm\xi u, \tag{50}$$

where we can take $\xi = \zeta^{1/2} = \exp(2\pi\imath/2k)$ with ζ having order k and $k|2n$. It follows that ϕ has order $2k$. In particular, after a change in indices

$$\begin{aligned} p(t) &= t(t-\alpha_1)(t-\zeta\alpha_1)\cdots(t-\zeta^{k-1}\alpha_1)\cdots(t-\alpha_{2n/k})\cdots(t-\zeta^{k-1}\alpha_{2n/k}) \\ &= t(t^k-\alpha_1^k)(t^k-\alpha_2^k)\cdots(t^k-\alpha_{2n/k}^k) \\ &= \sum_{q=0}^{\frac{2n}{k}}(-1)^q e_q(\alpha_1^k,\ldots,\alpha_{2n/k}^k)t^{2n-qk+1}, \end{aligned} \tag{51}$$

where $e_q(x_1, x_2, \ldots, x_{2n/k})$ is the elementary symmetric polynomial of degree q in $x_1, \ldots, x_{2n/k}$:

$$e_q(x_1, x_2, \ldots, x_{2n/k}) = \sum_{1\le j_1<j_2<\cdots<j_q\le 2n/k} x_{j_1}x_{j_2}\cdots x_{j_q}.$$

In this case $\langle\phi_\xi^+\rangle \cong C_{2k}$.

(2) *If in addition to the above, there exists $\beta \in S_{2n}$ such that $\alpha_i\alpha_{\beta(i)} = c^2$ for all i, then $\phi_\xi^\pm(t) = \zeta t$ and $\phi_\xi^\pm(u) = \pm\xi u$, and $\psi(t) = c^2t^{-1}$ and*

$$\psi_c^\pm(u) = \pm t^{-n-1}c^{n+1}u \quad \text{if } a_1 = \prod_{i=1}^{2n}\alpha_i = c^{2n}, \tag{a}$$

or

$$\psi_c^\pm(u) = \pm t^{-n-1}\imath c^{n+1}u \quad \text{if } a_1 = \prod_{i=1}^{2n}\alpha_i = -c^{2n}. \tag{b}$$

In this case

$$p(t) = \sum_{r=1}^{2n+1} a_r t^r,$$

where

$$a_k = \pm c^{2n-2k+2}a_{2n+2-k} \tag{52}$$

for $k = 1, \ldots, 2n+1$. Here the $\pm$ in (52) corresponds to the $\pm$ in $a_1 = \pm c^{2n}$.

PROOF. Case (1). Thus after a renaming of the indices we may assume $r = 1$ and we may write

$$\begin{aligned} p(t) &= t(t-\alpha_1)(t-\zeta\alpha_1)\cdots(t-\zeta^{k-1}\alpha_1)(t-\alpha_2)(t-\zeta\alpha_2)\cdots(t-\zeta^{k-1}\alpha_2)\cdots \\ &\quad\cdots(t-\alpha_{2n/k})\cdots(t-\zeta^{k-1}\alpha_{2n/k}) \\ &= t(t^k-\alpha_1^k)(t^k-\alpha_2^k)\cdots(t^k-\alpha_{2n/k}^k) \\ &= \sum_{q=0}^{2n/k}(-1)^q e_q(\alpha_1^k,\dots,\alpha_{2n/k}^k)t^{2n-qk+1}, \end{aligned}$$

where $e_q(x_1,\dots,x_{2n/k})$ are the elementary symmetric polynomial of degree q in $x_1,\dots,x_{2n/k}$

$$e_q(x_1,\dots,x_{2n/k}) = \sum_{1\le j_1<j_2<\dots<j_q\le 2n/k} x_{j_1}\cdots x_{j_q}.$$

For the second part we know $C_{2k}\cong\langle\phi_\xi^+\rangle\subseteq \operatorname{Aut}(R_2(p))$ for some $k|2n$ and we have

$$\begin{aligned} \psi_c^{\pm}(p(t)) &= \sum_{j=1}^{2n+1} a_j c^{2j}t^{-j} \\ &= t^{-2n-2}\sum_{j=1}^{2n+1} a_j c^{2j} t^{2n+2-j} \\ &= t^{-2n-2}\sum_{q=1}^{2n+1} a_{2n+2-q}c^{4n+4-2q}t^q, \end{aligned}$$

which we require to satisfy

$$\begin{aligned} \psi_c^{\pm}(u^2) = \psi_c^{\pm}(p(t)) &= t^{-2n-2}\sum_{q=1}^{2n+1} a_{2n+2-q}c^{4n+4-2q}t^q \\ = \psi_c^{\pm}(u)^2 = \pm c^{2n+2}t^{-2n-2}p(t) &= \pm c^{2n+2}t^{-2n-2}\left(\sum_{j=1}^{2n+1} a_j t^j\right). \end{aligned}$$

As a consequence (since $c\neq 0$), one has

$$a_j = \pm c^{2n-2j+2}a_{2n+2-j},$$

for $j = 1,\dots,2n+1$. Here $+$ is taken for case (a) and $-$ for case (b).

□

REMARK 6.5. *In the dihedral case with the roots $\zeta^r\alpha_i$, where $0\le r\le l-1$ and $i = 1,\dots,k$ so that $kl = 2n$, we simplify $|c^{2n}|$ to obtain*

$$|c^{2n}| = \left|\prod_{i=1}^k \alpha_i\right|^l = |a_1|,$$

and thus $c^2 = \omega\sqrt[n]{\left|\prod_{i=1}^k\alpha_i\right|^l}$, where $\omega^n = 1$.

For example if $\alpha_1 = 1$, $\alpha_2 = 1 + 2\imath$, $\alpha_3 = 1 + 3\imath$, $l = 4$, $n = 6$, *and* $k = 3$, *then for any* $\gamma \in S_{2n}$,

$$|c^2| = |\alpha_1 \alpha_{\gamma(1)}| = 1 \text{ or } \sqrt{5} \text{ or } \sqrt{10}.$$

Now for the dihedral group we would also have

$$|c^2| = \sqrt[n]{\left|\prod_{i=1}^{k} \alpha_i\right|^l} = \sqrt[6]{\sqrt{50}^4}.$$

But then $\sqrt[6]{\sqrt{50}^4} \neq |\alpha_1 \alpha_{\gamma(1)}|$ *for any* γ. *As a consequence, the automorphism group is* C_6, *and not* D_6.

From [**Skr88**], we know that for any automorphism ϕ of the associative algebra $R_2(p)$, one obtains an automorphism τ of the Lie algebra $\mathcal{R}_2(p) := \mathrm{Der}(R_2(p))$ through the equation

$$\tau(f(t)\partial) = \phi(f)(\phi \circ \partial \circ \phi^{-1}) \quad \text{for all } f \in R_2(p). \tag{53}$$

In addition, any Lie algebra automorphism of $\mathcal{R}_2(p)$ can be obtained from (53). Denote by $\tau_\zeta^\pm$ and $\sigma_c^\pm$ the Lie algebra automorphisms corresponding to the associative algebra automorphisms $\phi_\zeta^\pm$ and $\psi_c^\pm$ in Theorem 6.2 (1) and (2) respectively (if they indeed exist). For convenience, denote

$$\tau_\zeta = \tau_\zeta^+ \text{ and } \sigma_c = \sigma_c^+.$$

Let C_k be the cyclic group of order k and D_k be the dihedral group of order $2k$.

COROLLARY 6.6 ([**CGLZ17**], Corollary 16). *Let* $p(t) = t(t - \alpha_1)\cdots(t - \alpha_{2n})$, *with distinct roots.*

(1) *If* $\sigma_c^\pm$ *does not exist in* $\mathrm{Aut}(\mathcal{R}_2(p))$ *for any nonzero complex number* c, *then* $\mathrm{Aut}(\mathcal{R}_2(p))$ *is generated by the automorphism* τ_ζ^+ *of order* $2k$, *where* $k|2n$. *In otherwords we have*

$$\mathrm{Aut}(\mathcal{R}_2(p)) = \langle \tau_\zeta^+ \rangle \simeq C_{2k}.$$

(2) *If* $\sigma_c^\pm$ *exists in* $\mathrm{Aut}(\mathcal{R}_2(p))$ *for some nonzero complex number* c *with* $c^{2n} = a_1$, *then* $\mathrm{Aut}(\mathcal{R}_2(p))$ *is generated by* σ_c^+, *and some automorphism* τ_ζ^+ *of order* $2k$, *where* $k|2n$. *If* $k|n$, *then we have*

$$\mathrm{Aut}(\mathcal{R}_2(p)) = \langle \tau_\zeta^+, \sigma_c^+ \rangle \simeq D_{2k}.$$

PROOF. This follows from Theorem 6.2. □

7. The decomposition of the space of Kähler differentials modulo exact forms for D_{2k}

Let $r = 2n$, $R = R_2(p)$ and let $G := \mathrm{Aut}(R)$ be the groups in Corollary 6.6. For $\phi \in G$ and $\overline{rds} \in \Omega_R/dR$, the action of G on the Kähler differential is given by:

$$\phi(\overline{rds}) = \overline{\phi(r)d\phi(s)}. \tag{54}$$

First we note the following:

LEMMA 7.1. *For n even, the character table is given by the matrix*

$$M = \begin{array}{c} \\ 1 \\ \psi \\ \psi\phi \\ \phi \\ \vdots \\ \phi^k \\ \vdots \\ \phi^{n/2} \end{array} \begin{array}{c} \begin{array}{ccccccccc} \rho_1 & \rho_2 & \rho_3 & \rho_4 & \chi_1 & \cdots & \chi_h & \cdots & \chi_{(n/2)-1} \end{array} \\ \begin{pmatrix} 1 & 1 & 1 & 1 & 2 & \cdots & 2 & \cdots & 2 \\ 1 & -1 & 1 & -1 & 0 & \cdots & 0 & \cdots & 0 \\ 1 & -1 & -1 & 1 & 0 & \cdots & 0 & \cdots & 0 \\ 1 & 1 & -1 & -1 & 2\cos\left(\frac{2\pi}{n}\right) & \cdots & 2\cos\left(\frac{2\pi h}{n}\right) & \cdots & 2\cos\left(\frac{2\pi((n/2)-1)}{n}\right) \\ \vdots & \vdots & \vdots & \vdots & \vdots & \ddots & \vdots & \ddots & \vdots \\ 1 & 1 & (-1)^k & (-1)^k & 2\cos\left(\frac{2\pi k}{n}\right) & \cdots & 2\cos\left(\frac{2\pi hk}{n}\right) & \cdots & 2\cos\left(\frac{2\pi k((n/2)-1)}{n}\right) \\ \vdots & \vdots & \vdots & \vdots & \vdots & \ddots & \vdots & \ddots & \vdots \\ 1 & 1 & (-1)^{n/2} & (-1)^{n/2} & -2 & \cdots & 2(-1)^h & \cdots & 2(-1)^{(n/2)-1} \end{pmatrix} \end{array}.$$

So we have

$$M^{-1} = M^t \begin{pmatrix} \frac{1}{2n} & 0 & 0 & 0 & \cdots & 0 & \cdots & \cdots & 0 \\ 0 & \frac{n/2}{2n} & 0 & 0 & \cdots & 0 & \cdots & \cdots & 0 \\ 0 & 0 & \frac{n/2}{2n} & 0 & \cdots & 0 & \cdots & \cdots & 0 \\ 0 & 0 & 0 & \frac{2}{2n} & \cdots & 0 & \cdots & \cdots & 0 \\ \vdots & \vdots & \vdots & \vdots & \ddots & \vdots & \ddots & \vdots & \vdots \\ 0 & 0 & 0 & 0 & \cdots & \frac{2}{2n} & \cdots & \vdots & 0 \\ \vdots & \vdots & \vdots & \vdots & \ddots & \vdots & \ddots & \vdots & \vdots \\ 0 & 0 & 0 & 0 & \cdots & \vdots & \cdots & \frac{2}{2n} & 0 \\ 0 & 0 & 0 & 0 & \cdots & 0 & \cdots & 0 & \frac{1}{2n} \end{pmatrix}$$

$$= \begin{pmatrix} \frac{1}{2n} & \frac{1}{4} & \frac{1}{4} & \frac{1}{n} & \cdots & \frac{1}{n} & \cdots & \frac{1}{2n} \\ \frac{1}{2n} & -\frac{1}{4} & -\frac{1}{4} & \frac{1}{n} & \cdots & \frac{1}{n} & \cdots & \frac{1}{2n} \\ \frac{1}{2n} & \frac{1}{4} & -\frac{1}{4} & -\frac{1}{n} & \cdots & \frac{(-1)^k}{n} & \cdots & \frac{(-1)^{n/2}}{2n} \\ \frac{1}{2n} & -\frac{1}{4} & \frac{1}{4} & -\frac{1}{n} & \cdots & \frac{(-1)^k}{n} & \cdots & \frac{(-1)^{n/2}}{2n} \\ \frac{1}{n} & 0 & 0 & \frac{2}{n}\cos(2\pi/n) & \cdots & \frac{2}{n}\cos(2\pi h/n) & \cdots & -\frac{1}{n} \\ \vdots & \vdots & \vdots & \vdots & \ddots & \vdots & \ddots & \vdots \\ \frac{1}{n} & 0 & 0 & \frac{2}{n}\cos(2\pi h/n) & \cdots & \frac{2}{n}\cos(2\pi hk/n) & \cdots & \frac{(-1)^h}{n} \\ \vdots & \vdots & \vdots & \vdots & \ddots & \vdots & \ddots & \vdots \\ \frac{1}{n} & 0 & 0 & \frac{2}{n}\cos\left(2\pi\left((n/2)-1\right)/n\right) & \cdots & \frac{2}{n}\cos\left(2\pi k\left((n/2)-1\right)/n\right) & \cdots & \frac{(-1)^{(n/2)-1}}{n} \end{pmatrix}.$$

PROOF. Observe M is just the character table for D_n when n is even: from page 37 in [**Ser77**], we obtain the character table for $r = 2n$

	$(\phi_\zeta^+)^k$	$\psi_c^+(\phi_\zeta^+)^k$
ρ_1	1	1
ρ_2	1	-1
ρ_3	$(-1)^k$	$(-1)^k$
ρ_4	$(-1)^k$	$(-1)^{k+1}$
χ_h	$2\cos(2\pi hk/n)$	0

,

where $1 \le h < n/2$ for n even, ψ_c^+ is a reflection, and ϕ_ζ^+ is a rotation.

Let Ξ denote the set of conjugacy classes of the group D_n. Then from the orthogonality of the characters of the irreducible representations, we get the inverse matrix for M since one needs the following formula for any two irreducible representations π and ρ of D_n:

$$\sum_{\{g\}\in\Xi} \frac{|\{g\}|}{|D_n|}\chi_\pi(g)\overline{\chi_\rho(g)} = \begin{cases} 1 & \text{for } \pi \cong \rho, \\ 0 & \text{otherwise,} \end{cases} \tag{55}$$

(see page 260 in [**Ter99**]).

The distinct conjugacy classes of D_n via conjugation (for $n = 2\widehat{m}$ even) are:

$$\begin{aligned} &\{\mathrm{I}\},\{\phi_\zeta^+,(\phi_\zeta^+)^{-1}\},\dots,\{(\phi_\zeta^+)^j,(\phi_\zeta^+)^{-j}\},\dots,\{(\phi_\zeta^+)^{\frac{n}{2}-1},(\phi_\zeta^+)^{-(\frac{n}{2}-1)}\},\{(\phi_\zeta^+)^{\widehat{m}}\},\\ &\{\psi_c^+(\phi_\zeta^+)^\ell : \ell \text{ even}, 0\le \ell\le n-2\},\{\psi_c^+(\phi_\zeta^+)^p : p \text{ odd}, 1\le p\le n-1\}\end{aligned}$$

since the even dihedral group has nontrivial center (thus giving us one element orbits). □

So under an action by G, we decompose $\Omega_R/dR = Z(\widehat{\mathfrak{g}})$ into a direct sum of irreducible representations. Our goal in this section is to describe the module structure of Ω_R/dR into irreducibles under the action by G for a particular $R_2(p)$. Recall that $\mathcal{R}_2(p) = \mathrm{Der}(R_2(p))$.

THEOREM 7.2. *Let $p(t) = t(t-\alpha_1)\cdots(t-\alpha_{2n})$, where α_i are pairwisely distinct nonzero complex numbers.*

(1) *If $\sigma_c^\pm$ does not exist in $\mathrm{Aut}(\mathcal{R}_2(p))$ for any nonzero $c\in\mathbb{C}$, then $\mathrm{Aut}(\mathcal{R}_2(p)) \cong C_{2k}$ and the center for the universal central extension of $\mathfrak{g}\otimes R_2(p)$ decomposes as:*

$$\Omega_R/dR \cong U_0 \oplus \ldots \oplus U_{k-1}, \tag{56}$$

where $U_r = \bigoplus_{i\equiv r \mod k, 1\le i\le 2n} \mathbb{C}\omega_i$ for $r = 1,\dots,k-1$ is a sum of one-dimensional irreducible representation of C_{2k} with character $\chi_r(s) = \exp(2\pi\imath rs/2k)$, each occurring with multiplicity l and

$$U_0 = \mathbb{C}\omega_0 \oplus \bigoplus_{i=1}^{l} \mathbb{C}\omega_{ki}.$$

(2) *Assume $\sigma_c^\pm$ exists in $\mathrm{Aut}(\mathcal{R}_2(p))$ for some nonzero $c\in\mathbb{C}$, $c^{2n} = a_1$ and $k|n$. Then $\mathrm{Aut}(\mathcal{R}_2(p)) \cong D_{2k}$. Moreover, if k is also even then under the action of D_{2k} the center decomposes as:*

$$\Omega_R/dR \cong \mathbb{C}\omega_0 \oplus \bigoplus_{i=3}^{4} U_i^{\frac{(1-(-1)^k)n}{2k}} \oplus \bigoplus_{h=1}^{k-1} V_h^{\oplus\frac{(1-(-1)^h)n}{k}}. \tag{57}$$

where U_i, $i = 1,2,3,4$, are the irreducible one dimensional representations for D_{2k} with character ρ_i and V_h are the irreducible 2-dimensional representations for D_{2k} with character χ_h, $1\le h\le k-1$. Note $\mathbb{C}\omega_0$ and U_1 are the trivial representations.

When k is odd, the center decomposes as

$$\Omega_R/dR \cong \mathbb{C}\omega_0 \oplus \bigoplus_{i=3}^{4} U_i^{\oplus \Upsilon_i(\epsilon_i,\nu_i)} \oplus \bigoplus_{j=1}^{k-1} V_j^{\oplus \frac{(1-(-1)^j)n}{k}} \tag{58}$$

with

$$\Upsilon_i(\epsilon_i,\nu_i) = \frac{(1-(-1)^k)n}{2k}(\delta_{i,3}+\delta_{i,4}) + \frac{1-(-1)^n}{4}(\delta_{i,4}-\delta_{i,3}) + \frac{1}{2}(-1)^i \sum_{i=n+3}^{2n} c^{n+3-2i} P_{i-n-3,-i}.$$

COROLLARY 7.3. *When $\overline{\omega_i} = c^{-\frac{n+3-2i}{2}}\zeta^{-\frac{i}{2}}\omega_i$ for $1 \le i \le n+2$, we obtain that*

$$\overline{\omega_i} \text{ and } \overline{\omega_{n+3-i}}, \text{ where } 1 \le i \le (n+2)/2,$$

span a 2-dimensional irreducible representation for n even.

The following is a proof of Theorem 7.2.

PROOF. We will first prove (1). Recalling (48):

$$\phi(t) = \xi^2 t = \zeta t \text{ and } \phi(u) = \xi u$$

(so $\phi = \phi_\xi^+$ has order $2k$), the action of $\mathrm{Aut}_{\mathbb{C}}(R)$ shows that

$$\phi^j(\omega_0) = \phi^j(\overline{t^{-1}dt}) = \overline{\phi^j(t)^{-1}\, d\phi^j(t)} = \overline{\zeta^{-j}t^{-1}\, d(\zeta^j t)} = \overline{t^{-1}\, dt} = \omega_0$$

and

$$\begin{aligned}\phi^j(\omega_i) &= \phi^j(\overline{t^{-i}udt}) = \overline{\xi^{-2ij}t^{-i}\xi^j u\, d(\xi^{2j}t)} = \overline{\xi^{(3-i)j}t^{-i}u\, dt} = \xi^{(3-i)j}\omega_i \\ &= \exp(2\pi\imath(3-2i)j/2k)\omega_i = \exp(2\pi\imath(kl-2i+3)j/2k)\omega_i\end{aligned} \tag{59}$$

for all $0 \le j \le 2k$ and $0 \le i \le 2n$. Now the characters of the irreducible representations of C_{2k} are of the form $\chi_h(\phi^s) = \exp(2\pi\imath sh/2k)$ with $0 \le h \le 2k-1$. In order to figure out the multiplicities, we need to solve the number of solutions to

$$2s \equiv 2r \mod 2k$$

for $1 \le r \le s \le 2n$. In this case $2(s-r) = 2kd$, so $0 \le s-r = kd \le 2n-1 = kl-1$ for some integer d. Thus $s = r+kd$ where $0 \le d \le l-1$ and the multiplicity is l for each irreducible representation.

We conclude that the center Ω_R/dR decomposes into the direct sum of one-dimensional eigenspaces:

$$\Omega_R/dR \cong U_0 \oplus \ldots \oplus U_{k-1}, \tag{60}$$

where

$$U_r = \bigoplus_{i\equiv r \mod k, 1\le i\le 2n} \mathbb{C}\omega_i \quad \text{for } r = 1,\ldots,k-1,$$

a sum of one-dimensional irreducible representation of C_{2k} with character $\chi_r(s) = \exp(2\pi\imath rs/2k)$, each occurring with multiplicity l and

$$U_0 = \mathbb{C}\omega_0 \oplus \bigoplus_{i=1}^{l} \mathbb{C}\omega_{ki}.$$

where U_i are the one dimensional irreducible representations of D_{2k} with characters ρ_i, $i = 1,2,3,4$ and V_h are the irreducible representations with character χ_h, $1 \le h \le k-1$.

Next, we see that

$$\psi_c^+(\omega_0) = \psi_c^+(\overline{t^{-1}\,dt}) = \overline{c^{-2}t\,d(c^2t^{-1})} = \overline{t\,d(t^{-1})} = -\overline{t\cdot t^{-2}dt} = -\omega_0$$

and

$$\phi_\xi^+(\omega_0) = \phi_\xi^+(\overline{t^{-1}dt}) = \overline{\xi^{-2}t^{-1}d(\zeta t)} = \overline{t^{-1}dt} = \omega_0.$$

So ω_0 is a basis element for a one-dimensional irreducible representation under the action of D_{2k}.

Similarly, we have the rotations acting on ω_i as a scalar multiplication:

$$\begin{aligned}\phi_\xi^+(\omega_i) &= \phi_\xi^+(\overline{t^{-i}udt}) = \overline{\zeta^{-i}t^{-i}\xi u\zeta dt} = \overline{\xi^{3-2i}t^{-i}udt}\\ &= \xi^{3-2i}\omega_i\end{aligned} \tag{61}$$

and the reflections acting via:

$$\begin{aligned}\psi_c^+(\omega_i) &= \psi_c^+(\overline{t^{-i}udt}) = \overline{c^{-2i}t^it^{-n-1}c^{n+1}ud(c^2t^{-1})} = -\overline{c^{n-2i+3}t^{i-n-1}ut^{-2}dt}\\ &= -\overline{c^{n+3-2i}t^{-(n+3-i)}udt}\\ &= -c^{n+3-2i}\omega_{n+3-i} \quad \text{if } 1\le i\le n+2\end{aligned} \tag{62}$$

where we assumed $a_1 = c^n$.

We also have:

$$\begin{aligned}\psi_c^+(\omega_{n+3-i}) &= \psi_c^+(\overline{t^{-n-3+i}u\,dt}) = \overline{c^{-2n-6+2i}t^{n+3-i}t^{-n-1}c^{n+1}u\,d(c^2t^{-1})}\\ &= -\overline{c^{-n-3+2i}t^{-i}u\,dt}\\ &= -c^{-(n+3-2i)}\omega_i, \quad \text{if } 1\le i\le n+2\end{aligned}$$

i.e., $\sigma_c^+(-c^{n+3-2i}\omega_{n+3-i}) = \omega_i$.

Case 1. Let n be even (but different from 2). We see that for $1\le i\le \frac{n+2}{2}$, the 2-dimensional spaces $\mathbb{C}\omega_i\oplus\mathbb{C}\omega_{n+3-i}$ form irreducible D_{2k}-representations since the matrix representation for ϕ_ζ^+ and ψ_c^+ with respect to the basis ω_i and ω_{n+3-i} are:

$$\begin{aligned}\phi_\zeta^+|_{\{\omega_i,\omega_{n+3-i}\}} &= \begin{pmatrix}\zeta^{\frac{2n+3-2i}{2}} & 0\\ 0 & \zeta^{\frac{2i-3}{2}}\end{pmatrix} \text{ and}\\ \psi_c^+|_{\{\omega_i,\omega_{n+3-i}\}} &= \begin{pmatrix}0 & -c^{-(n+3-2i)}\\ -c^{n+3-2i} & 0\end{pmatrix},\end{aligned} \tag{63}$$

respectively, where $\mathrm{tr}(\phi_\zeta^+|_{\{\omega_i,\omega_{n+3-i}\}}) = \zeta^n$ and $\mathrm{tr}(\psi_c^+|_{\{\omega_i,\omega_{n+3-i}\}}) = 0$. It follows from Corollary 7.3 that we indeed have 2-dimensional irreducible representations.

For i between $n+3\le i\le 2n$,

$$\overline{t^{i-n-3}u\,dt} = \sum_{k=1}^{2n} P_{i-n-3,-k}\omega_k$$

by Equation (15). Thus for $n+3\le i\le 2n$, we have

$$\psi_c^+(\omega_i) = -c^{n+3-2i}\overline{t^{i-n-3}u\,dt} = -c^{n+3-2i}\sum_{k=1}^{2n} P_{i-n-3,-k}\omega_k. \tag{64}$$

Recall the recursion relations:

$$(2k+r+3)P_{l,-i} = -\sum_{j=1}^{r}(3j+2l-2r)a_jP_{l-r+j-1,-i} \tag{65}$$

for $l \geq 0$ with the initial condition $P_{-m,-i} = \delta_{-m,-i}$, $1 \leq i, m \leq r$. Now from Corollary 6.4 we have $a_j = 0$ unless $j = 1 + qk$ for some $0 \leq q \leq (2n)/k$. Hence we have

$$
\begin{aligned}
(2k+2n+3)P_{l,-i} &= -\sum_{j=1}^{2n}(3j+2l-4n)a_j P_{l-2n+j-1,-i} \\
&= -\sum_{q=0}^{\frac{2n}{k}-1}(3qk+2l-4n+3)a_{1+qk}P_{l-2n+qk,-i}
\end{aligned}
$$

so for a summand on the right to be nonzero we must have $l - 2n + qk = -i + ak$ for some $a \in \mathbb{Z}$. Or rather $l = -i + 2n + (a-q)k = -i + bk$ for some $b \in \mathbb{Z}$. Otherwise it might be that $P_{l,-i}$ is be nonzero for $l = -i + bk$.

In particular if $l = i - n - 3$, then $l \neq -i + bk$ for any $b \in \mathbb{Z}$ (otherwise $l \equiv i - 3 \mod k$ and $l \equiv -i \mod k$ gives us $i - 3 = -i \mod k$ and $2i = 3 + dk$ with k even). Hence $P_{i-n-3,-i} = 0$.

The matrix representation for (61) in basis $\{\omega_1, \dots, \omega_{2n}\}$ is

$$
\phi_\xi^+ = \begin{pmatrix}
\xi & 0 & 0 & 0 & 0 \\
0 & \xi^{-1} & 0 & 0 & 0 \\
0 & 0 & \xi^{-3} & 0 & 0 \\
\vdots & \vdots & \vdots & \ddots & \vdots \\
0 & 0 & 0 & 0 & \xi^{3-2n}
\end{pmatrix},
$$

which is traceless, while the matrix representation for (62) for $1 \leq i \leq n+2$ and (64) for $n+3 \leq i \leq 2n$ in $\{\omega_1, \dots, \omega_{2n}\}$ is

$$
\psi_c^+ = \begin{pmatrix}
0 & 0 & \cdots & 0 & -c^{-(n+1)} & -c^{-(n+3)}P_{0,-1} & \cdots & -c^{3-3n}P_{n-3,-1} \\
0 & 0 & \cdots & -c^{-(n-1)} & 0 & -c^{-(n+3)}P_{0,-2} & \cdots & -c^{3-3n}P_{n-3,-2} \\
\vdots & \vdots & \cdot^{\cdot^{\cdot}} & \vdots & \vdots & \vdots & \cdot^{\cdot^{\cdot}} & \vdots \\
0 & -c^{n-1} & \cdots & 0 & 0 & -c^{-(n+3)}P_{0,-n-1} & \cdots & -c^{3-3n}P_{n-3,-n-1} \\
-c^{n+1} & 0 & \cdots & 0 & 0 & -c^{-(n+3)}P_{0,-n-2} & \cdots & -c^{3-3n}P_{n-3,-n-2} \\
0 & 0 & \vdots & 0 & 0 & -c^{-(n+3)}P_{0,-n-3} & \vdots & -c^{3-3n}P_{n-3,-n-3} \\
\vdots & \vdots & \cdot^{\cdot^{\cdot}} & \vdots & \vdots & \vdots & \cdot^{\cdot^{\cdot}} & \vdots \\
0 & 0 & \cdots & 0 & 0 & -c^{-(n+3)}P_{0,-2n+2} & \cdots & -c^{3-3n}P_{n-3,-2n+2} \\
0 & 0 & \cdots & 0 & 0 & -c^{-(n+3)}P_{0,-2n+1} & \cdots & -c^{3-3n}P_{n-3,-2n+1} \\
0 & 0 & \cdots & 0 & 0 & -c^{-(n+3)}P_{0,-2n} & \cdots & -c^{3-3n}P_{n-3,-2n}
\end{pmatrix},
$$

which has trace

$$
-\sum_{i=n+3}^{2n} c^{n+3-2i}P_{i-n-3,-i} = 0
$$

since $k|n$ and k is even.

So for n even, the set of equations we need to solve is

$$\chi_{(\Omega_R/dR)/\mathbb{C}\omega_0} = n_1\rho_1 + n_2\rho_2 + n_3\rho_3 + n_4\rho_4 + \sum_{h=1}^{k-1} m_h\chi_h,$$

which are precisely,

$$\begin{aligned}
2n &= n_1 + n_2 + n_3 + n_4 + \sum_{h=1}^{k-1} 2m_h, \\
0 &= n_1 - n_2 + n_3 - n_4 \quad \text{for } \psi_c^+, \\
0 &= n_1 - n_2 - n_3 + n_4 \quad \text{for } \psi_c^+\phi_\xi^+, \\
0 &= n_1 + n_2 + (-1)^q n_3 + (-1)^q n_4 + \sum_{h=1}^{k-1} 2m_h \cos(2\pi hq/2k) \\
&\qquad\qquad\qquad\qquad \text{for} 1 \leq q \leq (2k/2) - 1 = k-1 \\
-2n &= n_1 + n_2 + (-1)^q n_3 + (-1)^q n_4 + \sum_{h=1}^{k-1} 2m_h(-1)^h.
\end{aligned}$$

In the above we used the fact that for $1 \leq j \leq k$ one has

$$\chi_{(\Omega_R/dR)/\mathbb{C}\omega_0}((\phi_\xi^+)^j) = \sum_{i=1}^{2n} \xi^{(3-2i)j} = \xi^{3j}\sum_{i=1}^{2n} \xi^{-2ij} = 2n\xi^{3j}\delta_{j,k} = -2n\delta_{j,k}$$

since

$$\begin{aligned}
0 = \xi^{2nj} - 1 &= (\xi^{2j} - 1)(\xi^{2j(n-1)} + \xi^{2j(n-2)} + \cdots + \xi^{2j} + 1) \\
&= (\xi^{2j} - 1)(\xi^{-2j} + \xi^{-4j} + \cdots + \xi^{2j} + 1) = (\xi^{2j} - 1)\left(\sum_{i=1}^{2n} \xi^{-2ij}\right).
\end{aligned}$$

If $1 \leq j < k$, then the left factor in the last equality is not zero so the sum must be zero. The set of equations above can be written as

$$M \begin{pmatrix} n_1 \\ n_2 \\ n_3 \\ n_4 \\ m_1 \\ \vdots \\ m_h \\ \vdots \\ m_{n/2-1} \end{pmatrix} = \begin{pmatrix} 2n \\ 0 \\ 0 \\ 0 \\ 0 \\ \vdots \\ 0 \\ 0 \\ -2n \end{pmatrix}.$$

Thus

$$\begin{pmatrix} n_1 \\ n_2 \\ n_3 \\ n_4 \\ m_1 \\ \vdots \\ m_h \\ \vdots \\ m_{n/2-1} \end{pmatrix} = M^{-1} \begin{pmatrix} 2n \\ 0 \\ 0 \\ 0 \\ 0 \\ \vdots \\ 0 \\ 0 \\ -2n \end{pmatrix}$$

$$= \begin{pmatrix} \frac{1}{4k} & \frac{1}{4} & \frac{1}{4} & \frac{1}{2k} & \cdots & \frac{1}{2k} & \cdots & \frac{1}{4k} \\ \frac{1}{4k} & -\frac{1}{4} & -\frac{1}{4} & \frac{1}{2k} & \cdots & \frac{1}{2k} & \cdots & \frac{1}{4k} \\ \frac{1}{4k} & \frac{1}{4} & -\frac{1}{4} & -\frac{1}{2k} & \cdots & \frac{(-1)^\ell}{2k} & \cdots & \frac{(-1)^k}{4k} \\ \frac{1}{4k} & -\frac{1}{4} & \frac{1}{4} & -\frac{1}{2k} & \cdots & \frac{(-1)^\ell}{2k} & \cdots & \frac{(-1)^k}{4k} \\ \frac{1}{2k} & 0 & 0 & \frac{2}{2k}\cos\left(\frac{2\pi}{2k}\right) & \cdots & \frac{2}{2k}\cos\left(\frac{2\pi h}{2k}\right) & \cdots & -\frac{1}{2k} \\ \vdots & \vdots & \vdots & \vdots & \ddots & \vdots & \ddots & \vdots \\ \frac{1}{2k} & 0 & 0 & \frac{2}{2k}\cos\left(\frac{2\pi h}{2k}\right) & \cdots & \frac{2}{2k}\cos\left(\frac{2\pi h\ell}{2k}\right) & \cdots & \frac{(-1)^h}{2k} \\ \vdots & \vdots & \vdots & \vdots & \ddots & \vdots & \ddots & \vdots \\ \frac{1}{2k} & 0 & 0 & \frac{2}{2k}\cos\left(\frac{2\pi\left(\frac{2k}{2}-1\right)}{2k}\right) & \cdots & \frac{2}{2k}\cos\left(\frac{2\pi\ell\left(\frac{2k}{2}-1\right)}{2k}\right) & \cdots & -\frac{(-1)^k}{2k} \end{pmatrix} \begin{pmatrix} 2n \\ 0 \\ 0 \\ 0 \\ 0 \\ 0 \\ \vdots \\ 0 \\ -2n \end{pmatrix}$$

$$= \begin{pmatrix} 0 \\ 0 \\ \frac{(1-(-1)^k)n}{2k} \\ \frac{(1-(-1)^k)n}{2k} \\ \frac{2n}{k} \\ 0 \\ \vdots \\ \frac{(1-(-1)^h)n}{k} \\ \vdots \\ \frac{(1+(-1)^k)n}{k} \end{pmatrix}.$$

In the case where $a_1 = c^{2n}$, $l = (2n)/k$ is even but k is odd, the multiplicities of irreducible representations are given by

$$\begin{pmatrix} n_1 \\ n_2 \\ n_3 \\ n_4 \\ m_1 \\ \vdots \\ m_h \\ \vdots \\ m_{n/2-1} \end{pmatrix} = M^{-1} \begin{pmatrix} 2n \\ -\frac{1-(-1)^n}{2} - \sum_{i=n+3}^{2n} c^{n+3-2i} P_{i-n-3,-i} \\ \frac{1-(-1)^n}{2} - \sum_{i=n+3}^{2n} \xi^{3-2i} c^{n+3-2i} P_{i-n-3,-i} \\ 0 \\ \vdots \\ 0 \\ -2n \end{pmatrix}$$

$$= \begin{pmatrix}
\frac{1}{4k} & \frac{1}{4} & \frac{1}{4} & \frac{1}{2k} & \cdots & \frac{1}{2k} & \cdots & \frac{1}{4k} \\
\frac{1}{4k} & -\frac{1}{4} & -\frac{1}{4} & \frac{1}{2k} & \cdots & \frac{1}{2k} & \cdots & \frac{1}{4k} \\
\frac{1}{4k} & \frac{1}{4} & -\frac{1}{4} & -\frac{1}{2k} & \cdots & \frac{(-1)^{\ell}}{2k} & \cdots & \frac{(-1)^{k}}{4k} \\
\frac{1}{4k} & -\frac{1}{4} & \frac{1}{4} & -\frac{1}{2k} & \cdots & \frac{(-1)^{\ell}}{2k} & \cdots & \frac{(-1)^{k}}{4k} \\
\frac{1}{2k} & 0 & 0 & \frac{2}{2k}\cos\left(\frac{2\pi}{2k}\right) & \cdots & \frac{2}{2k}\cos\left(\frac{2\pi h}{2k}\right) & \cdots & \frac{1}{2k} \\
\vdots & \vdots & \vdots & \vdots & \ddots & \vdots & \ddots & \vdots \\
\frac{1}{2k} & 0 & 0 & \frac{2}{2k}\cos\left(\frac{2\pi h}{2k}\right) & \cdots & \frac{2}{2k}\cos\left(\frac{2\pi h\ell}{2k}\right) & \cdots & \frac{(-1)^{h}}{2k} \\
\vdots & \vdots & \vdots & \vdots & \ddots & \vdots & \ddots & \vdots \\
\frac{1}{2k} & 0 & 0 & \frac{2}{2k}\cos\left(\frac{2\pi\left(\frac{2k}{2}-1\right)}{2k}\right) & \cdots & \frac{2}{2k}\cos\left(\frac{2\pi\ell\left(\frac{2k}{2}-1\right)}{2k}\right) & \cdots & \frac{(-1)^{k+1}}{2k}
\end{pmatrix}$$

$$\times \begin{pmatrix} 2n \\ -\frac{1-(-1)^n}{2} - \sum_{i=n+3}^{2n} c^{n+3-2i} P_{i-n-3,-i} \\ \frac{1-(-1)^n}{2} - \sum_{i=n+3}^{2n} \xi^{3-2i} c^{n+3-2i} P_{i-n-3,-i} \\ 0 \\ \vdots \\ 0 \\ -2n \end{pmatrix}$$

$$= \begin{pmatrix} -\frac{1}{4}\sum_{i=n+3}^{2n} c^{n+3-2i}P_{i-n-3,-i} - \frac{1}{4}\sum_{i=n+3}^{2n}\xi^{3-2i}c^{n+3-2i}P_{i-n-3,-i} \\ \frac{1}{4}\sum_{i=n+3}^{2n} c^{n+3-2i}P_{i-n-3,-i} + \frac{1}{4}\sum_{i=n+3}^{2n}\xi^{3-2i}c^{n+3-2i}P_{i-n-3,-i} \\ \frac{(1-(-1)^k)n}{2k} - \frac{1-(-1)^n}{4} - \frac{1}{4}\sum_{i=n+3}^{2n} c^{n+3-2i}P_{i-n-3,-i} + \frac{1}{4}\sum_{i=n+3}^{2n}\xi^{3-2i}c^{n+3-2i}P_{i-n-3,-i} \\ \frac{(1-(-1)^k)n}{2k} + \frac{1-(-1)^n}{4} + \frac{1}{4}\sum_{i=n+3}^{2n} c^{n+3-2i}P_{i-n-3,-i} - \frac{1}{4}\sum_{i=n+3}^{2n}\xi^{3-2i}c^{n+3-2i}P_{i-n-3,-i} \\ \frac{2n}{k} \\ \vdots \\ \frac{(1-(-1)^h)n}{k} \\ \vdots \\ \frac{(1+(-1)^k)n}{k} \end{pmatrix}.$$

Observe now that

$$0 \le n_1 = -\frac{1}{4}\sum_{i=n+3}^{2n} c^{n+3-2i}P_{i-n-3,-i} - \frac{1}{4}\sum_{i=n+3}^{2n}\xi^{3-2i}c^{n+3-2i}P_{i-n-3,-i}$$

$$0 \le n_2 = \frac{1}{4}\sum_{i=n+3}^{2n} c^{n+3-2i}P_{i-n-3,-i} + \frac{1}{4}\sum_{i=n+3}^{2n}\xi^{3-2i}c^{n+3-2i}P_{i-n-3,-i}$$

so that

$$\sum_{i=n+3}^{2n} c^{n+3-2i}P_{i-n-3,-i} = -\sum_{i=n+3}^{2n}\xi^{3-2i}c^{n+3-2i}P_{i-n-3,-i}.$$

□

We will now prove Corollary 7.3.

PROOF. Let n be even. We change the basis to

$$\overline{\omega_i} = c^{-\frac{n+3-2i}{2}}\zeta^{-\frac{i}{2}}\omega_i \text{ for } 1 \le i \le n+2$$

to obtain that we indeed have 2-dimensional irreducible representations. Since

$$\overline{\omega_{n+3-i}} = c^{\frac{n+3-2i}{2}}\zeta^{-\frac{n+3-i}{2}}\omega_{n+3-i} \text{ for } 1 \le i \le n+2,$$

we have

(i) $\phi_\zeta^+(\overline{\omega_i}) = c^{-\frac{n+3-2i}{2}}\zeta^{-\frac{2i}{4}}\zeta^{\frac{2n+3-2i}{2}}\omega_i = \zeta^{\frac{2n+3-2i}{2}}\overline{\omega_i}$,
(ii) $\phi_\zeta^+(\overline{\omega_{n+3-i}}) = c^{\frac{n+3-2i}{2}}\zeta^{-\frac{n+3-i}{2}}\zeta^{\frac{-3+2i}{4}}\omega_{n+3-i} = \zeta^{-\frac{2n+3-2i}{2}}\overline{\omega_{n+3-i}}$,
(iii) $\psi_c^+(\overline{\omega_i}) = -c^{-\frac{n+3-2i}{2}}\zeta^{-\frac{2i}{4}}c^{n+3-2i}\omega_{n+3-i} = \zeta^{\frac{2n+3-2i}{2}}\overline{\omega_{n+3-i}}$,
(iv) $\psi_c^+(\overline{\omega_{n+3-i}}) = -c^{\frac{n+3-2i}{2}}\zeta^{-\frac{n+3-i}{2}}c^{-\frac{n+3-2i}{2}}\omega_i = \zeta^{-\frac{2n+3-2i}{2}}\overline{\omega_i}$.

With respect to the basis $\{\overline{\omega_1}, \dots, \overline{\omega_{n+2}}\}$, this implies

$$\phi_\zeta^+\Big|_{\{\overline{\omega_i}, \overline{\omega_{n+3-i}}\}} = \begin{pmatrix} \zeta^{\frac{2n+3-2i}{2}} & 0 \\ 0 & \zeta^{-\frac{2n+3-2i}{2}} \end{pmatrix}$$

and

$$\psi_c^+\Big|_{\{\overline{\omega_i},\overline{\omega_{n+3-i}}\}} = \begin{pmatrix} 0 & \zeta^{-\frac{2n+3-2i}{2}} \\ \zeta^{\frac{2n+3-2i}{2}} & 0 \end{pmatrix},$$

which coincide with classical 2-dimensional irreducible representations for dihedral groups.

Now, let n be odd. With respect to the basis

$$\overline{\omega_i} = c^{-\frac{n+3-2i}{4}}\zeta^{-\frac{i}{2}}\omega_i \text{ for } 1 \le i \le n+2,$$

we have

$$\phi_\zeta^+\Big|_{\{\overline{\omega_i},\overline{\omega_{n+3-i}}\}} = \begin{pmatrix} \zeta^{\frac{2n+3-2i}{2}} & 0 \\ 0 & \zeta^{-\frac{2n+3-2i}{2}} \end{pmatrix}$$

and

$$\psi_c^+\Big|_{\{\overline{\omega_i},\overline{\omega_{n+3-i}}\}} = \begin{pmatrix} 0 & \zeta^{-\frac{2n+3-2i}{2}} \\ \zeta^{\frac{2n+3-2i}{2}} & 0 \end{pmatrix},$$

and we note that

$$\psi_c^+(\overline{\omega_{\frac{n+3}{2}}}) = -\overline{\omega_{\frac{n+3}{2}}} \quad \text{and} \quad \phi_\zeta^+(\overline{\omega_{\frac{n+3}{2}}}) = \zeta^{n/2}\overline{\omega_{\frac{n+3}{2}}} = -\overline{\omega_{\frac{n+3}{2}}}. \tag{66}$$

□

EXAMPLE 7.4. *In the case when $n = 3$ and $k = 3$ for $p(t) = t(t^3 - \alpha_1^3)(t^3 - \alpha_2^3)$,*

$$\psi_c^+ = \begin{pmatrix} 0 & 0 & 0 & 0 & -\frac{1}{c^2} & 0 \\ 0 & 0 & 0 & -\frac{1}{c} & 0 & 0 \\ 0 & 0 & -1 & 0 & 0 & 0 \\ 0 & -c & 0 & 0 & 0 & 0 \\ -c^2 & 0 & 0 & 0 & 0 & 0 \\ 0 & 0 & 0 & 0 & 0 & -\frac{\alpha_1^3\alpha_2^3}{c^6} \end{pmatrix}.$$

In this case $\alpha_1^3\alpha_2^3 = c^6$, the trace of ψ_c^+ equals -2, giving us multiplicities

$$n_1 = n_2 = n_3 = 0, n_4 = 2, m_1 = 2.$$

EXAMPLE 7.5. *For $n = 6$ and $k = 3$, we have*

$$\begin{aligned} p(t) &= t(t^3-\alpha_1^3)(t^3-\alpha_2^3)(t^3-\alpha_3^3)(t^3-\alpha_4^3) \\ &= t^{13} - \left(\alpha_1^3+\alpha_2^3+\alpha_3^3+\alpha_4^3\right)t^{10} + \left(\alpha_1^3\alpha_2^3+\alpha_1^3\alpha_3^3+\alpha_1^3\alpha_4^3+\alpha_2^3\alpha_3^3+\alpha_2^3\alpha_4^3+\alpha_3^3\alpha_4^3\right)t^7 \\ &\quad - \left(\alpha_1^3\alpha_2^3\alpha_3^3+\alpha_1^3\alpha_2^3\alpha_4^3+\alpha_1^3\alpha_3^3\alpha_4^3+\alpha_2^3\alpha_3^3\alpha_4^3\right)t^4 + \alpha_1^3\alpha_2^3\alpha_3^3\alpha_4^3 t \end{aligned}$$

Then ψ_c^+ *is*

$$\begin{pmatrix}
0 & 0 & 0 & 0 & 0 & 0 & 0 & -\frac{1}{c^7} & 0 & 0 & \Lambda_3 & 0 \\
0 & 0 & 0 & 0 & 0 & 0 & -\frac{1}{c^5} & 0 & 0 & \Lambda_2 & 0 & 0 \\
0 & 0 & 0 & 0 & 0 & -\frac{1}{c^3} & 0 & 0 & \Lambda_1 & 0 & 0 & \Theta_4 \\
0 & 0 & 0 & 0 & -\frac{1}{c} & 0 & 0 & 0 & 0 & 0 & \Theta_3 & 0 \\
0 & 0 & 0 & -c & 0 & 0 & 0 & 0 & 0 & \Theta_2 & 0 & 0 \\
0 & 0 & -c^3 & 0 & 0 & 0 & 0 & 0 & \Theta_1 & 0 & 0 & \Delta_4 \\
0 & -c^5 & 0 & 0 & 0 & 0 & 0 & 0 & 0 & 0 & \Delta_3 & 0 \\
-c^7 & 0 & 0 & 0 & 0 & 0 & 0 & 0 & 0 & \Delta_2 & 0 & 0 \\
0 & 0 & 0 & 0 & 0 & 0 & 0 & 0 & \Delta_1 & 0 & 0 & \Gamma_4 \\
0 & 0 & 0 & 0 & 0 & 0 & 0 & 0 & 0 & 0 & \Gamma_3 & 0 \\
0 & 0 & 0 & 0 & 0 & 0 & 0 & 0 & 0 & \Gamma_2 & 0 & 0 \\
0 & 0 & 0 & 0 & 0 & 0 & 0 & 0 & \Gamma_1 & 0 & 0 & \Psi
\end{pmatrix},$$

where

$$\begin{aligned}
\Lambda_1 &= -\frac{2\left(\alpha_1^3+\alpha_2^3+\alpha_3^3+\alpha_4^3\right)}{5c^9} = \frac{2a_{10}}{5c^9}, \\
\Lambda_2 &= -\frac{8\left(\alpha_1^3+\alpha_2^3+\alpha_3^3+\alpha_4^3\right)}{17c^{11}} = \frac{8a_{10}}{17c^{11}}, \\
\Lambda_3 &= -\frac{10\left(\alpha_1^3+\alpha_2^3+\alpha_3^3+\alpha_4^3\right)}{19c^{13}} = \frac{10a_{10}}{19c^{13}}, \\
\Theta_1 &= -\frac{\alpha_1^3\alpha_2^3+\alpha_1^3\alpha_3^3+\alpha_1^3\alpha_4^3+\alpha_2^3\alpha_3^3+\alpha_2^3\alpha_4^3+\alpha_3^3\alpha_4^3}{5c^9} = -\frac{a_7}{5c^9}, \\
\Theta_2 &= -\frac{\alpha_1^3\alpha_2^3+\alpha_1^3\alpha_3^3+\alpha_1^3\alpha_4^3+\alpha_2^3\alpha_3^3+\alpha_2^3\alpha_4^3+\alpha_3^3\alpha_4^3}{17c^{11}} = -\frac{a_7}{17c^{11}}, \\
\Theta_3 &= \frac{\alpha_1^3\alpha_2^3+\alpha_1^3\alpha_3^3+\alpha_1^3\alpha_4^3+\alpha_2^3\alpha_3^3+\alpha_2^3\alpha_4^3+\alpha_3^3\alpha_4^3}{19c^{13}} = \frac{a_7}{19c^{13}}, \\
\Theta_4 &= -\frac{8(\alpha_1^6+\alpha_2^6+\alpha_3^6+\alpha_4^6)+11\left(\alpha_1^3\alpha_2^3+\alpha_1^3\alpha_3^3+\alpha_1^3\alpha_4^3+\alpha_2^3\alpha_3^3+\alpha_2^3\alpha_4^3+\alpha_3^3\alpha_4^3\right)}{35c^{15}} \\
&= -\frac{8(a_{10}^2-2a_7)+11a_7}{35c^{15}} = -\frac{8a_{10}^2-5a_7}{35c^{15}}, \\
\Delta_1 &= \frac{4\left(\alpha_1^3\alpha_2^3\alpha_3^3+\alpha_1^3\alpha_2^3\alpha_4^3+\alpha_1^3\alpha_3^3\alpha_4^3+\alpha_2^3\alpha_3^3\alpha_4^3\right)}{5c^9} = -\frac{4a_4}{5c^9}, \\
\Delta_2 &= \frac{10\left(\alpha_1^3\alpha_2^3\alpha_3^3+\alpha_1^3\alpha_2^3\alpha_4^3+\alpha_1^3\alpha_3^3\alpha_4^3+\alpha_2^3\alpha_3^3\alpha_4^3\right)}{17c^{11}} = -\frac{10a_4}{17c^{11}}, \\
\Delta_3 &= \frac{8\left(\alpha_1^3\alpha_2^3\alpha_3^3+\alpha_1^3\alpha_2^3\alpha_4^3+\alpha_1^3\alpha_3^3\alpha_4^3+\alpha_2^3\alpha_3^3\alpha_4^3\right)}{19c^{13}} = -\frac{8a_4}{19c^{13}}, \\
\Delta_4 &= -\frac{2}{35c^{15}}\left(\alpha_1^3\alpha_2^3\alpha_3^3+\alpha_1^3\alpha_2^3\alpha_4^3+\alpha_1^3\alpha_3^3\alpha_4^3+\alpha_2^3\alpha_3^3\alpha_4^3\right. \\
&\quad +2\left(\alpha_1^6(\alpha_2^3+\alpha_3^3+\alpha_4^3)+\alpha_2^6(\alpha_1^3+\alpha_3^3+\alpha_4^3)\right. \\
&\quad \left.\left.+\alpha_3^6(\alpha_1^3+\alpha_2^3+\alpha_4^3)+\alpha_4^6(\alpha_1^3+\alpha_2^3+\alpha_3^3)\right)\right) \\
&= -\frac{2}{35c^{15}}\left(5a_4-2a_7a_{10}\right), \\
\Gamma_1 &= -\frac{7\alpha_1^3\alpha_2^3\alpha_3^3\alpha_4^3}{5c^9},
\end{aligned}$$

$$\Gamma_2 = -\frac{19\alpha_1^3\alpha_2^3\alpha_3^3\alpha_4^3}{17c^{11}},$$
$$\Gamma_3 = -\frac{17\alpha_1^3\alpha_2^3\alpha_3^3\alpha_4^3}{19c^{13}},$$
$$\Gamma_4 = \frac{1}{35c^{15}}\left(16\left(\alpha_1^6(\alpha_2^3\alpha_3^3+\alpha_2^3\alpha_4^3+\alpha_3^3\alpha_4^3)+\alpha_2^6(\alpha_1^3\alpha_3^3+\alpha_1^3\alpha_4^3+\alpha_3^3\alpha_4^3)\right.\right.$$
$$\left.\left.+\alpha_3^6(\alpha_1^3\alpha_2^3+\alpha_1^3\alpha_4^3+\alpha_2^3\alpha_4^3)+\alpha_4^6(\alpha_1^3\alpha_2^3+\alpha_1^3\alpha_3^3+\alpha_2^3\alpha_3^3)\right)+39\alpha_1^3\alpha_2^3\alpha_3^3\alpha_4^3\right)$$
$$= \frac{1}{35c^{15}}\left(16a_4a_{10}-25a_1\right),$$
$$\Psi = -\frac{4\alpha_1^3\alpha_2^3\alpha_3^3\alpha_4^3\left(\alpha_1^3+\alpha_2^3+\alpha_3^3+\alpha_4^3\right)}{5c^{15}} = -\frac{4\left(\alpha_1^3+\alpha_2^3+\alpha_3^3+\alpha_4^3\right)}{5c^3} = \frac{4a_{10}}{5c^3}.$$

By (52) *we have* $a_{10} = c^{-6}a_4$ *so that*

$$\operatorname{tr}\psi_c^+ = \Delta_1 + \Psi = -\frac{4a_4}{5c^9} + \frac{4a_{10}}{5c^3} = 0.$$

This implies that the multiplicities appearing are

$$n_1 = n_2 = 0, \quad n_3 = n_4 = 2 \text{ and } m_1 = 4, \quad m_2 = 0.$$

EXAMPLE 7.6. *When* $n = 9$ *and* $k = 3$, *we used Mathematica to get*

$$\operatorname{tr}(\psi_c^+) = -2 = -\operatorname{tr}(\psi_c^+\phi_\xi^+),$$

and hence

$$n_1 = n_2 = 0, n_3 = 2, n_4 = 4, m_1 = 6, m_2 = 0.$$

Acknowledgment

The second author acknowledges the College of Charleston for providing a productive working environment in May of 2016.

References

[Arb00] L. F. A. Arbogast, *Du calcul des derivations.*

[Bel28] E. T. Bell, *Partition polynomials*, Ann. of Math. (2) **29** (1927/28), no. 1-4, 38–46. MR1502817

[BGG93] E. Bujalance, J. M. Gamboa, and G. Gromadzki, *The full automorphism groups of hyperelliptic Riemann surfaces*, Manuscripta Math. **79** (1993), no. 3-4, 267–282. MR1223022

[Blo81] Spencer Bloch, *The dilogarithm and extensions of Lie algebras*, Algebraic K-theory, Evanston 1980 (Proc. Conf., Northwestern Univ., Evanston, Ill., 1980), Lecture Notes in Math., vol. 854, Springer, Berlin-New York, 1981, pp. 1–23. MR618298

[Bre94] Murray Bremner, *Universal central extensions of elliptic affine Lie algebras*, J. Math. Phys. **35** (1994), no. 12, 6685–6692. MR1303073

[Bre00] Thomas Breuer, *Characters and automorphism groups of compact Riemann surfaces*, London Mathematical Society Lecture Note Series, vol. 280, Cambridge University Press, Cambridge, 2000. MR1796706

[CF11] Ben Cox and Vyacheslav Futorny, *DJKM algebras I: their universal central extension*, Proc. Amer. Math. Soc. **139** (2011), no. 10, 3451–3460. MR2813377

[CGLZ14] Ben Cox, Xiangqian Guo, Rencai Lu, and Kaiming Zhao, *n-point Virasoro algebras and their modules of densities*, Commun. Contemp. Math. **16** (2014), no. 3, 1350047, 27. MR3211093

[CGLZ17] Ben Cox, Xiangqian Guo, Rencai Lu, and Kaiming Zhao, *Simple superelliptic Lie algebras*, Commun. Contemp. Math. **19** (2017), no. 3, 1650032, 22. MR3631928

[Cox08] Ben Cox, *Realizations of the four point affine Lie algebra* $\mathfrak{sl}(2, R) \oplus (\Omega_R/dR)$, Pacific J. Math. **234** (2008), no. 2, 261–289. MR2373448

[Cox16a] Ben Cox, *Module structure of the center of the universal central extension of a genus zero Krichever-Novikov algebra*, J. Algebra **467** (2016), 58–79. MR3545955

[Cox16b] Ben Cox, *On the universal central extension of hyperelliptic current algebras*, Proc. Amer. Math. Soc. **144** (2016), no. 7, 2825–2835. MR3487217

[CZ17] Ben Cox and Kaiming Zhao, *Certain families of polynomials arising in the study of hyperelliptic Lie algebras*, The Ramanujan Journal, to appear, arXiv:1602.01432. (2017).

[FdB55] C. F. Faà di Bruno, *Sullo sviluppo delle funzione*, Ann. di Scienze Matem. et Fisiche di Tortoloni **6** (1855), 479–480.

[FdB57] ———, *Note sur un nouvelle formule de calcul différentiel*, Quart. J. Math. **1** (1857), 359–360.

[FH91] William Fulton and Joe Harris, *Representation theory*, Graduate Texts in Mathematics, vol. 129, Springer-Verlag, New York, 1991. A first course; Readings in Mathematics. MR1153249

[Jor86] D. A. Jordan, *On the ideals of a Lie algebra of derivations*, J. London Math. Soc. (2) **33** (1986), no. 1, 33–39. MR829385

[Ser77] Jean-Pierre Serre, *Linear representations of finite groups*, Springer-Verlag, New York-Heidelberg, 1977. Translated from the second French edition by Leonard L. Scott; Graduate Texts in Mathematics, Vol. 42. MR0450380

[Sha03] Tanush Shaska, *Determining the automorphism group of a hyperelliptic curve*, Proceedings of the 2003 International Symposium on Symbolic and Algebraic Computation, ACM, New York, 2003, pp. 248–254. MR2035219

[Skr88] S. M. Skryabin, *Regular Lie rings of derivations* (Russian), Vestnik Moskov. Univ. Ser. I Mat. Mekh. **3** (1988), 59–62; English transl., Moscow Univ. Math. Bull. **43** (1988), no. 3, 56–58. MR966871

[Skr04] Serge Skryabin, *Degree one cohomology for the Lie algebras of derivations*, Lobachevskii J. Math. **14** (2004), 69–107. MR2035385

[Tay04] Richard Taylor, *Galois representations* (English, with English and French summaries), Ann. Fac. Sci. Toulouse Math. (6) **13** (2004), no. 1, 73–119. MR2060030

[Ter99] Audrey Terras, *Fourier analysis on finite groups and applications*, London Mathematical Society Student Texts, vol. 43, Cambridge University Press, Cambridge, 1999. MR1695775

Department of Mathematics, College of Charleston, Charleston, South Carolina 29424

Email address: coxbl@cofc.edu

Department of Mathematical Sciences, United States Military Academy, West Point, New York 10996

Email address: meeseongim@gmail.com

Contemporary Mathematics
Volume **713**, 2018
https://doi.org/10.1090/conm/713/14313

Classification of 5-dimensional complex nilpotent Leibniz algebras

Ismail Demir

ABSTRACT. Leibniz algebras are certain generalization of Lie algebras. In this paper we give the classification of 5−dimensional complex non-Lie nilpotent Leibniz algebras. We use the canonical forms for the congruence classes of matrices of bilinear forms to classify the case $\dim(A^2) = 3$ and $\dim(Leib(A)) = 1$ which can be applied to higher dimensions. The remaining cases are classified via direct method.

1. Introduction

Leibniz algebras introduced by Loday [**13**] are natural generalization of Lie algebras. Such algebras had been first considered by Bloh who called them D-algebras [**5**], considering their connections with derivations. A left (resp. right) Leibniz algebra A is a $\mathbb{C}$-vector space equipped with a bilinear product $[\,,\,] : A \times A \longrightarrow A$ such that the left (resp. right) multiplication operator is a derivation. A left Leibniz algebra is not necessarily a right Leibniz algebra. In this paper, we consider (left) Leibniz algebras following Barnes [**4**]. Any Lie algebra is a Leibniz algebra, but the converse is not true. For a Leibniz algebra A, we define the ideals $A^1 = A$ and $A^i = [A, A^{i-1}]$ for $i \in \mathbb{Z}_{\geq 2}$. The Leibniz algebra A is said be abelian if $A^2 = 0$. It is nilpotent of class c if $A^{c+1} = 0$ but $A^c \neq 0$ for some positive integer c. A subspace $Leib(A) = \text{span}\{[a,a] \mid a \in A\}$ is an important abelian ideal of A. A is a Lie algebra if and only if $Leib(A) = \{0\}$. The center of A is denoted by $Z(A) = \{x \in A \mid [a,x] = 0 = [x,a] \text{ for all a} \in \text{A}\}$. If A is a nilpotent Leibniz algebra of class c then $A^c \subseteq Z(A)$. A is nilpotent and $\dim(Leib(A)) = 1$ implies $Leib(A) \subseteq Z(A)$. We say that a Leibniz algebra A is split if it can be written as a direct sum of two nontrivial ideals. Otherwise A is said to be non-split. If A is a non-split Leibniz algebra then $Z(A) \subseteq A^2$. A $n-$dimensional Leibniz algebra A is said to be filiform Leibniz algebra if $\dim(A^i) = n - i$, for $2 \leq i \leq n$. In this paper we consider only non-split non-Lie Leibniz algebras over the field of complex numbers $\mathbb{C}$.

The classification of nilpotent Lie algebras is still unsolved and it is a very complicated problem. So far the complete classification of complex nilpotent Lie algebras of dimension $n \leq 7$ is given [**12**]. The lack of antisymmetry property in Leibniz algebras makes the problem of classification of non-Lie nilpotent Leibniz

2010 *Mathematics Subject Classification.* Primary 17A32.

algebras even harder. The complete classification of non-Lie nilpotent Leibniz algebras over $\mathbb{C}$ of dimension $n \leq 4$ is known (see [**1**], [**2**], [**6**], [**7**], [**9**], [**13**], [**16**]), and some partial results on higher dimensions (see [**8**], [**15**]). In this paper we give the classification of 5−dimensional non-Lie nilpotent Leibniz algebras using congruence classes of bilinear forms as in ([**7**], [**8**], [**9**]) for the classification of 5−dimensional non-Lie nilpotent Leibniz algebras with $\dim(A^2) = 3$ and $\dim(Leib(A)) = 1$. For the remaining cases we use direct method. Our approach of using congruence classes of bilinear forms can be used to classify n−dimensional Leibniz algebras with $\dim(A^2) = n - 2$ and $\dim(Leib(A)) = 1$ which we plan to pursue in future.

We give the following Lemmas which are very useful.

LEMMA 1.1. *Let A be a $n-$dimensional nilpotent Leibniz algebra with $\dim(Z(A)) = n - k$. If $\dim(Leib(A)) = 1$ then $\dim(A^2) \leq \frac{k^2-k+2}{2}$.*

PROOF. We have $Leib(A) \subseteq Z(A)$. Let $Leib(A) = \text{span}\{e_n\}$ and extend this to a basis $\{e_{k+1}, e_{k+2}, \ldots, e_{n-1}, e_n\}$ for $Z(A)$. Then the nonzero products in $A = \text{span}\{e_1, e_2, \ldots, e_k, e_{k+1}, \ldots, e_n\}$ are given by

$$[e_r, e_r] = \theta_r e_n, [e_i, e_j] = \sum_{t=1}^{n-1} \alpha_{ij}^t e_t + \beta_{ij} e_n, [e_j, e_i] = -\sum_{t=1}^{n-1} \alpha_{ij}^t e_t + \gamma_{ji} e_n.$$

for $1 \leq r, i, j \leq k, i \neq j$. Then $dim(A^2) \leq \{\text{number of } (i,j)\text{'s where } 1 \leq i < j \leq k\} + 1$. Note that the number of (i,j)'s where $1 \leq i < j \leq k$ is equal to $\frac{k^2-k}{2}$. Hence $\dim(A^2) \leq \frac{k^2-k+2}{2}$. □

LEMMA 1.2. *Let A be a $n-$dimensional nilpotent Leibniz algebra with $\dim(A^2) = n - k, \dim(Leib(A)) = 1$ and $\dim(A^3) = t$. Then*

(i): $n \leq t + \frac{k^2+k+2}{2}$
(ii): $n \leq t + \frac{k^2+k}{2}$ *if $Leib(A) \subseteq A^3$*

PROOF. First suppose $Leib(A) \subseteq A^3$. Let $Leib(A) = \text{span}\{e_n\}$. Then we can extend this to a basis $\{e_{n-t+1}, e_{n-t+2}, \ldots, e_{n-1}, e_n\}$ for A^3 and a basis $\{e_{k+1}, e_{k+2}, \ldots, e_{n-1}, e_n\}$ for A^2. Then the nonzero products in $A = \text{span}\{e_1, e_2, \ldots, e_k, e_{k+1}, \ldots, e_n\}$ are given by

$$[e_r, e_r] = \theta_r e_n, [e_i, e_j] = \sum_{m=k+1}^{n-1} \alpha_{ij}^t e_t + \beta_{ij} e_n, [e_j, e_i] = -\sum_{m=k+1}^{n-1} \alpha_{ij}^t e_t + \gamma_{ji} e_n,$$
$$[e_p, e_s] = \sum_{m=n-t+1}^{n-1} \alpha_{ij}^t e_t + \theta_{ij} e_n, [e_s, e_p] = -\sum_{m=n-t+1}^{n-1} \alpha_{ij}^t e_t + \delta_{ji} e_n,$$

for $1 \leq i < j \leq k, \quad k+1 \leq s \leq n-1$ and $1 \leq r, p \leq n$. Note that $dim(A^2) \leq \{\text{number of } (i,j)\text{'s where } 1 \leq i < j \leq k\} + \dim(A^3)$. Note that the number of (i,j)'s where $1 \leq i < j \leq k$ is equal to $\frac{k^2-k}{2}$. Hence $n \leq t + \frac{k^2+k}{2}$. This completes the proof of part (ii). Similarly we can prove part (i). In that case notice that $dim(A^2) \leq \{\text{number of } (i,j)\text{'s where } 1 \leq i < j \leq k\} + \dim(A^3) + 1$ because we may have $Leib(A) \not\subseteq A^3$. Therefore, we obtain $n \leq t + \frac{k^2+k+2}{2}$. □

2. Classification of nilpotent Leibniz algebras with $\dim(A^2) = 3$ and $\dim(Leib(A)) = 1$

Let A be a $n-$dimensional complex non-Lie nilpotent Leibniz algebra with $\dim(A^2) = n-2$ and $\dim(Leib(A)) = 1$. Choose $Leib(A) = \text{span}\{e_n\}$, and extend this to a basis $\{e_3, e_4, \ldots, e_{n-1}, e_n\}$ for A^2. Let V be a complementary subspace to A^2 in A so that $A = A^2 \oplus V$. Then for any $u, v \in V$, we have $[u, v] = \alpha_3 e_3 + \alpha_4 e_4 + \alpha_{n-1} e_{n-1} + \alpha_n e_n$ for some $\alpha_i \in \mathbb{C}$, $3 \le i \le n$. Define the bilinear form $f(\ ,\ ) : V \times V \to \mathbb{C}$ by $f(u, v) = \alpha_n$ for all $u, v \in V$.

Using ([**17**], Theorem 3.1), we choose a basis $\{e_1, e_2\}$ for V so that the matrix N of the bilinear form $f(\ ,\ ) : V \times V \to \mathbb{C}$ is one of the following:

$$(i)\begin{pmatrix} 0 & 1 \\ -1 & 0 \end{pmatrix},\quad (ii)\begin{pmatrix} 1 & 0 \\ 0 & 0 \end{pmatrix},\quad (iii)\begin{pmatrix} 1 & 0 \\ 0 & 1 \end{pmatrix},\quad (iv)\begin{pmatrix} 0 & 1 \\ -1 & 1 \end{pmatrix},\quad (v)\begin{pmatrix} 0 & 1 \\ c & 0 \end{pmatrix}$$

where $c \neq 1, -1$. However, since $e_n \in Leib(A)$ we observe that N cannot be the matrix (i). We can reduce the number of possible matrices to two using the following Lemma.

LEMMA 2.1. *The Leibniz algebras corresponding to the matrices (ii) and (iv) are isomorphic. Additionally, the Leibniz algebras corresponding to the matrices (iii) and (v) are isomorphic.*

PROOF. If N is the matrix (ii) we have the nontrivial products in A given by $[e_1, e_1] = e_n, [e_1, e_2] = \alpha_3 e_3 + \alpha_4 e_4 + \ldots + \alpha_{n-1} e_{n-1} = -[e_2, e_1]$ and $[e_i, e_j] = \beta_3 e_3 + \beta_4 e_4 + \ldots + \beta_n e_n$ for the rest of $i, j's, 1 \le i, j \le n-1$.
If N is the matrix (iv) we have the nontrivial products in A given by
$[e_2, e_2] = e_n, [e_1, e_2] = \alpha_3 e_3 + \alpha_4 e_4 + \ldots + \alpha_{n-1} e_{n-1} + e_n = -[e_2, e_1]$ and $[e_i, e_j] = \beta_3 e_3 + \beta_4 e_4 + \ldots + \beta_n e_n$ for the rest of $i, j's, 1 \le i, j \le n-1$. Here the base change $x_1 = e_2, x_2 = e_1, x_k = e_k, x_m = \alpha_m e_m + e_n, x_n = e_n$ for some m such that $\alpha_m \neq 0$, for k such that $3 \le k \le n-1$; shows that these two algebras are isomorphic.

If N is the matrix (iii) we have the nontrivial products in A given by
$[e_1, e_1] = e_n, [e_1, e_2] = \alpha_3 e_3 + \alpha_4 e_4 + \ldots + \alpha_{n-1} e_{n-1} = -[e_2, e_1], [e_2, e_2] = e_n$, and $[e_i, e_j] = \beta_3 e_3 + \beta_4 e_4 + \ldots + \beta_n e_n$ for the rest of $i, j's, 1 \le i, j \le n-1$.
If N is the matrix (v) we have the nontrivial products in A given by
$[e_1, e_2] = \alpha_3 e_3 + \alpha_4 e_4 + \ldots + \alpha_{n-1} e_{n-1} + e_n, [e_2, e_1] = -\alpha_3 e_3 - \alpha_4 e_4 - \ldots - \alpha_{n-1} e_{n-1} + c e_n$ and $[e_i, e_j] = \beta_3 e_3 + \beta_4 e_4 + \ldots + \beta_n e_n$ for the rest of $i, j's, 1 \le i, j \le n-1$. The base change $x_1 = e_1 + \frac{1}{2} e_2, x_2 = -ie_1 + \frac{i}{2} e_2, x_k = e_k, x_m = \alpha_m e_m + \frac{i}{2}(1-c) e_n, x_n = \frac{1}{2}(1+c) e_n$ for some m such that $\alpha_m \neq 0$, for k such that $3 \le k \le n-1$; shows that these two algebras are isomorphic. □

Let A be a $5-$dimensional complex non-split non-Lie nilpotent Leibniz algebra with $\dim(A^2) = 3$ and $\dim(Leib(A)) = 1$. Then $\dim(A^3) \le 2$. If $\dim(A^3) = 0$ then we have $A^2 = Z(A)$. Then from Lemma 1.1 we get $\dim(A^2) \le 2$, which is a contradiction. Hence $\dim(A^3) = 1$ or 2. First let $\dim(A^3) = 2$. Then $\dim(A^4) = 0$ or 1.

THEOREM 2.2. *Let A be a $5-$dimensional non-split non-Lie nilpotent Leibniz algebra with $\dim(A^2) = 3$, $\dim(A^3) = 2$, $\dim(A^4) = 1$, and $\dim(Leib(A)) = 1$. Then A is isomorphic to a Leibniz algebra spanned by $\{x_1, x_2, x_3, x_4, x_5\}$ with the nonzero products given by one of the following:*

$\mathcal{A}_1$: $[x_1, x_1] = x_5, [x_1, x_2] = x_3 = -[x_2, x_1], [x_1, x_3] = x_4 = -[x_3, x_1], [x_1, x_4] = x_5 = -[x_4, x_1]$.

$\mathcal{A}_2$: $[x_1, x_1] = x_5, [x_1, x_2] = x_3 = -[x_2, x_1], [x_1, x_3] = x_4 = -[x_3, x_1], [x_2, x_3] = x_5 = -[x_3, x_2], [x_1, x_4] = x_5 = -[x_4, x_1]$.
$\mathcal{A}_3$: $[x_1, x_1] = x_5, [x_1, x_2] = x_3 = -[x_2, x_1], [x_2, x_3] = x_4 = -[x_3, x_2], [x_2, x_4] = x_5 = -[x_4, x_2]$.
$\mathcal{A}_4$: $[x_1, x_1] = x_5, [x_1, x_2] = x_3 = -[x_2, x_1], [x_1, x_3] = x_5 = -[x_3, x_1], [x_2, x_3] = x_4 = -[x_3, x_2], [x_2, x_4] = x_5 = -[x_4, x_2]$.
$\mathcal{A}_5(\alpha)$: $[x_1, x_1] = x_5, [x_1, x_2] = x_3 = -[x_2, x_1], [x_2, x_2] = x_5, [x_1, x_3] = x_4 = -[x_3, x_1], [x_2, x_3] = \alpha x_5 = -[x_3, x_2], [x_1, x_4] = x_5 = -[x_4, x_1], \quad \alpha \in \mathbb{C}$.
$\mathcal{A}_6$: $[x_1, x_1] = x_5, [x_1, x_2] = x_3, [x_2, x_1] = -x_3 + x_5, [x_1, x_3] = x_4 = -[x_3, x_1], [x_1, x_4] = x_5 = -[x_4, x_1]$.
$\mathcal{A}_7$: $[x_1, x_1] = x_5, [x_1, x_2] = x_3, [x_2, x_1] = -x_3 + x_5, [x_1, x_3] = x_4 = -[x_3, x_1], [x_2, x_3] = x_5 = -[x_3, x_2], [x_1, x_4] = x_5 = -[x_4, x_1]$.

PROOF. We have $A^4 \subseteq Z(A) \subset A^2$. Assume $\dim(Z(A)) = 2$. Note that $A^3 \neq Z(A)$ since $A^4 \neq 0$. Take W such that $A^2 = A^3 \oplus W$. Then $W \subseteq Z(A)$. $A^3 = [A, A^2] = [A, A^3 \oplus W] = A^4$ which is a contradiction. Therefore $\dim(Z(A)) = 1$, and $A^4 = Z(A)$. Let $Leib(A) = Z(A) = A^4 = \text{span}\{e_5\}$. Extend this to bases $\{e_4, e_5\}$ and $\{e_3, e_4, e_5\}$ of A^3 and A^2, respectively. Take $V = \text{span}\{e_1, e_2\}$.

If N is the matrix (ii) then the nontrivial products in A given by
$[e_1, e_1] = e_5, [e_1, e_2] = \alpha_1 e_3 + \alpha_2 e_4 = -[e_2, e_1], [e_1, e_3] = \beta_1 e_4 + \beta_2 e_5, [e_3, e_1] = -\beta_1 e_4 + \beta_3 e_5, [e_2, e_3] = \beta_4 e_4 + \beta_5 e_5, [e_3, e_2] = -\beta_4 e_4 + \beta_6 e_5, [e_3, e_3] = \beta_7 e_5, [e_1, e_4] = \gamma_1 e_5, [e_4, e_1] = \gamma_2 e_5, [e_2, e_4] = \gamma_3 e_5, [e_4, e_2] = \gamma_4 e_5, [e_3, e_4] = \gamma_5 e_5, [e_4, e_3] = \gamma_6 e_5, [e_4, e_4] = \gamma_7 e_5$.
From Leibniz identities we get the following equations:

$$(2.1) \quad \begin{cases} \alpha_1(\beta_2 + \beta_3) + \alpha_2(\gamma_1 + \gamma_2) = 0 & \alpha_1(\beta_5 + \beta_6) + \alpha_2(\gamma_3 + \gamma_4) = 0 \\ \beta_4\gamma_1 = \alpha_1\beta_7 + \alpha_2\gamma_6 + \beta_1\gamma_3 & \alpha_1\gamma_5 + \alpha_2\gamma_7 = 0 \\ \beta_1(\gamma_1 + \gamma_2) = 0 & \beta_1\gamma_4 + \beta_4\gamma_1 + \alpha_1\beta_7 + \alpha_2\gamma_5 = 0 \\ \beta_1(\gamma_5 + \gamma_6) = 0 & \beta_1\gamma_7 = 0 \\ \beta_1\gamma_3 + \beta_4\gamma_2 - \alpha_1\beta_7 - \alpha_2\gamma_5 = 0 & \beta_4(\gamma_3 + \gamma_4) = 0 \\ \beta_4(\gamma_5 + \gamma_6) = 0 & \beta_4\gamma_7 = 0 \end{cases}$$

So we have $\gamma_5 = 0 = \gamma_6 = \gamma_7$.

Case 1: Let $\beta_4 = 0$. Then $\beta_1 \neq 0$ since $\dim(A^3) = 2$. So by (2.1) we obtain $\gamma_2 = -\gamma_1$ and $\beta_3 = -\beta_2$. If we assume $\gamma_4 \neq 0$ then we arrive contradiction using the equations (2.1). Hence $\gamma_4 = 0$. Then by (2.1) we have $\beta_7 = 0 = \gamma_3$ and $\beta_6 = -\beta_5$. If $\beta_5 = 0$ then the base change $x_1 = e_1, x_2 = \frac{1}{\alpha_1\beta_1\gamma_1}e_2, x_3 = frac1\alpha_1\beta_1\gamma_1(\alpha_1 e_3 + \alpha_2 e_4), x_4 = \frac{1}{\beta_1\gamma_1}(\beta_1 e_4 + \beta_2 e_5) + \frac{\alpha_2}{\alpha_1\beta_1}e_5, x_5 = e_5$ shows that A is isomorphic to the algebra $\mathcal{A}_1$. If $\beta_5 \neq 0$ then the base change $x_1 = \frac{\beta_5^{1/3}}{\alpha_1^{1/3}(\beta_1\gamma_1)^{2/3}}e_1, x_2 = \frac{1}{(\alpha_1^2\beta_1\beta_5\gamma_1)^{1/3}}e_2, x_3 = \frac{1}{\alpha_1\beta_1\gamma_1}(\alpha_1 e_3 + \alpha_2 e_4), x_4 = \frac{\beta_5^{1/3}}{\alpha_1^{1/3}(\beta_1\gamma_1)^{5/3}}(\beta_1 e_4 + \beta_2 e_5) + \frac{\alpha_2\beta_5^{1/3}}{\alpha_1^{4/3}\beta_1^{5/3}\gamma_1^{2/3}}e_5, x_5 = \frac{\beta_5^{2/3}}{\alpha_1^{2/3}(\beta_1\gamma_1)^{4/3}}e_5$ shows that A is isomorphic to the algebra $\mathcal{A}_2$.

Case 2: Let $\beta_4 \neq 0$. Then with the base change $x_1 = \beta_4 e_1 - \beta_1 e_2, x_2 = e_2, x_3 = e_3, x_4 = e_4, x_5 = \beta_4^2 e_5$ we can make $\beta_1 = 0$. Then by (2.1) we have $\beta_7 = 0 = \gamma_1 = \gamma_2, \beta_3 = -\beta_2, \beta_6 = -\beta_5$ and $\gamma_4 = -\gamma_3$.
If $\beta_2 = 0$ then the base change $x_1 = \alpha_1\beta_4\gamma_3 e_1, x_2 = e_2, x_3 = \alpha_1\beta_4\gamma_3(\alpha_1 e_3 + \alpha_2 e_4), x_4 = \alpha_1^2\beta_4\gamma_3(\beta_4 e_4 + \beta_5 e_5) + \alpha_1\alpha_2\beta_4\gamma_3^2 e_5, x_5 = (\alpha_1\beta_4\gamma_3)^2 e_5$ shows that A is isomorphic to the algebra $\mathcal{A}_3$. If $\beta_2 \neq 0$ then the base change $x_1 = \frac{\beta_4\gamma_3}{\alpha_1^2\beta_2^3}e_1, x_2 = \frac{1}{\alpha_1\beta_2}e_2, x_3 = \frac{\beta_4\gamma_3}{\alpha_1^3\beta_2^4}(\alpha_1 e_3 + \alpha_2 e_4), x_4 = \frac{\beta_4\gamma_3}{\alpha_1^3\beta_2^5}(\beta_4 e_4 + \beta_5 e_5) + \frac{\alpha_2\beta_4\gamma_3^2}{\alpha_1^4\beta_2^5}e_5, x_5 = (\frac{\beta_4\gamma_3}{\alpha_1^2\beta_2^3})^2 e_5$ shows that A is isomorphic to the algebra $\mathcal{A}_4$.

If N is the matrix (iii) then the nontrivial products in A given by
$[e_1,e_1]=e_5,[e_1,e_2]=\alpha_1e_3+\alpha_2e_4=-[e_2,e_1],[e_2,e_2]=e_5,[e_1,e_3]=\beta_1e_4+\beta_2e_5,$ $[e_3,e_1]=-\beta_1e_4+\beta_3e_5,[e_2,e_3]=\beta_4e_4+\beta_5e_5,[e_3,e_2]=-\beta_4e_4+\beta_6e_5,[e_3,e_3]=$ $\beta_7e_5,[e_1,e_4]=\gamma_1e_5,[e_4,e_1]=\gamma_2e_5,[e_2,e_4]=\gamma_3e_5,[e_4,e_2]=\gamma_4e_5,[e_3,e_4]=\gamma_5e_5,$ $[e_4,e_3]=\gamma_6e_5,[e_4,e_4]=\gamma_7e_5$.
From Leibniz identities we get the equations (2.1). So we have $\gamma_5=0=\gamma_6=\gamma_7$.

Case 1: Let $\beta_4=0$. Then $\beta_1\neq 0$ since $\dim(A^3)=2$. So by (2.1) we obtain $\gamma_2=-\gamma_1$ and $\beta_3=-\beta_2$. If we assume $\gamma_4\neq 0$ then we arrive contradiction using the equations (2.1). Hence $\gamma_4=0$. Then by (2.1) we have $\beta_7=0=\gamma_3$ and $\beta_6=-\beta_5$. Then the nontrivial products in A given by

$$(2.2)\quad [e_1,e_1]=e_5,[e_1,e_2]=\alpha_1e_3+\alpha_2e_4=-[e_2,e_1],[e_2,e_2]=e_5,$$
$$[e_1,e_3]=\beta_1e_4+\beta_2e_5=-[e_3,e_1],[e_2,e_3]=\beta_5e_5=-[e_3,e_2],[e_1,e_4]=\gamma_1e_5=-[e_4,e_1].$$

Then the base change $x_1=\frac{1}{(\alpha_1\beta_1\gamma_1)^{1/2}}e_1,x_2=\frac{1}{(\alpha_1\beta_1\gamma_1)^{1/2}}e_2,x_3=\frac{1}{\alpha_1\beta_1\gamma_1}(\alpha_1e_3+\alpha_2e_4),x_4=\frac{\alpha_1}{(\alpha_1\beta_1\gamma_1)^{3/2}}(\beta_1e_4+\beta_2e_5)+\frac{\alpha_2\gamma_1}{(\alpha_1\beta_1\gamma_1)^{3/2}}e_5,x_5=\frac{1}{\alpha_1\beta_1\gamma_1}e_5$ shows that A is isomorphic to the algebra $\mathcal{A}_5(\alpha)$.

Case 2: Let $\beta_4\neq 0$. Then by (2.1) we have $\beta_7=0$ and $\beta_4\gamma_1=\beta_1\gamma_3$. If $\beta_1^2+\beta_4^2\neq 0$ then the base change $x_1=\beta_1e_1+\beta_4e_2,x_2=\beta_4e_1-\beta_1e_2,x_3=e_3,x_4=e_4,x_5=(\beta_1^2+\beta_4^2)e_5$ shows that A is isomorphic to an algebra with the nonzero products given by (2.2). Hence A is isomorphic to $\mathcal{A}_5(\alpha)$. So let $\beta_1^2+\beta_4^2=0$. Take $\theta=-\frac{i(\beta_4\beta_2-\beta_1\beta_5)}{\beta_4}(\frac{\alpha_1}{2\beta_4\gamma_1})^{1/2}$. If $\beta_4\beta_2-\beta_1\beta_5=0$ then the base change $x_1=i(\frac{2\beta_4}{\alpha_1\beta_1^2\gamma_1})^{1/2}e_1,x_2=i(\frac{1}{2\alpha_1\beta_1^2\beta_4\gamma_1})^{1/2}(\beta_4e_1-\beta_1e_2),x_3=\frac{1}{\alpha_1\beta_1^2\gamma_1}(\alpha_1e_3+\alpha_2e_4-\beta_4e_5),x_4=\frac{i}{\beta_1^2\gamma_1}(\frac{2\beta_4}{\alpha_1\beta_1^2\gamma_1})^{1/2}(\beta_1e_4+\beta_2e_5)+\frac{i\alpha_2}{\alpha_1\beta_1^2}(\frac{2\beta_4}{\alpha_1\beta_1^2\gamma_1})^{1/2}e_5,x_5=-\frac{2\beta_4}{\alpha_1\beta_1^2\gamma_1}e_5$ shows that A is isomorphic to the algebra $\mathcal{A}_6$. If $\beta_4\beta_2-\beta_1\beta_5\neq 0$ then the base change $x_1=\frac{i(1-\theta)}{\theta}(\frac{2\beta_4}{\alpha_1\beta_1^2\gamma_1})^{1/2}e_1-\frac{i\beta_1(1+\theta)}{2\beta_4\theta}(\frac{2\beta_4}{\alpha_1\beta_1^2\gamma_1})^{1/2}e_2,x_2=\frac{i}{2\theta}(\frac{2\beta_4}{\alpha_1\beta_1^2\gamma_1})^{1/2}e_1-\frac{i\beta_1}{2\beta_4\theta}(\frac{2\beta_4}{\alpha_1\beta_1^2\gamma_1})^{1/2}e_2+\frac{1+\theta}{\beta_1^2\gamma_1\theta}e_3+\frac{\alpha_2(1+\theta)}{\alpha_1\beta_1^2\gamma_1\theta}e_4,x_3=-\frac{1}{\beta_1^2\gamma_1\theta}e_3-(\frac{\alpha_2}{\alpha_1\beta_1^2\gamma_1\theta}+\frac{i(1+\theta)}{\gamma_1\theta}(\frac{2\beta_4}{\alpha_1\beta_1^2\gamma_1})^{1/2})e_4+(\frac{\beta_4}{\alpha_1\beta_1^2\gamma_1\theta}-\frac{i\beta_2(1+\theta)}{\beta_1\gamma_1\theta}(\frac{2\beta_4}{\alpha_1\beta_1^2\gamma_1})^{1/2}-\frac{2\beta_4(1+\theta)^2}{\alpha_1\beta_1^2\gamma_1\theta})e_5,x_4=\frac{i}{\beta_1\gamma_1\theta}(\frac{2\beta_4}{\alpha_1\beta_1^2\gamma_1})^{1/2}(\beta_1e_4+\beta_2e_5),$ $x_5=\frac{2\beta_4}{\alpha_1\beta_1^2\gamma_1\theta}e_5$ shows that A is isomorphic to the algebra $\mathcal{A}_7$. □

THEOREM 2.3. *Let A be a $5-$dimensional non-split non-Lie nilpotent Leibniz algebra with* $\dim(A^2)=3$, $\dim(A^3)=2$, $\dim(A^4)=0$, *and* $\dim(Leib(A))=1$. *Then A is isomorphic to a Leibniz algebra spanned by $\{x_1,x_2,x_3,x_4,x_5\}$ with the nonzero products given by one of the following:*

$\mathcal{A}_8$: $[x_1,x_1]=x_5,[x_1,x_2]=x_3=-[x_2,x_1],[x_1,x_3]=x_4=-[x_3,x_1],[x_2,x_3]=x_5=-[x_3,x_2]$.

$\mathcal{A}_9$: $[x_1,x_1]=x_5,[x_1,x_2]=x_3=-[x_2,x_1],[x_1,x_3]=x_5=-[x_3,x_1],[x_2,x_3]=x_4=-[x_3,x_2]$.

$\mathcal{A}_{10}$: $[x_1,x_1]=x_5,[x_1,x_2]=x_3=-[x_2,x_1],[x_2,x_2]=x_5,[x_1,x_3]=x_4=-[x_3,x_1],$ $[x_2,x_3]=x_5=-[x_3,x_2]$.

$\mathcal{A}_{11}$: $[x_1,x_1]=x_5,[x_1,x_2]=x_3,[x_2,x_1]=-x_3+x_5,[x_1,x_3]=x_4=-[x_3,x_1],[x_2,x_3]=x_5=-[x_3,x_2]$.

PROOF. We have $A^3\subseteq Z(A)\subset A^2$. Then $A^3=Z(A)$. Let $Leib(A)=\mathrm{span}\{e_5\}$. Extend this to bases $\{e_4,e_5\}$ and $\{e_3,e_4,e_5\}$ of $A^3=Z(A)$ and A^2, respectively. Take $V=\mathrm{span}\{e_1,e_2\}$.

If N is the matrix (ii) then the nontrivial products in A given by
$[e_1,e_1]=e_5,[e_1,e_2]=\alpha_1e_3+\alpha_2e_4=-[e_2,e_1],[e_1,e_3]=\alpha_3e_4+\alpha_4e_5,[e_3,e_1]=-\alpha_3e_4+\alpha_5e_5,[e_2,e_3]=\beta_1e_4+\beta_2e_5,[e_3,e_2]=-\beta_1e_4+\beta_3e_5,[e_3,e_3]=\beta_4e_5$.
Then from Leibniz identities we get the equations $\beta_3=-\beta_2,\alpha_5=-\alpha_4$ and $\beta_4=0$. Also $\alpha_3\beta_2-\alpha_4\beta_1\neq 0$ since $\dim(A^3)=2$.

Suppose $\beta_1 = 0$. Then $\alpha_3, \beta_2 \neq 0$. Then the base change $x_1 = \alpha_1\beta_2 e_1, x_2 = e_2, x_3 = \alpha_1\beta_2(\alpha_1 e_3 + \alpha_2 e_4), x_4 = \alpha_1^3\beta_2^2(\alpha_3 e_4 + \alpha_4 e_5), x_5 = (\alpha_1\beta_2)^2 e_5$ shows that A is isomorphic to the algebra $\mathcal{A}_8$. Now suppose $\beta_1 \neq 0$. Then with the base change $x_1 = \beta_1 e_1 - \alpha_3 e_2, x_2 = e_2, x_3 = e_3, x_4 = e_4, x_5 = \beta_1^2 e_5$ we can make $\alpha_3 = 0$. So let $\alpha_3 = 0$. Then the base change $x_1 = e_1, x_2 = \frac{1}{\alpha_1\alpha_4} e_2, x_3 = \frac{1}{\alpha_1\alpha_4}(\alpha_1 e_3 + \alpha_2 e_4), x_4 = \frac{1}{\alpha_1\alpha_4^2}(\beta_1 e_4 + \beta_2 e_5), x_5 = e_5$ shows that A is isomorphic to the algebra $\mathcal{A}_9$.

If N is the matrix (iii) then the nontrivial products in A given by
$[e_1, e_1] = e_5, [e_1, e_2] = \alpha_1 e_3 + \alpha_2 e_4 = -[e_2, e_1], [e_2, e_2] = e_5, [e_1, e_3] = \alpha_3 e_4 + \alpha_4 e_5,$ $[e_3, e_1] = -\alpha_3 e_4 + \alpha_5 e_5, [e_2, e_3] = \beta_1 e_4 + \beta_2 e_5, [e_3, e_2] = -\beta_1 e_4 + \beta_3 e_5, [e_3, e_3] = \beta_4 e_5$. Then from Leibniz identities we get the equations $\beta_3 = -\beta_2, \alpha_5 = -\alpha_4$ and $\beta_4 = 0$. Also $\alpha_3\beta_2 - \alpha_4\beta_1 \neq 0$ since $\dim(A^3) = 2$.
Suppose $\beta_1 = 0$. Then the nontrivial products in A are the following:

$$(2.3) \quad \begin{aligned} &[e_1, e_1] = e_5, [e_1, e_2] = \alpha_1 e_3 + \alpha_2 e_4 = -[e_2, e_1], [e_2, e_2] = e_5, \\ &[e_1, e_3] = \alpha_3 e_4 + \alpha_4 e_5 = -[e_3, e_1], [e_2, e_3] = \beta_2 e_5 = -[e_3, e_2]. \end{aligned}$$

Then $\alpha_3, \beta_2 \neq 0$. Then the base change $x_1 = \frac{1}{\alpha_1\beta_2} e_1, x_2 = \frac{1}{\alpha_1\beta_2} e_2, x_3 = \frac{1}{\alpha_1^2\beta_2^2}(\alpha_1 e_3 + \alpha_2 e_4), x_4 = \frac{1}{\alpha_1^2\beta_2^3}(\alpha_3 e_4 + \alpha_4 e_5), x_5 = \frac{1}{\alpha_1^2\beta_2^2} e_5$ shows that A is isomorphic to the algebra $\mathcal{A}_{10}$. Now suppose $\beta_1 \neq 0$. If $\alpha_3^2 + \beta_1^2 \neq 0$ then the base change $x_1 = \alpha_3 e_1 + \beta_1 e_2, x_2 = \beta_1 e_1 - \alpha_3 e_2, x_3 = e_3, x_4 = e_4, x_5 = (\alpha_3^2 + \beta_1^2) e_5$ shows that A is isomorphic to an algebra with the nonzero products given by (2.3). Hence A is isomorphic to $\mathcal{A}_{10}$. So let $\alpha_3^2 + \beta_1^2 = 0$. Then the base change $x_1 = -\frac{4\beta_1^2}{\alpha_1\alpha_3(\alpha_4\beta_1 - \alpha_3\beta_2)} e_1, x_2 = -\frac{\beta_1(\beta_1 e_1 - \alpha_3 e_2)}{\alpha_1\alpha_3(\alpha_4\beta_1 - \alpha_3\beta_2)}, x_3 = \frac{8\beta_1^3}{(\alpha_1\alpha_3)^2(\alpha_4\beta_1 - \alpha_3\beta_2)^2}[-\alpha_1\alpha_3 e_3 - \alpha_2\alpha_3 e_4 + \beta_1 e_5], x_4 = \frac{32\beta_1^5}{(\alpha_1\alpha_3)^2(\alpha_4\beta_1 - \alpha_3\beta_2)^3}(\alpha_3 e_4 + \alpha_4 e_5),\ x_5 = \frac{16\beta_1^4}{(\alpha_1\alpha_3)^2(\alpha_4\beta_1 - \alpha_3\beta_2)^2} e_5$ shows that A is isomorphic to the algebra $\mathcal{A}_{11}$. □

Now let $\dim(A^3) = 1$. Then we have $A^3 \subseteq Z(A) \subset A^2$. Assume $\dim(Z(A)) = 1$. Then we get $A^3 = Z(A) = Leib(A)$, and from part **(ii)** in Lemma 1.2 we arrive $5 \leq 4$, which is a contradiction. Hence $\dim(Z(A)) = 2$.

THEOREM 2.4. *Let A be a 5−dimensional non-split non-Lie nilpotent Leibniz algebra with* $\dim(A^2) = 3$, $\dim(A^3) = 1$, *and* $\dim(Leib(A)) = 1$. *Then A is isomorphic to a Leibniz algebra spanned by $\{x_1, x_2, x_3, x_4, x_5\}$ with the nonzero products given by one of the following:*

$\mathcal{A}_{12}$: $[x_1, x_1] = x_5, [x_1, x_2] = x_3 = -[x_2, x_1], [x_1, x_3] = x_4 = -[x_3, x_1]$.
$\mathcal{A}_{13}$: $[x_1, x_2] = x_3 = -[x_2, x_1], [x_2, x_2] = x_5, [x_1, x_3] = x_4 = -[x_3, x_1]$.
$\mathcal{A}_{14}$: $[x_1, x_1] = x_5, [x_1, x_2] = x_3 = -[x_2, x_1], [x_2, x_2] = x_5, [x_1, x_3] = x_4 = -[x_3, x_1]$.
$\mathcal{A}_{15}$: $[x_1, x_1] = x_5, [x_1, x_2] = x_3, [x_2, x_1] = -x_3 + x_5, [x_1, x_3] = x_4 = -[x_3, x_1]$.

PROOF. Using Lemma 1.2 we see that $A^3 \neq Leib(A)$. Let $Leib(A) = \text{span}\{e_5\}$ and $A^3 = \text{span}\{e_4\}$. Then $Z(A) = \text{span}\{e_4, e_5\}$. Extend this to a basis $\{e_3, e_4, e_5\}$ of A^2. Take $V = \text{span}\{e_1, e_2\}$.

If N is the matrix (ii) then the nontrivial products in A given by
$[e_1, e_1] = e_5, [e_1, e_2] = \alpha_1 e_3 + \alpha_2 e_4 = -[e_2, e_1], [e_1, e_3] = \beta_1 e_4 = -[e_3, e_1], [e_2, e_3] = \beta_2 e_4 = -[e_3, e_2]$.
Suppose $\beta_2 = 0$. Then $\beta_1 \neq 0$ since $A^3 \neq 0$. Then the base change $x_1 = e_1, x_2 = e_2, x_3 = \alpha_1 e_3 + \alpha_2 e_4, x_4 = \alpha_1\beta_1 e_4, x_5 = e_5$ shows that A is isomorphic to the algebra $\mathcal{A}_{12}$. Now suppose $\beta_2 \neq 0$. Then with the base change $x_1 = \beta_2 e_1 - \beta_1 e_2, x_2 = e_2, x_3 = e_3, x_4 = e_4, x_5 = \beta_2^2 e_5$ we can make $\beta_1 = 0$. Then the base change $x_1 = e_2, x_2 = e_1, x_3 = -\alpha_1 e_3 - \alpha_2 e_4, x_4 = -\alpha_1\beta_2 e_4, x_5 = e_5$ shows that A is isomorphic to the algebra $\mathcal{A}_{13}$.

If N is the matrix (iii) then the nontrivial products in A given by $[e_1,e_1] = e_5, [e_1,e_2] = \alpha_1 e_3 + \alpha_2 e_4 = -[e_2,e_1], [e_2,e_2] = e_5, [e_1,e_3] = \beta_1 e_4 = -[e_3,e_1], [e_2,e_3] = \beta_2 e_4 = -[e_3,e_2]$.
Suppose $\beta_2 = 0$. Then $\beta_1 \neq 0$ since $A^3 \neq 0$. Then the base change $x_1 = e_1, x_2 = e_2, x_3 = \alpha_1 e_3 + \alpha_2 e_4, x_4 = \alpha_1\beta_1 e_4, x_5 = e_5$ shows that A is isomorphic to the algebra $\mathcal{A}_{14}$. Now suppose $\beta_2 \neq 0$. If $\beta_1^2+\beta_2^2 \neq 0$ then the base change $x_1 = \beta_1 e_1 + \beta_2 e_2, x_2 = \beta_2 e_1 - \beta_1 e_2, x_3 = -(\beta_1^2+\beta_2^2)(\alpha_1 e_3 + \alpha_2 e_4), x_4 = -\alpha_1(\beta_1^2+\beta_2^2)^2 e_4, x_5 = (\beta_1^2+\beta_2^2)e_5$ shows that A is isomorphic to the algebra $\mathcal{A}_3$ again. Now let $\beta_1^2+\beta_2^2 = 0$. Then the base change $x_1 = 2\beta_2 e_1, x_2 = \beta_2 e_1 - \beta_1 e_2, x_3 = -2\beta_1\beta_2(\alpha_1 e_3 + \alpha_2 e_4) + 2\beta_2^2 e_5, x_4 = -2\alpha_2\beta_1^2\beta_2^2 e_4, x_5 = 4\beta_2^2 e_5$ shows that A is isomorphic to the algebra $\mathcal{A}_{15}$. □

3. Classification of 5−dimensional complex nilpotent Leibniz algebras

Let A be a 5−dimensional complex non-split non-Lie nilpotent Leibniz algebra. Then $\dim(A^2) = 1, 2, 3$ or 4. The case $\dim(A^2) = 4$ can be done using Lemma 1 in [**3**].

THEOREM 3.1. *Let A be a 5−dimensional non-split non-Lie nilpotent Leibniz algebra with $\dim(A^2) = 4$. Then A is isomorphic to a Leibniz algebra spanned by $\{x_1, x_2, x_3, x_4, x_5\}$ with the nonzero products given by the following:*

$\mathcal{A}_{16}$: $[x_1,x_1] = x_2, [x_1,x_2] = x_3, [x_1,x_3] = x_4, [x_1,x_4] = x_5$.

The classification for the case $\dim(A^2) = 1$ is given in [**8**].

THEOREM 3.2. *Let A be a 5−dimensional non-split non-Lie nilpotent Leibniz algebra with $\dim(A^2) = 1$. Then A is isomorphic to a Leibniz algebra spanned by $\{x_1, x_2, x_3, x_4, x_5\}$ with the nonzero products given by one of the following:*

$\mathcal{A}_{17}(\alpha)$: $[x_1,x_4] = x_5, [x_2,x_3] = x_5, [x_2,x_4] = \alpha x_5, [x_3,x_2] = \alpha x_5, [x_4,x_1] = \alpha x_5, [x_4,x_2] = x_5, \quad \alpha \in \mathbb{C}\backslash\{-1,1\}$.
$\mathcal{A}_{18}$: $[x_1,x_4] = x_5, [x_2,x_3] = x_5, [x_2,x_4] = x_5 = -[x_4,x_2], [x_3,x_2] = x_5, [x_4,x_1] = x_5$.
$\mathcal{A}_{19}$: $[x_1,x_4] = x_5 = -[x_4,x_1], [x_2,x_3] = x_5 = -[x_3,x_2], [x_2,x_4] = x_5, [x_3,x_3] = x_5, [x_4,x_2] = x_5$.
$\mathcal{A}_{20}$: $[x_1,x_3] = x_5, [x_3,x_2] = x_5, [x_4,x_4] = x_5$.
$\mathcal{A}_{21}$: $[x_1,x_3] = x_5, [x_2,x_2] = x_5, [x_2,x_3] = x_5 = -[x_3,x_2], [x_3,x_1] = x_5, [x_4,x_4] = x_5$.
$\mathcal{A}_{22}$: $[x_1,x_2] = x_5 = -[x_2,x_1], [x_3,x_4] = x_5 = -[x_4,x_3], [x_4,x_4] = x_5$.
$\mathcal{A}_{23}(\alpha)$: $[x_1,x_2] = x_5 = -[x_2,x_1], [x_3,x_4] = x_5, [x_4,x_3] = \alpha x_5, \quad \alpha \in \mathbb{C}\backslash\{-1,1\}$.
$\mathcal{A}_{24}$: $[x_1,x_2] = x_5 = -[x_2,x_1], [x_2,x_2] = x_5, [x_3,x_4] = x_5 = -[x_4,x_3], [x_4,x_4] = x_5$
$\mathcal{A}_{25}(\alpha)$: $[x_1,x_2] = x_5 = -[x_2,x_1], [x_2,x_2] = x_5, [x_3,x_4] = x_5, [x_4,x_3] = \alpha x_5, \quad \alpha \in \mathbb{C}\backslash\{-1,1\}$.
$\mathcal{A}_{26}(\alpha,\beta)$: $[x_1,x_2] = x_5, [x_2,x_1] = \alpha x_5, [x_3,x_4] = x_5, [x_4,x_3] = \beta x_5, \quad \alpha,\beta \in \mathbb{C}\backslash\{-1,1\}$.
$\mathcal{A}_{27}$: $[x_1,x_2] = x_5 = -[x_2,x_1], [x_3,x_3] = x_5, [x_4,x_4] = x_5$.
$\mathcal{A}_{28}$: $[x_1,x_2] = x_5 = -[x_2,x_1], [x_2,x_2] = x_5, [x_3,x_3] = x_5, [x_4,x_4] = x_5$.
$\mathcal{A}_{29}(\alpha)$: $[x_1,x_2] = x_5, [x_2,x_1] = \alpha x_5, [x_3,x_3] = x_5, [x_4,x_4] = x_5 \quad \alpha \in \mathbb{C}\backslash\{-1,1\}$.
$\mathcal{A}_{30}$: $[x_1,x_1] = x_5, [x_2,x_2] = x_5, [x_3,x_3] = x_5, [x_4,x_4] = x_5$.

In Section 2, we give the classification of 5−dimensional non-Lie nilpotent Leibniz algebra with $\dim(A^2) = 3$ and $\dim(Leib(A)) = 1$. It is left to consider the case $\dim(A^2) = 2$ and $\dim(Leib(A)) = 1$ in order to finish the case $\dim(Leib(A)) = 1$. Notice that if $\dim(A^2) = 2$ then $\dim(A^3) = 1$ or 0.

THEOREM 3.3. *Let A be a 5−dimensional non-split non-Lie nilpotent Leibniz algebra with $\dim(A^2) = 2$ and $\dim(A^3) = 1 = \dim(Leib(A))$. Then A is isomorphic to a Leibniz algebra spanned by $\{x_1, x_2, x_3, x_4, x_5\}$ with the nonzero products given by one of the following:*

$\mathcal{A}_{31}$: $[x_1,x_2] = x_4 = -[x_2,x_1], [x_3,x_3] = x_5, [x_1,x_4] = x_5 = -[x_4,x_1]$.
$\mathcal{A}_{32}$: $[x_1,x_2] = x_4, [x_2,x_1] = -x_4 + x_5, [x_3,x_3] = x_5, [x_1,x_4] = x_5 = -[x_4,x_1]$.
$\mathcal{A}_{33}$: $[x_1,x_2] = x_4 = -[x_2,x_1], [x_2,x_2] = x_5, [x_3,x_3] = x_5, [x_1,x_4] = x_5 = -[x_4,x_1]$.
$\mathcal{A}_{34}$: $[x_1,x_2] = x_4, [x_2,x_1] = -x_4 + x_5, [x_2,x_2] = x_5, [x_3,x_3] = x_5, [x_1,x_4] = x_5 = -[x_4,x_1]$.
$\mathcal{A}_{35}$: $[x_1,x_1] = x_5, [x_1,x_2] = x_4 = -[x_2,x_1], [x_3,x_3] = x_5, [x_1,x_4] = x_5 = -[x_4,x_1]$.
$\mathcal{A}_{36}$: $[x_1,x_2] = x_4 = -[x_2,x_1], [x_1,x_3] = x_5, [x_1,x_4] = x_5 = -[x_4,x_1]$.

$\mathcal{A}_{37}$: $[x_1,x_2]=x_4=-[x_2,x_1],[x_2,x_2]=x_5,[x_1,x_3]=x_5,[x_1,x_4]=x_5=-[x_4,x_1]$.
$\mathcal{A}_{38}$: $[x_1,x_2]=x_4=-[x_2,x_1],[x_2,x_3]=x_5,[x_1,x_4]=x_5=-[x_4,x_1]$.
$\mathcal{A}_{39}$: $[x_1,x_1]=x_5,[x_1,x_2]=x_4=-[x_2,x_1],[x_2,x_3]=x_5,[x_1,x_4]=x_5=-[x_4,x_1]$.
$\mathcal{A}_{40}(\alpha)$: $[x_1,x_2]=x_4=-[x_2,x_1],[x_2,x_2]=\alpha x_5,[x_3,x_3]=x_5,[x_1,x_4]=x_5=-[x_4,x_1],\ \alpha\in\mathbb{C}$.
$\mathcal{A}_{41}(\alpha)$: $[x_1,x_2]=x_4,[x_2,x_1]=-x_4+x_5,[x_2,x_2]=\alpha x_5,[x_2,x_3]=x_5,[x_3,x_3]=x_5,$ $[x_1,x_4]=x_5=-[x_4,x_1],\quad \alpha\in\mathbb{C}$.
$\mathcal{A}_{42}$: $[x_1,x_1]=x_5,[x_1,x_2]=x_4=-[x_2,x_1],[x_2,x_2]=\frac{1}{4}x_5,[x_2,x_3]=x_5,[x_3,x_3]=x_5,$ $[x_1,x_4]=x_5=-[x_4,x_1]$.
$\mathcal{A}_{43}$: $[x_1,x_2]=x_4,[x_2,x_1]=-x_4+x_5,[x_3,x_2]=x_5=-[x_2,x_3],[x_1,x_4]=x_5=-[x_4,x_1]$.
$\mathcal{A}_{44}(\alpha)$: $[x_1,x_2]=x_4=-[x_2,x_1],[x_2,x_3]=\alpha x_5,[x_3,x_2]=x_5,[x_1,x_4]=x_5=-[x_4,x_1],\ \alpha\in\mathbb{C}\backslash\{-1\}$.
$\mathcal{A}_{45}(\alpha)$: $[x_1,x_2]=x_4,[x_2,x_1]=-x_4+x_5,[x_1,x_3]=x_5,[x_2,x_3]=\alpha x_5,[x_3,x_2]=x_5,$ $[x_1,x_4]=x_5=-[x_4,x_1],\quad \alpha\in\mathbb{C}$.
$\mathcal{A}_{46}$: $[x_1,x_1]=x_5,[x_1,x_2]=x_4=-[x_2,x_1],[x_3,x_2]=x_5=-[x_2,x_3],[x_1,x_4]=x_5=-[x_4,x_1]$.
$\mathcal{A}_{47}$: $[x_1,x_2]=x_4=-[x_2,x_1],[x_2,x_2]=x_5,[x_3,x_2]=x_5=-[x_2,x_3],[x_1,x_4]=x_5=-[x_4,x_1]$.
$\mathcal{A}_{48}$: $[x_1,x_2]=x_4,[x_2,x_1]=-x_4+x_5,[x_2,x_2]=x_5,[x_3,x_2]=x_5=-[x_2,x_3],[x_1,x_4]=x_5=-[x_4,x_1]$.
$\mathcal{A}_{49}$: $[x_1,x_2]=x_4=-[x_2,x_1],[x_2,x_2]=x_5,[x_1,x_3]=x_5,[x_3,x_2]=x_5=-[x_2,x_3],$ $[x_1,x_4]=x_5=-[x_4,x_1]$.

PROOF. Note that we have $A^3=Z(A)=Leib(A)$. Let $A^3=Z(A)=Leib(A)=\text{span}\{e_5\}$. Extend this to a basis $\{e_4,e_5\}$ of A^2. Then the nonzero products in $A=\text{span}\{e_1,e_2,e_3,e_4,e_5\}$ are given by
$[e_1,e_1]=\alpha_1e_5,[e_1,e_2]=\alpha_2e_4+\alpha_3e_5,[e_2,e_1]=-\alpha_2e_4+\alpha_4e_5,[e_2,e_2]=\alpha_5e_5,[e_1,e_3]=\beta_1e_4+\beta_2e_5,[e_3,e_1]=-\beta_1e_4+\beta_3e_5,[e_2,e_3]=\beta_4e_4+\beta_5e_5,[e_3,e_2]=-\beta_4e_4+\beta_6e_5,[e_3,e_3]=\beta_7e_5,[e_1,e_4]=\gamma_1e_5,[e_4,e_1]=\gamma_2e_5,[e_2,e_4]=\gamma_3e_5,[e_4,e_2]=\gamma_4e_5,[e_3,e_4]=\gamma_5e_5,[e_4,e_3]=\gamma_6e_5,[e_4,e_4]=\gamma_7e_7$.
Leibniz identities give the following equations:

$$\tag{3.1}\begin{cases}\alpha_2\gamma_2=-\alpha_2\gamma_1 & \alpha_2\gamma_4=-\alpha_2\gamma_3\\ \beta_4\gamma_1=\alpha_2\gamma_6+\beta_1\gamma_3 & \alpha_2\gamma_7=0\\ \beta_1\gamma_2=-\beta_1\gamma_1 & \beta_4\gamma_1+\beta_1\gamma_4+\alpha_2\gamma_5=0\\ \beta_1\gamma_6=-\beta_1\gamma_5 & \beta_1\gamma_3+\beta_4\gamma_2-\alpha_2\gamma_5=0\\ \beta_4\gamma_4=-\beta_4\gamma_3 & \beta_4\gamma_6=-\beta_4\gamma_5\\ \beta_1\gamma_7=0 & \beta_4\gamma_7=0\end{cases}$$

Note that $(\alpha_2,\beta_1,\beta_4)\neq(0,0,0)$ since $\dim(A^2)=2$. Then by (3.1) we have $\gamma_7=0$. Note that if $\beta_4\neq0$ and $\beta_1=0$(resp. $\beta_1\neq0$) then with the base change $x_1=e_2,x_2=e_1,x_3=e_3,x_4=e_4,x_5=e_5$(resp. $x_1=e_1,x_2=\beta_4e_1-\beta_1e_2,x_3=e_3,x_4=e_4,x_5=e_5$) we can make $\beta_4=0$. So let $\beta_4=0$. If $\beta_1\neq0$ and $\alpha_2=0$(resp. $\alpha_2\neq0$) then with the base change $x_1=e_1,x_2=e_3,x_3=e_2,x_4=e_4,x_5=e_5$(resp. $x_1=e_1,x_2=e_2,x_3=\beta_1e_2-\alpha_2e_3,x_4=e_4,x_5=e_5$) we can make $\beta_1=0$. So we can assume $\beta_1=0$. Then by (3.1) we have $\gamma_2=-\gamma_1,\gamma_4=-\gamma_3$ and $\gamma_6=0=\gamma_5$. If $\gamma_3\neq0$ and $\gamma_1=0$(resp. $\gamma_1\neq0$) then with the base change $x_1=e_2,x_2=e_1,x_3=e_3,x_4=e_4,x_5=e_5$(resp. $x_1=e_1,x_2=\gamma_3e_1-\gamma_1e_2,x_3=e_3,x_4=e_4,x_5=e_5$) we can make $\gamma_3=0$. So we can assume $\gamma_3=0$. Then $\gamma_1\neq0$ since $\dim(Z(A))=1$.

Case 1: Let $\beta_6=0$. Then we have the following products in A:

$$\tag{3.2}\begin{aligned}&[e_1,e_1]=\alpha_1e_5,[e_1,e_2]=\alpha_2e_4+\alpha_3e_5,[e_2,e_1]=-\alpha_2e_4+\alpha_4e_5,[e_2,e_2]=\alpha_5e_5,\\ &[e_1,e_3]=\beta_2e_5,[e_3,e_1]=\beta_3e_5,[e_2,e_3]=\beta_5e_5,[e_3,e_3]=\beta_7e_5,\\ &[e_1,e_4]=\gamma_1e_5=-[e_4,e_1].\end{aligned}$$

Without loss of generality we can assume $\beta_3 = 0$, because if $\beta_3 \neq 0$ then with the base change $x_1 = e_1, x_2 = e_2, x_3 = \gamma_1 e_3 + \beta_3 e_4, x_4 = e_4, x_5 = e_5$ we can make $\beta_3 = 0$.

Case 1.1: Let $\beta_5 = 0$.

Case 1.1.1: Let $\beta_2 = 0$. Then $\beta_7 \neq 0$ since $\dim(Z(A)) = 1$. Hence we have the following products in A:

(3.3)
$$[e_1, e_1] = \alpha_1 e_5, [e_1, e_2] = \alpha_2 e_4 + \alpha_3 e_5, [e_2, e_1] = -\alpha_2 e_4 + \alpha_4 e_5, [e_2, e_2] = \alpha_5 e_5,$$
$$[e_3, e_3] = \beta_7 e_5, [e_1, e_4] = \gamma_1 e_5 = -[e_4, e_1].$$

Case 1.1.1.1: Let $\alpha_1 = 0$. Then we have the following products in A:

(3.4)
$$[e_1, e_2] = \alpha_2 e_4 + \alpha_3 e_5, [e_2, e_1] = -\alpha_2 e_4 + \alpha_4 e_5, [e_2, e_2] = \alpha_5 e_5,$$
$$[e_3, e_3] = \beta_7 e_5, [e_1, e_4] = \gamma_1 e_5 = -[e_4, e_1].$$

- If $\alpha_5 = 0 = \alpha_3 + \alpha_4$ then the base change $x_1 = e_1, x_2 = \frac{\beta_7}{\alpha_2\gamma_1} e_2, x_3 = e_3, x_4 = \frac{\beta_7}{\alpha_2\gamma_1}(\alpha_2 e_4 + \alpha_3 e_5), x_5 = \beta_7 e_5$ shows that A is isomorphic to $\mathcal{A}_{31}$.
- If $\alpha_5 = 0$ and $\alpha_3 + \alpha_4 \neq 0$ then the base change $x_1 = \frac{\alpha_3+\alpha_4}{\alpha_2\gamma_1} e_1, x_2 = \frac{\alpha_2\beta_7\gamma_1}{(\alpha_3+\alpha_4)^2} e_2, x_3 = e_3, x_4 = \frac{\beta_7}{\alpha_3+\alpha_4}(\alpha_2 e_4 + \alpha_3 e_5), x_5 = \beta_7 e_5$ shows that A is isomorphic to $\mathcal{A}_{32}$.
- If $\alpha_5 \neq 0$ and $\alpha_3 + \alpha_4 = 0$ then the base change $x_1 = e_1, x_2 = \frac{\alpha_2\gamma_1}{\alpha_5} e_2, x_3 = \frac{\alpha_2\gamma_1}{(\alpha_5\beta_7)^{1/2}} e_3, x_4 = \frac{\alpha_2\gamma_1}{\alpha_5}(\alpha_2 e_4 + \alpha_3 e_5), x_5 = \frac{(\alpha_2\gamma_1)^2}{\alpha_5} e_5$ shows that A is isomorphic to $\mathcal{A}_{33}$.
- If $\alpha_5 \neq 0$ and $\alpha_3 + \alpha_4 \neq 0$ then the base change $x_1 = \frac{\alpha_3+\alpha_4}{\alpha_2\gamma_1} e_1, x_2 = \frac{(\alpha_3+\alpha_4)^2}{\alpha_2\alpha_5\gamma_1} e_2, x_3 = \frac{(\alpha_3+\alpha_4)^2}{\alpha_2\gamma_1(\alpha_5\beta_7)^{1/2}} e_3, x_4 = \frac{(\alpha_3+\alpha_4)^3}{(\alpha_2\gamma_1)^2\alpha_5}(\alpha_2 e_4 + \alpha_3 e_5)$, $x_5 = \frac{(\alpha_3+\alpha_4)^4}{(\alpha_2\gamma_1)^2\alpha_5} e_5$ shows that A is isomorphic to $\mathcal{A}_{34}$.

Case 1.1.1.2: Let $\alpha_1 \neq 0$. If $(\alpha_3 + \alpha_4, \alpha_5) \neq (0, 0)$ then the base change $x_1 = xe_1 + e_2, x_2 = e_2, x_3 = e_3, x_4 = e_4, x_5 = e_5$ (where $\alpha_1 x^2 + (\alpha_3 + \alpha_4)x + \alpha_5 = 0$) shows that A is isomorphic to an algebra with the nonzero products given by (3.4). Hence A is isomorphic to $\mathcal{A}_{31}, \mathcal{A}_{32}, \mathcal{A}_{33}$ or $\mathcal{A}_{34}$. So let $\alpha_3 + \alpha_4 = 0 = \alpha_5$. Then the base change $x_1 = e_1, x_2 = \frac{\alpha_1}{\alpha_2\gamma_1} e_2, x_3 = (\frac{\alpha_1}{\beta_7})^{1/2} e_3, x_4 = \frac{\alpha_1}{\alpha_2\gamma_1}(\alpha_2 e_4 + \alpha_3 e_5), x_5 = \alpha_1 e_5$ shows that A is isomorphic to $\mathcal{A}_{35}$.

Case 1.1.2: Let $\beta_2 \neq 0$. If $\beta_7 \neq 0$ then the base change $x_1 = 2\beta_7 e_1 - \beta_2 e_3, x_2 = e_2, x_3 = -\frac{2\gamma_1}{\beta_2} e_3 + e_4, x_4 = e_4, x_5 = e_5$ shows that A is isomorphic to an algebra with the nonzero products given by (3.3). Hence A is isomorphic to $\mathcal{A}_{31}, \mathcal{A}_{32}, \mathcal{A}_{33}, \mathcal{A}_{34}$ or $\mathcal{A}_{35}$. So let $\beta_7 = 0$. Note that if $\alpha_1 \neq 0$ then with the base change $x_1 = \beta_2 e_1 - \alpha_1 e_3, x_2 = e_2, x_3 = e_3, x_4 = e_4, x_5 = e_5$ we can make $\alpha_1 = 0$. So we can assume $\alpha_1 = 0$. Take $\theta = \frac{\alpha_3+\alpha_4}{\alpha_2\gamma_1}$. The base change $y_1 = e_1, y_2 = e_2, y_3 = \frac{\alpha_2\gamma_1}{\beta_2} e_3, y_4 = \alpha_2 e_4 - \alpha_4 e_5, y_5 = \alpha_2\gamma_1 e_5$ shows that A is isomorphic to the following algebra:

$$[y_1, y_2] = y_4 + \theta y_5, [y_2, y_1] = -y_4, [y_2, y_2] = \frac{\alpha_5}{\alpha_2\gamma_1} y_5, [y_1, y_3] = y_5,$$
$$[y_1, y_4] = y_5 = -[y_4, y_1].$$

Without loss of generality we can assume $\theta = 0$ because if $\theta \neq 0$ then with the base change $x_1 = y_1, x_2 = y_2 - \theta y_3, x_3 = y_3, x_4 = y_4, x_5 = y_5$ we can make $\theta = 0$.

- If $\alpha_5 = 0$ then the base change $x_1 = y_1, x_2 = y_2, x_3 = y_3, x_4 = y_4, x_5 = y_5$ shows that A is isomorphic to $\mathcal{A}_{36}$.
- If $\alpha_5 \neq 0$ then the base change $x_1 = y_1, x_2 = \frac{\alpha_2\gamma_1}{\alpha_5} y_2, x_3 = \frac{\alpha_2\gamma_1}{\alpha_5} y_3, x_4 = \frac{\alpha_2\gamma_1}{\alpha_5} y_4, x_5 = \frac{\alpha_2\gamma_1}{\alpha_5} y_5$ shows that A is isomorphic to $\mathcal{A}_{37}$.

Case 1.2: Let $\beta_5 \neq 0$. Without loss of generality we can assume $\beta_2 = 0$. This is because if $\beta_2 \neq 0$ then with the base change $x_1 = \beta_5 e_1 - \beta_2 e_2, x_2 = e_2, x_3 = e_3, x_4 = e_4, x_5 = e_5$ we can make $\beta_2 = 0$.

Case 1.2.1: Let $\beta_7 = 0$. If $\alpha_5 \neq 0$ then with the base change $x_1 = e_1, x_2 = \beta_5 e_2 - \alpha_5 e_3, x_3 = e_3, x_4 = e_4, x_5 = e_5$ we can make $\alpha_5 = 0$. So assume $\alpha_5 = 0$.

- If $\alpha_1 = 0$ then the base change $x_1 = e_1 - \frac{\alpha_3+\alpha_4}{\beta_5} e_3, x_2 = e_2, x_3 = \frac{\alpha_2\gamma_1}{\beta_5} e_3, x_4 = \alpha_2 e_4 + \alpha_3 e_5, x_5 = \alpha_2\gamma_1 e_5$ shows that A is isomorphic to $\mathcal{A}_{38}$.
- If $\alpha_1 \neq 0$ then the base change $x_1 = -e_1 + \frac{\alpha_3+\alpha_4}{\beta_5} e_3, x_2 = \frac{\alpha_1}{\alpha_2\gamma_1} e_2, x_3 = \frac{\alpha_2\gamma_1}{\beta_5} e_3, x_4 = -\frac{\alpha_1}{\alpha_2\gamma_1}(\alpha_2 e_4 + \alpha_3 e_5), x_5 = \alpha_1 e_5$ shows that A is isomorphic to $\mathcal{A}_{39}$.

Case 1.2.2: Let $\beta_7 \neq 0$.

Case 1.2.2.1: Let $\alpha_1 = 0$. Then we have the following products in A:

$$
\begin{aligned}
(3.5)\quad [e_1, e_2] = \alpha_2 e_4 + \alpha_3 e_5, [e_2, e_1] &= -\alpha_2 e_4 + \alpha_4 e_5, [e_2, e_2] = \alpha_5 e_5, \\
[e_2, e_3] = \beta_5 e_5, [e_3, e_3] &= \beta_7 e_5, [e_1, e_4] = \gamma_1 e_5 = -[e_4, e_1].
\end{aligned}
$$

- If $\alpha_3 + \alpha_4 = 0$ then the base change $x_1 = (\frac{\beta_5}{\alpha_2\gamma_1})^{1/2} e_1, x_2 = \frac{\beta_7}{\beta_5} e_2, x_3 = e_3, x_4 = (\frac{\beta_5}{\alpha_2\gamma_1})^{1/2} \frac{\beta_7}{\beta_5}(\alpha_2 e_4 + \alpha_3 e_5), x_5 = \beta_7 e_5$ shows that A is isomorphic to $\mathcal{A}_{40}(\alpha)$.
- If $\alpha_3+\alpha_4 \neq 0$ then the base change $x_1 = \frac{\alpha_3+\alpha_4}{\alpha_2\gamma_1} e_1, x_2 = \frac{(\alpha_3+\alpha_4)^2\beta_7}{\alpha_2\gamma_1\beta_5^2} e_2, x_3 = \frac{(\alpha_3+\alpha_4)^2}{\alpha_2\gamma_1\beta_5} e_3, x_4 = \frac{(\alpha_3+\alpha_4)^3\beta_7}{(\alpha_2\gamma_1\beta_5)^2}(\alpha_2 e_4 + \alpha_3 e_5), x_5 = \frac{(\alpha_3+\alpha_4)^4\beta_7}{(\alpha_2\gamma_1\beta_5)^2} e_5$ shows that A is isomorphic to $\mathcal{A}_{41}(\alpha)$.

Case 1.2.2.2: Let $\alpha_1 \neq 0$. If $(\alpha_3+\alpha_4, 4\alpha_5\beta_7 - \beta_5^2) \neq (0, 0)$ then the base change $x_1 = \frac{x\beta_7}{\gamma_1} e_1 - \frac{2\beta_7}{\beta_5} e_2 + e_3, x_2 = e_2, x_3 = xe_3 + e_4, x_4 = e_4, x_5 = e_5$ (where $\frac{\alpha_1\beta_7^2}{\gamma_1^2} x^2 - \frac{2(\alpha_3+\alpha_4)\beta_7^2}{\beta_5\gamma_1} x + \frac{\beta_7(4\alpha_5\beta_7-\beta_5^2)}{\beta_5^2} = 0$) shows that A is isomorphic to an algebra with the nonzero products given by (3.5). Hence A is isomorphic to $\mathcal{A}_{40}(\alpha)$ or $\mathcal{A}_{41}(\alpha)$. So let $\alpha_3 + \alpha_4 = 0 = 4\alpha_5\beta_7 - \beta_5^2$. Then the base change $x_1 = (\frac{\alpha_1}{\beta_7})^{1/2} \frac{\beta_5}{\alpha_2\gamma_1} e_1, x_2 = \frac{\alpha_1}{\alpha_2\gamma_1} e_2, x_3 = \frac{\alpha_1\beta_5}{\alpha_2\gamma_1\beta_7} e_3, x_4 = (\frac{\alpha_1}{\beta_7})^{1/2} \frac{\beta_5\alpha_1}{(\alpha_2\gamma_1)^2}(\alpha_2 e_4 + \alpha_3 e_5), x_5 = \frac{(\alpha_1\beta_5)^2}{(\alpha_2\gamma_1)^2\beta_7} e_5$ shows that A is isomorphic to $\mathcal{A}_{42}$.

Case 2: Let $\beta_6 \neq 0$. Without loss of generality we can assume $\beta_3 = 0$, because if $\beta_3 \neq 0$ then with the base change $x_1 = \beta_6 e_1 - \beta_3 e_2, x_2 = e_2, x_3 = e_3, x_4 = e_4, x_5 = e_5$ we can make $\beta_3 = 0$. Note that if $\beta_7 \neq 0$ then the base change $x_1 = e_1, x_2 = \beta_7 e_2 - \beta_6 e_3, x_3 = e_3, x_4 = e_4, x_5 = e_5$ shows that A is isomorphic to an algebra with the nonzero products given by (3.2). Hence A is isomorphic to $\mathcal{A}_{31}, \mathcal{A}_{32}, \ldots, \mathcal{A}_{41}(\alpha)$ or $\mathcal{A}_{42}$. So let $\beta_7 = 0$.

Case 2.1: Let $\alpha_5 = 0$. Then we have the following products in A:

(3.6)
$$[e_1, e_1] = \alpha_1 e_5, [e_1, e_2] = \alpha_2 e_4 + \alpha_3 e_5, [e_2, e_1] = -\alpha_2 e_4 + \alpha_4 e_5, [e_1, e_3] = \beta_2 e_5,$$
$$[e_2, e_3] = \beta_5 e_5, [e_3, e_2] = \beta_6 e_5, [e_1, e_4] = \gamma_1 e_5 = -[e_4, e_1].$$

Case 2.1.1: Let $\alpha_1 = 0$. Then we have the following products in A:

(3.7) $$[e_1, e_2] = \alpha_2 e_4 + \alpha_3 e_5, [e_2, e_1] = -\alpha_2 e_4 + \alpha_4 e_5, [e_1, e_3] = \beta_2 e_5,$$
$$[e_2, e_3] = \beta_5 e_5, [e_3, e_2] = \beta_6 e_5, [e_1, e_4] = \gamma_1 e_5 = -[e_4, e_1].$$

- If $\beta_2 = 0$ and $\beta_5 + \beta_6 = 0$ then $\alpha_3 + \alpha_4 \neq 0$ since $Leib(A) \neq 0$. Then the base change $x_1 = \frac{\alpha_3+\alpha_4}{\alpha_2\gamma_1} e_1, x_2 = (\frac{\alpha_2\gamma_1}{\alpha_3+\alpha_4})^2 e_2, x_3 = \frac{(\alpha_3+\alpha_4)^2}{\alpha_2\gamma_1\beta_6} e_3, x_4 = \frac{\alpha_2\gamma_1}{\alpha_3+\alpha_4}(\alpha_2 e_4 + \alpha_3 e_5), x_5 = \alpha_2\gamma_1 e_5$ shows that A is isomorphic to $\mathcal{A}_{43}$.
- If $\beta_2 = 0$ and $\beta_5 + \beta_6 \neq 0$ then the base change $x_1 = e_1 - \frac{\alpha_3+\alpha_4}{\beta_5+\beta_6} e_3, x_2 = e_2, x_3 = \frac{\alpha_2\gamma_1}{\beta_6} e_3, x_4 = \alpha_2 e_4 + (\alpha_3 - \frac{(\alpha_3+\alpha_4)\beta_6}{\beta_5+\beta_6}) e_5, x_5 = \alpha_2\gamma_1 e_5$ shows that A is isomorphic to $\mathcal{A}_{44}(\alpha)$.
- If $\beta_2 \neq 0, \beta_5+\beta_6 = 0$ and $\alpha_3+\alpha_4 = 0$ then the base change $x_1 = \beta_6 e_1, x_2 = \beta_2 e_2 - \alpha_2\beta_6\gamma_1 e_3, x_3 = \alpha_2\beta_6\gamma_1 e_3, x_4 = \beta_2\beta_6(\alpha_2 e_4 + (\alpha_3 - \alpha_2\beta_6\gamma_1) e_5), x_5 = \alpha_2\beta_2\beta_6^2\gamma_1 e_5$ shows that A is isomorphic to $\mathcal{A}_{45}(-1)$.
- If $\beta_2 \neq 0, \beta_5 + \beta_6 = 0$ and $\alpha_3 + \alpha_4 \neq 0$ then the base change $x_1 = \frac{\alpha_3+\alpha_4}{\alpha_2\gamma_1} e_1, x_2 = \frac{(\alpha_3+\alpha_4)\beta_2}{\alpha_2\beta_6\gamma_1} e_2, x_3 = \frac{(\alpha_3+\alpha_4)^2}{\alpha_2\beta_6\gamma_1} e_3, x_4 = \frac{(\alpha_3+\alpha_4)^2\beta_2}{\alpha_2^2\beta_6\gamma_1^2}(\alpha_2 e_4 + \alpha_3 e_5)$, $x_5 = \frac{(\alpha_3+\alpha_4)^3\beta_2}{\alpha_2^2\beta_6\gamma_2^2} e_5$ shows that A is isomorphic to $\mathcal{A}_{45}(-1)$.
- If $\beta_2 \neq 0, \beta_5 + \beta_6 \neq 0$ and $\alpha_3 + \alpha_4 = 0$ then the base change $x_1 = \frac{\beta_6}{\beta_2} e_1 - \frac{\beta_6}{\beta_5+\beta_6} e_2, x_2 = e_2, x_3 = \frac{\alpha_2\beta_6\gamma_1}{\beta_2^2} e_3 - \frac{\alpha_2\beta_6^2}{\beta_2(\beta_5+\beta_6)} e_4, x_4 = \frac{\beta_6}{\beta_2}(\alpha_2 e_4 + \alpha_3 e_5), x_5 = \frac{\alpha_2\beta_6^2\gamma_1}{\beta_2^2} e_5$ shows that A is isomorphic to $\mathcal{A}_{44}(\alpha)$.
- If $\beta_2 \neq 0, \beta_5 + \beta_6 \neq 0$ and $\alpha_3 + \alpha_4 \neq 0$ then the base change $x_1 = \frac{\alpha_3+\alpha_4}{\alpha_2\gamma_1} e_1, x_2 = \frac{(\alpha_3+\alpha_4)\beta_2}{\alpha_2\beta_6\gamma_1} e_2, x_3 = \frac{(\alpha_3+\alpha_4)^2}{\alpha_2\beta_6\gamma_1} e_3, x_4 = \frac{(\alpha_3+\alpha_4)^2\beta_2}{\alpha_2^2\beta_6\gamma_1^2}(\alpha_2 e_4 + \alpha_3 e_5)$, $x_5 = \frac{(\alpha_3+\alpha_4)^3\beta_2}{\alpha_2^2\beta_6\gamma_2^2} e_5$ shows that A is isomorphic to $\mathcal{A}_{45}(\alpha)(\alpha \in \mathbb{C}\backslash\{-1\})$.

Case 2.1.2: Let $\alpha_1 \neq 0$.

- If $\beta_2 \neq 0$ then the base change $x_1 = \beta_2 e_1 - \alpha_1 e_3, x_2 = e_2, x_3 = e_3, x_4 = e_4, x_5 = e_5$ shows that A is isomorphic to an algebra with the nonzero products given by (3.7). Hence A is isomorphic to $\mathcal{A}_{43}, \mathcal{A}_{44}(\alpha)$ or $\mathcal{A}_{45}(\alpha)$.
- If $\beta_2 = 0$ and $\alpha_3 + \alpha_4 \neq 0$ then the base change $x_1 = (\alpha_3 + \alpha_4) e_1 - \alpha_1 e_2, x_2 = e_2, x_3 = (\alpha_3 + \alpha_4)\gamma_1 e_3 - \alpha_1\beta_6 e_4, x_4 = e_4, x_5 = e_5$ shows that A is isomorphic to an algebra with the nonzero products given by (3.7). Hence A is isomorphic to $\mathcal{A}_{43}, \mathcal{A}_{44}(\alpha)$ or $\mathcal{A}_{45}(\alpha)$.
- If $\beta_2 = 0, \alpha_3 + \alpha_4 = 0$ and $\beta_5 + \beta_6 \neq 0$ then the base change $x_1 = e_1 + e_2 - \frac{\alpha_1}{\beta_5+\beta_6} e_3, x_2 = e_2, x_3 = \gamma_1 e_3 + \beta_6 e_4, x_4 = e_4, x_5 = e_5$ shows that A is isomorphic to an algebra with the nonzero products given by (3.7). Hence A is isomorphic to $\mathcal{A}_{43}, \mathcal{A}_{44}(\alpha)$ or $\mathcal{A}_{45}(\alpha)$.
- If $\beta_2 = 0, \alpha_3 + \alpha_4 = 0$ and $\beta_5 + \beta_6 = 0$ then the base change $x_1 = e_1, x_2 = \frac{\alpha_1}{\alpha_2\gamma_1} e_2, x_3 = \frac{\alpha_2\gamma_1}{\beta_6} e_3, x_4 = \frac{\alpha_1}{\alpha_2\gamma_1}(\alpha_2 e_4 + \alpha_3 e_5), x_5 = \alpha_1 e_5$ shows that A is isomorphic to $\mathcal{A}_{46}$.

Case 2.2: Let $\alpha_5 \neq 0$. If $\beta_5+\beta_6 \neq 0$ then the base change $x_1 = e_1, x_2 = (\beta_5+\beta_6)e_2-\alpha_5 e_3, x_3 = e_3, x_4 = e_4, x_5 = e_5$ shows that A is isomorphic to an algebra with the nonzero products given by (3.6). Hence A is isomorphic to $\mathcal{A}_{43}, \mathcal{A}_{44}(\alpha), \mathcal{A}_{45}(\alpha)$ or $\mathcal{A}_{46}$. So let $\beta_5+\beta_6 = 0$. Note that if $\alpha_1 \neq 0$ and $\beta_2 = 0$[resp. $\beta_2 \neq 0$] then with the base change $x_1 = xe_1+e_2, x_2 = e_2, x_3 = e_3+\frac{\beta_6}{x\gamma_1}e_4, x_4 = e_4, x_5 = e_5$ (where $\alpha_1 x^2+(\alpha_3+\alpha_4)x+\alpha_5 = 0$)[resp. $x_1 = \beta_2 e_1-\alpha_1 e_3, x_2 = e_2, x_3 = e_3, x_4 = e_4, x_5 = e_5$] we can make $\alpha_1 = 0$. So assume $\alpha_1 = 0$.

- If $\beta_2 = 0$ and $\alpha_3+\alpha_4 = 0$ then the base change $x_1 = e_1, x_2 = \frac{\alpha_2\gamma_1}{\alpha_5}e_2, x_3 = \frac{\alpha_2\gamma_1}{\beta_6}e_3, x_4 = \frac{\alpha_2\gamma_1}{\alpha_5}(\alpha_2 e_4+\alpha_3 e_5), x_5 = \frac{(\alpha_2\gamma_1)^2}{\alpha_5}e_5$ shows that A is isomorphic to $\mathcal{A}_{47}$.
- If $\beta_2 = 0$ and $\alpha_3+\alpha_4 \neq 0$ then the base change $x_1 = \frac{\alpha_3+\alpha_4}{\alpha_2\gamma_1}e_1, x_2 = \frac{(\alpha_3+\alpha_4)^2}{\alpha_2\alpha_5\gamma_1}e_2, x_3 = \frac{(\alpha_3+\alpha_4)^2}{\alpha_2\beta_6\gamma_1}e_3, x_4 = \frac{(\alpha_3+\alpha_4)^3}{(\alpha_2\gamma_1)^2\alpha_5}(\alpha_2 e_4+\alpha_3 e_5), x_5 = \frac{(\alpha_3+\alpha_4)^4}{(\alpha_2\gamma_1)^2\alpha_5}e_5$ shows that A is isomorphic to $\mathcal{A}_{48}$.
- If $\beta_2 \neq 0$ then the base change $x_1 = \frac{\alpha_5\beta_2}{\alpha_2\beta_6\gamma_1}e_1, x_2 = \frac{\alpha_5\beta_2^2}{\alpha_2\beta_6^2\gamma_1}e_2-\frac{(\alpha_3+\alpha_4)\alpha_5\beta_2}{\alpha_2\beta_6^2\gamma_1}e_3, x_3 = \frac{(\alpha_5\beta_2)^2}{\alpha_2\beta_6^3\gamma_1}e_3, x_4 = \frac{\alpha_5^2\beta_2^3}{\alpha_2\beta_6^3\gamma_1^2}e_4-\frac{\alpha_5^2\alpha_4\beta_2^3}{(\alpha_2\gamma_1)^2\beta_6^3}e_5, x_5 = \frac{\alpha_5^3\beta_2^4}{(\alpha_2\gamma_1)^2\beta_6^4}e_5$ shows that A is isomorphic to $\mathcal{A}_{49}$.

□

THEOREM 3.4. *Let A be a 5−dimensional non-split non-Lie nilpotent Leibniz algebra with $\dim(A^2) = 2, \dim(A^3) = 0$ and $\dim(Leib(A)) = 1$. Then A is isomorphic to a Leibniz algebra spanned by $\{x_1, x_2, x_3, x_4, x_5\}$ with the nonzero products given by one of the following:*

$\mathcal{A}_{50}$: $[x_1,x_2] = x_4, [x_2,x_1] = -x_4+x_5, [x_3,x_3] = x_5$.
$\mathcal{A}_{51}$: $[x_1,x_2] = x_4 = -[x_2,x_1], [x_3,x_1] = x_5$.
$\mathcal{A}_{52}$: $[x_1,x_2] = x_4, [x_2,x_1] = -x_4+x_5, [x_3,x_1] = x_5, [x_3,x_3] = x_5$.
$\mathcal{A}_{53}(\alpha)$: $[x_1,x_2] = x_4 = -[x_2,x_1], [x_1,x_3] = x_5, [x_3,x_1] = \alpha x_5, \quad \alpha \in \mathbb{C}\backslash\{-1\}$.
$\mathcal{A}_{54}$: $[x_1,x_2] = x_4, [x_2,x_1] = -x_4+x_5, [x_1,x_3] = x_5 = -[x_3,x_1]$.
$\mathcal{A}_{55}(\alpha)$: $[x_1,x_2] = x_4 = -[x_2,x_1], [x_1,x_3] = x_5, [x_3,x_1] = \alpha x_5, [x_3,x_3] = x_5, \quad \alpha \in \mathbb{C}$.
$\mathcal{A}_{56}(\alpha)$: $[x_1,x_2] = x_4, [x_2,x_1] = -x_4+\alpha x_5, [x_3,x_1] = x_5, [x_3,x_2] = x_5, [x_3,x_3] = x_5, \quad \alpha \in \mathbb{C}\backslash\{0\}$.
$\mathcal{A}_{57}$: $[x_1,x_2] = x_4, [x_2,x_1] = -x_4+x_5, [x_3,x_1] = x_5, [x_3,x_2] = x_5$.
$\mathcal{A}_{58}$: $[x_1,x_2] = x_4 = -[x_2,x_1], [x_1,x_3] = x_5, [x_3,x_2] = x_5$.
$\mathcal{A}_{59}$: $[x_1,x_2] = x_4, [x_2,x_1] = -x_4+x_5, [x_1,x_3] = x_5, [x_3,x_2] = x_5$.
$\mathcal{A}_{60}$: $[x_1,x_2] = x_4, [x_2,x_1] = -x_4+x_5, [x_1,x_3] = x_5, [x_3,x_1] = x_5, [x_3,x_2] = x_5, [x_3,x_3] = x_5$.
$\mathcal{A}_{61}(\alpha)$: $[x_1,x_2] = x_4, [x_2,x_1] = -x_4+x_5, [x_1,x_3] = x_5, [x_3,x_1] = \alpha x_5, [x_2,x_3] = x_5, [x_3,x_2] = \alpha x_5, \quad \alpha \in \mathbb{C}$.
$\mathcal{A}_{62}$: $[x_1,x_2] = x_4 = -[x_2,x_1], [x_2,x_2] = x_5, [x_2,x_3] = x_5 = -[x_3,x_2]$.
$\mathcal{A}_{63}$: $[x_1,x_2] = x_4 = -[x_2,x_1], [x_2,x_2] = x_5, [x_1,x_3] = x_5 = -[x_3,x_1]$.

PROOF. Let $Leib(A) = \text{span}\{e_5\}$. Extend this to a basis of $\{e_4, e_5\}$ of $A^2 = Z(A)$. Then the nonzero products in $A = \text{span}\{e_1, e_2, e_3, e_4, e_5\}$ are given by $[e_1,e_1] = \alpha_1 e_5, [e_1,e_2] = \alpha_2 e_4+\alpha_3 e_5, [e_2,e_1] = -\alpha_2 e_4+\alpha_4 e_5, [e_2,e_2] = \alpha_5 e_5, [e_1,e_3] = \beta_1 e_4+\beta_2 e_5, [e_3,e_1] = -\beta_1 e_4+\beta_3 e_5, [e_2,e_3] = \beta_4 e_4+\beta_5 e_5, [e_3,e_2] = -\beta_4 e_4+\beta_6 e_5, [e_3,e_3] = \beta_7 e_5$.
Without loss of generality we can assume $\beta_4 = 0$, because if $\beta_4 \neq 0$ and $\beta_1 = 0$(resp. $\beta_1 \neq 0$) then with the base change $x_1 = e_2, x_2 = e_1, x_3 = e_3, x_4 = e_4, x_5 = e_5$(resp. $x_1 = e_1, x_2 = \beta_4 e_1-\beta_1 e_2, x_3 = e_3, x_4 = e_4, x_5 = e_5$) we can make $\beta_4 = 0$. Note that if $\beta_1 \neq 0$ and $\alpha_2 = 0$(resp. $\alpha_2 \neq 0$) then with the base change $x_1 = e_1, x_2 = e_3, x_3 = e_2, x_4 = e_4, x_5 = e_5$(resp. $x_1 = e_1, x_2 = e_2, x_3 = \beta_1 e_2-\alpha_2 e_3, x_4 = e_4, x_5 = e_5$) we can make $\beta_1 = 0$. So let $\beta_1 = 0$. Then $\alpha_2 \neq 0$ since $\dim(A^2) = 2$. If $\alpha_1 \neq 0$ and $(\alpha_3+\alpha_4, \alpha_5) \neq (0,0)$[resp. $\alpha_3+\alpha_4 = 0 = \alpha_5$] then with the base change $x_1 = xe_1+e_2, x_2 = e_2, x_3 = e_3, x_4 = e_4, x_5 = e_5$ (where

$\alpha_1 x^2 + (\alpha_3 + \alpha_4)x + \alpha_5 = 0)$[resp. $x_1 = e_2, x_2 = e_1, x_3 = e_3, x_4 = e_4, x_5 = e_5$] we can make $\alpha_1 = 0$. So we can assume $\alpha_1 = 0$.

Case 1: Let $\alpha_5 = 0$. Then we have the following products in A:

(3.8)
$$[e_1, e_2] = \alpha_2 e_4 + \alpha_3 e_5, [e_2, e_1] = -\alpha_2 e_4 + \alpha_4 e_5, [e_1, e_3] = \beta_2 e_5, [e_3, e_1] = \beta_3 e_5,$$
$$[e_2, e_3] = \beta_5 e_5, [e_3, e_2] = \beta_6 e_5, [e_3, e_3] = \beta_7 e_5.$$

Case 1.1: Let $\beta_5 = 0$. Then we have the following products in A:

(3.9)
$$[e_1, e_2] = \alpha_2 e_4 + \alpha_3 e_5, [e_2, e_1] = -\alpha_2 e_4 + \alpha_4 e_5, [e_1, e_3] = \beta_2 e_5, [e_3, e_1] = \beta_3 e_5,$$
$$[e_3, e_2] = \beta_6 e_5, [e_3, e_3] = \beta_7 e_5.$$

Case 1.1.1: Let $\beta_6 = 0$. Then we have the following products in A:

(3.10)
$$[e_1, e_2] = \alpha_2 e_4 + \alpha_3 e_5, [e_2, e_1] = -\alpha_2 e_4 + \alpha_4 e_5, [e_1, e_3] = \beta_2 e_5, [e_3, e_1] = \beta_3 e_5,$$
$$[e_3, e_3] = \beta_7 e_5.$$

Case 1.1.1.1: Let $\beta_2 = 0$. Then we have the following products in A:

(3.11)
$$[e_1, e_2] = \alpha_2 e_4 + \alpha_3 e_5, [e_2, e_1] = -\alpha_2 e_4 + \alpha_4 e_5, [e_3, e_1] = \beta_3 e_5, [e_3, e_3] = \beta_7 e_5.$$

Case 1.1.1.1.1: Let $\beta_3 = 0$. Then $\beta_7 \neq 0$ since $\dim(Z(A)) = 2$. Note that $\alpha_3 + \alpha_4 \neq 0$ since A is non-split. Then the base change $x_1 = \frac{\beta_7}{\alpha_3+\alpha_4} e_1, x_2 = e_2, x_3 = e_3, x_4 = \frac{\beta_7}{\alpha_3+\alpha_4}(\alpha_2 e_4 + \alpha_3 e_5), x_5 = \beta_7 e_5$ shows that A is isomorphic to $\mathcal{A}_{50}$.

Case 1.1.1.1.2: Let $\beta_3 \neq 0$.

- If $\beta_7 = 0$ then the base change $x_1 = e_1, x_2 = \beta_3 e_2 - (\alpha_3 + \alpha_4)e_3, x_3 = e_3, x_4 = \beta_3(\alpha_2 e_4 + \alpha_3 e_5), x_5 = \beta_3 e_5$ shows that A is isomorphic to $\mathcal{A}_{51}$.
- If $\beta_7 \neq 0$ and $\alpha_3 + \alpha_4 = 0$ then the base change $x_1 = \beta_7 e_1 - \beta_3 e_3, x_2 = e_2, x_3 = -\beta_3 e_3, x_4 = \beta_7(\alpha_2 e_4 + \alpha_3 e_5), x_5 = \beta_3^2 \beta_7 e_5$ shows that A is isomorphic to $\mathcal{A}_{55}(0)$.
- If $\beta_7 \neq 0$ and $\alpha_3 + \alpha_4 \neq 0$ then the base change $x_1 = \beta_7 e_1, x_2 = \frac{\beta_3^2}{\alpha_3+\alpha_4} e_2, x_3 = \beta_3 e_3, x_4 = \frac{\beta_3^2 \beta_7}{\alpha_3+\alpha_4}(\alpha_2 e_4 + \alpha_3 e_5), x_5 = \beta_3^2 \beta_7 e_5$ shows that A is isomorphic to $\mathcal{A}_{52}$.

Case 1.1.1.2: Let $\beta_2 \neq 0$.

- If $\beta_7 = 0$ and $\beta_2 + \beta_3 \neq 0$ then the base change $x_1 = e_1, x_2 = (\beta_2 + \beta_3)e_2 - (\alpha_3 + \alpha_4)e_3, x_3 = e_3, x_4 = \alpha_2(\beta_2 + \beta_3)e_4 + (\alpha_3\beta_3 - \alpha_4\beta_2)e_5, x_5 = \beta_2 e_5$ shows that A is isomorphic to $\mathcal{A}_{53}(\alpha)$.
- If $\beta_7 = 0$ and $\beta_2 + \beta_3 = 0$ then $\alpha_3 + \alpha_4 \neq 0$ since $Leib(A) \neq 0$. Then the base change $x_1 = e_1, x_2 = \beta_2 e_2, x_3 = (\alpha_3 + \alpha_4)e_3, x_4 = \beta_2(\alpha_2 e_4 + \alpha_3 e_5), x_5 = \beta_2(\alpha_3 + \alpha_4)e_5$ shows that A is isomorphic to $\mathcal{A}_{54}$.
- If $\beta_7 \neq 0$ and $\alpha_3 + \alpha_4 = 0$ then the base change $x_1 = \beta_7 e_1, x_2 = e_2, x_3 = \beta_2 e_3, x_4 = \beta_7(\alpha_2 e_4 + \alpha_3 e_5), x_5 = \beta_2^2 \beta_7 e_5$ shows that A is isomorphic to $\mathcal{A}_{55}(\alpha)$.
- If $\beta_7 \neq 0$ and $\alpha_3 + \alpha_4 \neq 0$ then the base change $x_1 = -\beta_7 e_1 - \frac{\beta_2\beta_3}{\alpha_3+\alpha_4} e_2 + \beta_2 e_3, x_2 = e_2, x_3 = e_3, x_4 = e_4, x_5 = e_5$ shows that A is isomorphic to an

algebra with the nonzero products given by (3.11). Hence A is isomorphic to $\mathcal{A}_{50}, \mathcal{A}_{51}, \mathcal{A}_{52}$ or $\mathcal{A}_{55}(0)$.

Case 1.1.2: Let $\beta_6 \neq 0$.
Case 1.1.2.1: Let $\beta_2 = 0$. Then we have the following products in A:

(3.12)
$$[e_1, e_2] = \alpha_2 e_4 + \alpha_3 e_5, [e_2, e_1] = -\alpha_2 e_4 + \alpha_4 e_5, [e_3, e_1] = \beta_3 e_5, [e_3, e_2] = \beta_6 e_5, \\ [e_3, e_3] = \beta_7 e_5.$$

- If $\beta_3 = 0$ then the base change $x_1 = e_2, x_2 = e_1, x_3 = e_3, x_4 = e_4, x_5 = e_5$ shows that A is isomorphic to an algebra with the nonzero products given by (3.10). Hence A is isomorphic to $\mathcal{A}_{50}, \mathcal{A}_{51}, \dots, \mathcal{A}_{54}$ or $\mathcal{A}_{55}(\alpha)$.
- If $\beta_3 \neq 0$ and $\alpha_3 + \alpha_4 = 0$ then the base change $x_1 = e_1, x_2 = \beta_6 e_1 - \beta_3 e_2, x_3 = e_3, x_4 = e_4, x_5 = e_5$ shows that A is isomorphic to an algebra with the nonzero products given by (3.10). Hence A is isomorphic to $\mathcal{A}_{50}, \mathcal{A}_{51}, \dots, \mathcal{A}_{54}$ or $\mathcal{A}_{55}(\alpha)$.
- If $\beta_3 \neq 0, \alpha_3 + \alpha_4 \neq 0$ and $\beta_7 \neq 0$ then the base change $x_1 = \beta_6 e_1, x_2 = \beta_3 e_2, x_3 = \frac{\beta_3 \beta_6}{\beta_7} e_3, x_4 = \beta_3 \beta_6(\alpha_2 e_4 + \alpha_3 e_5), x_5 = \frac{(\beta_3\beta_6)^2}{\beta_7} e_5$ shows that A is isomorphic to $\mathcal{A}_{56}(\alpha)$.
- If $\beta_3 \neq 0, \alpha_3 + \alpha_4 \neq 0$ and $\beta_7 = 0$ then the base change $x_1 = \beta_6 e_1, x_2 = \beta_3 e_2, x_3 = (\alpha_3 + \alpha_4) e_3, x_4 = \beta_3 \beta_6 (\alpha_2 e_4 + \alpha_3 e_5), x_5 = \beta_3 \beta_6 (\alpha_3 + \alpha_4) e_5$ shows that A is isomorphic to $\mathcal{A}_{57}$.

Case 1.1.2.2: Let $\beta_2 \neq 0$.
Case 1.1.2.2.1: Let $\beta_7 = 0$. Note that if $\beta_3 \neq 0$ then with the base change $x_1 = \beta_6 e_1 - \beta_3 e_2 + \frac{\beta_3(\alpha_3+\alpha_4)}{\beta_2} e_3, x_2 = e_2, x_3 = e_3, x_4 = e_4, x_5 = e_5$ we can make $\beta_3 = 0$. So we can assume $\beta_3 = 0$.

- If $\alpha_3 + \alpha_4 = 0$ then the base change $x_1 = \beta_6 e_1, x_2 = \beta_2 e_2, x_3 = e_3, x_4 = \beta_2 \beta_6 (\alpha_2 e_4 + \alpha_3 e_5), x_5 = \beta_2 \beta_6 e_5$ shows that A is isomorphic to $\mathcal{A}_{58}$.
- If $\alpha_3 + \alpha_4 \neq 0$ then the base change $x_1 = \beta_6^2 e_1, x_2 = \beta_2 \beta_6 e_2, x_3 = \beta_6(\alpha_3 + \alpha_4) e_3, x_4 = \beta_2 \beta_6^3 (\alpha_2 e_4 + \alpha_3 e_5), x_5 = (\alpha_3 + \alpha_4) \beta_2 \beta_6^3 e_5$ shows that A is isomorphic to $\mathcal{A}_{59}$.

Case 1.1.2.2.2: Let $\beta_7 \neq 0$. Take $\theta = \beta_7(\alpha_3 + \alpha_4) - \beta_2 \beta_6$.

- If $\theta \neq 0$ then the base change $x_1 = \beta_7 e_1 + \frac{\beta_2 \beta_3 \beta_7}{\theta} e_2 - \beta_2 e_3, x_2 = e_2, x_3 = e_3, x_4 = e_4, x_5 = e_5$ shows that A is isomorphic to an algebra with the nonzero products given by (3.12). Hence A is isomorphic to $\mathcal{A}_{50}, \mathcal{A}_{51}, \dots, \mathcal{A}_{56}(\alpha)$ or $\mathcal{A}_{57}$.
- If $\theta = 0$ and $\beta_3 = 0$ then the base change $x_1 = \beta_7 e_1 - \beta_2 e_3, x_2 = e_2, x_3 = e_3, x_4 = e_4, x_5 = e_5$ shows that A is isomorphic to an algebra with the nonzero products given by (3.12). Hence A is isomorphic to $\mathcal{A}_{50}, \mathcal{A}_{51}, \dots, \mathcal{A}_{56}(\alpha)$ or $\mathcal{A}_{57}$.
- If $\theta = 0$ and $\beta_3 \neq 0$ then the base change $x_1 = \beta_6 e_1 + (\beta_2 - \beta_3) e_2, x_2 = \beta_2 e_2, x_3 = \frac{\beta_2 \beta_6}{\beta_7} e_3, x_4 = \beta_2 \beta_6 (\alpha_2 e_4 + \alpha_3 e_5), x_5 = \frac{(\beta_2 \beta_6)^2}{\beta_7} e_5$ shows that A is isomorphic to $\mathcal{A}_{60}$.

Case 1.2: Let $\beta_5 \neq 0$.

- If $\beta_2 = 0$ then the base change $x_1 = e_2, x_2 = e_1, x_3 = e_3, x_4 = e_4, x_5 = e_5$ shows that A is isomorphic to an algebra with the nonzero products given by (3.9). Hence A is isomorphic to $\mathcal{A}_{50}, \mathcal{A}_{51}, \ldots, \mathcal{A}_{59}$ or $\mathcal{A}_{60}$.
- If $\beta_2 \neq 0$ and $\alpha_3 + \alpha_4 = 0$ then the base change $x_1 = e_1, x_2 = \beta_5 e_1 - \beta_2 e_2, x_3 = e_3, x_4 = e_4, x_5 = e_5$ shows that A is isomorphic to an algebra with the nonzero products given by (3.9). Hence A is isomorphic to $\mathcal{A}_{50}, \mathcal{A}_{51}, \ldots, \mathcal{A}_{59}$ or $\mathcal{A}_{60}$.
- If $\beta_2 \neq 0, \alpha_3 + \alpha_4 \neq 0$ and $(\beta_7, \beta_5\beta_3 - \beta_2\beta_6) \neq (0,0)$ then the base change $x_1 = e_1, x_2 = xe_1 - \frac{x\beta_2+\beta_7}{\beta_5}e_2 + e_3, x_3 = e_3, x_4 = e_4, x_5 = e_5$(where $(\alpha_3 + \alpha_4)\beta_2 x^2 + ((\alpha_3 + \alpha_4)\beta_7 - \beta_3\beta_5 + \beta_2\beta_6)x + \beta_6\beta_7 = 0$) shows that A is isomorphic to an algebra with the nonzero products given by (3.9). Hence A is isomorphic to $\mathcal{A}_{50}, \mathcal{A}_{51}, \ldots, \mathcal{A}_{59}$ or $\mathcal{A}_{60}$.
- If $\beta_2 \neq 0, \alpha_3 + \alpha_4 \neq 0$ and $(\beta_7, \beta_5\beta_3 - \beta_2\beta_6) = (0,0)$ hen the base change $x_1 = \beta_5 e_1, x_2 = \beta_2 e_2, x_3 = (\alpha_3 + \alpha_4)e_3, x_4 = \beta_2\beta_5(\alpha_2 e_4 + \alpha_3 e_5), x_5 = \beta_2\beta_5(\alpha_3 + \alpha_4)e_5$ shows that A is isomorphic to $\mathcal{A}_{61}(\alpha)$.

Case 2: Let $\alpha_5 \neq 0$. Note that if $(\beta_5 + \beta_6, \beta_7) \neq (0,0)$ then the base change $x_1 = e_1, x_2 = xe_2 + e_3, x_3 = e_3, x_4 = e_4, x_5 = e_5$(where $\alpha_5 x^2 + (\beta_5 + \beta_6)x + \beta_7 = 0$) shows that A is isomorphic to an algebra with the nonzero products given by (3.8). Hence A is isomorphic to $\mathcal{A}_{50}, \mathcal{A}_{51}, \ldots, \mathcal{A}_{60}$ or $\mathcal{A}_{61}(\alpha)$. So let $\beta_5 + \beta_6 = 0 = \beta_7$. Furthermore, if $\alpha_3 + \alpha_4 \neq 0$ then the base change $x_1 = e_1, x_2 = \alpha_5 e_1 - (\alpha_3 + \alpha_4)e_2, x_3 = e_3, x_4 = e_4, x_5 = e_5$ shows that A is isomorphic to an algebra with the nonzero products given by (3.8). Hence A is isomorphic to $\mathcal{A}_{50}, \mathcal{A}_{51}, \ldots, \mathcal{A}_{60}$ or $\mathcal{A}_{61}(\alpha)$. So we can assume $\alpha_3 + \alpha_4 = 0$. If $\beta_2 + \beta_3 \neq 0$ then the base change $x_1 = e_1, x_2 = -\frac{\alpha_5}{\beta_2+\beta_3}e_1 + e_2 + e_3, x_3 = e_3, x_4 = e_4, x_5 = e_5$ shows that A is isomorphic to an algebra with the nonzero products given by (3.8). Hence A is isomorphic to $\mathcal{A}_{50}, \mathcal{A}_{51}, \ldots, \mathcal{A}_{60}$ or $\mathcal{A}_{61}(\alpha)$. So let $\beta_2 + \beta_3 = 0$.

Case 2.1: Let $\beta_2 = 0$. Then $\beta_5 \neq 0$ since $\dim(Z(A)) = 2$. Then the base change $x_1 = e_1, x_2 = e_2, x_3 = \frac{\alpha_5}{\beta_5}e_3, x_4 = \alpha_2 e_4 + \alpha_3 e_5, x_5 = \alpha_5 e_5$ shows that A is isomorphic to $\mathcal{A}_{62}$.

Case 2.2: Let $\beta_2 \neq 0$. Without loss of generality we can assume $\beta_5 = 0$ because if $\beta_5 \neq 0$ then with the base change $x_1 = e_1, x_2 = \beta_5 e_1 - \beta_2 e_2, x_3 = e_3, x_4 = e_4, x_5 = e_5$ we can make $\beta_5 = 0$. Then the base change $x_1 = \frac{\alpha_5}{\beta_2}e_1, x_2 = e_2, x_3 = e_3, x_4 = \frac{\alpha_5}{\beta_2}(\alpha_2 e_4 + \alpha_3 e_5), x_5 = \alpha_5 e_5$ shows that A is isomorphic to $\mathcal{A}_{63}$. □

Now we need to consider the case $\dim(Leib(A)) \neq 1$. We only give the proof of the classification of 5−dimensional filiform Leibniz algebras and compare it with the classification given in [**15**]. We skip the proofs of the remaining cases since they are similar. The detailed proofs of the remaining cases can be seen in [**11**].

Let A be a 5−dimensional filiform Leibniz algebra. Then $\dim(A^2) = 3, \dim(A^3) = 2$ and $\dim(A^4) = 1$. Hence using $0 \neq A^4 \subseteq A^3$ we get $A^4 = Z(A)$. Assume $\dim(Leib(A)) = 2$. Take W such that $A^2 = Leib(A) \oplus W$. If $W = Z(A)$ then $A^3 = Leib(A)$ since $Leib(A)$ is an ideal. If $W \neq Z(A)$ and $W \not\subseteq A^3$ then $A^3 = Leib(A)$. Furthermore if $W \neq Z(A)$ and $W \subseteq A^3$ then $[A, W] \subseteq [A, A^3] = A^4 = Z(A) \subseteq Leib(A)$. Then $A^3 = [A, A^2] \subseteq Leib(A)$, so $A^3 = Leib(A)$. In all cases we get $A^3 = Leib(A)$. Hence $A^3 = Leib(A)$. Let $Z(A) = A^4 = \mathrm{span}\{e_5\}$. Extend this to bases of $\{e_4, e_5\}$ and $\{e_3, e_4, e_5\}$ of $Leib(A) = A^3$ and A^2, respectively. Then the nonzero products in $A = \mathrm{span}\{e_1, e_2, e_3, e_4, e_5\}$ are given by

$[e_1, e_1] = \alpha_1 e_4 + \alpha_2 e_5, [e_1, e_2] = \alpha_3 e_3 + \alpha_4 e_4 + \alpha_5 e_5, [e_2, e_1] = -\alpha_3 e_3 + \beta_1 e_4 + \beta_2 e_5,$ $[e_2, e_2] = \beta_3 e_4 + \beta_4 e_5, [e_1, e_3] = \beta_5 e_4 + \beta_6 e_5, [e_3, e_1] = \gamma_1 e_4 + \gamma_2 e_5, [e_2, e_3] = \gamma_3 e_4 + \gamma_4 e_5, [e_3, e_2] = \gamma_5 e_4 + \gamma_6 e_5, [e_3, e_3] = \theta_1 e_4 + \theta_2 e_5, [e_1, e_4] = \theta_3 e_5, [e_2, e_4] = \theta_4 e_5,$ $[e_3, e_4] = \theta_5 e_5.$

Leibniz identities give the following equations:

$$(3.13) \qquad \begin{cases} \gamma_1 = -\beta_5 & \alpha_3(\beta_6 + \gamma_2) + \alpha_1\theta_4 - \beta_1\theta_3 = 0 \\ \gamma_5 = -\gamma_3 & \alpha_3(\gamma_4 + \gamma_6) + \alpha_4\theta_4 - \beta_3\theta_3 = 0 \\ \theta_1 = 0 & \alpha_3\theta_2 + \beta_5\theta_4 - \gamma_3\theta_3 \\ \theta_5 = 0 & \gamma_1\theta_3 = 0 \\ \gamma_5\theta_3 = \alpha_3\theta_2 & \gamma_1\theta_4 = -\alpha_3\theta_2 \\ \gamma_5\theta_4 = 0 & \end{cases}$$

Suppose $\theta_4 = 0$. Then $\theta_3 \neq 0$ since $\dim(Z(A)) = 1$. By (3.13) we have $\theta_2 = 0 = \gamma_5 = \gamma_3 = \gamma_1 = \beta_5$. Then $\dim(A^3) = 1$ which is a contradiction. Now suppose $\theta \neq 0$. If $\theta_3 = 0$ then by (3.13) we have $\theta_2 = 0 = \gamma_1 = \gamma_5 = \gamma_3 = \beta_5$. This implies that $\dim(A^3) = 1$, contradiction. If $\theta_3 \neq 0$ then by (3.13) we have $\gamma_1 = 0 = \gamma_5 = \gamma_3 = \beta_5$. Then again we get $\dim(A^3) = 1$, contradiction. Hence our assumption was wrong. Then $\dim(Leib(A)) = 3$.

THEOREM 3.5. *Let A be a 5−dimensional non-split non-Lie nilpotent Leibniz algebra with $\dim(A^2) = 3 = \dim(Leib(A)), \dim(A^3) = 2$ and $\dim(A^4) = 1$. Then A is isomorphic to a Leibniz algebra spanned by $\{x_1, x_2, x_3, x_4, x_5\}$ with the nonzero products given by one of the following:*

$\mathcal{A}_{64}$: $[x_1, x_2] = x_3, [x_1, x_3] = x_4, [x_1, x_4] = x_5.$
$\mathcal{A}_{65}$: $[x_1, x_2] = x_3, [x_2, x_1] = x_5, [x_1, x_3] = x_4, [x_1, x_4] = x_5.$
$\mathcal{A}_{66}$: $[x_1, x_2] = x_3, [x_2, x_2] = x_5, [x_1, x_3] = x_4, [x_1, x_4] = x_5.$
$\mathcal{A}_{67}$: $[x_1, x_2] = x_3, [x_2, x_1] = x_5, [x_2, x_2] = x_5, [x_1, x_3] = x_4, [x_1, x_4] = x_5.$
$\mathcal{A}_{68}$: $[x_1, x_2] = x_3, [x_2, x_2] = x_4, [x_1, x_3] = x_4, [x_2, x_3] = x_5, [x_1, x_4] = x_5.$
$\mathcal{A}_{69}$: $[x_1, x_2] = x_3, [x_2, x_1] = x_5, [x_2, x_2] = x_4, [x_1, x_3] = x_4, [x_2, x_3] = x_5, [x_1, x_4] = x_5.$
$\mathcal{A}_{70}(\alpha)$: $[x_1, x_2] = x_3, [x_2, x_1] = \alpha x_5, [x_2, x_2] = x_4 + x_5, [x_1, x_3] = x_4, [x_2, x_3] = x_5, [x_1, x_4] = x_5, \quad \alpha \in \mathbb{C}.$
$\mathcal{A}_{71}$: $[x_1, x_1] = x_3, [x_2, x_1] = x_5, [x_1, x_3] = x_4, [x_1, x_4] = x_5.$
$\mathcal{A}_{72}$: $[x_1, x_1] = x_3, [x_2, x_2] = x_5, [x_1, x_3] = x_4, [x_1, x_4] = x_5.$
$\mathcal{A}_{73}(\alpha)$: $[x_1, x_1] = x_3, [x_2, x_1] = x_4, [x_2, x_2] = \alpha x_5, [x_1, x_3] = x_4, [x_2, x_3] = x_5, [x_1, x_4] = x_5, \quad \alpha \in \mathbb{C}.$
$\mathcal{A}_{74}$: $[x_1, x_1] = x_3, [x_2, x_1] = x_4 + x_5, [x_2, x_2] = 2x_5, [x_1, x_3] = x_4, [x_2, x_3] = x_5, [x_1, x_4] = x_5.$

PROOF. Let $A^4 = Z(A) = \text{span}\{e_5\}$. Extend this to bases $\{e_4, e_5\}$ and $\{e_3, e_4, e_5\}$ of A^3 and A^2, respectively. Then the nonzero products in $A = \text{span}\{e_1, e_2, e_3, e_4, e_5\}$ are given by

$$[e_1, e_1] = \alpha_1 e_5 + \alpha_2 e_4 + \alpha_3 e_5, [e_1, e_2] = \alpha_4 e_3 + \alpha_5 e_4 + \alpha_6 e_5, [e_2, e_1] = \beta_1 e_3 + \beta_2 e_4 + \beta_3 e_5,$$
$$[e_2, e_2] = \beta_4 e_3 + \beta_5 e_4 + \beta_6 e_5, [e_1, e_3] = \gamma_1 e_4 + \gamma_2 e_5, [e_2, e_3] = \gamma_3 e_4 + \gamma_4 e_5,$$
$$[e_1, e_4] = \gamma_5 e_5, [e_2, e_4] = \gamma_6 e_5.$$

Leibniz identities give the following equations:

$$(3.14) \qquad \begin{cases} \beta_1\gamma_1 = \alpha_1\gamma_3 & \beta_1\gamma_2 + \beta_2\gamma_5 = \alpha_1\gamma_4 + \alpha_2\gamma_6 \\ \beta_4\gamma_1 = \alpha_4\gamma_3 & \beta_4\gamma_2 + \beta_5\gamma_5 = \alpha_4\gamma_4 + \alpha_5\gamma_6 \\ \gamma_3\gamma_5 = \gamma_1\gamma_6 & \end{cases}$$

We can assume $\gamma_3 = 0$, because if $\gamma_3 \neq 0$ and $\gamma_1 = 0$(resp. $\gamma_1 \neq 0$) then with the base change $x_1 = e_2, x_2 = e_1, x_3 = e_3, x_4 = e_4, x_5 = e_5$(resp. $x_1 = e_1, x_2 =$

$\gamma_3 e_1 - \gamma_1 e_2, x_3 = e_3, x_4 = e_4, x_5 = e_5)$ we can make $\gamma_3 = 0$. Then $\gamma_1 \neq 0$ since $\dim(A^3) = 2$. So by (3.14) we have $\beta_1 = 0 = \beta_4 = \gamma_6$. Then $\gamma_5 \neq 0$ since $\dim(Z(A)) = 1$.

Case 1: Let $\alpha_1 = 0$. Then $\alpha_4 \neq 0$ since $\dim(A^2) = 3$. Then from (3.14) we get $\beta_2 = 0$. Hence we have the following products in A:

$$(3.15)\quad [e_1, e_1] = \alpha_2 e_4 + \alpha_3 e_5, [e_1, e_2] = \alpha_4 e_3 + \alpha_5 e_4 + \alpha_6 e_5, [e_2, e_1] = \beta_3 e_5,$$
$$[e_2, e_2] = \beta_5 e_4 + \beta_6 e_5, [e_1, e_3] = \gamma_1 e_4 + \gamma_2 e_5, [e_2, e_3] = \gamma_4 e_5, [e_1, e_4] = \gamma_5 e_5.$$

If $\alpha_2 \neq 0$ then with the base change $x_1 = \gamma_1 e_1 - \alpha_2 e_3, x_2 = e_2, x_3 = e_3, x_4 = e_4, x_5 = e_5$ we can make $\alpha_2 = 0$. So let $\alpha_2 = 0$. Also if $\alpha_3 \neq 0$ then with the base change $x_1 = \gamma_5 e_1 - \gamma_3 e_4, x_2 = e_2, x_3 = e_3, x_4 = e_4, x_5 = e_5$ we can make $\alpha_3 = 0$. So we can assume $\alpha_3 = 0$.

Case 1.1: Let $\gamma_4 = 0$. Then by (3.14) we have $\beta_5 = 0$.

- If $\beta_6 = 0 = \beta_3$ then $x_1 = e_1, x_2 = e_2, x_3 = \alpha_4 e_3 + \alpha_5 e_4 + \alpha_6 e_5, x_4 = \alpha_4(\gamma_1 e_4 + \gamma_2 e_5), x_5 = \alpha_4 \gamma_1 \gamma_5 e_5$ shows that A is isomorphic to $\mathcal{A}_{64}$.
- If $\beta_6 = 0$ and $\beta_3 \neq 0$ then the base change $x_1 = (\frac{\beta_3}{\alpha_4\gamma_1\gamma_5})^{1/2} e_1, x_2 = e_2, x_3 = (\frac{\beta_3}{\alpha_4\gamma_1\gamma_5})^{1/2}(\alpha_4 e_3 + \alpha_5 e_4 + \alpha_6 e_5), x_4 = \frac{\beta_3}{\gamma_1\gamma_5}(\gamma_1 e_4 + \gamma_2 e_5), x_5 = \beta_3(\frac{\beta_3}{\alpha_4\gamma_1\gamma_5})^{1/2} e_5$ shows that A is isomorphic to $\mathcal{A}_{65}$.
- If $\beta_6 \neq 0$ and $\beta_3 = 0$ then the base change $x_1 = e_1, x_2 = \frac{\alpha_4\gamma_1\gamma_5}{\beta_6} e_2, x_3 = \frac{\alpha_4\gamma_1\gamma_5}{\beta_6}(\alpha_4 e_3 + \alpha_5 e_4 + \alpha_6 e_5), x_4 = \frac{\alpha_4^2\gamma_1\gamma_5}{\beta_6}(\gamma_1 e_4 + \gamma_2 e_5), x_5 = \frac{(\alpha_4\gamma_1\gamma_5)^2}{\beta_6} e_5$ shows that A is isomorphic to $\mathcal{A}_{66}$.
- If $\beta_6 \neq 0$ and $\beta_3 \neq 0$ then the base change $x_1 = (\frac{\beta_3}{\alpha_4\gamma_1\gamma_5})^{1/2} e_1, x_2 = \frac{(\beta_3)^{3/2}}{\beta_6(\alpha_4\gamma_1\gamma_5)^{1/2}} e_2, x_3 = \frac{\beta_3^2}{\alpha_4\beta_6\gamma_1\gamma_5}(\alpha_4 e_3 + \alpha_5 e_4 + \alpha_6 e_5), x_4 = \frac{\beta_3^{5/2}}{\alpha_4^{1/2}\beta_6\gamma_1^{3/2}\gamma_5^{3/2}}(\gamma_1 e_4 + \gamma_2 e_5), x_5 = \frac{\beta_3^3}{\alpha_4\beta_6\gamma_1\gamma_5} e_5$ shows that A is isomorphic to $\mathcal{A}_{67}$.

Case 1.2: Let $\gamma_4 \neq 0$. Then $\beta_5 = \frac{\alpha_4\gamma_4}{\gamma_5}$ from (3.14). Take $\theta = \frac{\alpha_4^2\gamma_1(\beta_6\gamma_1 - \beta_5\gamma_2)}{\beta_5^3\gamma_5}$. The base change $y_1 = e_1, y_2 = \frac{\alpha_4\gamma_1}{\beta_5} e_2, y_3 = \frac{\alpha_4\gamma_1}{\beta_5}(\alpha_4 e_3 + \alpha_5 e_4 + \alpha_6 e_5), y_4 = \frac{\alpha_4^2\gamma_1}{\beta_5}(\gamma_1 e_4 + \gamma_2 e_5), y_5 = \frac{\alpha_4^2\gamma_1^2\gamma_5}{\beta_5} e_5$ shows that A is isomorphic to the following algebra:

$$[y_1, y_2] = y_3, [y_2, y_1] = \frac{\beta_3}{\alpha_4\gamma_1\gamma_5} y_5, [y_2, y_2] = y_4 + \theta y_5, [y_1, y_3] = y_4, [y_2, y_3] = y_5,$$
$$[y_1, y_4] = y_5.$$

- If $\theta = 0$ and $\beta_3 = 0$ then the base change $x_1 = y_1, x_2 = y_2, x_3 = y_3, x_4 = y_4, x_5 = y_5$ shows that A is isomorphic to $\mathcal{A}_{68}$.
- If $\theta = 0$ and $\beta_3 \neq 0$ then the base change $x_1 = (\frac{\beta_3}{\alpha_4\gamma_1\gamma_5})^{1/2} y_1, x_2 = \frac{\beta_3}{\alpha_4\gamma_1\gamma_5} y_2, x_3 = (\frac{\beta_3}{\alpha_4\gamma_1\gamma_5})^{3/2} y_3, x_4 = (\frac{\beta_3}{\alpha_4\gamma_1\gamma_5})^2 y_4, x_5 = (\frac{\beta_3}{\alpha_4\gamma_1\gamma_5})^{5/2} y_5$ shows that A is isomorphic to $\mathcal{A}_{69}$.
- If $\theta \neq 0$ then the base change $x_1 = \theta y_1, x_2 = \theta^2 y_2, x_3 = \theta^3 y_3, x_4 = \theta^4 y_4, x_5 = \theta^5 y_5$ shows that A is isomorphic to $\mathcal{A}_{70}(\alpha)$.

Case 2: Let $\alpha_1 \neq 0$. If $\alpha_4 \neq 0$ the base change $x_1 = \alpha_4 e_1 - \alpha_1 e_2, x_2 = e_2, x_3 = e_3, x_4 = e_4, x_5 = e_5$ shows that A is isomorphic to an algebra with the nonzero products given by (3.15). Hence A is isomorphic to $\mathcal{A}_{64}, \mathcal{A}_{65}, \ldots, \mathcal{A}_{69}$ or $\mathcal{A}_{70}(\alpha)$. So let $\alpha_4 = 0$. Then from (3.14) we have $\beta_5 = 0$. Note that here if $\alpha_5 \neq 0$ then with the base change $x_1 = e_1, x_2 = \gamma_1 e_2 - \alpha_5 e_3, x_3 = e_3, x_4 = e_4, x_5 = e_5$ we can make $\alpha_5 = 0$. Then we can assume $\alpha_5 = 0$. If $\alpha_6 \neq 0$ then with the base change

$x_1 = e_1, x_2 = \gamma_5 e_2 - \alpha_6 e_4, x_3 = e_3, x_4 = e_4, x_5 = e_5$ we can make $\alpha_6 = 0$. Then we can assume $\alpha_6 = 0$.

Case 2.1: Let $\gamma_4 = 0$. Then by (3.14) we have $\beta_2 = 0$. Note that if $\beta_3 = 0 = \beta_6$ then A is split. So let $(\beta_3, \beta_6) \neq (0,0)$. If $\beta_6 = 0$ then the base change $x_1 = e_1, x_2 = \frac{\alpha_1\gamma_1\gamma_5}{\beta_3} e_2, x_3 = \alpha_1 e_3 + \alpha_2 e_4 + \alpha_3 e_5, x_4 = \alpha_1(\gamma_1 e_4 + \gamma_2 e_5), x_5 = \alpha_1\gamma_1\gamma_5 e_5$ shows that A is isomorphic to $\mathcal{A}_{71}$. Now suppose $\beta_6 \neq 0$. Without loss of generality we can assume $\beta_3 = 0$ because if $\beta_3 \neq 0$ then with the base change $x_1 = \beta_6 e_1 - \beta_3 e_2, x_2 = e_2 + \frac{\beta_3}{\gamma_5} e_4, x_3 = e_3, x_4 = e_4, x_5 = e_5$ we can make $\beta_3 = 0$. Then the base change $x_1 = e_1, x_2 = \sqrt{\frac{\alpha_1\gamma_1\gamma_5}{\beta_6}} e_2, x_3 = \alpha_1 e_3 + \alpha_2 e_4 + \alpha_3 e_5, x_4 = \alpha_1(\gamma_1 e_4 + \gamma_2 e_5), x_5 = \alpha_1\gamma_1\gamma_5 e_5$ shows that A is isomorphic to $\mathcal{A}_{72}$.

Case 2.2: Let $\gamma_4 \neq 0$. Then $\beta_2 = \frac{\alpha_1\gamma_4}{\gamma_5}$ from (3.14). Take $\theta = \frac{\beta_3\gamma_1 - \beta_2\gamma_2}{\beta_2\gamma_1\gamma_5}$. The base change $y_1 = e_1, y_2 = \frac{\alpha_1\gamma_1}{\beta_2} e_2, y_3 = \alpha_1 e_3 + \alpha_2 e_4 + \alpha_3 e_5, y_4 = \alpha_1(\gamma_1 e_4 + \gamma_2 e_5), y_5 = \alpha_1\gamma_1\gamma_5 e_5$ shows that A is isomorphic to the following algebra:

$$[y_1, y_1] = y_3, [y_2, y_1] = y_4 + \theta y_5, [y_2, y_2] = \frac{\alpha_1\beta_6\gamma_1}{\beta_2^2\gamma_5} y_5, [y_1, y_3] = y_4, [y_2, y_3] = y_5,$$
$$[y_1, y_4] = y_5.$$

- If $\theta = 0$ then the base change $x_1 = y_1, x_2 = y_2, x_3 = y_3, x_4 = y_4, x_5 = y_5$ shows that A is isomorphic to $\mathcal{A}_{73}(\alpha)$.
- If $\theta \neq 0$ and $\frac{\alpha_1\beta_6\gamma_1}{\beta_2^2\gamma_5} \neq 2$ then with suitable change of basis (w.s.c.o.b.) A is isomorphic to $\mathcal{A}_{73}(\alpha)$.
- If $\theta \neq 0$ and $\frac{\alpha_1\beta_6\gamma_1}{\beta_2^2\gamma_5} = 2$ then the base change $x_1 = \theta y_1, x_2 = \theta^2 y_2, x_3 = \theta^2 y_3, x_4 = \theta^3 y_4, x_6 = \theta^4 y_5$ shows that A is isomorphic to $\mathcal{A}_{74}$.

□

Note that we obtain the complete classification of 5−dimensional filiform Leibniz algebras from Theorem 2.2 and Theorem 3.5. We compare our classification with the classification given in [**15**]. They obtained the isomorphism classes in the classes FLb_5, SLb_5 and TLb_5. It can be seen that $\dim(Leib(FLb_5)) = 3 = \dim(Leib(SLb_5))$ and $\dim(Leib(TLb_5)) = 1$ or 0. The classification of FLb_5 and SLb_5 given in [**15**] completely agrees with Theorem 3.5. However, we find some redundancy in the classification of TLb_5 since $L(2,1,0) \cong L(0,1,0)$ and $L(2,1,1) \cong L(0, \frac{1}{8}, \frac{1}{8})$. Also they missed the isomorphism classes $\mathcal{A}_2$ and $\mathcal{A}_4$ listed in Theorem 2.2.

Let $\dim(A^2) = 3, \dim(A^3) = 2$ and $A^4 = 0$. Then we have $A^3 = Z(A)$. Assume $\dim(Leib(A)) = 3$. Let $A^3 = Z(A) = \text{span}\{e_4, e_5\}$. Extend this to a basis $\{e_3, e_4, e_5\}$ of $Leib(A) = A^2$. Then the nonzero products in $A = \text{span}\{e_1, e_2, e_3, e_4, e_5\}$ are given by

$[e_1, e_1] = \alpha_1 e_3 + \alpha_2 e_4 + \alpha_3 e_5, [e_1, e_2] = \alpha_4 e_3 + \alpha_5 e_4 + \alpha_6 e_5, [e_2, e_1] = \beta_1 e_3 + \beta_2 e_4 + \beta_3 e_5, [e_2, e_2] = \beta_4 e_3 + \beta_5 e_4 + \beta_6 e_5, [e_1, e_3] = \gamma_1 e_4 + \gamma_2 e_4, [e_2, e_3] = \gamma_3 e_4 + \gamma_4 e_5$.
From the Leibniz identities we get the following equations:

$$\begin{cases} \beta_1\gamma_1 = \alpha_1\gamma_3 & \beta_1\gamma_2 = \alpha_1\gamma_4 \\ \beta_4\gamma_1 = \alpha_4\gamma_3 & \beta_4\gamma_2 = \alpha_4\gamma_4 \end{cases} \tag{3.16}$$

Suppose $\gamma_3 = 0$. Then $\gamma_1, \gamma_4 \neq 0$ since $\dim(A^3) = 2$. From (3.16) we have $\beta_1 = 0 = \beta_4 = \alpha_1 = \alpha_4$, which is a contradiction since $\dim(A^2) = 3$. Now suppose $\gamma_3 \neq 0$. If $\beta_1 = 0 = \beta_4$ then by (3.16) we get $\alpha_1 = 0 = \alpha_4$, contradiction. If

$\beta_1 \neq 0$(resp. $\beta_4 \neq 0$) then by (3.16) we have $\gamma_3\gamma_2 - \gamma_1\gamma_4 = 0$ that contradicts with the fact that $\dim(A^3) = 2$. Hence our assumption was wrong. Therefore $\dim(Leib(A)) = 2$.

THEOREM 3.6. *Let A be a 5−dimensional non-split non-Lie nilpotent Leibniz algebra with* $\dim(A^2) = 3, \dim(A^3) = 2 = \dim(Leib(A))$ *and* $A^4 = 0$. *Then A is isomorphic to a Leibniz algebra spanned by* $\{x_1, x_2, x_3, x_4, x_5\}$ *with the nonzero products given by one of the following:*

$\mathcal{A}_{75}(\alpha)$: $[x_1, x_2] = x_3, [x_2, x_1] = -x_3 + x_4, [x_2, x_2] = \alpha x_5, [x_1, x_3] = x_4 = -[x_3, x_1], [x_2, x_3] = x_5 = -[x_3, x_2], \quad \alpha \in \mathbb{C}\backslash\{0\}$.

$\mathcal{A}_{76}(\alpha)$: $[x_1, x_1] = x_5, [x_1, x_2] = x_3, [x_2, x_1] = -x_3 + x_4, [x_2, x_2] = \alpha x_5, [x_1, x_3] = x_4 = -[x_3, x_1], [x_2, x_3] = x_5 = -[x_3, x_2], \quad \alpha \in \mathbb{C}$.

$\mathcal{A}_{77}(\alpha)$: $[x_1, x_1] = \alpha x_5, [x_1, x_2] = x_3, [x_2, x_1] = -x_3 + x_4 + x_5, [x_1, x_3] = x_4 = -[x_3, x_1], [x_2, x_3] = x_5 = -[x_3, x_2], \quad \alpha \in \mathbb{C}\backslash\{0\}$.

$\mathcal{A}_{78}(\alpha)$: $[x_1, x_1] = \alpha x_5, [x_1, x_2] = x_3, [x_2, x_1] = -x_3 + x_4 + x_5, [x_2, x_2] = x_5, [x_1, x_3] = x_4 = -[x_3, x_1], [x_2, x_3] = x_5 = -[x_3, x_2], \quad \alpha \in \mathbb{C}\backslash\{0\}$.

$\mathcal{A}_{79}(\alpha)$: $[x_1, x_1] = \alpha x_5, [x_1, x_2] = x_3, [x_2, x_1] = -x_3 + x_4 + x_5, [x_2, x_2] = -\frac{1}{2}x_5, [x_1, x_3] = x_4 = -[x_3, x_1], [x_2, x_3] = x_5 = -[x_3, x_2], \quad \alpha \in \mathbb{C}\backslash\{-\frac{1}{6}, 0\}$.

$\mathcal{A}_{80}(\alpha)$: $[x_1, x_1] = \alpha x_5, [x_1, x_2] = x_3 = -[x_2, x_1], [x_2, x_2] = x_4 + x_5, [x_1, x_3] = x_4 = -[x_3, x_1], [x_2, x_3] = x_5 = -[x_3, x_2], \quad \alpha \in \mathbb{C}\backslash\{-\frac{4}{27}, 0\}$.

$\mathcal{A}_{81}(\alpha, \beta)$: $[x_1, x_1] = \alpha x_5, [x_1, x_2] = x_3, [x_2, x_1] = -x_3 + x_5, [x_2, x_2] = x_4 + \beta x_5, [x_1, x_3] = x_4 = -[x_3, x_1], [x_2, x_3] = x_5 = -[x_3, x_2], \quad \alpha \in \mathbb{C}\backslash\{0\}, \beta \in \mathbb{C}, 4\alpha\beta \neq 1, 8\alpha\beta^3 - 2\beta^2 + 1 \neq 0, 16\alpha\beta^3 \neq 1 + 6\beta^2 \pm \sqrt{4\beta^2 + 12\beta + 1}, -27\alpha\beta \neq 9\beta^2 + 2\beta^4 \pm 2\sqrt{\beta^2(3+\beta^2)^3}$.

$\mathcal{A}_{82}(\alpha, \beta, \gamma)$: $[x_1, x_1] = \alpha x_5, [x_1, x_2] = x_3, [x_2, x_1] = -x_3 + x_4 + \beta x_5, [x_2, x_2] = x_4 + \gamma x_5, [x_1, x_3] = x_4 = -[x_3, x_1], [x_2, x_3] = x_5 = -[x_3, x_2], \quad \alpha, \beta, \gamma \in \mathbb{C}$.

$\mathcal{A}_{83}(\alpha, \beta)$: $[x_1, x_1] = x_4 + \alpha x_5, [x_1, x_2] = x_3, [x_2, x_1] = -x_3 + \beta x_5, [x_2, x_2] = x_5, [x_1, x_3] = x_4 = -[x_3, x_1], [x_2, x_3] = x_5 = -[x_3, x_2], \quad \alpha, \beta \in \mathbb{C}$.

Let $\dim(A^2) = 3$ and $\dim(A^3) = 1$. Then we have $A^3 \subseteq Z(A) \subsetneq A^2$. Note that $A^2 \neq Z(A)$ since $A^3 \neq 0$. Hence $\dim(Z(A)) = 1$ or 2. First we consider the case $\dim(Z(A)) = 2$. Note that since $Leib(A) \subseteq A^2$ we have $2 \leq \dim(Leib(A)) \leq 3$.

THEOREM 3.7. *Let A be a 5−dimensional non-split non-Lie nilpotent Leibniz algebra with* $\dim(A^2) = 3$, $\dim(A^3) = 1$, $\dim(Z(A)) = 2 = \dim(Leib(A))$ *and* $Leib(A) \neq Z(A)$. *Then A is isomorphic to a Leibniz algebra spanned by* $\{x_1, x_2, x_3, x_4, x_5\}$ *with the nonzero products given by one of the following:*

$\mathcal{A}_{84}$: $[x_1, x_2] = x_3 + x_4, [x_2, x_1] = -x_3, [x_1, x_4] = x_5$.
$\mathcal{A}_{85}$: $[x_1, x_2] = x_3 + x_4, [x_2, x_1] = -x_3, [x_2, x_2] = x_5, [x_1, x_4] = x_5$.
$\mathcal{A}_{86}$: $[x_1, x_1] = x_4, [x_1, x_2] = x_3 = -[x_2, x_1], [x_1, x_4] = x_5$.
$\mathcal{A}_{87}$: $[x_1, x_1] = x_4, [x_1, x_2] = x_3 = -[x_2, x_1], [x_2, x_2] = x_5, [x_1, x_4] = x_5$.

THEOREM 3.8. *Let A be a 5−dimensional non-split non-Lie nilpotent Leibniz algebra with* $\dim(A^2) = 3$, $\dim(A^3) = 1$, $\dim(Z(A)) = 2 = \dim(Leib(A))$ *and* $Leib(A) = Z(A)$. *Then A is isomorphic to a Leibniz algebra spanned by* $\{x_1, x_2, x_3, x_4, x_5\}$ *with the nonzero products given by one of the following:*

$\mathcal{A}_{88}$: $[x_1, x_1] = x_5, [x_1, x_2] = x_3, [x_2, x_1] = -x_3 + x_4, [x_1, x_3] = x_5 = -[x_3, x_1]$.
$\mathcal{A}_{89}$: $[x_1, x_2] = x_3, [x_2, x_1] = -x_3 + x_4, [x_2, x_2] = x_5, [x_1, x_3] = x_5 = -[x_3, x_1]$.
$\mathcal{A}_{90}$: $[x_1, x_1] = x_5, [x_1, x_2] = x_3, [x_2, x_1] = -x_3 + x_4, [x_2, x_2] = x_5, [x_1, x_3] = x_5 = -[x_3, x_1]$.
$\mathcal{A}_{91}$: $[x_1, x_2] = x_3, [x_2, x_1] = -x_3 + x_5, [x_2, x_2] = x_4, [x_1, x_3] = x_5 = -[x_3, x_1]$.
$\mathcal{A}_{92}$: $[x_1, x_1] = x_5, [x_1, x_2] = x_3 = -[x_2, x_1], [x_2, x_2] = x_4, [x_1, x_3] = x_5 = -[x_3, x_1]$.
$\mathcal{A}_{93}$: $[x_1, x_1] = x_5, [x_1, x_2] = x_3, [x_2, x_1] = -x_3 + x_5, [x_2, x_2] = x_4, [x_1, x_3] = x_5 = -[x_3, x_1]$.
$\mathcal{A}_{94}(\alpha)$: $[x_1, x_1] = x_5, [x_1, x_2] = x_3, [x_2, x_1] = -x_3 + x_4 + \alpha x_5, [x_2, x_2] = x_4, [x_1, x_3] = x_5 = -[x_3, x_1], \quad \alpha \in \mathbb{C}$.
$\mathcal{A}_{95}$: $[x_1, x_1] = x_4, [x_1, x_2] = x_3, [x_2, x_1] = -x_3 + x_5, [x_1, x_3] = x_5 = -[x_3, x_1]$.
$\mathcal{A}_{96}$: $[x_1, x_1] = x_4, [x_1, x_2] = x_3 = -[x_2, x_1], [x_2, x_2] = x_5, [x_1, x_3] = x_5 = -[x_3, x_1]$.

THEOREM 3.9. *Let A be a 5−dimensional non-split non-Lie nilpotent Leibniz algebra with* $\dim(A^2) = 3 = \dim(Leib(A))$, $\dim(A^3) = 1$ *and* $\dim(Z(A)) = 2$. *Then A is isomorphic to a Leibniz algebra spanned by* $\{x_1, x_2, x_3, x_4, x_5\}$ *with the nonzero products given by one of the following:*

$\mathcal{A}_{97}$: $[x_1, x_1] = x_3, [x_2, x_1] = x_4, [x_1, x_3] = x_5$.
$\mathcal{A}_{98}$: $[x_1, x_1] = x_3, [x_2, x_1] = x_4, [x_2, x_2] = x_5, [x_1, x_3] = x_5$.
$\mathcal{A}_{99}$: $[x_1, x_1] = x_3, [x_2, x_2] = x_4, [x_1, x_3] = x_5$.
$\mathcal{A}_{100}$: $[x_1, x_1] = x_3, [x_2, x_1] = x_5, [x_2, x_2] = x_4, [x_1, x_3] = x_5$.

$\mathcal{A}_{101}$: $[x_1,x_1]=x_3,[x_2,x_1]=x_4,[x_2,x_2]=x_4,[x_1,x_3]=x_5$.
$\mathcal{A}_{102}$: $[x_1,x_1]=x_3,[x_2,x_1]=x_4+x_5,[x_2,x_2]=x_4,[x_1,x_3]=x_5$.
$\mathcal{A}_{103}$: $[x_1,x_1]=x_3,[x_1,x_2]=x_4,[x_2,x_1]=x_5,[x_1,x_3]=x_5$.
$\mathcal{A}_{104}(\alpha)$: $[x_1,x_1]=x_3,[x_1,x_2]=x_4,[x_2,x_1]=\alpha x_4,[x_1,x_3]=x_5, \quad \alpha\in\mathbb{C}\backslash\{-1\}$.
$\mathcal{A}_{105}(\alpha)$: $[x_1,x_1]=x_3,[x_1,x_2]=x_4,[x_2,x_1]=\alpha x_4,[x_2,x_2]=x_5,[x_1,x_3]=x_5, \quad \alpha\in\mathbb{C}\backslash\{-1\}$.
$\mathcal{A}_{106}$: $[x_1,x_2]=x_3,[x_2,x_2]=x_4,[x_1,x_3]=x_5$.
$\mathcal{A}_{107}$: $[x_1,x_2]=x_3,[x_2,x_1]=x_5,[x_2,x_2]=x_4,[x_1,x_3]=x_5$.
$\mathcal{A}_{108}$: $[x_1,x_2]=x_3,[x_2,x_1]=x_4,[x_2,x_2]=x_4,[x_1,x_3]=x_5$.
$\mathcal{A}_{109}$: $[x_1,x_2]=x_3,[x_2,x_1]=x_4+x_5,[x_2,x_2]=x_4,[x_1,x_3]=x_5$.
$\mathcal{A}_{110}$: $[x_1,x_1]=x_4,[x_1,x_2]=x_3,[x_1,x_3]=x_5$.
$\mathcal{A}_{111}$: $[x_1,x_1]=x_4,[x_1,x_2]=x_3,[x_2,x_1]=x_5,[x_1,x_3]=x_5$.
$\mathcal{A}_{112}$: $[x_1,x_1]=x_4,[x_1,x_2]=x_3,[x_2,x_2]=x_5,[x_1,x_3]=x_5$.
$\mathcal{A}_{113}$: $[x_1,x_1]=x_4,[x_1,x_2]=x_3,[x_2,x_1]=x_5,[x_2,x_2]=x_5,[x_1,x_3]=x_5$.
$\mathcal{A}_{114}$: $[x_1,x_1]=x_4,[x_1,x_2]=x_3,[x_2,x_1]=x_4,[x_1,x_3]=x_5$.
$\mathcal{A}_{115}$: $[x_1,x_1]=x_4,[x_1,x_2]=x_3,[x_2,x_1]=x_4,[x_2,x_2]=x_5,[x_1,x_3]=x_5$.
$\mathcal{A}_{116}(\alpha)$: $[x_1,x_1]=x_4,[x_1,x_2]=x_3,[x_2,x_1]=\alpha x_4,[x_2,x_2]=x_4,[x_1,x_3]=x_5, \quad \alpha\in\mathbb{C}$.
$\mathcal{A}_{117}(\alpha)$: $[x_1,x_1]=x_4,[x_1,x_2]=x_3,[x_2,x_1]=\alpha x_4+x_5,[x_2,x_2]=x_4,[x_1,x_3]=x_5, \quad \alpha\in\mathbb{C}$.

Now let $\dim(Z(A))=1$. Then $\dim(Leib(A))=2$ or 3.

THEOREM 3.10. *Let A be a 5−dimensional non-split non-Lie nilpotent Leibniz algebra with $\dim(A^2)=3$, $\dim(A^3)=1=\dim(Z(A))$ and $\dim(Leib(A))=2$. Then A is isomorphic to a Leibniz algebra spanned by $\{x_1,x_2,x_3,x_4,x_5\}$ with the nonzero products given by one of the following:*

$\mathcal{A}_{118}$: $[x_1,x_2]=-x_3+x_4,[x_2,x_1]=x_3,[x_2,x_3]=x_5=-[x_3,x_2],[x_1,x_4]=x_5$.
$\mathcal{A}_{119}$: $[x_1,x_2]=-x_3+x_4,[x_2,x_1]=x_3,[x_2,x_2]=x_5,[x_2,x_3]=x_5=-[x_3,x_2],[x_1,x_4]=x_5$.
$\mathcal{A}_{120}(\alpha)$: $[x_1,x_2]=-x_3=[x_2,x_1],[x_2,x_2]=x_4,[x_2,x_3]=-\alpha x_5,[x_3,x_2]=(\alpha-1)x_5,[x_1,x_4]=x_5, \quad \alpha\in\mathbb{C}$.
$\mathcal{A}_{121}(\alpha)$: $[x_1,x_2]=-x_3+x_4,[x_2,x_1]=x_3,[x_2,x_2]=x_4,[x_2,x_3]=-\alpha x_5,[x_3,x_2]=(\alpha-1)x_5,[x_1,x_4]=x_5, \quad \alpha\in\mathbb{C}$.
$\mathcal{A}_{122}$: $[x_1,x_2]=x_3,[x_2,x_1]=-x_3+x_4,[x_3,x_1]=x_5,[x_1,x_4]=x_5$.
$\mathcal{A}_{123}$: $[x_1,x_2]=x_3,[x_2,x_1]=-x_3+x_4,[x_2,x_2]=x_5,[x_3,x_1]=x_5,[x_1,x_4]=x_5$.
$\mathcal{A}_{124}$: $[x_1,x_2]=x_3,[x_2,x_1]=-x_3+x_4,[x_3,x_1]=x_5,[x_2,x_3]=x_5=-[x_3,x_2],[x_1,x_4]=x_5$.
$\mathcal{A}_{125}$: $[x_1,x_2]=x_3,[x_2,x_1]=-x_3+x_4,[x_2,x_2]=x_5,[x_3,x_1]=x_5,[x_2,x_3]=x_5=-[x_3,x_2],[x_1,x_4]=x_5$.
$\mathcal{A}_{126}(\alpha)$: $[x_1,x_2]=x_3,[x_2,x_1]=-x_3+x_4,[x_2,x_2]=x_4,[x_3,x_1]=x_5,[x_2,x_3]=\alpha x_5,[x_3,x_2]=(1-\alpha)x_5,[x_1,x_4]=x_5, \quad \alpha\in\mathbb{C}\backslash\{0\}$.
$\mathcal{A}_{127}$: $[x_1,x_1]=x_4,[x_1,x_2]=x_3=-[x_2,x_1],[x_2,x_3]=x_5=-[x_3,x_2],[x_1,x_4]=x_5$.
$\mathcal{A}_{128}$: $[x_1,x_1]=x_4,[x_1,x_2]=x_3=-[x_2,x_1],[x_2,x_2]=x_5,[x_2,x_3]=x_5=-[x_3,x_2],[x_1,x_4]=x_5$.

THEOREM 3.11. *Let A be a 5−dimensional non-split non-Lie nilpotent Leibniz algebra with $\dim(A^2)=3=\dim(Leib(A))$ and $\dim(A^3)=1=\dim(Z(A))$. Then A is isomorphic to a Leibniz algebra spanned by $\{x_1,x_2,x_3,x_4,x_5\}$ with the nonzero products given by one of the following:*

$\mathcal{A}_{129}(\alpha)$: $[x_1,x_1]=x_4,[x_1,x_2]=\alpha x_4,[x_2,x_1]=x_3,[x_2,x_3]=x_5,[x_1,x_4]=x_5, \quad \alpha\in\mathbb{C}$.
$\mathcal{A}_{130}$: $[x_1,x_1]=x_4,[x_1,x_2]=-x_4,[x_2,x_1]=-x_3,[x_2,x_2]=x_3,[x_2,x_3]=x_5,[x_1,x_4]=x_5$.
$\mathcal{A}_{131}(\alpha,\beta)$: $[x_1,x_1]=x_4,[x_1,x_2]=\alpha x_4,[x_2,x_1]=\beta x_3,[x_2,x_2]=x_3,[x_2,x_3]=x_5,[x_1,x_4]=x_5, \quad \alpha,\beta\in\mathbb{C},\alpha\beta\neq 1$.
$\mathcal{A}_{132}(\alpha)$: $[x_1,x_2]=x_3,[x_2,x_1]=\alpha x_3,[x_2,x_2]=x_4,[x_2,x_3]=x_5,[x_1,x_4]=x_5, \quad \alpha\in\mathbb{C}\backslash\{-1,0\}$.
$\mathcal{A}_{133}(\alpha,\beta)$: $[x_1,x_1]=x_4,[x_1,x_2]=x_3+\alpha x_4,[x_2,x_1]=\beta x_3,[x_2,x_2]=x_4,[x_2,x_3]=x_5,[x_1,x_4]=x_5, \quad \alpha\in\mathbb{C},\beta\in\mathbb{C}\backslash\{-1\}$.
$\mathcal{A}_{134}(\alpha,\beta,\gamma)$: $[x_1,x_1]=\alpha x_4,[x_1,x_2]=x_3+\beta x_4,[x_2,x_1]=\gamma x_3,[x_2,x_2]=x_4,[x_2,x_3]=x_5,[x_1,x_4]=x_5, \quad \alpha,\beta,\gamma\in\mathbb{C}$.
$\mathcal{A}_{135}(\alpha,\beta)$: $[x_1,x_1]=x_3+\alpha x_4,[x_1,x_2]=x_3+\beta x_4,[x_2,x_1]=-x_3+x_4,[x_2,x_2]=x_4,[x_2,x_3]=x_5,[x_1,x_4]=x_5, \quad \alpha,\beta\in\mathbb{C}$.

Let $\dim(A^2)=3$ and $A^3=0$. Then we have $A^2=Z(A)$. Then since $Leib(A)\subseteq A^2$ we have $2\leq\dim(Leib(A))\leq 3$.

THEOREM 3.12. *Let A be a 5−dimensional non-split non-Lie nilpotent Leibniz algebra with $\dim(A^2)=3$, $\dim(A^3)=0$ and $\dim(Leib(A))=2$. Then A is isomorphic to a Leibniz algebra spanned by $\{x_1,x_2,x_3,x_4,x_5\}$ with the nonzero products given by one of the following:*

$\mathcal{A}_{136}$: $[x_1,x_1]=x_4,[x_1,x_2]=x_3,[x_2,x_1]=-x_3+x_4+x_5,[x_2,x_2]=x_5$.
$\mathcal{A}_{137}$: $[x_1,x_1]=x_4,[x_1,x_2]=x_3=-[x_2,x_1],[x_2,x_2]=x_5$.

THEOREM 3.13. *Let A be a 5−dimensional non-split non-Lie nilpotent Leibniz algebra with* $\dim(A^2) = 3$, $\dim(A^3) = 0$ *and* $\dim(Leib(A)) = 3$. *Then A is isomorphic to a Leibniz algebra spanned by* $\{x_1, x_2, x_3, x_4, x_5\}$ *with the nonzero products given by one of the following:*

$\mathcal{A}_{138}(\alpha)$: $[x_1, x_1] = x_4, [x_1, x_2] = \alpha x_4 + x_5, [x_2, x_1] = x_3, [x_2, x_2] = x_5, \quad \alpha \in \mathbb{C}$.
$\mathcal{A}_{139}$: $[x_1, x_1] = x_4, [x_1, x_2] = x_3, [x_2, x_1] = x_3, [x_2, x_2] = x_5$.

Let $\dim(A^2) = 2$ and $\dim(A^3) = 1$. Then we have $A^3 = Z(A)$.

THEOREM 3.14. *Let A be a 5−dimensional non-split non-Lie nilpotent Leibniz algebra with* $\dim(A^2) = 2 = \dim(Leib(A))$ *and* $\dim(A^3) = 1$. *Then A is isomorphic to a Leibniz algebra spanned by* $\{x_1, x_2, x_3, x_4, x_5\}$ *with the nonzero products given by one of the following:*

$\mathcal{A}_{140}$: $[x_1, x_2] = x_4, [x_3, x_1] = x_5, [x_1, x_4] = x_5$.
$\mathcal{A}_{141}$: $[x_1, x_2] = x_4, [x_2, x_2] = x_5, [x_3, x_1] = x_5, [x_1, x_4] = x_5$.
$\mathcal{A}_{142}$: $[x_1, x_2] = x_4, [x_2, x_3] = x_5, [x_1, x_4] = x_5$.
$\mathcal{A}_{143}$: $[x_1, x_2] = x_4, [x_2, x_3] = x_5, [x_3, x_1] = x_5, [x_1, x_4] = x_5$.
$\mathcal{A}_{144}$: $[x_1, x_1] = x_4, [x_2, x_2] = x_5, [x_3, x_1] = x_5, [x_1, x_4] = x_5$.
$\mathcal{A}_{145}$: $[x_1, x_1] = x_4, [x_2, x_3] = x_5, [x_1, x_4] = x_5$.
$\mathcal{A}_{146}$: $[x_1, x_1] = x_4, [x_2, x_3] = x_5, [x_3, x_1] = x_5, [x_1, x_4] = x_5$.
$\mathcal{A}_{147}(\alpha)$: $[x_1, x_1] = x_4, [x_2, x_3] = \alpha x_5, [x_3, x_2] = x_5, [x_1, x_4] = x_5, \quad \alpha \in \mathbb{C}\backslash\{0\}$.
$\mathcal{A}_{148}$: $[x_1, x_2] = x_4, [x_2, x_1] = x_5, [x_3, x_2] = x_5, [x_1, x_4] = x_5$.
$\mathcal{A}_{149}(\alpha)$: $[x_1, x_2] = x_4, [x_2, x_3] = \alpha x_5, [x_3, x_2] = x_5, [x_1, x_4] = x_5, \quad \alpha \in \mathbb{C}$.
$\mathcal{A}_{150}(\alpha)$: $[x_1, x_2] = x_4, [x_3, x_1] = x_5, [x_2, x_3] = \alpha x_5, [x_3, x_2] = x_5, [x_1, x_4] = x_5, \quad \alpha \in \mathbb{C}$.
$\mathcal{A}_{151}$: $[x_1, x_2] = x_4, [x_2, x_1] = x_5, [x_3, x_1] = x_5, [x_3, x_2] = x_5, [x_1, x_4] = x_5$.
$\mathcal{A}_{152}$: $[x_1, x_1] = x_4, [x_2, x_2] = x_5, [x_2, x_3] = x_5 = -[x_3, x_2], [x_1, x_4] = x_5$.
$\mathcal{A}_{153}$: $[x_1, x_2] = x_4, [x_2, x_2] = x_5, [x_2, x_3] = x_5 = -[x_3, x_2], [x_1, x_4] = x_5$.
$\mathcal{A}_{154}$: $[x_1, x_1] = x_4, [x_1, x_2] = x_4, [x_2, x_2] = x_5, [x_2, x_3] = x_5 = -[x_3, x_2], [x_1, x_4] = x_5$.
$\mathcal{A}_{155}$: $[x_1, x_2] = x_4, [x_3, x_3] = x_5, [x_1, x_4] = x_5$.
$\mathcal{A}_{156}$: $[x_1, x_2] = x_4, [x_2, x_3] = x_5, [x_3, x_3] = x_5, [x_1, x_4] = x_5$.
$\mathcal{A}_{157}(\alpha)$: $[x_1, x_2] = x_4, [x_2, x_2] = x_5, [x_2, x_3] = \alpha x_5, [x_3, x_3] = x_5, [x_1, x_4] = x_5, \quad \alpha \in \mathbb{C}$.
$\mathcal{A}_{158}$: $[x_1, x_2] = x_4, [x_2, x_1] = x_5, [x_3, x_3] = x_5, [x_1, x_4] = x_5$.
$\mathcal{A}_{159}$: $[x_1, x_2] = x_4, [x_2, x_1] = x_5, [x_2, x_3] = x_5, [x_3, x_3] = x_5, [x_1, x_4] = x_5$.
$\mathcal{A}_{160}(\alpha)$: $[x_1, x_2] = x_4, [x_2, x_1] = x_5, [x_2, x_2] = x_5, [x_2, x_3] = \alpha x_5, [x_3, x_3] = x_5, [x_1, x_4] = x_5, \quad \alpha \in \mathbb{C}$.

Let $\dim(A^2) = 2$ and $A^3 = 0$. Then we have $A^2 = Z(A)$.

THEOREM 3.15. *Let A be a 5−dimensional non-split non-Lie nilpotent Leibniz algebra with* $\dim(A^2) = 2 = \dim(Leib(A))$ *and* $A^3 = 0$. *Then A is isomorphic to a Leibniz algebra spanned by* $\{x_1, x_2, x_3, x_4, x_5\}$ *with the nonzero products given by one of the following:*

$\mathcal{A}_{161}$: $[x_1, x_1] = x_5, [x_1, x_2] = x_4, [x_3, x_3] = x_5$.
$\mathcal{A}_{162}$: $[x_1, x_1] = x_5, [x_1, x_2] = x_4, [x_2, x_1] = x_5, [x_3, x_3] = x_5$.
$\mathcal{A}_{163}(\alpha)$: $[x_1, x_1] = x_5, [x_1, x_2] = x_4, [x_2, x_1] = \alpha x_5, [x_2, x_2] = x_5, [x_3, x_3] = x_5, \quad \alpha \in \mathbb{C}$.
$\mathcal{A}_{164}(\alpha)$: $[x_1, x_2] = x_4, [x_2, x_1] = \alpha x_4 + x_5, [x_3, x_3] = x_5, \quad \alpha \in \mathbb{C}\backslash\{-1\}$.
$\mathcal{A}_{165}(\alpha)$: $[x_1, x_2] = x_4, [x_2, x_1] = \alpha x_4, [x_2, x_2] = x_5, [x_3, x_3] = x_5, \quad \alpha \in \mathbb{C}\backslash\{-1\}$.
$\mathcal{A}_{166}(\alpha)$: $[x_1, x_2] = x_4, [x_2, x_1] = \alpha x_4 + x_5, [x_2, x_2] = x_5, [x_3, x_3] = x_5, \quad \alpha \in \mathbb{C}\backslash\{-1\}$.
$\mathcal{A}_{167}(\alpha, \beta)$: $[x_1, x_1] = x_5, [x_1, x_2] = x_4, [x_2, x_1] = \alpha x_4 + \beta x_5, [x_2, x_2] = x_5, [x_3, x_3] = x_5, \alpha \in \mathbb{C}\backslash\{-1, 0\}, \beta \in \mathbb{C}$.
$\mathcal{A}_{168}$: $[x_1, x_2] = x_5, [x_2, x_2] = x_4, [x_3, x_3] = x_5$.
$\mathcal{A}_{169}(\alpha)$: $[x_1, x_2] = \alpha x_5, [x_2, x_1] = x_5, [x_2, x_2] = x_4, [x_3, x_3] = x_5, \quad \alpha \in \mathbb{C}$.
$\mathcal{A}_{170}$: $[x_1, x_1] = x_5, [x_1, x_2] = x_5, [x_2, x_2] = x_4, [x_3, x_3] = x_5$.
$\mathcal{A}_{171}$: $[x_1, x_2] = x_4 + x_5 = -[x_2, x_1], [x_2, x_2] = x_4, [x_3, x_3] = x_5$.
$\mathcal{A}_{172}$: $[x_1, x_2] = x_4 + x_5, [x_2, x_1] = -x_4, [x_2, x_2] = x_4, [x_3, x_3] = x_5$.
$\mathcal{A}_{173}(\alpha, \beta)$: $[x_1, x_2] = x_4 + \alpha x_5, [x_2, x_1] = -x_4 + \beta x_5, [x_2, x_2] = x_4, [x_3, x_3] = x_5, \quad \alpha, \beta \in \mathbb{C}$.
$\mathcal{A}_{174}$: $[x_1, x_2] = x_4, [x_2, x_3] = x_5$.
$\mathcal{A}_{175}$: $[x_1, x_1] = x_5, [x_1, x_2] = x_4, [x_2, x_3] = x_5$.
$\mathcal{A}_{176}(\alpha)$: $[x_1, x_2] = \alpha x_4, [x_2, x_1] = x_4, [x_2, x_3] = x_5, \quad \alpha \in \mathbb{C}\backslash\{-1\}$.
$\mathcal{A}_{177}(\alpha)$: $[x_1, x_1] = x_5, [x_1, x_2] = \alpha x_4, [x_2, x_1] = x_4, [x_2, x_3] = x_5, \quad \alpha \in \mathbb{C}\backslash\{-1\}$.
$\mathcal{A}_{178}$: $[x_1, x_1] = x_4, [x_1, x_2] = x_5, [x_2, x_3] = x_5$.
$\mathcal{A}_{179}$: $[x_1, x_1] = x_4, [x_1, x_2] = x_4, [x_2, x_3] = x_5$.
$\mathcal{A}_{180}$: $[x_1, x_1] = x_4, [x_1, x_2] = x_4 + x_5, [x_2, x_3] = x_5$.
$\mathcal{A}_{181}$: $[x_1, x_1] = x_4, [x_1, x_2] = x_4 + x_5, [x_2, x_1] = x_4, [x_2, x_3] = x_5$.
$\mathcal{A}_{182}(\alpha)$: $[x_1, x_1] = x_4, [x_1, x_2] = \alpha x_4, [x_2, x_1] = x_4, [x_2, x_3] = x_5, \quad \alpha \in \mathbb{C}\backslash\{0, 1\}$.
$\mathcal{A}_{183}(\alpha)$: $[x_1, x_1] = x_4, [x_1, x_2] = \alpha x_4 + x_5, [x_2, x_1] = x_4, [x_2, x_3] = x_5, \quad \alpha \in \mathbb{C}\backslash\{0, 1\}$.
$\mathcal{A}_{184}(\alpha)$: $[x_1, x_2] = x_4, [x_2, x_2] = \alpha x_5, [x_2, x_3] = x_5, [x_3, x_3] = x_5, \quad \alpha \in \mathbb{C}$.
$\mathcal{A}_{185}(\alpha)$: $[x_1, x_2] = x_4, [x_2, x_1] = x_5, [x_2, x_2] = \alpha x_5, [x_2, x_3] = x_5, [x_3, x_3] = x_5, \quad \alpha \in \mathbb{C}$.

$\mathcal{A}_{186}(\alpha,\beta)$: $[x_1,x_1]=x_5,[x_1,x_2]=x_4,[x_2,x_1]=\alpha x_5,[x_2,x_2]=\beta x_5,[x_2,x_3]=x_5,[x_3,x_3]=x_5, \quad \alpha,\beta\in\mathbb{C}$.
$\mathcal{A}_{187}(\alpha,\beta)$: $[x_1,x_2]=\alpha x_4,[x_2,x_1]=x_4,[x_2,x_2]=\beta x_5,[x_2,x_3]=x_5,[x_3,x_3]=x_5$, $\alpha\in\mathbb{C}\backslash\{-1,0\},\beta\in\mathbb{C}$.
$\mathcal{A}_{188}(\alpha,\beta)$: $[x_1,x_2]=\alpha x_4+x_5,[x_2,x_1]=x_4,[x_2,x_2]=\beta x_5,[x_2,x_3]=x_5,[x_3,x_3]=x_5$, $\alpha\in\mathbb{C}\backslash\{-1\},\beta\in\mathbb{C}$.
$\mathcal{A}_{189}(\alpha,\beta,\gamma)$: $[x_1,x_1]=x_5,[x_1,x_2]=\alpha x_4+\beta x_5,[x_2,x_1]=x_4,[x_2,x_2]=\gamma x_5,[x_2,x_3]=x_5,[x_3,x_3]=x_5, \quad \alpha\in\mathbb{C}\backslash\{-1\},\beta,\gamma\in\mathbb{C}$.
$\mathcal{A}_{190}(\alpha)$: $[x_1,x_1]=x_4,[x_1,x_2]=x_5,[x_2,x_2]=\alpha x_5,[x_2,x_3]=x_5,[x_3,x_3]=x_5, \quad \alpha\in\mathbb{C}\backslash\{0\}$.
$\mathcal{A}_{191}$: $[x_1,x_1]=x_4,[x_2,x_1]=x_5,[x_2,x_3]=x_5,[x_3,x_3]=x_5$.
$\mathcal{A}_{192}(\alpha,\beta,\gamma)$: $[x_1,x_1]=x_4,[x_1,x_2]=x_4+\alpha x_5,[x_2,x_1]=\beta x_5,[x_2,x_2]=\gamma x_5,[x_2,x_3]=x_5,[x_3,x_3]=x_5, \quad \alpha,\beta,\gamma\in\mathbb{C}$.
$\mathcal{A}_{193}(\alpha,\beta,\gamma)$: $[x_1,x_1]=x_4,[x_1,x_2]=x_4+\alpha x_5,[x_2,x_1]=x_4+\beta x_5,[x_2,x_2]=\gamma x_5,[x_2,x_3]=x_5,[x_3,x_3]=x_5, \quad \alpha,\beta,\gamma\in\mathbb{C},(\alpha,\beta)\neq(0,0)$.
$\mathcal{A}_{194}(\alpha,\beta,\gamma,\theta)$: $[x_1,x_1]=x_4,[x_1,x_2]=\alpha x_4+\beta x_5,[x_2,x_1]=x_4+\gamma x_5,[x_2,x_2]=\theta x_5,[x_2,x_3]=x_5,[x_3,x_3]=x_5, \quad \alpha\in\mathbb{C}\backslash\{0,1\},\beta,\gamma,\theta\in\mathbb{C},(\beta,\gamma,\theta)\neq(0,0,0)$.
$\mathcal{A}_{195}$: $[x_1,x_2]=x_5,[x_2,x_2]=x_4,[x_2,x_3]=x_5$.
$\mathcal{A}_{196}$: $[x_1,x_2]=x_4=-[x_2,x_1],[x_2,x_2]=x_4,[x_2,x_3]=x_5$.
$\mathcal{A}_{197}$: $[x_1,x_1]=x_5,[x_1,x_2]=x_4=-[x_2,x_1],[x_2,x_2]=x_4,[x_2,x_3]=x_5$.
$\mathcal{A}_{198}$: $[x_1,x_1]=x_5,[x_2,x_1]=ix_5,[x_2,x_2]=x_4,[x_2,x_3]=x_5,[x_3,x_3]=x_5$.
$\mathcal{A}_{199}(\alpha,\beta)$: $[x_1,x_1]=\alpha x_5,[x_1,x_2]=x_4,[x_2,x_1]=-x_4+\beta x_5,[x_2,x_2]=x_4,[x_2,x_3]=x_5,[x_3,x_3]=x_5, \quad \alpha,\beta\in\mathbb{C}$.
$\mathcal{A}_{200}(\alpha)$: $[x_1,x_1]=x_5,[x_2,x_1]=x_4,[x_2,x_3]=\alpha x_5,[x_3,x_2]=x_5, \quad \alpha\in\mathbb{C}$.
$\mathcal{A}_{201}$: $[x_1,x_1]=x_5,[x_2,x_1]=x_4,[x_2,x_2]=x_5,[x_2,x_3]=x_5=-[x_3,x_2]$.
$\mathcal{A}_{202}(\alpha)$: $[x_2,x_1]=x_4,[x_1,x_3]=x_5,[x_2,x_3]=\alpha x_5,[x_3,x_2]=x_5, \quad \alpha\in\mathbb{C}$.
$\mathcal{A}_{203}(\alpha)$: $[x_2,x_1]=x_4,[x_2,x_2]=x_5,[x_1,x_3]=x_5,[x_2,x_3]=\alpha x_5,[x_3,x_2]=x_5, \quad \alpha\in\mathbb{C}$.
$\mathcal{A}_{204}(\alpha,\beta)$: $[x_1,x_2]=x_4,[x_2,x_1]=\alpha x_4,[x_2,x_3]=\beta x_5,[x_3,x_2]=x_5, \quad \alpha,\beta\in\mathbb{C}\backslash\{-1\}$.
$\mathcal{A}_{205}(\alpha,\beta)$: $[x_1,x_1]=x_5,[x_1,x_2]=x_4,[x_2,x_1]=\alpha x_4,[x_2,x_3]=\beta x_5,[x_3,x_2]=x_5, \quad \alpha\in\mathbb{C}\backslash\{-1\},\beta\in\mathbb{C}$.
$\mathcal{A}_{206}(\alpha)$: $[x_1,x_2]=x_4,[x_2,x_1]=\alpha x_4+x_5,[x_2,x_3]=\alpha x_5,[x_3,x_2]=x_5, \quad \alpha\in\mathbb{C}\backslash\{-1\}$.
$\mathcal{A}_{207}(\alpha)$: $[x_1,x_1]=x_5,[x_1,x_2]=x_4,[x_2,x_1]=\alpha x_4,[x_2,x_3]=\alpha x_5,[x_3,x_2]=x_5, \quad \alpha\in\mathbb{C}\backslash\{-1\}$.
$\mathcal{A}_{208}(\alpha,\beta)$: $[x_1,x_2]=x_4,[x_2,x_1]=\alpha x_4,[x_1,x_3]=x_5,[x_2,x_3]=\beta x_5,[x_3,x_2]=x_5, \quad \alpha\in\mathbb{C}\backslash\{-1\},\beta\in\mathbb{C}$.
$\mathcal{A}_{209}(\alpha,\beta)$: $[x_1,x_2]=x_4,[x_2,x_1]=\alpha x_4+x_5,[x_1,x_3]=x_5,[x_2,x_3]=\beta x_5,[x_3,x_2]=x_5, \quad \alpha,\beta\in\mathbb{C}\backslash\{-1\}$.
$\mathcal{A}_{210}(\alpha)$: $[x_1,x_2]=x_4,[x_2,x_1]=\alpha x_4,[x_2,x_2]=x_5,[x_2,x_3]=x_5=-[x_3,x_2], \quad \alpha\in\mathbb{C}\backslash\{-1\}$.
$\mathcal{A}_{211}(\alpha)$: $[x_1,x_1]=x_5,[x_1,x_2]=x_4,[x_2,x_1]=\alpha x_4,[x_2,x_2]=x_5,[x_2,x_3]=x_5=-[x_3,x_2]$, $\alpha\in\mathbb{C}\backslash\{-1\}$.
$\mathcal{A}_{212}(\alpha)$: $[x_1,x_2]=x_4,[x_2,x_1]=\alpha x_4,[x_2,x_2]=x_5,[x_1,x_3]=x_5,[x_2,x_3]=x_5=-[x_3,x_2]$, $\alpha\in\mathbb{C}\backslash\{-1\}$.
$\mathcal{A}_{213}(\alpha)$: $[x_1,x_1]=x_4,[x_1,x_3]=x_5,[x_2,x_3]=\alpha x_5,[x_3,x_2]=x_5, \quad \alpha\in\mathbb{C}\backslash\{0,1\}$.
$\mathcal{A}_{214}$: $[x_1,x_1]=x_4,[x_2,x_1]=x_5,[x_1,x_3]=x_5,[x_3,x_2]=x_5$.
$\mathcal{A}_{215}(\alpha,\beta)$: $[x_1,x_1]=x_4,[x_2,x_1]=x_4,[x_1,x_3]=\alpha x_5,[x_2,x_3]=\beta x_5,[x_3,x_2]=x_5, \quad \alpha\in\mathbb{C}\backslash\{0\},\beta\in\mathbb{C}$.
$\mathcal{A}_{216}$: $[x_1,x_1]=x_4,[x_2,x_1]=x_4+x_5,[x_2,x_3]=x_5=-[x_3,x_2]$.
$\mathcal{A}_{217}(\alpha,\beta)$: $[x_1,x_1]=x_4,[x_2,x_1]=x_4+x_5,[x_1,x_3]=\alpha x_5,[x_2,x_3]=\beta x_5,[x_3,x_2]=x_5$, $\alpha\in\mathbb{C},\beta\in\mathbb{C}\backslash\{-1\}$.
$\mathcal{A}_{218}(\alpha,\beta,\gamma)$: $[x_1,x_1]=x_4,[x_1,x_2]=x_4,[x_2,x_1]=\alpha x_4,[x_1,x_3]=\beta x_5,[x_2,x_3]=\gamma x_5,[x_3,x_2]=x_5, \quad \alpha,\beta\in\mathbb{C}\backslash\{0\},\gamma\in\mathbb{C}$.
$\mathcal{A}_{219}$: $[x_1,x_1]=x_4,[x_1,x_2]=x_4+x_5,[x_1,x_3]=x_5,[x_3,x_2]=x_5$.
$\mathcal{A}_{220}(\alpha,\beta)$: $[x_1,x_1]=x_4,[x_1,x_2]=x_4+x_5,[x_2,x_1]=\alpha x_4,[x_1,x_3]=(1+\beta)x_5,[x_2,x_3]=\beta x_5,[x_3,x_2]=x_5, \quad \alpha,\beta\in\mathbb{C},\alpha(1+\beta)\neq\beta$.
$\mathcal{A}_{221}(\alpha,\beta)$: $[x_1,x_1]=x_4,[x_1,x_2]=x_4+x_5,[x_2,x_1]=x_4,[x_1,x_3]=\alpha x_5,[x_2,x_3]=\beta x_5,[x_3,x_2]=x_5, \quad \alpha,\beta\in\mathbb{C},\alpha\neq1+\beta,\alpha\neq\beta$.
$\mathcal{A}_{222}(\alpha,\beta)$: $[x_1,x_1]=x_4,[x_1,x_2]=x_4+x_5,[x_2,x_1]=\alpha x_4,[x_1,x_3]=\beta x_5,[x_3,x_2]=x_5, \quad \alpha,\beta\in\mathbb{C}\backslash\{0,1\}$.
$\mathcal{A}_{223}(\alpha,\beta,\gamma)$: $[x_1,x_1]=x_4,[x_1,x_2]=x_4+x_5,[x_2,x_1]=\alpha x_4,[x_1,x_3]=\beta x_5,[x_2,x_3]=\gamma x_5,[x_3,x_2]=x_5, \quad \alpha\in\mathbb{C}\backslash\{0,1\},\beta\in\mathbb{C},\gamma\in\mathbb{C}\backslash\{0\},\beta\neq1+\gamma,\alpha\beta\neq\gamma,(\alpha+1)\beta\neq\gamma$.
$\mathcal{A}_{224}(\alpha)$: $[x_1,x_1]=x_4,[x_1,x_2]=x_4+\frac{1}{\alpha}x_5,[x_2,x_1]=x_5,[x_2,x_3]=\alpha x_5,[x_3,x_2]=x_5, \quad \alpha\in\mathbb{C}\backslash\{-1,0\}$.
$\mathcal{A}_{225}$: $[x_1,x_1]=x_4,[x_1,x_2]=x_4,[x_2,x_1]=x_4+x_5,[x_3,x_2]=x_5$.
$\mathcal{A}_{226}(\alpha)$: $[x_1,x_1]=x_4,[x_1,x_2]=x_4,[x_2,x_1]=x_4+x_5,[x_1,x_3]=\alpha x_5,[x_2,x_3]=\alpha x_5,[x_3,x_2]=x_5, \quad \alpha\in\mathbb{C}\backslash\{0\}$.
$\mathcal{A}_{227}(\alpha)$: $[x_1,x_1]=x_4,[x_1,x_2]=x_4,[x_2,x_1]=\alpha x_4+x_5,[x_3,x_2]=x_5, \quad \alpha\in\mathbb{C}\backslash\{0,1\}$.
$\mathcal{A}_{228}(\alpha)$: $[x_1,x_1]=x_4,[x_1,x_2]=x_4,[x_2,x_1]=\frac{\alpha}{1+\alpha}x_4+x_5,[x_1,x_3]=(1+\alpha)x_5,[x_2,x_3]=\alpha x_5,[x_3,x_2]=x_5, \quad \alpha\in\mathbb{C}\backslash\{-1,0,1\}$.
$\mathcal{A}_{229}$: $[x_1,x_1]=x_4,[x_2,x_1]=x_5,[x_2,x_2]=x_5,[x_2,x_3]=x_5=-[x_3,x_2]$.
$\mathcal{A}_{230}(\alpha)$: $[x_1,x_1]=x_4,[x_2,x_1]=x_4+\alpha x_5,[x_2,x_2]=x_5,[x_2,x_3]=x_5=-[x_3,x_2], \quad \alpha\in\mathbb{C}\backslash\{0,1\}$.
$\mathcal{A}_{231}(\alpha)$: $[x_1,x_1]=x_4,[x_2,x_1]=x_4+\alpha x_5,[x_2,x_2]=x_5,[x_1,x_3]=x_5,[x_2,x_3]=x_5=-[x_3,x_2], \quad \alpha\in\mathbb{C}\backslash\{0\}$.

$\mathcal{A}_{232}(\alpha)$: $[x_1,x_1]=x_4,[x_1,x_2]=x_4,[x_2,x_1]=x_4+\alpha x_5,[x_2,x_2]=x_5,[x_2,x_3]=x_5=-[x_3,x_2],\quad \alpha\in\mathbb{C}\backslash\{0,\frac{1}{2}\}$.

$\mathcal{A}_{233}(\alpha)$: $[x_1,x_1]=x_4,[x_1,x_2]=x_4,[x_2,x_1]=x_4+\alpha x_5,[x_2,x_2]=x_5,[x_1,x_3]=x_5,$ $[x_2,x_3]=x_5=-[x_3,x_2],\quad \alpha\in\mathbb{C}\backslash\{\frac{1}{2}\}$.

$\mathcal{A}_{234}(\alpha,\beta)$: $[x_1,x_1]=x_4,[x_1,x_2]=x_4,[x_2,x_1]=\alpha x_4+\beta x_5,[x_2,x_2]=x_5,[x_2,x_3]=x_5=-[x_3,x_2],\quad \alpha\in\mathbb{C}\backslash\{0,1\},\beta\in\mathbb{C}$.

$\mathcal{A}_{235}(\alpha)$: $[x_1,x_1]=x_4,[x_1,x_2]=x_4,[x_2,x_1]=\frac{\alpha-1}{\alpha}x_4,[x_2,x_2]=x_5,[x_1,x_3]=\alpha x_5,$ $[x_2,x_3]=x_5=-[x_3,x_2],\quad \alpha\in\mathbb{C}\backslash\{0,1\}$.

$\mathcal{A}_{236}(\alpha)$: $[x_1,x_1]=x_5,[x_1,x_2]=x_4=-[x_2,x_1],[x_2,x_2]=x_4,[x_2,x_3]=\alpha x_5,[x_3,x_2]=x_5,\quad \alpha\in\mathbb{C}$.

$\mathcal{A}_{237}(\alpha)$: $[x_1,x_2]=x_4=-[x_2,x_1],[x_2,x_2]=x_4,[x_1,x_3]=\alpha x_5,[x_3,x_2]=x_5,\quad \alpha\in\mathbb{C}\backslash\{0\}$.

$\mathcal{A}_{238}(\alpha)$: $[x_1,x_2]=x_4,[x_2,x_1]=-x_4+x_5,[x_2,x_2]=x_4,[x_1,x_3]=\alpha x_5,[x_3,x_2]=x_5,\quad \alpha\in\mathbb{C}\backslash\{0,1\}$.

$\mathcal{A}_{239}$: $[x_1,x_1]=x_5,[x_1,x_2]=x_4=-[x_2,x_1],[x_2,x_2]=x_4,[x_1,x_3]=x_5,[x_3,x_2]=x_5$.

$\mathcal{A}_{240}$: $[x_1,x_2]=x_4,[x_2,x_1]=x_5,[x_2,x_3]=x_4$.

$\mathcal{A}_{241}(\alpha,\beta)$: $[x_1,x_2]=x_4+\alpha x_5,[x_2,x_1]=\beta x_4,[x_2,x_2]=x_5,[x_1,x_3]=x_5,[x_2,x_3]=x_4,$ $\alpha\in\mathbb{C},\beta\in\mathbb{C}\backslash\{0\}$.

$\mathcal{A}_{242}(\alpha)$: $[x_1,x_2]=x_4,[x_2,x_1]=\alpha x_4+(\alpha+1)x_5,[x_2,x_3]=x_4,[x_3,x_2]=x_5,\quad \alpha\in\mathbb{C}$.

$\mathcal{A}_{243}$: $[x_1,x_2]=x_4+2x_5,[x_2,x_1]=-x_5,[x_2,x_2]=x_5,[x_2,x_3]=x_4,[x_3,x_2]=x_5$.

$\mathcal{A}_{244}(\alpha,\beta)$: $[x_1,x_2]=x_4,[x_2,x_1]=\alpha x_5,[x_2,x_2]=\beta x_5,[x_1,x_3]=x_5,[x_2,x_3]=x_4,[x_3,x_2]=x_5,\quad \alpha,\beta\in\mathbb{C}$.

$\mathcal{A}_{245}(\alpha,\beta,\gamma)$: $[x_1,x_2]=x_4+\alpha x_5,[x_2,x_1]=\beta x_4,[x_2,x_2]=\gamma x_5,[x_1,x_3]=x_5,[x_2,x_3]=x_4,$ $[x_3,x_2]=x_5,\quad \alpha,\gamma\in\mathbb{C},\beta\in\mathbb{C}\backslash\{0\}$.

$\mathcal{A}_{246}$: $[x_1,x_2]=x_4,[x_1,x_3]=x_5,[x_3,x_1]=x_5,[x_2,x_3]=x_4$.

$\mathcal{A}_{246}$: $[x_1,x_2]=x_4+x_5,[x_1,x_3]=-x_5,[x_3,x_1]=x_5,[x_2,x_3]=x_4$.

$\mathcal{A}_{247}(\alpha,\beta)$: $[x_1,x_2]=x_4+x_5,[x_2,x_1]=\alpha x_4,[x_1,x_3]=\beta x_5,[x_3,x_1]=x_5,[x_2,x_3]=x_4,$ $\alpha\in\mathbb{C},\beta\in\mathbb{C}\backslash\{-1\}$.

$\mathcal{A}_{248}(\alpha,\beta)$: $[x_1,x_2]=x_4+\alpha x_5,[x_2,x_1]=\beta x_4+x_5,[x_3,x_1]=x_5,[x_2,x_3]=x_4,\quad \alpha\in\mathbb{C}\backslash\{0\},\beta\in\mathbb{C},\alpha\beta\neq 1$.

$\mathcal{A}_{249}(\alpha,\beta,\gamma)$: $[x_1,x_2]=x_4+\alpha x_5,[x_2,x_1]=\beta x_4+x_5,[x_1,x_3]=\gamma x_5,[x_3,x_1]=x_5,[x_2,x_3]=x_4,\quad \alpha,\beta\in\mathbb{C},\gamma\in\mathbb{C}\backslash\{0\},\alpha\neq\gamma$.

$\mathcal{A}_{250}(\alpha)$: $[x_1,x_2]=x_4+\alpha x_5,[x_2,x_1]=-x_4+\frac{-1-\alpha^2}{\alpha}x_5,[x_2,x_2]=x_5,[x_3,x_1]=x_5,[x_2,x_3]=x_4,\quad \alpha\in\mathbb{C}\backslash\{0\}$.

$\mathcal{A}_{251}(\alpha,\beta,\gamma)$: $[x_1,x_2]=x_4+\alpha x_5,[x_2,x_1]=\beta x_4+\gamma x_5,[x_2,x_2]=x_5,[x_3,x_1]=x_5,[x_2,x_3]=x_4,\quad \alpha,\beta,\gamma\in\mathbb{C},\alpha\gamma-\alpha^2\beta+1\neq 0$.

$\mathcal{A}_{252}(\alpha,\beta,\gamma)$: $[x_1,x_2]=x_4+\alpha x_5,[x_2,x_1]=\beta x_4,[x_2,x_2]=x_5,[x_1,x_3]=\gamma x_5,[x_3,x_1]=x_5,$ $[x_2,x_3]=x_4,\quad \alpha,\beta\in\mathbb{C},\gamma\in\mathbb{C}\backslash\{0\}$.

$\mathcal{A}_{253}(\alpha,\beta)$: $[x_1,x_2]=x_4-\frac{2}{\alpha}x_5,[x_2,x_1]=\beta x_4+\alpha x_5,[x_2,x_2]=x_5,[x_1,x_3]=-\frac{1}{\alpha^2}x_5,[x_3,x_1]=x_5,[x_2,x_3]=x_4,\quad \alpha\in\mathbb{C}\backslash\{-1,0,1\},\beta\in\mathbb{C}\backslash\{0\}$.

$\mathcal{A}_{254}(\alpha,\beta,\gamma)$: $[x_1,x_2]=x_4+2\beta\gamma x_5,[x_2,x_1]=\alpha x_4+\beta x_5,[x_2,x_2]=x_5,[x_1,x_3]=\gamma x_5,[x_3,x_1]=x_5,[x_2,x_3]=x_4,\quad \alpha\in\mathbb{C},\beta,\gamma\in\mathbb{C}\backslash\{0\},\beta^2\gamma\neq -1$.

$\mathcal{A}_{255}(\alpha,\beta)$: $[x_1,x_2]=x_4+\alpha x_5,[x_2,x_1]=\beta x_5,[x_2,x_2]=x_5,[x_1,x_3]=-x_5,[x_3,x_1]=x_5,$ $[x_2,x_3]=x_4,\quad \alpha\in\mathbb{C},\beta\in\mathbb{C}\backslash\{0\},\alpha\neq -2\beta,\beta^2+\alpha\beta+1\neq 0$.

$\mathcal{A}_{256}(\alpha,\beta,\gamma,\theta)$: $[x_1,x_2]=x_4+\alpha x_5,[x_2,x_1]=\beta x_4+\gamma x_5,[x_2,x_2]=x_5,[x_1,x_3]=\theta x_5,[x_3,x_1]=x_5,[x_2,x_3]=x_4,\quad \alpha,\beta\in\mathbb{C},\gamma\in\mathbb{C}\backslash\{0\},\theta\in\mathbb{C}\backslash\{-1,0\},\alpha\neq 2\gamma\theta$.

$\mathcal{A}_{257}(\alpha)$: $[x_1,x_2]=x_4,[x_1,x_3]=\alpha x_5,[x_2,x_3]=x_4,[x_3,x_3]=x_5,\quad \alpha\in\mathbb{C}\backslash\{0\}$.

$\mathcal{A}_{258}(\alpha)$: $[x_1,x_2]=x_4+x_5,[x_2,x_1]=\alpha x_4,[x_1,x_3]=-x_5,[x_3,x_1]=\alpha x_5,[x_2,x_3]=x_4,$ $[x_3,x_3]=x_5,\quad \alpha\in\mathbb{C}\backslash\{0\}$.

$\mathcal{A}_{259}(\alpha,\beta)$: $[x_1,x_2]=x_4+x_5,[x_2,x_1]=\alpha x_4,[x_1,x_3]=\beta x_5,[x_3,x_1]=\alpha x_5,[x_2,x_3]=x_4,$ $[x_3,x_3]=x_5,\quad \alpha\in\mathbb{C}\backslash\{0,1\},\beta\in\mathbb{C}\backslash\{-1,0\},\alpha\neq\beta$.

$\mathcal{A}_{260}(\alpha,\beta)$: $[x_1,x_2]=x_4+x_5,[x_2,x_1]=\alpha x_4,[x_1,x_3]=\alpha x_5,[x_3,x_1]=\beta x_5,[x_2,x_3]=x_4,$ $[x_3,x_3]=x_5,\quad \alpha\in\mathbb{C}\backslash\{0\},\alpha\neq\beta$.

$\mathcal{A}_{261}(\alpha,\beta,\gamma)$: $[x_1,x_2]=x_4+x_5,[x_2,x_1]=\alpha x_4,[x_1,x_3]=\beta x_5,[x_3,x_1]=\gamma x_5,[x_2,x_3]=x_4,$ $[x_3,x_3]=x_5,\quad \alpha,\beta,\gamma\in\mathbb{C},\alpha\neq\beta,\alpha\neq\gamma$.

$\mathcal{R}_1$: $[x_1,x_2]=x_4+\alpha x_5,[x_2,x_1]=\beta x_4+x_5,[x_1,x_3]=\gamma x_5,[x_3,x_1]=\theta x_5,[x_2,x_3]=x_4,$ $[x_3,x_3]=x_5,\quad \alpha,\beta,\gamma,\theta\in\mathbb{C}$.

$\mathcal{R}_2$: $[x_1,x_2]=x_4+\alpha x_5,[x_2,x_1]=\beta x_4+\gamma x_5,[x_1,x_3]=\theta x_5,[x_3,x_1]=\delta x_5,[x_2,x_3]=x_4,$ $[x_3,x_3]=x_5,\quad \alpha,\beta,\gamma,\theta,\delta\in\mathbb{C}$.

$\mathcal{R}_3$: $[x_1,x_2]=x_4+\alpha x_5,[x_2,x_1]=\beta x_4+\gamma x_5,[x_2,x_2]=\theta x_5,[x_1,x_3]=\delta x_5,[x_3,x_1]=\lambda x_5,$ $[x_2,x_3]=x_4,[x_3,x_3]=x_5,\quad \alpha,\beta,\gamma,\theta,\delta,\lambda\in\mathbb{C}$.

$\mathcal{R}_4$: $[x_1,x_1]=x_5,[x_1,x_2]=x_4+\alpha x_5,[x_2,x_1]=\beta x_5,[x_2,x_2]=\gamma x_5,[x_1,x_3]=\theta x_5=-[x_3,x_1],[x_2,x_3]=x_4,[x_3,x_2]=\delta x_5,\quad \alpha,\beta,\gamma,\theta,\delta\in\mathbb{C}$.

$\mathcal{R}_5$: $[x_1,x_1]=x_4,[x_1,x_2]=\alpha x_5,[x_2,x_1]=\beta x_5,[x_2,x_2]=\gamma x_5,[x_1,x_3]=\theta x_5,[x_3,x_1]=\delta x_5,$ $[x_2,x_3]=x_4,[x_3,x_2]=\lambda x_5,[x_3,x_3]=\mu x_5,\quad \alpha,\beta,\gamma,\theta,\delta,\lambda,\mu\in\mathbb{C}$.

$\mathcal{R}_6$: $[x_1,x_1]=x_4+x_5,[x_1,x_2]=\alpha x_5,[x_2,x_1]=\beta x_5,[x_2,x_2]=\gamma x_5,[x_1,x_3]=\theta x_5=-[x_3,x_1],[x_2,x_3]=x_4,[x_3,x_2]=\delta x_5,\quad \alpha,\beta,\gamma,\theta,\delta\in\mathbb{C}$.

$\mathcal{R}_7$: $[x_1,x_1]=x_4+\alpha x_5,[x_1,x_2]=\beta x_4+\gamma x_5,[x_2,x_1]=\theta x_5,[x_2,x_2]=\delta x_5,[x_1,x_3]=\lambda x_5,$ $[x_2,x_3]=x_4=-[x_3,x_2],\quad \alpha,\beta,\gamma,\theta,\delta,\lambda\in\mathbb{C}$.

$\mathcal{R}_8$: $[x_1,x_1]=x_4+\alpha x_5,[x_1,x_2]=\beta x_4+\gamma x_5,[x_2,x_1]=\theta x_5,[x_1,x_3]=\delta x_5,[x_2,x_3]=x_4+x_5,$ $[x_3,x_2]=-x_4,\quad \alpha,\beta,\gamma,\theta,\delta\in\mathbb{C}$.

$\mathcal{R}_9$: $[x_1,x_1]=x_4+\alpha x_5, [x_1,x_2]=\beta x_4+\gamma x_5, [x_2,x_2]=\theta x_5, [x_1,x_3]=\delta x_5, [x_3,x_1]=x_5, [x_2,x_3]=x_4+\lambda x_5, [x_3,x_2]=-x_4, \quad \alpha,\beta,\gamma,\theta,\delta,\lambda\in\mathbb{C}$.

$\mathcal{R}_{10}$: $[x_1,x_1]=x_4+\alpha x_5, [x_1,x_2]=\beta x_4+\gamma x_5, [x_2,x_1]=\theta x_5, [x_1,x_3]=\delta x_5, [x_3,x_1]=\lambda x_5, [x_2,x_3]=x_4+\mu x_5, [x_3,x_2]=-x_4, [x_3,x_3]=x_5, \quad \alpha,\beta,\gamma,\theta,\delta,\lambda,\mu\in\mathbb{C}$.

$\mathcal{R}_{11}$: $[x_1,x_1]=x_4+\alpha x_5, [x_1,x_2]=\beta x_4+\gamma x_5, [x_2,x_1]=\theta x_5, [x_2,x_2]=\delta x_5, [x_2,x_3]=\lambda x_4+\mu x_5, [x_3,x_2]=x_4, \quad \alpha,\beta,\gamma,\theta,\delta,\lambda,\mu\in\mathbb{C}$.

$\mathcal{R}_{12}$: $[x_1,x_1]=x_4, [x_1,x_2]=\alpha x_4+\beta x_5, [x_2,x_1]=\gamma x_5, [x_2,x_2]=\theta x_5, [x_1,x_3]=x_5, [x_2,x_3]=\delta x_4+\lambda x_5, [x_3,x_2]=x_4, \quad \alpha,\beta,\gamma,\theta,\delta,\lambda\in\mathbb{C}$.

$\mathcal{R}_{13}$: $[x_1,x_1]=x_4+\alpha x_5, [x_1,x_2]=\beta x_4+\gamma x_5, [x_2,x_1]=\theta x_5, [x_2,x_2]=\delta x_5, [x_1,x_3]=\lambda x_5, [x_3,x_1]=x_5, [x_2,x_3]=\mu x_4+\omega x_5, [x_3,x_2]=x_4, \quad \alpha,\beta,\gamma,\theta,\delta,\lambda,\mu,\omega\in\mathbb{C}$.

$\mathcal{R}_{14}$: $[x_1,x_1]=x_4+\alpha x_5, [x_1,x_2]=\beta x_4+\gamma x_5, [x_2,x_1]=\theta x_5, [x_2,x_2]=\delta x_5, [x_1,x_3]=\lambda x_5, [x_2,x_3]=\mu x_4+\omega x_5, [x_3,x_2]=x_4, [x_3,x_3]=x_5, \quad \alpha,\beta,\gamma,\theta,\delta,\lambda,\mu,\omega\in\mathbb{C}$.

$\mathcal{R}_{15}$: $[x_1,x_1]=\alpha x_5, [x_1,x_2]=\beta x_4+\gamma x_5, [x_2,x_1]=x_4+\theta x_5, [x_2,x_2]=\delta x_5, [x_1,x_3]=x_4+\lambda x_5, [x_3,x_1]=\mu x_5, [x_2,x_3]=\omega x_5, [x_3,x_2]=x_4, [x_3,x_3]=\varphi x_5, \quad \alpha,\beta,\gamma,\theta,\delta,\lambda,\mu,\omega,\varphi\in\mathbb{C}$.

Throughout this work, we use Mathematica program implementing Algorithm 2.6 given in [**6**] which determines if given two Leibniz algebras are isomorphic. However we note that for some difficult cases in Theorem 3.15, this program cannot decide whether given two Leibniz algebras are isomorphic. Note that in Theorem 3.15 we obtain 101 distinct isomorphism classes $\mathcal{A}_{161}, \mathcal{A}_{162}, \ldots, \mathcal{A}_{261}(\alpha,\beta,\gamma)$; and additional 15 algebras $\mathcal{R}_1,\ldots,\mathcal{R}_{15}$ that are not distinct and can be isomorphic to the isomorphism classes $\mathcal{A}_{161}, \mathcal{A}_{162}, \ldots, \mathcal{A}_{261}(\alpha,\beta,\gamma)$.

Acknowledgments

This work is part of author's Ph.D. Thesis submitted to North Carolina State University. The author thanks to his advisors, Dr. Kailash Misra and Dr. Ernest Stitzinger, for their guidance throughout his Ph.D.

References

[1] S. Albeverio, B. A. Omirov, and I. S. Rakhimov, *Classification of 4-dimensional nilpotent complex Leibniz algebras*, Extracta Math. **21** (2006), no. 3, 197–210. MR2332070

[2] Sh. A. Ayupov and B. A. Omirov, *On 3-dimensional Leibniz algebras* (English, with Russian and Uzbek summaries), Uzbek. Mat. Zh. **1** (1999), 9–14. MR1799250

[3] Sh. A. Ayupov and B. A. Omirov, *On some classes of nilpotent Leibniz algebras* (Russian, with Russian summary), Sibirsk. Mat. Zh. **42** (2001), no. 1, 18–29, i; English transl., Siberian Math. J. **42** (2001), no. 1, 15–24. MR1830788

[4] D. W. Barnes, *Some theorems on Leibniz algebras*, Comm. Algebra **39** (2011), no. 7, 2463–2472. MR2821724

[5] A. Bloh, *On a generalization of the concept of Lie algebra* (Russian), Dokl. Akad. Nauk SSSR **165** (1965), 471–473. MR0193114

[6] J. M. Casas, M. A. Insua, M. Ladra, and S. Ladra, *An algorithm for the classification of 3-dimensional complex Leibniz algebras*, Linear Algebra Appl. **436** (2012), no. 9, 3747–3756. MR2900750

[7] I. Demir, K. C. Misra, and E. Stitzinger, *On some structures of Leibniz algebras*, Recent advances in representation theory, quantum groups, algebraic geometry, and related topics, Contemp. Math., vol. 623, Amer. Math. Soc., Providence, RI, 2014, pp. 41–54. MR3288621

[8] I. Demir, K. C. Misra, and E. Stitzinger, *Classification of some solvable Leibniz algebras*, Algebr. Represent. Theory **19** (2016), no. 2, 405–417. MR3489103

[9] I. Demir, K. C. Misra, and E. Stitzinger, *On classification of four-dimensional nilpotent Leibniz algebras*, Comm. Algebra **45** (2017), no. 3, 1012–1018. MR3573356

[10] I. Demir, *Classification of 5-Dimensional Complex Nilpotent Leibniz Algebras*, ProQuest LLC, Ann Arbor, MI, 2016. Thesis (Ph.D.)–North Carolina State University. MR3653708

[11] I. Demir, *Classification of 5−Dimensional Complex Nilpotent Leibniz Algebras*, arxiv:1706.00951.

[12] Gong, M.-P., Classification of nilpotent Lie algebras of dimension 7. *PhD Thesis*, University of Waterloo, Waterloo, Canada, 1998.

[13] J.-L. Loday, *Une version non commutative des algèbres de Lie: les algèbres de Leibniz* (French), Enseign. Math. (2) **39** (1993), no. 3-4, 269–293. MR1252069

[14] J.-L. Loday and T. Pirashvili, *Universal enveloping algebras of Leibniz algebras and (co)homology*, Math. Ann. **296** (1993), no. 1, 139–158. MR1213376

[15] I. S. Rakhimov and U. D. Bekbaev, *On isomorphisms and invariants of finite dimensional complex filiform Leibniz algebras*, Comm. Algebra **38** (2010), no. 12, 4705–4738. MR2805139

[16] Rikhsiboev, I.M., Rakhimov I.S. Classification of three dimensional complex Leibniz algebras. *AIP Conference Proc.*, 2012, 1450:358-362, DOI: 10.1063/1.4724168.

[17] Teran, F.D. Canonical forms for congruence of matrices: a tribute to H. W. Turnbull and A. C. Aitken. 2nd meeting on Linear Algebra, Matrix Analysis and Applications(ALAMA 2010), Valencia(Spain), 2010.

The Ministry of National Education of Turkey, Konak/Izmir, Turkey 35220
Email address: idemir@ncsu.edu

Contemporary Mathematics
Volume **713**, 2018
https://doi.org/10.1090/conm/713/14314

Gelfand-Tsetlin modules of $\mathfrak{sl}(3)$ in the principal block

Vyacheslav Futorny, Dimitar Grantcharov, and Luis Enrique Ramirez

ABSTRACT. We provide complete description of the simple modules in the principal block of the category of Gelfand-Tsetlin modules of $\mathfrak{sl}(3)$. In addition, we prove that every such module is a subquotient of a localization of a highest weight module.

1. Introduction

The category of Gelfand-Tsetlin modules of $\mathfrak{g} = \mathfrak{sl}(n)$ is an important category of modules that plays important role in many areas of mathematics and theoretical physics. Gelfand-Tsetlin modules are those that have generalized eigenspace decompositions over a certain maximal commutative subalgebra (*Gelfand-Tsetlin subalgebra*) Γ of the universal enveloping algebra of $\mathfrak{g}$. The concept of a Gelfand-Tsetlin module generalizes the construction of classical Gelfand-Tsetlin bases that have been introduced in [**11**]. The general theory of Gelfand-Tsetlin modules was developed in [**1**], [**2**], [**4**], [**5**], [**6**], [**7**], [**8**], [**12**], [**13**], [**14**], [**15**], [**17**], [**18**], [**19**], [**21**], among others.

Throughout the paper, for $n \geq 2$, $T_n(\mathbb{C})$ will stand for the space consisting of the following *Gelfand-Tsetlin tableau of height n*:

$$\begin{array}{ccccccccc}
\boxed{v_{n1}} & & \boxed{v_{n2}} & & \cdots & & \boxed{v_{n,n-1}} & & \boxed{v_{nn}} \\
& \boxed{v_{n-1,1}} & & & \cdots & & & \boxed{v_{n-1,n-1}} & \\
& & \cdots & & \cdots & & \cdots & & \\
& & & \boxed{v_{21}} & & \boxed{v_{22}} & & & \\
& & & & \boxed{v_{11}} & & & &
\end{array}$$

There is a natural bijection $\mathbb{C}^{\frac{n(n+1)}{2}} \to T_n(\mathbb{C})$ defined by mapping an element

$$v = (v_{n1}, ..., v_{nn} | v_{n-1,1}, ..., v_{n-1,n-1} | \cdots | v_{21}, v_{22} | v_{11})$$

in $\mathbb{C}^{\frac{n(n+1)}{2}}$ to the Gelfand-Tsetlin tableau $T(v)$ written above.

2010 *Mathematics Subject Classification.* Primary 17B10.
Key words and phrases. Gelfand-Tsetlin modules, Gelfand-Tsetlin basis, tableaux realization.

For a fixed tableau $T(v)$ in $T_n(\mathbb{C})$ we define the lattice of tableaux containing $T(v)$ as follows:

$$T(v)_{\mathbb{Z}} := \{T(v+z) \mid z=(z_{ij}) \in \mathbb{Z}^{\frac{n(n+1)}{2}}, z_{nk}=0, k=1,\dots,n\}.$$

Henceforth, we will consider the elements z in $\mathbb{Z}^{\frac{n(n-1)}{2}}$ as elements in $\mathbb{Z}^{\frac{n(n+1)}{2}}$ by adding zeros as n first entries.

In what follows, we define $V(T(v))$ to be the complex vector space with basis the set $T(v)_{\mathbb{Z}}$, i.e. $V(T(v)) = \bigoplus_{T(v')\in T(v)_{\mathbb{Z}}} \mathbb{C}T(v')$. Note that if $v', v'' \in T(v)_{\mathbb{Z}}$, then $T(v'+v'') \neq T(v') + T(v'')$ in $V(T(v))$ (even if $v'+v'' \in T(v)_{\mathbb{Z}}$).

A problem dating back to the work of Gelfand and Graev, [**10**], is to define a Gelfand-Tsetlin $\mathfrak{gl}(n)$-module (or, equivalently, $\mathfrak{sl}(n)$-module) structure on the space (or certain subspaces) of $V(T(v))$. This problem has been partially solved in a series of recent works, [**6**], [**7**], [**22**], [**23**], [**24**], [**25**]. In the case $n=3$, it is enough to consider only two cases for v: generic v and 1-singular v. Details in that case are provided in Section 3. For better visualization of the modules, we work with $\mathfrak{sl}(3)$ instead of $\mathfrak{gl}(3)$.

An important feature of the module $V(T(v))$ is that for $n=3$ it is *universal*, i.e. it contains all simple objects of the corresponding block as subquotients, see Theorem 3.8. The situation is especially interesting in the principal block as this block has the maximum number of nonisomorphic simple objects.

For the purpose of this paper, the principal block $\mathcal{GT}_0$ is defined as follows. To each $v \in \mathbb{C}^{\frac{n(n+1)}{2}}$ we associate the maximal ideal m_v of the Gelfand-Tsetlin subalgebra Γ generated by $c_{ij} - \gamma_{ij}(v)$, $1 \leq j \leq i \leq n$, where c_{ij} are the generators of Γ, and γ_{ij} are symmetric polynomials defined in (2). A natural subcategory of the category $\mathcal{GT}(n)$ of all Gelfand-Tsetlin $\mathfrak{sl}(n)$-modules related to Γ is the category $\mathcal{GT}_v(n)$ consisting of all V such that

$$\mathrm{Supp}_{GT}(V) \subseteq \{\mathsf{m}_{v'} \mid T(v') \in T(v)_{\mathbb{Z}}\},$$

where $\mathrm{Supp}_{GT}(V)$ is the Gelfand-Tsetlin support of V, see Definition 2.3. If $\bar{v}_{nk} = -k+1$ and all remaining $\bar{v}_{ij}$ are zero, then $\mathcal{GT}_{\bar{v}}(n)$ is called *the principal block* of the category of Gelfand-Tsetlin modules and will be denoted by $\mathcal{GT}_0$.

The purpose of this paper is to give an explicit description of all simple modules in $\mathcal{GT}_0$ for $n=3$ and to show that every such module is a subquotient of a localized highest weight module where the localizing set is generated by a single root element. These results provide convenient descriptions of every simple object of $\mathcal{GT}_0$.

The organization of the paper is the following. In Section 2 we collect some definitions and basic properties of Gelfand-Tsetlin modules of $\mathfrak{gl}(n)$ and $\mathfrak{sl}(n)$. In Section 3 we recall the description of the Gelfand-Tsetlin structure of $V(T(v))$ for a generic v and 1-singular v. Section 4 is devoted to the description of the simple objects of the principal block and their parameterization in terms of subsets of $\mathbb{Z}^3$. In the last section we use localization to prove that all simple modules described in Section 4 are isomorphic to subquotients of localized highest weight modules.

2. Gelfand-Tsetlin modules for $\mathfrak{gl}(n)$

Throughout the paper $\mathfrak{gl}(n)$ stands for the Lie algebra of all $n \times n$ complex matrices and $\mathfrak{sl}(n)$ is the Lie subalgebra of $\mathfrak{gl}(n)$ consisting of all traceless matrices. Let for $m \leqslant n$, $\mathfrak{gl}_m$ be the Lie subalgebra of $\mathfrak{gl}(n)$ spanned by $\{E_{ij} \mid i,j=1,\dots,m\}$, where E_{ij} is the (i,j)th elementary matrix. We have the following chain

$$\mathfrak{gl}_1 \subset \mathfrak{gl}_2 \subset \ldots \subset \mathfrak{gl}_n .$$

It induces the chain $U_1 \subset U_2 \subset \ldots \subset U_n$ for the universal enveloping algebras $U_m = U(\mathfrak{gl}_m)$, $1 \leq m \leq n$. Let Z_m be the center of U_m. Then Z_m is the polynomial algebra in the m variables $\{c_{mk} \,|\, k = 1, \ldots, m\}$,

$$c_{mk} = \sum_{(i_1,\ldots,i_k)\in\{1,\ldots,m\}^k} E_{i_1 i_2} E_{i_2 i_3} \ldots E_{i_k i_1}. \tag{1}$$

The *(standard) Gelfand-Tsetlin subalgebra* Γ of U ([**5**,**11**]) is generated by $\{Z_m \,|\, m = 1, \ldots, n\}$. The algebra Γ is a polynomial algebra in $\dfrac{n(n+1)}{2}$ variables $\{c_{ij} \,|\, 1 \leqslant j \leqslant i \leqslant n\}$.

For $i = 1, \ldots, n$ denote by S_i the i-th symmetric group and set $G := S_n \times \cdots \times S_1$.

Let Λ be the polynomial algebra in variables $\{v_{ij} \mid 1 \leqslant j \leqslant i \leqslant n\}$ and let $\imath : \Gamma \longrightarrow \Lambda$ be the embedding given by $\imath(c_{mk}) = \gamma_{mk}(v)$ where

$$\gamma_{mk}(v) = \sum_{i=1}^{m} (v_{mi} + m - 1)^k \prod_{j \neq i}^{m} \left(1 - \frac{1}{v_{mi} - v_{mj}}\right). \tag{2}$$

REMARK 2.1. *For each $1 \leq m \leq n$, the $\gamma_{mk}(v)$ is a symmetric polynomial of degree k in variables $v_{m1}, \ldots, v_{mm}$, and $\{\gamma_{m1}(v), \ldots, \gamma_{mm}(v)\}$ generate the algebra of symmetric polynomials in variables $v_{m1}, \ldots, v_{mm}$. As a result, the image of $\imath$ coincides with the subalgebra of $G-$invariant polynomials in Λ which we identify with Γ (see* [**26**] *for details).*

DEFINITION 2.2. *Any algebra homomorphism $\chi : \Gamma \to \mathbb{C}$ will be called* Gelfand-Tsetlin character. *Equivalently, χ is a Gelfand-Tsetlin character if χ is in* $\operatorname{Specm}\Gamma$, *the set of maximal ideals of* Γ.

DEFINITION 2.3. *A finitely generated U-module M is called a* Gelfand-Tsetlin module (with respect to Γ) *provided that the restriction $M|_\Gamma$ is a direct sum of Γ-modules:*

$$M|_\Gamma = \bigoplus_{\chi\in\Gamma^*} M_\chi, \tag{3}$$

where $M_\chi = \{v \in M \mid$ for each $\gamma \in \Gamma$, $\exists k \in \mathbb{Z}_{\geq 0}$ such that $(\gamma - \chi(\gamma))^k v = 0\}$. Equivalently, M is a Gelfand-Tsetlin module if

$$M = \bigoplus_{\mathsf{m}\in\operatorname{Specm}\Gamma} M_{\mathsf{m}},$$

where $M_{\mathsf{m}} = \{v \in M \mid \mathsf{m}^k v = 0$ for some $k \geq 0\}$.

DEFINITION 2.4. *Given a Gelfand-Tsetlin module M, the* Gelfand-Tsetlin support *of M is the set* $\operatorname{Supp}_{GT}(M) := \{\chi \in \Gamma^* \mid M_\chi \neq 0\}$, *equivalently,* $\operatorname{Supp}_{GT}(M) := \{\mathsf{m} \in \operatorname{Specm}\Gamma \mid M_{\mathsf{m}} \neq 0\}$, *and* $\dim M_\chi$ *(respectively,* $\dim M_{\mathsf{m}}$*) is called the* Gelfand-Tsetlin χ-multiplicity (resp., m-multiplicity) *of M.*

2.1. Gelfand-Tsetlin modules and weight modules of $\mathfrak{sl}(n)$. Let Γ be a Gelfand-Tsetlin subalgebra of $\mathfrak{gl}(n)$. Consider the natural projection $\pi : \mathfrak{gl}(n) \to \mathfrak{sl}(n)$ which naturally extends to an epimorphism $\bar{\pi} : U(\mathfrak{gl}(n)) \to U(\mathfrak{sl}(n))$. Then the image $\bar{\pi}(\Gamma)$ of Γ is called the *standard Gelfand-Tsetlin subalgebra* of $\mathfrak{sl}(n)$ and with a slight abuse of notation is also denoted by Γ. It is a maximal commutative subalgebra of $U(\mathfrak{sl}(n))$ which is isomorphic to a polynomial ring in $\frac{n(n+1)}{2} - 1$ variables.

If Γ is the standard Gelfand-Tsetlin subalgebra of $\mathfrak{sl}(n)$ then we denote by $\mathcal{GT}(n) = \mathcal{GT}_\Gamma(n)$ the category of all Gelfand-Tsetlin $\mathfrak{sl}(n)$-modules with respect to Γ. As explained in the introduction, $\mathcal{GT}_v(n)$ is the full subcategory of $\mathcal{GT}(n)$ consisting of all V such that $\mathrm{Supp}_{GT}(V) \subseteq \{\mathsf{m}_{v'} \mid T(v') \in T(v)_{\mathbb{Z}}\}$, where $\mathsf{m}_{v'}$ is the ideal of Γ generated by $c_{ij} - \gamma_{ij}(v')$, $1 \leq j \leq i \leq n$. We write $\mathcal{GT}_0$ for $\mathcal{GT}_{\bar{v}}(n)$ when $\bar{v} = (0, -1, ..., -n+1|0, ..., 0|...|0)$.

Let $\mathfrak{h}_{\mathfrak{gl}(n)}$ denote the subalgebra of $\mathfrak{gl}(n)$ spanned by E_{ii}, $i = 1, \ldots, n$. We call this algebra *the standard Cartan subalgebra of* $\mathfrak{gl}(n)$. By $\mathfrak{h} = \mathfrak{h}_{\mathfrak{sl}(n)}$ we denote the standard Cartan subalgebra $\pi(\mathfrak{h}_{\mathfrak{gl}(n)})$ of $\mathfrak{sl}(n)$, namely $\mathfrak{h} = \mathfrak{h}_{\mathfrak{gl}(n)} \cap \mathfrak{sl}(n)$. Recall that an $\mathfrak{sl}(n)$-module M is called a *generalized weight module* if

$$M = \bigoplus_{\lambda \in \mathfrak{h}^*} M_\lambda,$$

where $M_\lambda = \{m \in M \mid \text{ for each } h \in \mathfrak{h}, \exists k \in \mathbb{Z}_{\geq 0} \text{ such that } (h - \lambda(h))^k m = 0\}$.

We call M_λ the *λ-weight space* of M, the set $\{\lambda \in \mathfrak{h}^* \mid M_\lambda \neq 0\}$ *the weight support of* M, and $\dim M_\lambda$ the *λ-weight multiplicity* of M. If $\mathfrak{h}$ acts semisimply on a generalized weight module M, then M is called a *weight module.* It is easy to show that a simple generalized weight module is a weight module.

It is clear that every simple Gelfand-Tsetlin module is a weight module, possibly with infinite weight multiplicities.

The advantage of working with Gelfand-Tsetlin $\mathfrak{sl}(n)$-modules instead of with Gelfand-Tsetlin $\mathfrak{gl}(n)$-modules is that in the case $n = 3$ one can easily picture the weight support of the modules (being a lattice in $\mathbb{R}^2$, not in $\mathbb{R}^3$) and better visualize the relations between them, see Section 4.

3. Gelfand-Tsetlin modules with tableaux realization for $\mathfrak{gl}(n)$

The classical examples of Gelfand-Tsetlin modules are the irreducible finite-dimensional modules. In this case, the tableaux basis of these modules is also an eigenbasis for the action of the generators of Γ and $\dim(M_\chi) \leq 1$ for each $\chi \in \Gamma^*$, see [**26**]. For infinite-dimensional Gelfand-Tsetlin modules, the situation is more complicated already in the case of $\mathfrak{sl}(3)$. In this section we collect some results about the tableaux realization of more general Gelfand-Tsetlin modules. In particular, we provide such realization for the $\mathfrak{sl}(3)$-module $V(T(v))$ for any v.

3.1. Irreducible finite-dimensional modules as Gelfand-Tsetlin modules. In this subsection we recall the classical result of Gelfand and Tsetlin which gives an explicit tableaux basis for each irreducible finite-dimensional $\mathfrak{gl}(n)$-module.

DEFINITION 3.1. *A Gelfand-Tsetlin tableau $T(v)$ of height n is called* standard *if $v_{ki} - v_{k-1,i} \in \mathbb{Z}_{\geq 0}$ and $v_{k-1,i} - v_{k,i+1} \in \mathbb{Z}_{>0}$ for all $1 \leq i \leq k \leq n-1$.*

Note that, for sake of convenience, the second condition above is slightly different from the original condition in [**11**].

THEOREM 3.2 ([**11**]). *Let $L(\lambda)$ be the irreducible finite-dimensional module over $\mathfrak{gl}(n)$ of highest weight $\lambda = (\lambda_1, \ldots, \lambda_n)$. Then there exists a basis of $L(\lambda)$ consisting of all standard tableaux $T(v) = T(v_{ij})$ with fixed top row $v_{nj} = \lambda_j - j + 1$, $j = 1, \ldots, n$. Moreover, the action of the generators of $\mathfrak{gl}(n)$ on $L(\lambda)$ is given by the* Gelfand-Tsetlin formulas*:*

$$E_{k,k+1}(T(v)) = -\sum_{i=1}^{k} \left(\frac{\prod_{j=1}^{k+1}(v_{ki} - v_{k+1,j})}{\prod_{j\neq i}^{k}(v_{ki} - v_{kj})} \right) T(v + \delta^{ki}),$$

$$E_{k+1,k}(T(v)) = \sum_{i=1}^{k} \left(\frac{\prod_{j=1}^{k-1}(v_{ki} - v_{k-1,j})}{\prod_{j\neq i}^{k}(v_{ki} - v_{kj})} \right) T(v - \delta^{ki}),$$

$$E_{kk}(T(v)) = \left(k - 1 + \sum_{i=1}^{k} v_{ki} - \sum_{i=1}^{k-1} v_{k-1,i} \right) T(v),$$

where $\delta^{ij} \in \mathbb{Z}^{\frac{n(n+1)}{2}}$ is defined by $(\delta^{ij})_{ij} = 1$ and all other $(\delta^{ij})_{k\ell}$ are zero. If the new tableau $T(v \pm \delta^{ki})$ is not standard, then the corresponding summand of $E_{k,k+1}(T(v))$ or $E_{k+1,k}(T(v))$ is zero by definition. Furthermore, for $s \leq r$,

$$c_{rs}(T(v)) = \gamma_{rs}(v)T(v),$$

where $\gamma_{rs}(v)$ are defined in (2).

The formulas above are called *Gelfand-Tsetlin formulas* for $\mathfrak{gl}(n)$.

REMARK 3.3. *We can easily formulate the theorem above in the case of $\mathfrak{sl}(n)$. Namely, we use the same Gelfand-Tsetlin formulas for the generators $E_{r,r+1}$ or $E_{r+1,r}$, and the action of the generators of the Cartan subalegbra is given by $h_i(T(v)) := (E_{ii} - E_{i+1,i+1})(T(v))$. We also fix the action of the central element $E_{11} + \ldots + E_{nn}$ to be trivial.*

3.2. Generic Gelfand-Tsetlin modules. Observing that the coefficients in the Gelfand-Tsetlin formulas in Theorem 3.2 are rational functions on the entries of the tableaux, it is natural to define an $\mathfrak{sl}(n)$-module structure on modules whose tableaux basis satisfy the property that the denominators of the rational functions are nonzero. This leads to the following definition.

DEFINITION 3.4. *A Gelfand-Tsetlin tableau $T(v) \in T_n(\mathbb{C})$ (or, an element v of $\mathbb{C}^{\frac{n(n+1)}{2}}$) is called* generic *if $v_{ki} - v_{kj} \notin \mathbb{Z}$ for all $1 \leq i \neq j \leq k \leq n-1$.*

In [**5**], it is proven that if v is generic then the space $V(T(v))$ with basis $T(v)_{\mathbb{Z}}$ has an $\mathfrak{sl}(n)$-module structure. In this way we have a natural example of a generic Gelfand-Tsetlin module all Gelfand-Tsetlin multiplicities of which are 1. Further properties of the generic module $V(T(v))$ and its subquotients are obtained in [**6**].

3.3. Singular Gelfand-Tsetlin modules.

DEFINITION 3.5. *A vector $v \in \mathbb{C}^{\frac{n(n+1)}{2}}$ will be called* singular *if it is not generic, i.e. if there exist $1 \leq s < t \leq r \leq n-1$ such that $v_{rs} - v_{rt} \in \mathbb{Z}$. The vector v will be called* 1-singular *if there exist k, i, j with $1 \leq i < j \leq k \leq n-1$ such that $v_{ki} - v_{kj} \in \mathbb{Z}$ and $v_{rs} - v_{rt} \notin \mathbb{Z}$ for all $(r, s, t) \neq (k, i, j)$. We also call such 1-singular vector (k, i, j)*-singular.

We note that for $n=3$ every singular vector is a 1-singular vector. Generalized tableaux realization of modules $V(T(\bar{v}))$ associated with 1-singular vectors $\bar{v}$ is studied in [**7**]. In what follows we summarize the main details of this construction.

From now on we fix (i,j,k) such that $1 \leq i < j \leq k \leq n-1$ and a (k,i,j)-singular vector $\bar{v}$. Also, from now on, τ will denote the element $(\tau[n],\ldots,\tau[2],\tau[1])$ in G such that $\tau[k]$ is the transposition (i,j) and all other $\tau[t]$ are Id.

In addition to the issue with the zero denominators in the Gelfand-Tsetlin formulas in Theorem 3.2, the singular case faces another challenge. The formulas involving $T(\bar{v}+z)$ and $T(\bar{v}+\tau(z))$ are identical and hence it is natural to impose the condition $T(\bar{v}+z) = T(\bar{v}+\tau(z))$. To deal with the zero denominators issue, we formally introduce new tableaux $\mathcal{D}T(\bar{v}+z')$ for every $z' \in \mathbb{Z}^{\frac{n(n-1)}{2}}$ subject to the relations $\mathcal{D}T(\bar{v}+z') + \mathcal{D}T(\bar{v}+\tau(z')) = 0$. We call $\mathcal{D}T(\bar{v}+z')$ *the derivative Gelfand-Tsetlin tableau* associated with z'.

DEFINITION 3.6. *Define the* (generalized) Gelfand-Tsetlin tableau associated *to w to be*

$$Tab(\bar{v}+w) := \begin{cases} T(\bar{v}+w), & \text{if } w_{ki}-w_{kj} \leq 0 \\ \mathcal{D}T(\bar{v}+w), & \text{if } w_{ki}-w_{kj} > 0. \end{cases}$$

Since $\bar{v}$ is fixed, for simplicity we will write $Tab(w) = Tab(\bar{v}+w)$. One easily checks that the set $\mathcal{B}(T(v)) := \left\{ Tab(w) \mid w \in \mathbb{Z}^{\frac{n(n-1)}{2}} \right\}$ is linearly independent and the space with basis $\mathcal{B}(T(v))$ is naturally isomorphic to the space $T(v)_{\mathbb{Z}}$ introduced in Introduction. For this reason, with a slight abuse of notation we assume that

$$V(T(\bar{v})) = \bigoplus_{Tab(w)\in\mathcal{B}(T(v))} \mathbb{C}(Tab(w))$$

for the fixed (k,i,j)-singular vector $\bar{v}$.

For a vector v of variables v_{rs} and a rational function $f = f(v)$ which is smooth on the hyperplane $v_{ki}-v_{kj}=0$ we can define a "differential operator" $\mathcal{D}^{\bar{v}}$ on products of f and a generic tableau $T(v+z)$, as follows:

$$\mathcal{D}^{\bar{v}}(fT(v+z)) = \mathcal{D}^{\bar{v}}(f)T(\bar{v}+z) + f(\bar{v})\mathcal{D}T(\bar{v}+z),$$

where $\mathcal{D}^{\bar{v}}(f) = \frac{1}{2}\left(\frac{\partial f}{\partial v_{ki}} - \frac{\partial f}{\partial v_{kj}}\right)(\bar{v})$. We then extend $\mathcal{D}^{\bar{v}}$ linearly. The operator $\mathcal{D}^{\bar{v}}$ may be defined in rigorous terms; see §4.2 in [**6**], but to avoid technicalities we skip this rigorous definition.

THEOREM 3.7 ([**7**], Theorems 4.11 and 4.12). *If $\bar{v}$ is as above and v is generic, $V(T(\bar{v}))$ is an 1-singular Gelfand-Tsetlin $\mathfrak{sl}(n)$-module, with action of the generators of $\mathfrak{sl}(n)$ given by the formulas:*

$$E_{rs}(T(\bar{v}+z)) = \mathcal{D}^{\bar{v}}((v_{ki}-v_{kj})E_{rs}(T(v+z))),$$
$$E_{rs}(\mathcal{D}T(\bar{v}+z')) = \mathcal{D}^{\bar{v}}(E_{rs}(T(v+z'))),$$

while the action of the generators of Γ is given by the formulas:

$$c_{rs}(T(\bar{v}+z)) = \gamma_{rs}(\bar{v}+z)T(\bar{v}+z)$$

$$c_{rs}(\mathcal{D}T(\bar{v}+z')) = \gamma_{rs}(\bar{v}+z')\mathcal{D}T(\bar{v}+z') + \mathcal{D}^{\bar{v}}(\gamma_{rs}(v+z'))T(\bar{v}+z')$$

for any $z,z' \in \mathbb{Z}^{\frac{n(n-1)}{2}}$ with $z' \neq \tau(z')$.

The following result plays central role in this paper.

THEOREM 3.8. *Every simple object of the principal block $\mathcal{GT}_0$ of $\mathfrak{sl}(3)$ is a subquotient of $V(T(\bar{v}))$ where $\bar{v} = (0, -1, -2|0, 0|0)$.*

The theorem holds in fact for any Gelfand-Tsetlin block $\mathcal{GT}_v(3)$ of $\mathfrak{sl}(3)$ and it was proven in [**7**] (see Theorem 7.2) in all cases except one. The remaining case was completed in [**8**] (see Main Theorem).

4. Irreducible Gelfand-Tsetlin modules in the principal block

Until the end of the paper we fix $\bar{v} = (0, -1, -2; 0, 0; 0)$. Hence for any $z = (m, n, k) \in \mathbb{Z}^3$, $T(\bar{v} + z)$ is the following Gelfand-Tsetlin tableau:

$$\begin{array}{ccccc} \boxed{0} & & \boxed{-1} & & \boxed{-2} \\ & \boxed{m} & & \boxed{n} & \\ & & \boxed{k} & & \end{array}$$

In this section we describe the simple subquotients (with multiplicities) and the Loewy decomposition of $V(T(\bar{v}))$. It turns out that $V(T(\bar{v}))$ has 32 irreducible subquotients $\{M_i\}_{i=1}^{32}$, twenty of which are nonisomorphic. The proofs of these statements will be provided in [**9**].

4.1. Simple subquotients of $V(T(\bar{v}))$. The module $V(T(\bar{v}))$ has length 32 and its simple subquotients (with multiplicities) M_i can be separated in four groups:

(i) Two 1-dimensional (trivial) modules.
(ii) Twenty highest weight modules whose weight multiplicities are 1.
(iii) Eight simple Verma modules (with unbounded set of weight multiplicities).
(iv) Two weight modules with infinite weight multiplicities.

From the above description, looking at the nonisomorphic simple subquotients of $V(T(\bar{v}))$, we easily obtain a complete description of the simple objects in the principal Gelfand-Tsetlin block $\mathcal{GT}_0$ of $\mathfrak{sl}(3)$:

(a) One 1-dimensional (trivial) module.
(b) Twelve highest weight modules whose weight multiplicities are 1.
(c) Six simple Verma modules.
(d) One weight module with infinite weight multiplicities.

Recall that for $n = 3$, we have that

$$\mathcal{B}(T(\bar{v})) = \{T(\bar{v} + (m, n, k)), \mathcal{D}T(\bar{v} + (m', n', k')) \mid m \leq n \text{ and } m' > n'\}.$$

Also, recall that $w \in \mathbb{Z}^3$, the tableau associated to w is

$$Tab(w) := \begin{cases} T(\bar{v} + w), & \text{if } w_{21} \leq w_{22} \\ \mathcal{D}T(\bar{v} + w), & \text{if } w_{21} > w_{22}. \end{cases}$$

In particular, the map $w \mapsto Tab(w)$ defines a bijection $Tab : \mathbb{Z}^3 \to \mathcal{B}(T(\bar{v}))$. Henceforth the modules M_i will be defined by their corresponding sets in $\mathbb{Z}^3$, equivalently, by their bases in $\mathcal{B}(T(\bar{v}))$. We should note that all 32 subsets $\mathcal{B}_i$ of $\mathbb{Z}^3$ that define M_i, $1 \leq i \leq 32$, are defined by a set of inequalities of the form $a \leq b$ or $a < b$ where a, b are elements in the set $\{m, n, k, 0, -1, -2\}$. With this in mind we have the following definition.

Definition 4.1. *Let $\mathcal{B}$ be a convex subset of $\mathbb{Z}^3$ such that there is a subquotient of $V(T(\bar{v}))$ whose basis is $Tab(\mathcal{B})$ (such $\mathcal{B}$ is a union of some the 32 subsets $\mathcal{B}_i$). Then this subquotient is unique and is denoted by $M(\mathcal{B})$. If $M(\mathcal{B})$ is simple, it will be denoted by $L(\mathcal{B})$.*

We now describe the sets $\mathcal{B}_i$ that define all irreducible subquotients $M_i = L(\mathcal{B}_i)$ of $V(T(\bar{v}))$. For convenience, the modules listed in one row are isomorphic. It is worth noting that the map $\tau : (m,n,k) \to (n,m,k)$ induces all isomorphisms between simple subquotients of $V(T(\bar{v}))$.

(i) Two 1-dimensional (trivial) modules. In this case there is one pair of isomorphic 1-dimensional modules:

Module	Basis	Module	Basis
M_8	$L\left(\begin{cases} m=-1 \\ n=k=0 \end{cases}\right)$	M_{20}	$L\left(\begin{cases} n=-1 \\ m=k=0 \end{cases}\right)$

(ii) Twenty weight modules with weight multiplicities 1. There are eight pairs of isomorphic modules and four more modules in the list:

Module	Basis	Module	Basis
M_2	$L\left(\begin{cases} k \le m \le -2 \\ n=0 \end{cases}\right)$	M_9	$L\left(\begin{cases} k \le n \le -2 \\ m=0 \end{cases}\right)$
M_3	$L\left(\begin{cases} m \le -2 \\ n=-1 \\ m<k\le -1 \end{cases}\right)$	M_{12}	$L\left(\begin{cases} n \le -2 \\ m=-1 \\ n<k\le -1 \end{cases}\right)$
M_6	$L\left(\begin{cases} m \le -2 \\ n=0 \\ m<k\le 0 \end{cases}\right)$	M_{15}	$L\left(\begin{cases} n \le -2 \\ m=0 \\ n<k\le 0 \end{cases}\right)$
M_{10}	$L\left(\begin{cases} m \le -2 \\ n=0 \\ 0<k \end{cases}\right)$	M_{22}	$L\left(\begin{cases} n \le -2 \\ m=0 \\ 0<k \end{cases}\right)$
M_{11}	$L\left(\begin{cases} m=-1 \\ 0<n \\ k\le -1 \end{cases}\right)$	M_{23}	$L\left(\begin{cases} n=-1 \\ 0<m \\ k\le -1 \end{cases}\right)$
M_{16}	$L\left(\begin{cases} m=-1 \\ 0<n \\ -1<k\le n \end{cases}\right)$	M_{27}	$L\left(\begin{cases} n=-1 \\ 0<m \\ -1<k\le m \end{cases}\right)$
M_{21}	$L\left(\begin{cases} m=0 \\ 0<k\le n \end{cases}\right)$	M_{30}	$L\left(\begin{cases} n=0 \\ 0<k\le m \end{cases}\right)$
M_{24}	$L\left(\begin{cases} m=-1 \\ 0<n<k \end{cases}\right)$	M_{31}	$L\left(\begin{cases} n=-1 \\ 0<m<k \end{cases}\right)$

Module	Basis
M_7	$L\left(\begin{cases}-1\leq n\leq 0\\ m=-1\\ k\leq -1\end{cases}\bigcup\begin{cases}k\leq n<-1\\ m=-1\end{cases}\right)$
M_{14}	$L\left(\begin{cases}m=0\\ 0\leq n\\ k\leq 0\end{cases}\bigcup\begin{cases}m=0\\ n=-1\\ k\leq -1\end{cases}\right)$
M_{17}	$L\left(\begin{cases}-1\leq n\leq 0<k\\ m=-1\end{cases}\bigcup\begin{cases}m=-1\\ n\leq -2\\ -1<k\end{cases}\right)$
M_{26}	$L\left(\begin{cases}m=0\\ 0\leq n<k\end{cases}\bigcup\begin{cases}m=0\\ n=-1\\ 0<k\end{cases}\right)$

The following picture describes the weight support of the above listed modules. Recall that modules in the same row are isomorphic, and that the weight multiplicities of all modules are one.

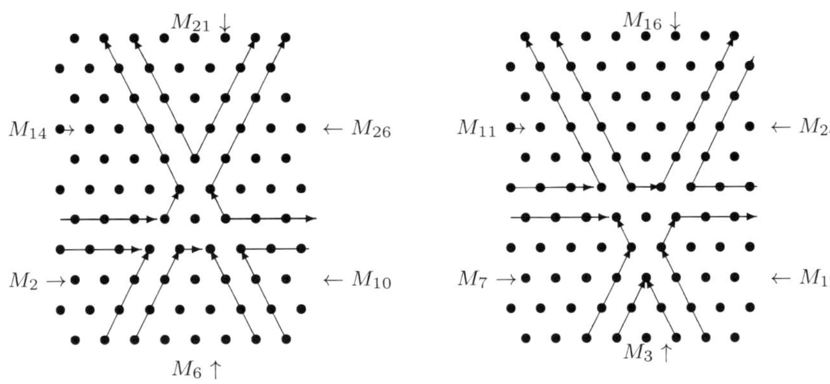

Note that the action of the Cartan subalgebra on each M_i can be recovered from the Gelfand-Tsetlin formulas. In particular, the weight corresponding to the tableau $Tab(m,n,k)$ is $(2k-(m+n+1),2(m+n+1)-k)$. In the pictures above, the point in the middle is (0,0) and corresponds to the trivial (i.e. the finite-dimensional) module. The horizontal lines have equations to $y=-\frac{1}{2}x$, $y=-\frac{1}{2}(x+3)$, and $y=-\frac{1}{2}(x-3)$, while the diagonal lines have equations $y=x$, $y=x+3$, $y=x-3$, $y=-2x$, $y=-2(x+3)$ and $y=-2(x-3)$.

(iii) Eight simple Verma modules (with unbounded set of weight multiplicities). There are two pairs of isomorphic modules and four more modules in the list.

Module	Basis	Module	Basis
M_5	$L\left(\begin{cases}k\leq m\leq -2\\ 0<n\end{cases}\right)$	M_{18}	$L\left(\begin{cases}k\leq n\leq -2\\ 0<m\end{cases}\right)$
M_{19}	$L\left(\begin{cases}0<n<k\\ m\leq -2\end{cases}\right)$	M_{29}	$L\left(\begin{cases}0<m<k\\ n\leq -2\end{cases}\right)$

Module	Basis
M_1	$L\left(\begin{cases} k\le m\le -2 \\ n=-1 \end{cases} \bigcup \begin{cases} m\le n \\ k\le n\le -2 \end{cases} \bigcup \begin{cases} n<m\le -2 \\ k\le n \end{cases}\right)$
M_4	$L\left(\begin{cases} m\le n \\ m\le -2 \\ n\le -1 \\ n<k \end{cases} \bigcup \begin{cases} n<m \\ m\le -2 \\ n<k \end{cases}\right)$
M_{28}	$L\left(\begin{cases} m\le n \\ 0<m \\ k\le n \end{cases} \bigcup \begin{cases} n<m \\ 0<m \\ -1<n \\ k\le n \end{cases}\right)$
M_{32}	$L\left(\begin{cases} m\le n \\ 0<m \\ n<k \end{cases} \bigcup \begin{cases} n<m \\ 0<n \\ n<k \end{cases} \bigcup \begin{cases} 0<m<k \\ n=0 \end{cases}\right)$

The weight supports are listed below. For reader's convenience the weight multiplicities are included as well.

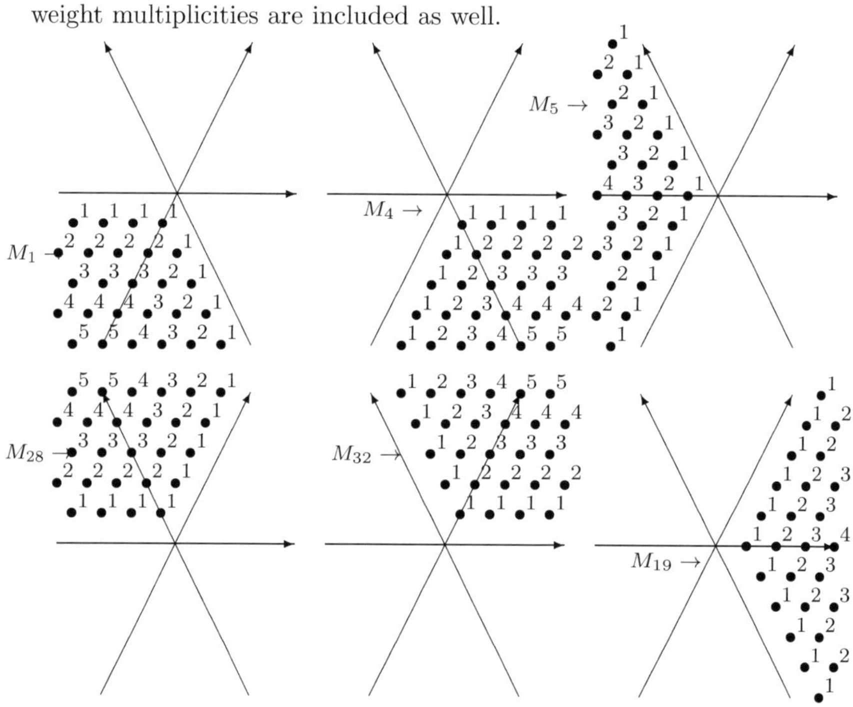

(iv) Two weight modules with infinite weight multiplicities. In this case we have one pair of isomorphic modules:

Module	Basis	Module	Basis
M_{13}	$L\left(\begin{cases} m\le -2 \\ 0<n \\ m<k\le n \end{cases}\right)$	M_{25}	$L\left(\begin{cases} n\le -2 \\ 0<m \\ n<k\le m \end{cases}\right)$

Note that the support of this module is the whole plane.

4.2. Loewy decomposition of $V(T(\bar{v}))$. The Loewy decomposition (the first module in the list being the socle) of $V(T(\bar{v}))$ is the following:
M_1, $M_2 \oplus M_3$, $M_4 \oplus M_5 \oplus M_6 \oplus M_7$, $M_8 \oplus M_9 \oplus M_{10} \oplus M_{11} \oplus M_{12} \oplus M_{13}$, $M_{14} \oplus M_{15} \oplus M_{16} \oplus M_{17} \oplus M_{18} \oplus M_{19}$, $M_{20} \oplus M_{21} \oplus M_{22} \oplus M_{23} \oplus M_{24} \oplus M_{25}$, $M_{26} \oplus M_{27} \oplus M_{28} \oplus M_{29}$, $M_{30} \oplus M_{31}$, M_{32}.

5. E_{21}-localization of Gelfand-Tsetlin modules in the principal block

5.1. Irreducible Gelfand-Tsetlin modules and localization functors. In this section $U = U(\mathfrak{sl}(3))$. We recall the definition of the localization functor on U-modules. For details we refer the reader to [**3**] and [**16**].

For every root α of $\mathfrak{sl}(3)$ the multiplicative set $\mathbf{F}_\alpha := \{f_\alpha^n \mid n \in \mathbb{Z}_{\geq 0}\} \subset U$ satisfies Ore's localizability conditions because ad f_α acts locally nilpotent on U. Let $\mathcal{D}_\alpha U$ be the localization of U relative to $\mathbf{F}_\alpha$. For every weight module M we denote by $\mathcal{D}_\alpha M$ the *α–localization of* M, defined as $\mathcal{D}_\alpha M = \mathcal{D}_\alpha U \otimes_U M$. If f_α is injective on M, then M can be naturally viewed as a submodule of $\mathcal{D}_\alpha M$. Furthermore, if f_α is injective on M, then it is bijective on M if and only if $\mathcal{D}_\alpha M = M$. Note that the localization functor is exact.

In the case when f_α acts injectively on M, set $\mathcal{QD}_\alpha M := \mathcal{D}_\alpha M/M$. Also, if $\alpha = \epsilon_i - \epsilon_j$ we will write $\mathcal{D}_{ji}$ and $\mathcal{QD}_{ji}$ for $\mathcal{D}_\alpha$ and $\mathcal{QD}_\alpha$, respectively. In this section we will work exclusively with $\alpha = \epsilon_2 - \epsilon_1$.

DEFINITION 5.1. *For a region $\mathcal{B}$ denote by $\mathcal{B}+t\delta^{11}$ the region $\{(m,n,k) \mid (m,n,k-t) \in \mathcal{B}\}$, and $\mathcal{B} + \mathbb{N}\delta^{11} = \bigcup_{t=0}^{\infty}(\mathcal{B} + t\delta^{11})$.*

REMARK 5.2. *Since the action of E_{21} on the basis vectors of $V(T(\bar{v}))$ is given by $E_{21}(Tab(w)) = Tab(w - \delta^{11})$, it follows that the module $M(\mathcal{B})$ is E_{21}-injective if and only if $\mathcal{B} \subseteq \mathcal{B}+\delta^{11}$. Moreover, $M(\mathcal{B})$ is E_{21}-surjective if and only if $\mathcal{B}+\delta^{11} \subseteq \mathcal{B}$. Therefore, $M(\mathcal{B})$ is E_{21}-bijective if and only if $\mathcal{B} = \mathcal{B} + \delta^{11}$.*

PROPOSITION 5.3. *Let $\mathcal{B} \subset \mathbb{Z}^3$ be such that E_{21} acts injectively on the irreducible Gelfand-Tsetlin module $L(\mathcal{B})$. Then $\mathcal{D}_{12}L(\mathcal{B}) \simeq M(\mathcal{B} + \mathbb{N}\delta^{11})$ and $\mathcal{QD}_{12}L(\mathcal{B}) \simeq M((\mathcal{B} + \mathbb{N}\delta^{11}) \setminus \mathcal{B})$.*

PROOF. We first note that the module $V(T(\bar{v}))$ is E_{21}-bijective (see Remark 5.2). In particular $\mathcal{D}_{12}V(T(\bar{v})) = V(T(\bar{v}))$. Also, since the localization functor is exact, we have that $\mathcal{D}_{12}M(\mathcal{B})$ is a subquotient of $V(T(\bar{v}))$. Hence, $\mathcal{D}_{12}M(\mathcal{B}) = M(\mathcal{B}')$ for some $\mathcal{B}'$ in $\mathbb{Z}^3$. Now using again Remark 5.2 we see that $\mathcal{D}_{12}M(\mathcal{B}) = M(\mathcal{B} + \mathbb{N}\delta^{11})$. This together with the facts that $M(\mathcal{B}) = L(\mathcal{B})$ and $M(\mathcal{B} + \mathbb{N}\delta^{11})/M(\mathcal{B}) = M((\mathcal{B} + \mathbb{N}\delta^{11}) \setminus \mathcal{B})$ completes the proof. □

EXAMPLE 5.4. *Using Proposition 5.3 we describe the module M_{13} with infinite weight multiplicities as a subquotient of the localized highest weight module $\mathcal{D}_{12}M_5$ as follows. The module $\mathcal{D}_{12}M_5$ has length 3: M_5 is a submodule, M_{13} is a submodule of $\mathcal{QD}_{12}(M_5)$ and $M_{19} \simeq \mathcal{QD}_{12}(M_5)/M_{13}$. This in terms of subsets of $\mathbb{Z}^3$ can be expressed as follows: $\mathcal{B}_5 + \mathbb{N}\delta^{11} = \mathcal{B}_5 \sqcup \mathcal{B}_{13} \sqcup \mathcal{B}_{19}$, and, more precisely:*

$$\left(\begin{cases} m \leq -2 \\ 0 < n \end{cases}\right) = \left(\begin{cases} m \leq -2 \\ 0 < n \\ k \leq m \end{cases}\right) \bigcup \left(\begin{cases} m \leq -2 \\ 0 < n \\ m < k \leq n \end{cases}\right) \bigcup \left(\begin{cases} m \leq -2 \\ 0 < n \\ n < k \end{cases}\right)$$

5.2. Simple objects of $\mathcal{GT}_0$ in terms of E_{21}-localization. We will list all simple objects in the principal block as subquotients of localized highest weight modules. We first list all simple highest weight modules (relative to different Borel subalegbras) that are E_{21}-injective:

$$M_1,\ M_2 \simeq M_9,\ M_5 \simeq M_{18},\ M_7,\ M_{11} \simeq M_{23},\ M_{14},\ M_{28}.$$

We next apply Proposition 5.3 to obtain all remaining simple objects of $\mathcal{GT}_0$ as subquotients of E_{21}-localization of the above modules. By $\mathrm{soc}(M)$ we denote the socle of the module M.

Module	Subquotient of E_{21}-localization
$M_3 \simeq M_{12}$	$\mathrm{soc}(\mathcal{QD}_{12}(M_7)/M_8) \simeq \mathrm{soc}(\mathcal{QD}_{12}(M_1))$
M_4	$\mathcal{QD}_{12}(M_1)/M_3$
$M_6 \simeq M_{15}$	$\mathrm{soc}(\mathcal{QD}_{12}(M_2)) \simeq \mathrm{soc}(\mathcal{QD}_{12}(M_9))$
$M_8 \simeq M_{20}$	$\mathrm{soc}(\mathcal{QD}_{12}(M_7)/M_3) \simeq \mathrm{soc}(\mathcal{QD}_{12}(M_{14})/M_{21})$
M_{10}	$\mathcal{QD}_{12}(M_2)/M_6$
M_{13}	$Soc(\mathcal{QD}_{12}(M_5))$
M_{16}	$\mathrm{soc}(\mathcal{QD}_{12}(M_{11}))$
M_{17}	$\mathcal{QD}_{12}(M_7)/(M_8 \oplus M_{12})$
M_{19}	$\mathcal{QD}_{12}(M_5)/M_{13}$
$M_{21} \simeq M_{30}$	$\mathrm{soc}(\mathcal{QD}_{12}(M_{14})/M_{20}) \simeq \mathrm{soc}(\mathcal{QD}_{12}(M_{28}))$
M_{22}	$\mathcal{QD}_{12}(M_9)/M_{15}$
M_{24}	$\mathcal{QD}_{12}(M_{11})/M_{16}$
M_{25}	$\mathrm{soc}(\mathcal{QD}_{12}(M_{18}))$
M_{26}	$\mathcal{QD}_{12}(M_{14})/(M_{20} \oplus M_{21})$
M_{27}	$\mathrm{soc}(\mathcal{QD}_{12}(M_{23}))$
M_{29}	$\mathcal{QD}_{12}(M_{18})/M_{25}$
M_{31}	$\mathcal{QD}_{12}(M_{23})/M_{27}$
M_{32}	$\mathcal{QD}_{12}(M_{28})/M_{21}$

COROLLARY 5.5. *Every simple Gelfand-Tsetlin module in the principal block is isomorphic to a subquotient of the localization $\mathcal{D}_{12}(L)$ of an E_{21}-injective simple highest weight module L.*

Acknowledgments

V.F. was supported in part by CNPq grant (301320/2013-6) and by Fapesp grant (2014/09310-5). D.G. was supported in part by Simons Collaboration Grant 358245. We also would like to thank Christophe Breuil for asking a question that initiated the writing of this paper.

References

[1] M. Colarusso and S. Evens, *On algebraic integrability of Gelfand-Zeitlin fields*, Transform. Groups **15** (2010), no. 1, 46–71. MR2600695

[2] M. Colarusso and S. Evens, *The Gelfand-Zeitlin integrable system and K-orbits on the flag variety*, Symmetry: representation theory and its applications, Progr. Math., vol. 257, Birkhäuser/Springer, New York, 2014, pp. 85–119. MR3363008

[3] V. V. Deodhar, *On a construction of representations and a problem of Enright*, Invent. Math. **57** (1980), no. 2, 101–118. MR567193

[4] Yu. A. Drozd, S. A. Ovsienko, and V. M. Futornyĭ, *Irreducible weighted* sl(3)*-modules* (Russian), Funktsional. Anal. i Prilozhen. **23** (1989), no. 3, 57–58; English transl., Funct. Anal. Appl. **23** (1989), no. 3, 217–218 (1990). MR1026989

[5] Yu. A. Drozd, V. M. Futorny, and S. A. Ovsienko, *Harish-Chandra subalgebras and Gel′fand-Zetlin modules*, Finite-dimensional algebras and related topics (Ottawa, ON, 1992), NATO Adv. Sci. Inst. Ser. C Math. Phys. Sci., vol. 424, Kluwer Acad. Publ., Dordrecht, 1994, pp. 79–93. MR1308982

[6] V. Futorny, D. Grantcharov, and L. E. Ramirez, *Irreducible generic Gelfand-Tsetlin modules of* $\mathfrak{gl}(n)$, SIGMA Symmetry Integrability Geom. Methods Appl. **11** (2015), Paper 018, 13. MR3322336

[7] V. Futorny, D. Grantcharov, and L. E. Ramirez, *Singular Gelfand-Tsetlin modules of* $\mathfrak{gl}(n)$, Adv. Math. **290** (2016), 453–482. MR3451929

[8] V. Futorny, D. Grantcharov, L. E. Ramirez, Drinfeld category and the classification of singular Gelfand-Tsetlin $\mathfrak{gl}_n$-modules, Int. Math. Res. Notices, https://doi.org/10.1093/imrn/rnx159.

[9] V. Futorny, D. Grantcharov, L.E. Ramirez, Classification of irreducible Gelfand-Tsetlin modules for $\mathfrak{sl}(3)$, in progress.

[10] I. M. Gel′fand and M. I. Graev, *Finite-dimensional irreducible representations of the unitary and complete linear group and special functions associated with them* (Russian), Izv. Akad. Nauk SSSR Ser. Mat. **29** (1965), 1329–1356. MR0201568

[11] I. M. Gel′fand and M. L. Cetlin, *Finite-dimensional representations of the group of unimodular matrices* (Russian), Doklady Akad. Nauk SSSR (N.S.) **71** (1950), 825–828. MR0035774

[12] M. I. Graev, *Infinite-dimensional representations of the Lie algebra* $\mathfrak{gl}(n, \mathbb{C})$ *related to complex analogs of the Gelfand-Tsetlin patterns and general hypergeometric functions on the Lie group* $\mathrm{GL}(n, \mathbb{C})$, Acta Appl. Math. **81** (2004), no. 1-3, 93–120. MR2069333

[13] M. I. Graev, *A continuous analogue of Gel′fand-Tsetlin schemes and a realization of the principal series of irreducible unitary representations of the group* $\mathrm{GL}(n, \mathbb{C})$ *in the space of functions on the manifold of these schemes* (Russian), Dokl. Akad. Nauk **412** (2007), no. 2, 154–158; English transl., Dokl. Math. **75** (2007), no. 1, 31–35. MR2451955

[14] B. Kostant and N. Wallach, *Gelfand-Zeitlin theory from the perspective of classical mechanics. I*, Studies in Lie theory, Progr. Math., vol. 243, Birkhäuser Boston, Boston, MA, 2006, pp. 319–364. MR2214253

[15] B. Kostant and N. Wallach, *Gelfand-Zeitlin theory from the perspective of classical mechanics. II*, The unity of mathematics, Progr. Math., vol. 244, Birkhäuser Boston, Boston, MA, 2006, pp. 387–420. MR2181811

[16] O. Mathieu, *Classification of irreducible weight modules* (English, with English and French summaries), Ann. Inst. Fourier (Grenoble) **50** (2000), no. 2, 537–592. MR1775361

[17] V. Mazorchuk, *Tableaux realization of generalized Verma modules*, Canad. J. Math. **50** (1998), no. 4, 816–828. MR1638623

[18] V. Mazorchuk, *On categories of Gelfand-Zetlin modules*, Noncommutative structures in mathematics and physics (Kiev, 2000), NATO Sci. Ser. II Math. Phys. Chem., vol. 22, Kluwer Acad. Publ., Dordrecht, 2001, pp. 299–307. MR1893473

[19] S. Ovsienko, *Finiteness statements for Gelfand-Zetlin modules*, Third International Algebraic Conference in the Ukraine (Ukrainian), Natsīonal. Akad. Nauk Ukraïni, Īnst. Mat., Kiev, 2002, pp. 323–338. MR2210503

[20] L. E. Ramirez, *Combinatorics of irreducible Gelfand-Tsetlin* $\mathfrak{sl}(3)$*-modules*, Algebra Discrete Math. **14** (2012), no. 2, 276–296. MR3099975

[21] L. E. Ramirez, Classificaçâo dos $sl(3)$-módulos de Gelfand-Tsetlin irredutíveis, PhD Thesis, Universidade de São Paulo (2013).

[22] L.E. Ramirez, P. Zadunaisky, Gelfand-Tsetlin modules over $\mathfrak{gl}(n, \mathbb{C})$ with arbitrary characters, arXiv:1705.10731.

[23] E. Vishnyakova, Geometric approach to 1-singular Gelfand-Tsetlin $\mathfrak{gl}_n$-modules, arXiv:1704.00170.

[24] E. Vishnyakova, Geometric approach to p-singular Gelfand-Tsetlin $\mathfrak{gl}_n$-modules, arXiv:1705.05793.

[25] P. Zadunaisky, *A new way to construct 1-singular Gelfand-Tsetlin modules*, Algebra Discrete Math. **23** (2017), no. 1, 180–193. MR3634504

[26] D. P. Želobenko, *Compact Lie groups and their representations*, American Mathematical Society, Providence, R.I., 1973. Translated from the Russian by Israel Program for Scientific Translations; Translations of Mathematical Monographs, Vol. 40. MR0473098

Instituto de Matemática e Estatística, Universidade de São Paulo, São Paulo SP, Brazil

Email address: futorny@ime.usp.br

Department of Mathematics, University of Texas at Arlington, Arlington, Texas 76019

Email address: grandim@uta.edu

Universidade Federal do ABC, Santo André SP, Brazil

Email address: luis.enrique@ufabc.edu.br

Contemporary Mathematics
Volume **713**, 2018
https://doi.org/10.1090/conm/713/14378

Fusion rings revisited

Jiuzu Hong

Abstract. In this note we describe a general elementary procedure to attach a fusion ring to any Kac-Moody algebra of affine type. In the case of untwisted affine algebras, they are usual fusion rings in the literature. In the case of twisted affine algebras, they are exactly the twisted fusion rings defined previously by the author via tracing out diagram automorphisms on conformal blocks for appropriate simply-laced Lie algebras. We also relate the fusion ring to the modular S-matrix for any Kac-Moody algebra of affine type.

1. Kac-Moody algebras of affine types

Let A be a generalized Cartan matrix of affine type of order $n+1$ (and rank n). We denote by $a_i(i = 0, 1, \cdots, n)$ the labelling on the vertices of Dynkin diagram associated to A ([**Ka**, §4.8, Table Aff 1, Table Aff 2]). We denote by $\check{a}_i(i = 0, 1, \cdots, n)$ the labelling of the vertices of the affine Dynkin diagram associated to the transpose A^t of A, which is obtained rom the Dynkin diagram of A by reversing all arrows and keeping the same enumeration of vertices.

Let $\mathfrak{g}(A)$ denote the Kac-Moody algebra associated to A with Cartan subalgebra $\mathfrak{h}$, roots $\alpha_i(i = 0, 1, \cdots n)$ and coroots $\check{\alpha}_i(i = 0, 1, \cdots, n)$. Note that $A = (\langle \alpha_i, \check{\alpha}_j \rangle)$,where $\langle \cdot, \cdot \rangle$ is the natural pairing between $\mathfrak{h}$ and its dual $\mathfrak{h}^*$. Let $e_i, f_i\ (i = 0, 1, \cdots, n)$ denote the Chevalley generators of $\mathfrak{g}(A)$. The following relations hold:

$$[e_i, f_j] = \delta_{ij}\check{\alpha}_i, [h, e_i] = \alpha_i(h)e_i, \text{ and } [h, f_i] = -\alpha_i(h)f_i,$$

for any $i, j = 0, \cdots, n$ and $h \in \mathfrak{h}$. Let $\mathfrak{g}'(A)$ denote the derived subalgebra $[\mathfrak{g}(A), \mathfrak{g}(A)]$ of $\mathfrak{g}(A)$. Note that $\mathfrak{g}'(A)$ is generated by $e_i, f_i(i = 0, \cdots, n)$. Fix an element $d \in \mathfrak{h}$ such that

$$\langle \alpha_i, d \rangle = 0 \text{ for } i = 1, \cdots, \quad \langle \alpha_0, d \rangle = 1.$$

Then $\mathfrak{g}(A) = \mathfrak{g}'(A) + \mathbb{C}d$, and $\mathfrak{h}$ is spanned by $\check{\alpha}_i(i = 0, \cdots, n)$ and d.

Put

$$K = \sum_{i=0}^{n} \check{a}_i\check{\alpha}_i.$$

Then K is the central element of $\mathfrak{g}(A)$.

2010 *Mathematics Subject Classification.* Primary 17B67, 81T40, 20G44.

Key words and phrases. Affine Lie algebra, affine Weyl group, conformal blocks, fusion ring, modular S -matrix, Verlinde formula.

This work was partially supported by the Simons Foundation collaboration grant 524406.

Let $\Lambda_i (i=0,\cdots,n)$ denote the fundamental weights of $\mathfrak{g}(A)$. The space $\mathfrak{h}^*$ is spanned by $\alpha_i, i=0,\cdots,n$ and Λ_0. There is a special imaginary root $\delta\in\mathfrak{h}^*$ given by $\delta=\sum_{i=0}^n a_i\alpha_i$. Let $(\cdot|\cdot)$ denote the symmetric normalized bilinear form on $\mathfrak{h}$ such that:

$$\begin{cases}(\check{\alpha}_i|\check{\alpha}_j)=a_j\check{a}_j a_{ij} \quad (i,j=0,\cdots,n)\\ (\check{\alpha}_i|d)=0 \quad i=1,\cdots,n\\ (\check{\alpha}_0|d)=a_0; \quad (d|d)=0.\end{cases}$$

Let $\kappa:\mathfrak{h}\to\mathfrak{h}^*$ be the induced isomorphism from $(\cdot|\cdot)$. Then

$$\kappa(\check{\alpha}_i)=\frac{a_i}{\check{a}_i}\alpha_i; \kappa(K)=\delta, \kappa(d)=a_0\Lambda_0.$$

We denote by $\mathring{\mathfrak{g}}$ the subalgebra of $\mathfrak{g}(A)$ generated by $e_i, f_i\,(i=1,\cdots,n)$. Let $\mathring{\mathfrak{h}}$ be the span of $\check{\alpha}_i\,(i=1,\cdots,n)$. Then $\mathring{\mathfrak{g}}$ is a simple Lie algebra with the Cartan subalgebra $\mathring{\mathfrak{h}}$. Let $\check{h}$ denote the dual Coxeter number of $\mathfrak{g}(A)$. We denote by $\mathring{A}$ the Cartan matrix of $\mathring{\mathfrak{g}}$.

We will use $X_N^{(r)}$ to denote the generalized Cartan matrix of affine type. When $r=1$, $X_N^{(1)}$ is untwisted; otherwise $X_N^{(r)}$ is twisted.

The following is the table for the associated $\mathring{A}$ and $\check{h}$ when A is of untwisted affine type.

(1)

A	$A_n^{(1)}$	$B_n^{(1)}$	$C_n^{(1)}$	$D_n^{(1)}$	$E_6^{(1)}$	$E_7^{(1)}$	$E_8^{(1)}$	$F_4^{(1)}$	$G_2^{(1)}$
$\mathring{A}$	A_n	B_n	C_n	D_n	E_6	E_7	E_8	F_4	G_2
$\check{h}$	n+1	2n-1	n+1	2n-2	12	18	30	9	4

In the above table, $n\geq 1$ when $A=A_n^{(1)}$, $n\geq 3$ when $A=B_n^{(1)}$, $n\geq 2$ when $A=C_n^{(1)}$, and $n\geq 4$ when $A=D_n^{(1)}$.

The following is the table for the associated $\mathring{A}$ and $\check{h}$ when A is of twisted affine type.

(2)

A	$A_{2n}^{(2)}$	$A_{2n-1}^{(2)}$	$D_{n+1}^{(2)}$	$E_6^{(2)}$	$D_4^{(3)}$
$\mathring{A}$	C_n	C_n	B_n	F_4	G_2
$\check{h}$	2n+1	2n	2n	12	6

In the above table, $n\geq 2$ when $A=A_{2n}^{(2)}$, $n\geq 3$ when $A=A_{2n-1}^{(2)}$, and $n\geq 3$ when $A=D_{n+1}^{(2)}$.

2. Fusion rings associated to Kac-Moody algebras of affine types

Let $\mathring{P}\subset\mathring{\mathfrak{h}}^*$ (resp. $\mathring{Q}\subset\mathring{\mathfrak{h}}^*$) be the weight lattice (resp. root lattice) associated to $\mathring{\mathfrak{g}}$. Let $\check{\mathring{P}}$ denote the dual lattice of $\mathring{Q}$, i.e.

$$\check{\mathring{P}}=\{\check{\mu}\in\mathring{\mathfrak{h}}\,|\,\langle\lambda,\check{\mu}\rangle\in\mathbb{Z}, \text{ for any } \lambda\in\mathring{Q}\}.$$

Similarly let $\check{\mathring{Q}}$ denote the dual lattice of $\mathring{P}$. Let $\mathring{P}^+$ (resp. $\check{\mathring{P}}^+$) be the set of dominant weights (resp. dominant coweights) of $\mathring{\mathfrak{g}}$. Let $\mathring{\Phi}$ (resp. $\mathring{\Phi}^+$) denote the set of roots (resp. positive roots) of $\mathring{\mathfrak{g}}$.

Let $\mathring{\rho}$ (resp. $\check{\mathring{\rho}}$) denote the summation of fundamental weights (resp. coweights) of $\mathfrak{g}$. Recall the symmetric bilinear form $(\cdot|\cdot)$ on $\mathfrak{h}$. It restricts to the subspace $\mathring{\mathfrak{h}}$, and it induces the isomorphism $\kappa:\mathring{\mathfrak{h}}\simeq\mathring{\mathfrak{h}}^*$.

Put

$$M = \begin{cases} \kappa(\check{\mathring{Q}}) & \text{if } A = X_N^{(1)} \text{ or } A = A_{2n}^{(2)} \\ \mathring{Q} & \text{otherwise} \end{cases} . \tag{3}$$

Let W (resp. $\mathring{W}$) denote the Weyl group of $\mathfrak{g}(A)$ (resp. $\mathring{\mathfrak{g}}$). Then W is naturally isomorphic to $\mathring{W} \ltimes M$ (cf.[**Ka**, Proposition 6.5]).

Let θ be the highest root of $\mathring{\mathfrak{g}}$. Set

$$\check{\theta} = \begin{cases} \kappa^{-1}(\theta) & \text{if } A = X_N^{(1)} \text{ or } A = A_{2n}^{(2)} \\ \text{highest coroot of } \mathring{\mathfrak{g}} & \text{otherwise.} \end{cases} \tag{4}$$

In fact when $A = X_N^{(1)}$, $\check{\theta}$ is the highest short coroot of $\mathring{\mathfrak{g}}$; when $A = A_{2n}^{(2)}$, $\check{\theta}$ is twice of the highest short coroot of $\mathring{\mathfrak{g}}$.

Given any $k \in \mathbb{N}$. We put

$$P_k = \{\lambda \in \mathring{P}^+ \mid \langle \lambda, \check{\theta} \rangle \leq k\}. \tag{5}$$

We also put

$$\check{P}_k = \{\check{\lambda} \in \check{\mathring{P}}^+ \mid \langle \check{\lambda}, \theta \rangle \leq k\},$$

(we only need it when $A = X_N^{(r)}$ with $r > 1$ and $A \neq A_{2n}^{(2)}$).

Let $\mathring{T}$ denote the torus $\mathrm{Hom}(\mathring{P}, \mathbb{C}^\times)$. We define a finite subset Σ_k of $\mathring{T}$. When $A = X_N^{(1)}$ or $A = A_{2n}^{(2)}$, we set

$$\Sigma_k = \{e^{\frac{2\pi i}{k+\check{h}}(\mathring{\rho}+\lambda|\cdot)} \in \mathring{T} \mid \lambda \in P_k\}. \tag{6}$$

When $A = X_N^{(r)}$ with $r > 1$ and $A \neq A_{2n}^{(2)}$, we set

$$\Sigma_k = \{e^{\frac{2\pi i}{k+h}\langle \check{\mathring{\rho}}+\check{\lambda}, \cdot\rangle} \in \mathring{T} \mid \check{\lambda} \in \check{P}_k\}.$$

Let $R_k(A)$ denote the commutative ring of $\mathbb{C}$-valued functions on the finite set Σ_k. We introduce an involution $*$ on the set P_k by sending λ to $-w_0(\lambda)$ where w_0 is the longest element in $\mathring{W}$.

For any $w \in \mathring{W}$, let $\ell(w)$ be the length of w. For any $\lambda \in P_k$ and $t \in \Sigma_k$, put

$$J_\lambda(t) := \sum_{w \in \mathring{W}} (-1)^{\ell(w)} w(\lambda + \mathring{\rho})(t), \tag{7}$$

and $\mathring{\chi}_\lambda(t) := \frac{J_\lambda(t)}{J_0(t)}$.

Let $\mathring{V}_\lambda$ denote the irreducible representation of $\mathring{\mathfrak{g}}$ of highest weight λ. By Weyl character formula $\mathring{\chi}_\lambda(t)$ is the trace of t on the representation $\mathring{V}_\lambda$. Therefore we get a collection of well-defined functions $\{\mathring{\chi}_\lambda \mid \lambda \in P_k\}$ on Σ_k.

Set $\Delta(t) = |J_0(t)|^2$. We introduce an Hermitian form $(\cdot, \cdot)$ on $R_k(A)$ as follows, for any $f, g \in R_k(A)$

$$(f, g) := \frac{1}{|\mathring{P}/(k + \check{h})M|} \sum_{t \in \Sigma_k} f(t)\overline{g(t)}\Delta(t). \tag{8}$$

THEOREM 2.1. (1) *The Hermitian form $(\cdot, \cdot)$ on $R_k(A)$ is an inner product.*

(2) *The set of functions $\{\mathring{\chi}_\lambda \mid \lambda \in P_k\}$ is an orthonormal basis of the ring $R_k(A)$ with respect to the inner product $(\cdot, \cdot)$.*

As a ring of functions on a finite set, $R_k(A)$ is semisimple. For any $\lambda, \mu \in P_k$, there exist unique coefficients $c_{\lambda\mu}^{\nu}$ such that

$$\mathring{\chi}_\lambda \cdot \mathring{\chi}_\mu = \sum_{\nu \in P_k} c_{\lambda\mu}^{\nu} \mathring{\chi}_\nu,$$

where $c_{\lambda\mu}^{\nu} \in \mathbb{C}$. Observe that $\mathring{\chi}_{\lambda^*}(t) = \overline{\mathring{\chi}_\lambda(t)}$ for any $\lambda \in P_k$ and $t \in \Sigma_k$. Then an immediate consequence of Theorem 2.1 is that the coefficients $c_{\lambda\mu}^{\nu}$ can be computed by the following formula

$$c_{\lambda\mu}^{\nu} = \frac{1}{|\mathring{P}/(k+\check{h})M|} \sum_{t \in \Sigma_k} \mathring{\chi}_\lambda(t)\mathring{\chi}_\mu(t)\mathring{\chi}_{\nu^*}(t)\Delta(t). \tag{9}$$

We call the ring $R_k(A)$ together with the basis $\{\mathring{\chi}_\lambda \mid \lambda \in P_k\}$ the fusion ring associated to the Kac-Moody algebra $\mathfrak{g}(A)$, and we call $c_{\lambda\mu}^{\nu}$ the fusion coefficients of $R_k(A)$.

The following theorem asserts the integrability of fusion coefficients, and the stabilization when the level is sufficient large.

THEOREM 2.2. *For any $\lambda, \mu, \nu \in P_k$, we have*

(1) *The coefficients $c_{\lambda\mu}^{\nu} \in \mathbb{Z}$.*

(2) *If $\langle \lambda + \mu + \nu, \check{\theta} \rangle \leq 2k$, then*

$$c_{\lambda\mu}^{\nu} = \dim \operatorname{Hom}_{\mathring{\mathfrak{g}}}(\mathring{V}_\nu, \mathring{V}_\lambda \otimes \mathring{V}_\mu).$$

The following theorem is an incarnation of [**Ho2**, Theorem 1.6] in terms of fusion rings.

THEOREM 2.3. *The fusion rings $R_{2k+1}(A_{2n}^{(2)})$ and $R_k(C_n^{(1)})$ are isomorphic.*

Theorem 2.1, Theorem 2.2 and Theorem 2.3 are basically the consequence of Verlinde formula for dimension of conformal blocks (cf.[**Be**]), and the Verlinde formula for the trace of diagram automorphism on conformal blocks [**Ho2**] by the author. In the rest of note we will explain uniformly how they are deduced.

When $A = X_N^{(1)}$, the theory of fusion rings and conformal blocks can be dated back to [**Fa**, **Be**, **TUY**, **V**]. The coefficients $c_{\lambda\mu}^{\nu}$ are always nonnegative, since they can be interpreted as the dimension of conformal blocks.

When $A = X_N^{(r)}$ with $r > 1$, the corresponding theory of fusion rings and its relation with usual conformal blocks is essentially explained in [**Ho2**], although the role of twisted affine Lie algebras was not clearly clarified there. We expect that the fusion coefficients $c_{\lambda\mu}^{\nu}$ are all nonnegative. For example by Theorem 2.3 the fusion coefficients for $A_{2n}^{(2)}$ are always nonnegative when the level is odd. We expect that there is an appropriate theory of twisted conformal blocks whose dimensions would correspond to the fusion coefficients for twisted affine Lie algebras. We also hope that there will be a theory of fusion category for twisted affine Lie algebra in the flavor of [**Fi**].

3. Twining formula and Verlinde formula

3.1. Twining formula for tensor invariant space. For each $\mathfrak{g}(A)$ with $A = X_N^{(r)}$, we attach a unique pair $(\dot{\mathfrak{g}}, \sigma)$ where $\dot{\mathfrak{g}}$ is a simple Lie algebra and σ is a diagram automorphism on $\dot{\mathfrak{g}}$ (possibly trivial) of order r. Let $\mathring{A}$ (resp. $\dot{A}$) denote the Cartan matrix of $\mathring{\mathfrak{g}}$ (resp. $\dot{\mathfrak{g}}$). The Cartan matrix $\dot{A}$ is attached as follows. If A is of untwisted affine type, then we take $\dot{A} = \mathring{A}$; if A is of twisted affine type, we have the following table:

(10)

A	$A_{2n}^{(2)}$	$A_{2n-1}^{(2)}$	$D_{n+1}^{(2)}$	$E_6^{(2)}$	$D_4^{(3)}$
$\mathring{A}$	C_n	C_n	B_n	F_4	G_2
$\dot{A}$	A_{2n}	D_{n+1}	A_{2n-1}	E_6	D_4

.

The simple Lie algebra $\mathring{\mathfrak{g}}$ is the orbit Lie algebra of the pair $(\dot{\mathfrak{g}}, \sigma)$ in the sense of [**Ho2**, Section 2.1].

Let $\mathring{I}$ (resp. $\dot{I}$) denote the set of vertices of Dynkin diagram of $\mathring{\mathfrak{g}}$ (resp. $\dot{\mathfrak{g}}$). We will denote by $\dot{P}$ (resp. $\dot{Q}$) the weight lattice (resp. root lattice) of $\dot{\mathfrak{g}}$, and we denote by $\dot{P}^+$ the set of dominant weights. Let $\{\dot{\omega}_j \mid j \in \dot{I}\}$ denote the fundamental weights of $\dot{\mathfrak{g}}$. There is a bijection $p : \mathring{I} \simeq \dot{I}/\sigma$, where $\dot{I}/\sigma$ denotes the set of σ-orbits on $\dot{I}$. Moreover there is a bijection $\iota : \mathring{P} \simeq \dot{P}^\sigma$ and a projection $\breve{\iota} : \check{\dot{Q}} \to \check{\mathring{Q}}$ such that

(1) $\iota(\mathring{\omega}_i) = \sum_{j \in p(i)} \dot{\omega}_i$ for any $i \in \mathring{I}$.

(2) $\langle \lambda, \breve{\iota}(\breve{\beta}) \rangle = \langle \iota(\lambda), \beta \rangle$, for any $\lambda \in \mathring{P}$ and $\breve{\beta} \in \check{\dot{Q}}$.

For any $\lambda \in \mathring{P}^+$, let $\dot{V}_\lambda$ denote the irreducible representation of $\dot{\mathfrak{g}}$ of highest weight $\iota(\lambda) \in \dot{P}^+$. There is a unique operator $\sigma : \dot{V}_\lambda \to \dot{V}_\lambda$ such that

(1) $\sigma(x \cdot v) = \sigma(x) \cdot \sigma(v)$ for any $x \in \dot{\mathfrak{g}}$ and $v \in \dot{V}$,

(2) $\sigma(v_\lambda) = v_\lambda$ where v_λ is the highest weight vector of $\dot{V}_\lambda$.

For any tuple $\vec{\lambda} = (\lambda_1, \cdots, \lambda_m)$ of elements in $\mathring{P}^+$, let $\dot{V}_{\vec{\lambda}}^{\dot{\mathfrak{g}}}$ (resp. $\mathring{V}_{\vec{\lambda}}^{\mathring{\mathfrak{g}}}$) denote the tensor invariant space

$$(\dot{V}_{\lambda_1} \otimes \cdots \otimes \dot{V}_{\lambda_m})^{\dot{\mathfrak{g}}} \text{ (resp. } (\mathring{V}_{\lambda_1} \otimes \cdots \otimes \mathring{V}_{\lambda_m})^{\mathring{\mathfrak{g}}}).$$

From each operator σ on $\dot{V}_{\lambda_i}$, there exists an operator σ (we still use the same notation) acting diagonally on $\dot{V}_{\vec{\lambda}}^{\dot{\mathfrak{g}}}$. The following twining formula was proved in [**HS**, Theorem 1.1].

THEOREM 3.1. *We have the equality* $\operatorname{tr}(\sigma | \dot{V}_{\vec{\lambda}}^{\dot{\mathfrak{g}}}) = \dim \mathring{V}_{\vec{\lambda}}^{\mathring{\mathfrak{g}}}$, *where* $\operatorname{tr}(\sigma | \dot{V}_{\vec{\lambda}}^{\dot{\mathfrak{g}}})$ *is the trace of* σ *on* $\dot{V}_{\vec{\lambda}}^{\dot{\mathfrak{g}}}$.

The twining formula for the weight spaces between representations of $\dot{\mathfrak{g}}$ and $\mathring{\mathfrak{g}}$ was first discovered by Jantzen (cf. [**Ho1**, **Ja**]). Theorem 3.1 can actually be deduced from Jantzen formula (cf.[**Ho2**, Section 5.1]).

3.2. Verlinde formula. Let $\mathbb{C}((t))$ be the field of Laurent series. Let $\hat{\mathcal{L}}(\dot{\mathfrak{g}})$ be the associated affine Lie algebra $\mathfrak{g}((t)) \oplus \mathbb{C}K \oplus \mathbb{C}d$ (cf.[**Ka**, §7]). The diagram automorphism σ still acts on $\hat{\mathcal{L}}(\dot{\mathfrak{g}})$. Note that if A is of untwisted affine type, then $\mathfrak{g}(A) \simeq \hat{\mathcal{L}}(\dot{\mathfrak{g}})$. If A is of twisted affine type, then the Kac-Moody algebra $\mathfrak{g}(A)$ is the orbit Lie algebra of $(\hat{\mathcal{L}}(\dot{\mathfrak{g}}), \sigma)$, and in fact there is also a twining formula for the weight spaces between representations of $\hat{\mathcal{L}}(\dot{\mathfrak{g}})$ and $\mathfrak{g}(A)$, which is due to Fuchs-Schellekens-Schweigert[**FSS**].

Let $\check{\dot{\theta}}$ denote the highest short coroot of $\dot{\mathfrak{g}}$. Let $\dot{P}_k$ denote the subset of $\dot{P}^+$ consisting of $\lambda \in \dot{P}^+$ such that $\langle \lambda, \check{\dot{\theta}} \rangle \leq k$. For any $\lambda \in \dot{P}_k$, let $\dot{\mathcal{H}}_{\lambda + k\dot{\Lambda}_0}$ denote the irreducible integrable representation of $\hat{\mathcal{L}}(\dot{\mathfrak{g}})$ of highest weight $\lambda + k\dot{\Lambda}_0$, where $\dot{\Lambda}_0$

is the fundamental weight of $\hat{\mathcal{L}}(\dot{\mathfrak{g}})$ corresponding to the vertex 0 in the extended Dynkin diagram of $\dot{\mathfrak{g}}$.

LEMMA 3.2. *The isomorphism* $\iota : \mathring{P} \simeq \dot{P}^\sigma$ *restricts to the bijection* $\iota : P_k \simeq (\dot{P}_k)^\sigma$.

PROOF. It is enough to show that for any $\lambda \in P_k$, $\iota(\lambda) \in \dot{P}_k$. By [**Ho2**, Lemma 2.1], $\check{\iota}(\check{\dot{\theta}}) = \check{\theta}$. For any $\lambda \in P_k$, we have

$$\langle \iota(\lambda), \check{\dot{\theta}} \rangle = \langle \lambda, \check{\iota}(\check{\dot{\theta}}) \rangle = \langle \lambda, \check{\theta} \rangle \leq k.$$

Hence $\iota(\lambda) \in \dot{P}_k$. □

Given an m-pointed smooth projective curve $(C, \vec{p})$ over $\mathbb{C}$, on each point p_i we associate a dominant weight $\lambda_i \in \dot{P}_k$ and an irreducible integral representation $\dot{\mathcal{H}}_{\lambda_i + k\dot{\Lambda}}$ an affine Lie algebra $\hat{\mathcal{L}}(\dot{\mathfrak{g}})$ depending on the point p_i. Let $\dot{\mathfrak{g}}(C\backslash\vec{p})$ be the space of $\dot{\mathfrak{g}}$-valued regular functions on $C\backslash\vec{p}$. The space $\dot{\mathfrak{g}}(C\backslash\vec{p})$ is naturally a Lie algebra induced from $\dot{\mathfrak{g}}$. The Lie algebra $\dot{\mathfrak{g}}(C\backslash\vec{p})$ acts on

$$\dot{\mathcal{H}}_{\vec{\lambda}} := \dot{\mathcal{H}}_{\lambda_1 + k\dot{\Lambda}_0} \otimes \cdots \otimes \dot{\mathcal{H}}_{\lambda_m + k\dot{\Lambda}_0}$$

naturally. The space $\dot{V}_{\vec{\lambda}}(C, \vec{p})$ of conformal blocks associated to $\vec{p}$ and $\vec{\lambda}$ is defined as the coinvariant space of $\dot{\mathcal{H}}_{\vec{\lambda}}$ with respect to the action of $\dot{\mathfrak{g}}(C\backslash\vec{p})$:

$$\dot{V}_{\vec{\lambda}}(C, \vec{p}) := \dot{\mathcal{H}}_{\vec{\lambda}} / \dot{\mathfrak{g}}(C\backslash\vec{p})\dot{\mathcal{H}}_{\vec{\lambda}}.$$

The trace $\mathrm{tr}(\sigma|\dot{V}_{\vec{\lambda}}(C, \vec{p}))$ is independent of the choice of $\vec{p}$ (cf.[**Ho2**, Section 3]). For convenience we write $\mathrm{tr}(\sigma|\dot{V}_{\vec{\lambda}}(C))$ by ignoring $\vec{p}$.

LEMMA 3.3. *We attach* $\lambda, \mu \in \dot{P}_k$ *to two distinct points on the projective line* $\mathbb{P}^1$. *Then we have*

$$\mathrm{tr}(\sigma|\dot{V}_{\lambda,\mu}(\mathbb{P}^1)) = \delta_{\lambda,\mu^*}.$$

PROOF. cf.[**Ho2**, Lemma 3.9]. □

Let $\check{\dot{h}}$ denote the dual Coxeter number of $\hat{\mathcal{L}}(\dot{\mathfrak{g}})$. By comparing Table (1), (2) and (10), we can see that $\check{\dot{h}} = \check{h}$.

LEMMA 3.4. *If* $A = X_N^{(1)}$ *or* $A = A_{2n}^{(2)}$, *then*

$$\kappa(\check{\mathring{Q}}) = \begin{cases} Q_l & A = X_N^{(1)} \\ \frac{1}{2}Q_l & A = A_{2n}^{(2)} \end{cases}.$$

PROOF. It follows from the discussions in [**Ka**, §6.5]. □

Recall the torus $\mathring{T} = \mathrm{Hom}(\mathring{P}, \mathbb{C})$. Set

$$T_k := \{t \in \mathring{T} \mid \alpha(t) = 1, \forall \alpha \in (k + \check{h})M\}.$$

Then T_k is a finite subgroup of $\mathring{T}$. An element $t \in \mathring{T}$ is regular if $\alpha(t) \neq 1$ for any $\alpha \in \mathring{\Phi}$. Equivalently an element $t \in \mathring{T}$ is regular if and only if the stabilizer group of $\mathring{W}$ at t is trivial. Let T_k^{reg} denote the set of regular elements in T_k.

The following is the general Verlinde formula for the trace of σ on the space of conformal blocks.

THEOREM 3.5. *Let $(C, \vec{p})$ be an m-pointed smooth projective curve of genus g. Given a tuple $\vec{\lambda} = (\lambda_1, \lambda_2, \cdots, \lambda_m)$ of elements in P_k. We have the following formula*

$$\operatorname{tr}(\sigma|\dot{V}_{\vec{\lambda}}(C)) = |T_k|^{g-1} \sum_{t \in T_k^{\mathrm{reg}}/\mathring{W}} \mathring{\chi}_{\lambda_1}(t) \cdots \mathring{\chi}_{\lambda_m}(t) \Delta(t)^{1-g}. \tag{11}$$

PROOF. When $A = X_N^{(1)}$, it is just the usual Verlinde formula(cf.[**Be**]). When $A = X_N^{(r)}$ with $r > 1$, it is the main theorem of [**Ho2**]. Lemma 3.4 ensures the data in the formulas of [**Be, Ho2**] matches with the note. □

In particular if $C = \mathbb{P}^1$ and $\vec{\lambda} = (\lambda, \mu, \nu)$ with $\lambda, \mu, \nu \in P_k$, we have

$$\operatorname{tr}(\sigma|\dot{V}_{\vec{\lambda}}(C)) = \frac{1}{|T_k|} \sum_{t \in T_k^{\mathrm{reg}}/\mathring{W}} \mathring{\chi}_\lambda(t) \mathring{\chi}_\mu(t) \mathring{\chi}_\nu(t) \Delta(t). \tag{12}$$

Let W_k denote the affine Weyl group $\mathring{W} \ltimes (k + \check{h})M$. We define the dotted action of W_k on $\mathring{P}$,

$$w \star \lambda = w(\lambda + \mathring{\rho}) - \mathring{\rho}, \text{ for any } w \in W_k.$$

Let $W_k^\dagger$ denote the set consisting of the minimal representatives of the left cosets of $\mathring{W}$ in W_k. For any $\lambda \in P_k$, $w \star \lambda \in \mathring{P}^+$ if and only if $w \in W_k^\dagger$.

The following is a general version of Kac-Walton formula.

THEOREM 3.6. *For any $\lambda, \mu, \nu \in P_k$, we have*

$$\operatorname{tr}(\sigma|\dot{V}_{\lambda,\mu,\nu}(\mathbb{P}^1)) = \sum_{w \in W_k^\dagger} (-1)^{\ell(w)} \dim \mathring{V}^{\mathring{\mathfrak{g}}}_{\lambda,\mu,w\star\nu}.$$

PROOF. When $A = X_N^{(1)}$ it is just the usual Kac-Walton formula (cf.[**Ka**, Ex.13.35]). When $A = X_N^{(r)}$ with $r > 1$, it follows from [**Ho2**, Theorem 5.11] and Theorem 3.1. □

4. Proofs

4.1. Proof of Theorem 2.1.

4.1.1. *Inner product.*

LEMMA 4.1. *We have the following equality*

$$\check{h} = \langle \mathring{\rho}, \check{\theta} \rangle + 1$$

PROOF. When $A = X_N^{(1)}$, $\check{\theta}$ is the highest short coroot of $\mathring{\mathfrak{g}}$. When $A = A_{2n}^{(2)}$, $\check{\theta}$ is twice of the highest short coroot of $\mathring{\mathfrak{g}}$. Then it is easy to check that $\check{h} = (\mathring{\rho}, \check{\theta}) + 1$ by comparing the numerical labels in the tables of [**Ka**, §4.8] and Tables (1) (2).

When $A = X_N^{(r)}$ with $r > 1$ and $A \neq A_{2n}^{(2)}$, $\langle \mathring{\rho}, \check{\theta} \rangle + 1$ is the Coxeter number of dual root system of $\mathring{\mathfrak{g}}$. By comparing the table in [**Ka**, §6.1], it is also easy to see the equality. □

To prove the Hermitian form defined in (8) is an inner product, it is enough to show that $\Delta(t) > 0$ for any $t \in \Sigma_k$. By Weyl denominator formula

$$\Delta(t) = \prod_{\alpha \in \mathring{\Phi}} (1 - \alpha(t)) = |\prod_{\alpha \in \mathring{\Phi}^+} (1 - \alpha(t))|^2.$$

When $A = X_N^{(1)}$ or $A = A_{2n}^{(2)}$, for any $\lambda \in P_k$ and $\alpha \in \mathring{\Phi}^+$, we have $0 \le (\lambda|\alpha) \le k$. By Lemma 4.1, $(\mathring{\rho}|\theta) = \langle \mathring{\rho}, \check{\theta} \rangle = \check{h} - 1$. Since θ is the highest root of $\mathring{\mathfrak{g}}$, for any $\alpha \in \mathring{\Phi}^+$ we have

$$0 < (\mathring{\rho}|\alpha) \le (\mathring{\rho}|\theta) = \check{h} - 1.$$

In particular

$$0 < (\lambda + \mathring{\rho}|\alpha) \le k + \check{h} - 1.$$

It shows that $\alpha(t)$ can not be equal to 1. Hence $\Delta(t) > 0$ for any $t \in \Sigma_k$.

When $A = X_N^{(r)}$ with $r > 1$ and $A \ne A_{2n}^{(2)}$, for any $\check{\lambda} \in \check{P}_k$, we have $0 \le \langle \alpha, \check{\lambda} \rangle \le k$. By Lemma 4.1, for any $\alpha \in \mathring{\Phi}^+$, we always have

$$0 < \langle \alpha, \mathring{\check{\rho}} + \check{\lambda} \rangle < k + \check{h}.$$

Hence in this case, we also have $\Delta(t) > 0$ for any $t \in \Sigma_k$.

4.1.2. *Orthonormal basis.* Recall that T_k is the group consisting of $t \in \mathring{T}$ such that $\alpha(t) = 1$ for any $\alpha \in (k + \check{h})M$. The group T_k is naturally isomorphic to

$$\begin{cases} \mathring{P}/(k+\check{h})M & \text{if } A = X_N^{(1)} \text{ or } A = A_{2n}^{(2)} \\ \mathring{\check{P}}/(k+\check{h})\mathring{\check{Q}} & \text{otherwise} \end{cases} \tag{13}$$

LEMMA 4.2. $|T_k| = |\mathring{P}/(k+\check{h})M|$.

PROOF. When $A = X_N^{(1)}$ or $A = A_{2n}^{(2)}$, it simply follows from the natural isomorphism (13).

When $A = X_N^{(r)}$ with $r > 1$ and $A \ne A_{2n}^{(2)}$, $\mathring{\check{P}}/(k+\check{h})\mathring{\check{Q}}$ is Pontryagin dual to $\mathring{P}/(k+\check{h})\mathring{Q}$. Then the lemma follows from the natural isomorphism (13). □

We recall the finite set Σ_k in $\mathring{T}$. It is clear that Σ_k is a subset of T_k^{reg}.

When $A = X_N^{(1)}$ or $A = A_{2n}^{(2)}$, let $\eta_1 : P_k \to \Sigma_k$ be the map by assigning $\lambda \in P_k$ to $t_\lambda := e^{\frac{2\pi i}{k+\check{h}}(\mathring{\rho}+\lambda|\cdot)} \in \Sigma_k$. When $A = X_N^{(r)}$ with $r > 1$ and $A \ne A_{2n}^{(2)}$, let $\eta_2 : \check{P}_k \to \Sigma_k$ be the map assigning $\check{\lambda} \in \check{P}_k$ to $t_{\check{\lambda}} := e^{\frac{2\pi i}{k+\check{h}}\langle \mathring{\check{\rho}}+\check{\lambda},\cdot\rangle}$.

LEMMA 4.3. *The maps η_1, η_2 defined above are bijections.*

PROOF. We first assume that $A = X_N^{(1)}$ or $A = A_{2n}^{(2)}$. It is enough to show that η_1 is injective, equivalent it is enough to show that $i_1 : P_k \to \mathring{P}/(k+\check{h})M$ the map given by

$$\lambda \mapsto \lambda + \mathring{\rho} \mod (k+\check{h})M$$

is injective. For any $\lambda \in (k+\check{h})M$ and $w \in \mathring{W}$, we always have

$$\langle \lambda, w(\check{\theta}) \rangle \in (k+\check{h})\mathbb{Z}. \tag{14}$$

For any $\lambda_1, \lambda_2 \in P_k$, we have $0 \le \langle \lambda_i, \check{\theta} \rangle \le k$. It also follows that for any $w \in \mathring{W}$,

$$-k \le \langle \lambda_i, w(\check{\theta}) \rangle \le k. \tag{15}$$

If $i(\lambda_1) = i(\lambda_2)$, equivalently $\lambda_1 - \lambda_2 \in (k+\check{h})M$, then by (14) and (15) we have

$$\langle \lambda_1 - \lambda_2, w(\check{\theta}) \rangle = 0,$$

for any $w \in \mathring{W}$. Since $\{w(\check{\theta}) \,|\, w \in \mathring{W}\}$ spans $\mathring{\mathfrak{h}}$, it follows that $\lambda_1 = \lambda_2$. It proves the injectivity of i_1.

When $A = X_N^{(r)}$ with $r > 1$ and $A \neq A_{2n}^{(2)}$, we consider the map $i_2 : \check{P}_k \to \check{\mathring{P}}/(k+\check{h})\check{\mathring{Q}}$ given by

$$\check{\lambda} \mapsto \check{\lambda} + \check{\mathring{\rho}} \mod (k+\check{h})\check{\mathring{Q}}.$$

In this case the lemma can be proved similarly. □

LEMMA 4.4. *When $A = X_N^{(r)}$ with $r > 1$ and $A \neq A_{2n}^{(2)}$, we have $|P_k| = |\check{P}_k|$.*

PROOF. When $A \neq A_{2n-1}^{(2)}$ and $A \neq D_{n+1}^{(2)}$, it is clear that $|P_k| = |\check{P}_k|$ since the dual Lie algebra of $\mathring{\mathfrak{g}}$ is isomorphic to $\mathring{\mathfrak{g}}$.

When $A = A_{2n-1}^{(2)}$ or $A = D_{n+1}^{(2)}$, we define a map $\epsilon : \mathring{P} \to \check{\mathring{P}}$ given by $\omega_i \mapsto \check{\omega}_{\ell-i}$ where ℓ is the rank of $\mathring{\mathfrak{g}}$. Note that ϵ induces the dual map $\check{\epsilon} : \mathring{Q} \to \check{\mathring{Q}}$. Observe that $\check{\epsilon}(\theta) = \check{\theta}$, where $\check{\theta}$ is the highest coroot of $\mathring{\mathfrak{g}}$. It is now clear that ϵ defines a bijection $P_k \simeq \check{P}_k$. Hence the lemma is proved. □

COROLLARY 4.5. *For any generalized Cartan matrix A of affine type, we have $|\Sigma_k| = |P_k|$.*

PROOF. It follows from Lemma 4.2 and Lemma 4.3. □

LEMMA 4.6. *Σ_k is the fundamental set of T_k^{reg} with respect to the action of the Weyl group $\mathring{W}$, i.e. for any $t \in T_k^{\mathrm{reg}}$ there exists a unique element $t_0 \in \Sigma_k$ and a unique $w \in \mathring{W}$ such that $w(t_0) = t$.*

PROOF. By Lemma 4.3, we are reduced to consider the action of the Weyl group $\mathring{W}$ on $\mathring{P}/(k+\check{h})M$.

When $A = X_N^{(1)}$ or $A = A_{2n}^{(2)}$, it follows from the fact that the set $\{\mathring{\rho}+\lambda \mid \lambda \in P_k\}$ is exactly those integral weights sitting in the interior of fundamental alcove with respect to the action of the affine Weyl group $\mathring{W} \ltimes (k+\check{h})M$ on $\mathring{P} \otimes \mathbb{R}$.

When $A = X_N^{(r)}$ with $r > 1$ and $A \neq A_{2n}^{(2)}$, it follows from the fact that the set $\{\check{\mathring{\rho}} + \check{\lambda} \mid \lambda \in \check{P}_k\}$ is those integral coweights sitting in the interior of fundamental alcove with respect to the action of the affine Weyl group $\mathring{W} \ltimes (k+\check{h})\check{\mathring{Q}}$ on $\check{\mathring{P}} \otimes \mathbb{R}$. □

From this lemma, we immediately get $\Sigma_k = T_k^{\mathrm{reg}}/\mathring{W}$. By Theorem 3.5, Lemma 4.2 and Lemma 4.6, Lemma 3.3 is equivalent to

$$\frac{1}{|\mathring{P}/(k+\check{h})M|} \sum_{t \in \Sigma_k} \mathring{\chi}_\lambda(t)\mathring{\chi}_{\mu^*}(t)\Delta(t) = \delta_{\lambda\mu}, \tag{16}$$

for any $\lambda, \mu \in P_k$. Recall the definition of the Hermitian form $(\cdot,\cdot)$ in (8), the formula (16) implies that $\{\mathring{\chi}_\lambda \mid \lambda \in P_k\}$ is an orthonormal subset of $R_k(A)$.

In the end in view of Corollary 4.5, the dimension $R_k(A)$ is equal to the cardinality of P_k. It follows that $\{\mathring{\chi}_\lambda \mid \lambda \in P_k\}$ is indeed an orthonormal basis of $R_k(A)$ with respect to the inner product $(\cdot,\cdot)$.

4.2. Proof of Theorem 2.2. From the proof of Theorem 2.1, the Verlinde formula (Theorem 3.5) implies that

$$c_{\lambda\mu}^{\nu} = \mathrm{tr}(\sigma | \dot{V}_{\lambda,\mu,\nu^*}^{\mathring{\mathfrak{g}}}), \tag{17}$$

for any $\lambda, \mu, \nu \in P_k$. By Kac-Walton formula (Theorem 3.6) we have

$$c^{\nu}_{\lambda\mu} = \sum_{w \in W_k^{\dagger}} (-1)^{\ell(w)} \dim \mathring{V}^{\mathring{\mathfrak{g}}}_{\lambda,\mu,w(\mu^*)}. \tag{18}$$

It follows that for any $\lambda, \mu, \nu \in P_k$, $c^{\nu}_{\lambda\mu} \in \mathbb{Z}$.

We now proceed to prove the stabilization of fusion coefficients. We first observe that $\langle \lambda^*, \check{\theta} \rangle = \langle \lambda, \check{\theta} \rangle$ for any $\lambda \in \mathring{P}$, since $\check{\theta}^* = \check{\theta}$. Hence the condition $\langle \lambda + \mu + \nu, \check{\theta} \rangle \leq 2k$ is equivalent to the condition $\langle \lambda + \mu + \nu^*, \check{\theta} \rangle \leq 2k$.

When $A = X_N^{(1)}$, the stabilization follows from [**Be**, Proposition 4.3].

When $A = X_N^{(r)}$ with $r > 1$, $\dot{\mathfrak{g}}$ is simply-laced, and λ, μ, ν satisfies the following inequality

$$\langle \iota(\lambda) + \iota(\mu) + \iota(\nu)^*, \check{\dot{\theta}} \rangle \leq 2k.$$

Again in view of [**Be**, Proposition 4.3] we have

$$\dot{V}_{\lambda,\mu,\nu^*}(\mathbb{P}^1) \simeq (\dot{V}_\lambda \otimes \dot{V}_\mu \otimes \dot{V}_{\nu^*})_{\dot{\mathfrak{g}}}.$$

In particular it follows that

$$\mathrm{tr}(\sigma | \dot{V}_{\lambda,\mu,\nu^*}(\mathbb{P}^1)) = \mathrm{tr}(\sigma | (\dot{V}_\lambda \otimes \dot{V}_\mu \otimes \dot{V}_{\nu^*})_{\dot{\mathfrak{g}}}) = \dim(\mathring{V}_\lambda \otimes \mathring{V}_\mu \otimes \mathring{V}_{\nu^*})_{\mathring{\mathfrak{g}}},$$

where the second equality follows from Theorem 3.1, since the tensor invariant space and tensor co-invariant space of $\dot{\mathfrak{g}}$ and $\mathring{\mathfrak{g}}$ are naturally isomorphic.

4.3. Proof of Theorem 2.3. Let $A_1 = A_{2n}^{(2)}$ and $A_2 = C_n^{(1)}$. The underlying simple Lie algebras $\mathring{\mathfrak{g}}_1$ and $\mathring{\mathfrak{g}}_2$ are both isomorphic to the simple Lie algebra of type C_n. We use notation $\mathring{\mathfrak{g}}$ for type C_n simple Lie algebra to identity $\mathring{\mathfrak{g}}_1$ and $\mathring{\mathfrak{g}}_2$. Let M_1 (resp. M_2) be the lattice in $\mathring{\mathfrak{h}}^*$ associated to the generalized Cartan matrix A_1 (resp. A_2) of level $2k+1$ (resp. k). Then we have

$$M_1 = (2k+1+2n+1)\frac{1}{2}\mathring{Q}_l = (k+n+1)\mathring{Q}_l = (k+n+1)\mathring{Q}_l = M_2.$$

Let $(\cdot|\cdot)_i$ be the inner product on $\mathring{\mathfrak{h}}$ associated to A_i for each $i = 1, 2$. Note that $(\cdot|\cdot)_1 = 2(\cdot|\cdot)$. Let $P_k(A_i)$ be the finite set of dominant weights associated to A_i as in (5) for each $i = 1, 2$. In particular we have

$$P_{2k+1}(A_1) = \{\lambda \in \mathring{P} \mid (\lambda|\theta)_1 \leq 2k+1\}$$

and

$$P_k(A_2) = \{\lambda \in \mathring{P} \mid (\lambda|\theta)_2 \leq k\}.$$

It is now not difficult to see that

$$P_{2k+1}(A_1) = P_k(A_2).$$

We set $P_k := P_k(A_2) = P_{2k+1}(A_1)$. Then the fusion ring $R_{2k+1}(A_1)$ and $R_k(A_2)$ have bases indexed by the same set P_k.

Let $\Sigma_k(A_i)$ be the finite set of $\mathring{T}$ defined as in (6) for each $i = 1, 2$. We observe that

$$\Sigma_{2k+1}(A_1) = \Sigma_k(A_2),$$

since for any $\lambda \in P_k$ we have

$$e^{\frac{2\pi i}{2k+1+2n+1}(\lambda+\mathring{\rho}|\cdot)_1} = e^{\frac{2\pi i}{k+n+1}(\lambda+\mathring{\rho}|\cdot)_2}.$$

From Formula (9), we see that the fusion rings $R_{2k+1}(A_1)$ and $R_k(A_2)$ are indeed isomorphic.

5. Modular S-matrix

The relation between the fusion ring and modular S-matrix is well-known when the genearlized Cartan matrix is of untwisted affine type, see details in [**Ka**, §13.8, Ex.13.34,13.35]. In the case of twisted affine case, we use the fusion rings that we have just defined to relate to the S-matrix for twisted affine algebras (cf.[**Ka**, §13.9]). For convenience we still treat all generalized Cartan matrices of affine type generally.

We first assume $A = X_N^{(r)}$ with $r > 1$ but $A \neq A_{2n}^{(2)}$. For any A, we attach the adjacent Cartan matrix A' as follows:

(19)

A	$A_{2n-1}^{(2)}$	$D_{n+1}^{(2)}$	$E_6^{(2)}$	$D_4^{(3)}$
A'	$D_{n+1}^{(2)}$	$A_{2n+1}^{(2)}$	$E_6^{(2)}$	$D_4^{(3)}$

We define a linear isomorphism $\epsilon : \mathring{\mathfrak{h}}'^* \simeq \mathring{\mathfrak{h}}$ such that

(1) If $A = A_{2n-1}^{(2)}$ or $A = D_{n+1}^{(2)}$, then $\epsilon(\omega_i') = \check{\omega}_{n-i}$ for any $i = 1, \cdots, n$;
(2) otherwise $\epsilon(\omega_i') = \check{\omega}_i$ for any $i = 1, \cdots, n$,

where $\check{\omega}_i$ denotes the fundamental coweight of $\mathring{\mathfrak{g}}$.

Let $\check{\theta}'$ be an element in the coroot lattice of $\mathring{\mathfrak{g}}'$ defined similarly as $\check{\theta}$ in (4), and let P_k' be similarly defined as P_k in (5).

LEMMA 5.1. *The map ϵ restricts to a bijection $\epsilon : P_k' \simeq \check{P}_k$.*

PROOF. The root system of $\mathring{\mathfrak{g}}'$ is dual to $\mathring{\mathfrak{g}}$. The lemma is clear by looking at the Dynkin diagram of $\mathring{\mathfrak{g}}$ and $\mathring{\mathfrak{g}}'$. □

We now identify $\mathring{\mathfrak{h}}$ and $\mathring{\mathfrak{h}}'$ via the isomorphism $\varsigma := \epsilon^* \circ \kappa$ from $\mathring{\mathfrak{h}}$ to $\mathring{\mathfrak{h}}'$, where ϵ^* is the dual operator of $\epsilon : \mathring{\mathfrak{h}}'^* \simeq \mathring{\mathfrak{h}}$.

When $A = X_N^{(1)}$ or $A = A_{2n}^{(2)}$, we take $A' = A$ and take ς to be the identity on $\mathring{\mathfrak{h}}$.

When $A = X_N^{(r)}$ with $r > 1$ and $A \neq A_{2n}^{(2)}$, for any $\mu' \in P_k$, let $t_{\mu'}$ denote the element in Σ_k associated to $\epsilon(\mu') \in \check{P}_k$. When $A = X_N^{(1)}$ or $A = A_{2n}^{(2)}$, for any $\mu' \in P_k' = P_k$, we denote by $t_{\mu'}$ the element in Σ associated to $\mu' \in P_k$. Put

$$S_{\lambda,\mu'} = i^{|\mathring{\Phi}^+|} |\mathring{P}/(k+\check{h})M|^{-1/2} J_\lambda(t_{\mu'}), \tag{20}$$

where i is the imaginary unit.

The matrix $S = (S_{\lambda,\mu'})$ is indexed by $P_k \times P_k'$. We call it the modular matrix from A to A'. Let S^t (resp. $\bar{S}^t$) be the transpose (resp. conjugate transpose) of S. Then S^t (resp. $\bar{S}^t$) is a matrix indexed by $P_k' \times P_k$.

LEMMA 5.2. *$S\bar{S}^t$ is the identify matrix indexed by $P_k \times P_k$.*

PROOF. Note that the number of positive roots of type B_n is equal to the number of positive roots of type C_n (cf.[**Hu**, §12.2, Table 1]). It is easy to see that $i^{|\mathring{\Phi}^+|} = i^{|\mathring{\Phi}'^+|}$ for any A, where $\mathring{\Phi}'^+$ is the set of positive roots of $\mathring{\mathfrak{g}}'$. Hence the lemma is equivalent to that $\{\mathring{\chi}_\lambda \,|\, \lambda \in P_k\}$ is an orthonormal basis of $R_k(A)$ with respect to the inner product $(\cdot,\cdot)$ defined in (8). □

Note that $(A')' = A$. Let S' be the modular matrix $(S_{\lambda',\mu})$ from A' to A.

LEMMA 5.3. *$S^t = S'$.*

PROOF. When $A = X_N^{(1)}$ or $A = A_{2n}^{(2)}$, the lemma is clear since $J_\lambda(t_\mu) = J_\mu(t_\lambda)$. We now assume $A = X_N^{(r)}$ with $r > 1$ and $A \neq A_{2n}^{(2)}$. Note that $M = \mathring{Q}$ (resp. $M' = \mathring{Q}'$). moreover

$$\check{h}' = \check{h}, \quad \epsilon(\mathring{P}') = \check{\mathring{P}}, \text{ and } \epsilon(\mathring{Q}') = \check{\mathring{Q}}.$$

It follows that ϵ induces the isomorphism

$$\mathring{P}'/(k + \check{h}')\mathring{Q}' \simeq \check{\mathring{P}}/(k + \check{h})\check{\mathring{Q}}.$$

From (7), we observe that $J_\lambda(t_{\mu'}) = J_{\mu'}(t_\lambda)$ for any $\lambda \in P_k$ and $\mu' \in P'_k$. It concludes the Lemma. $\square$

For any dominant integral weight Λ of $\mathfrak{g}(A)$ of level k, we denote by $\mathcal{H}_\Lambda$ the irreducible integrable representation of highest weight Λ. Let χ_Λ denote the normalized character of $\mathfrak{g}(A)$, i.e. $\chi_\Lambda = e^{-m_\Lambda \delta} \mathrm{ch}_{\mathcal{H}_\Lambda}$, where $\mathrm{ch}_{\mathcal{H}_\Lambda}$ is the character of $\mathcal{H}_\Lambda$ and the number m_Λ is the so-called modular anomaly (cf.[**Ka**, §12.7.5]). Let $V(A, k)$ be the vector space spanned by the normalized characters χ_Λ of level k. Then the matrix S gives a linear transformation from $V(A, k)$ to $V(A', k)$. For precise transformation formula, see [**Ka**, Theorem 13.8, 13.9]. In the literature the formula (20) is called Kac-Peterson formula (cf.[**Ka**, **Ka1**]) for the modular S-matrix as the linear transformation from $V(A, k)$ to $V(A', k)$.

Acknowledgments

This note originated from the comments of I. Cherednik on the work [**Ho2**]. The elementary definition of fusion rings presented in this note should be well-known in the case of untwisted affine Lie algebra, but the author is not able to find an appropriate reference.

References

[Be] A. Beauville, *Conformal blocks, fusion rules and the Verlinde formula*, Proceedings of the Hirzebruch 65 Conference on Algebraic Geometry (Ramat Gan, 1993), Israel Math. Conf. Proc., vol. 9, Bar-Ilan Univ., Ramat Gan, 1996, pp. 75–96. MR1360497

[Fa] G. Faltings, *A proof for the Verlinde formula*, J. Algebraic Geom. **3** (1994), no. 2, 347–374. MR1257326

[Fi] M. Finkelberg, *An equivalence of fusion categories*, Geom. Funct. Anal. **6** (1996), no. 2, 249–267. MR1384612

[FS] J. Fuchs and C. Schweigert, *The action of outer automorphisms on bundles of chiral blocks*, Comm. Math. Phys. **206** (1999), no. 3, 691–736. MR1721887

[FSS] J. Fuchs, B. Schellekens, and C. Schweigert, *From Dynkin diagram symmetries to fixed point structures*, Comm. Math. Phys. **180** (1996), no. 1, 39–97. MR1403859

[Ho1] J. Hong, *Mirković-Vilonen cycles and polytopes for a symmetric pair*, Represent. Theory **13** (2009), 19–32. MR2480386

[Ho2] J. Hong, Conformal blocks, Verlinde formula and diagram automorphisms. arXiv:1610.02975.

[HS] J. Hong and L. Shen, *Tensor invariants, saturation problems, and Dynkin automorphisms*, Adv. Math. **285** (2015), 629–657. MR3406511

[Hu] J. E. Humphreys, *Introduction to Lie algebras and representation theory*, Springer-Verlag, New York-Berlin, 1972. Graduate Texts in Mathematics, Vol. 9. MR0323842

[Ja] J. C. Jantzen, *Darstellungen halbeinfacher algebraischer Gruppen und zugeordnete kontravariante Formen*, Bonn. Math. Schr. **67** (1973), v+124. MR0401935

[Ka] V. G. Kac, *Infinite-dimensional Lie algebras*, 3rd ed., Cambridge University Press, Cambridge, 1990. MR1104219

[Ka1] V. G. Kac and D. H. Peterson, *Infinite-dimensional Lie algebras, theta functions and modular forms*, Adv. in Math. **53** (1984), no. 2, 125–264. MR750341

[TUY] A. Tsuchiya, K. Ueno, and Y. Yamada, *Conformal field theory on universal family of stable curves with gauge symmetries*, Integrable systems in quantum field theory and statistical mechanics, Adv. Stud. Pure Math., vol. 19, Academic Press, Boston, MA, 1989, pp. 459–566. MR1048605

[V] E. Verlinde, *Fusion rules and modular transformations in 2D conformal field theory*, Nuclear Phys. B **300** (1988), no. 3, 360–376. MR954762

Department of Mathematics, University of North Carolina at Chapel Hill, Chapel Hill, North Carolina 27599-3250

Email address: jiuzu@email.unc.edu

Contemporary Mathematics
Volume **713**, 2018
https://doi.org/10.1090/conm/713/14316

Kostant-Lusztig $\mathbb{A}$-bases of multiparameter quantum groups

Naihuan Jing, Kailash C. Misra, and Hiroyuki Yamane

Abstract. We study the Kostant-Lusztig $\mathbb{A}$-base of the multiparameter quantum groups. To simplify calculations, especially for G_2-type, we utilize the duality of the pairing of the universal R-matrix.

1. Introduction

Quantum enveloping algebras $U_q(\mathfrak{g})$ [**3**, **13**] of the Kac-Moody algebras $\mathfrak{g}$ are one of the important classes of quantum groups. Quantum enveloping algebras and their integrable highest weight representations enjoy favorable properties, among which the canonical bases of Lusztig [**18**] or global crystal bases of Kashiwara [**15**] are the most prominent ones. Existence of such canonical bases has also been established for quantum enveloping algebras $U_q(\mathfrak{g})$ of Borcherds' generalized Kac-Moody algebras $\mathfrak{g}$ [**14**].

On the other hand, quantum enveloping algebras have been extended to Nichols algebras of diagonal types, which include multiparameter quantum enveloping algebras as examples, in particular, one-parameter quantum enveloping algebras $U_q(\mathfrak{g})$ in Kac-Moody types. In [**1**] Andruskiuwitch and Schneider proved that finite dimensional pointed Hopf algebras with finite abelian group (with order > 7) of group-like elements are essentially Lusztig's small quantum groups and their variants. Furthermore, Heckenburger classified the Nicholas algebras with arithmetic root data [**7**] and proved results similar to quantum enveloping algebras (see also [**5**]).

Lusztig [**17**] introduced the Kostant-Lusztig $\mathbb{A}$-base of the quantum groups. The $\mathbb{A}$-base and the finite dimensional Hopf algbebra of the quantum group at root of unity have been key figures in study of Lusztig conjectures, (see [**12**] for history). In this paper we establish the Kostant-Lusztig $\mathbb{A}$-forms for the multiparameter quantum groups and construct invariant bases for each factor of the triangular decomposition. Our general result is based on the structure theory for multiparameter quantum groups and information on the lower rank cases, most notably the case of G_2. The recent work of Fan and Li [**4**] on two-parameter quantum algebras has

---

2010 *Mathematics Subject Classification.* Primary 17B37,17B10; Secondary 81R50.

The first author was partially supported by National Natural Science Foundation of China grant # 11531004 and Simons Foundation grant # 523868.

The second author was partially supported by Simons Foundation grant # 307555.

The third author was partially supported by JSPS Grand-in-Aid for Scientific Research (C), 16K05095.

made us believe that one should be able to use our $\mathbb{A}$-forms to construct canonical bases for the multiparameter quantum groups.

2. Generalized quantum groups

In this section we recall definitions and known results about multiparameter quantum groups. We will use the following notations throughout this paper. The ring of real numbers and integers will be denoted by $\mathbb{R}$ and $\mathbb{Z}$ respectively. For x, $y \in \mathbb{R}$, let $\mathbb{J}_{x,y} := \{z \in \mathbb{Z} | x \leq z \leq y\}$ and $\mathbb{J}_{x,\infty} := \{z \in \mathbb{Z} | x \leq z\}$. Then $\mathbb{N} = \mathbb{J}_{1,\infty}$ and $\mathbb{Z}_{\geq 0} := \mathbb{J}_{0,\infty}$. Let $\mathbb{K}$ be a field of characteristic zero and $\mathbb{K}^\times := \mathbb{K} \setminus \{0\}$. For x, $y \in \mathbb{K}$ and $r \in \mathbb{Z}_{\geq 0}$, we denote $(r)_x := \sum_{k=0}^{r-1} x^k$, $(r)_x! := \prod_{k=1}^{r}(k)_x$, $(r; x, y) := 1 - x^{r-1}y$ and $(r; x, y)! := \prod_{k=1}^{r}(k; x, y)$.

2.1. Definition of generalized quantum groups. Let $\theta \in \mathbb{N}$ and $I := \mathbb{J}_{1,\theta}$. Let V be a θ-dimensional $\mathbb{R}$-linear space with a basis $\{v_i | i \in I\}$. Let $V_{\mathbb{Z}} := \oplus_{i \in I}\mathbb{Z}v_i$, so $V_{\mathbb{Z}}$ is a rank-θ free $\mathbb{Z}$-module (or a free abelian group). Let $\chi : V_{\mathbb{Z}} \times V_{\mathbb{Z}} \to \mathbb{K}^\times$ be a map such that $\chi(x + y, z) = \chi(x, z)\chi(y, z)$ and $\chi(x, y + z) = \chi(x, y)\chi(x, z)$ hold for all x, y, $z \in V_{\mathbb{Z}}$. We call such χ *a bi-character.*

DEFINITION 2.1. Let $\pi : I \to V_{\mathbb{Z}}$ be a map such that $\pi(I)$ is a $\mathbb{Z}$-base of $V_{\mathbb{Z}}$. The generalized quantum group $U = U(\chi, \pi)$ is the unique associative $\mathbb{K}$-algebra (with 1) satisfying the following conditions (i)-(v).

(i) As a $\mathbb{K}$-algebra, $U(\chi, \pi)$ is generated by K_λ, L_λ, $(\lambda \in V_{\mathbb{Z}})$, E_i, F_i $(i \in I)$.

(ii) The following equations hold. Let $q_{ij} := \chi(\pi(i), \pi(j))$.

$$\begin{array}{l} K_0 = 1,\ K_\lambda K_\mu = K_{\lambda+\mu}, \\ L_0 = 1,\ L_\lambda L_\mu = L_{\lambda+\mu},\ K_\lambda L_\mu = L_\mu K_\lambda, \\ K_{\pi(i)} E_j K_{\pi(i)}^{-1} = q_{ij} E_j,\ K_{\pi(i)} F_j K_{\pi(i)}^{-1} = q_{ij}^{-1} F_j, \\ L_{\pi(i)} E_j L_{\pi(i)}^{-1} = q_{ji}^{-1} E_j,\ L_{\pi(i)} F_j L_{\pi(i)}^{-1} = q_{ji} F_j, \\ [E_i, F_j] = \delta_{ij}(-K_{\pi(i)} + L_{\pi(i)}) \end{array}$$

(iii) There are subspaces U_λ $(\lambda \in V_{\mathbb{Z}})$ such that $U = \oplus_{\lambda \in V_{\mathbb{Z}}} U_\lambda$, $U_\lambda U_\mu \subset U_{\lambda+\mu}$, $K_{\pi(i)}^{\pm 1} \in U_0$, $L_{\pi(i)}^{\pm 1} \in U_0$, $E_i \in U_{\pi(i)}$, $F_i \in U_{-\pi(i)}$

(iv) Let U^0 be the $\mathbb{K}$-subalgebra of U generated by $K_\lambda L_\mu$ $(\lambda, \mu \in V_{\mathbb{Z}})$. Let U^+ (resp. U^-) be the $\mathbb{K}$-subalgebra of U generated by E_i (resp. F_i) $(i \in I)$. Then the elements $K_\lambda L_\mu$ $(\lambda, \mu \in V_{\mathbb{Z}})$ form a $\mathbb{K}$-basis of U^0 and we have the $\mathbb{K}$-linear isomorphism $\mathbf{m} : U^+ \otimes U^0 \otimes U^- \to U$ defined by $X \otimes Y \otimes Z \mapsto XYZ$.

(v) There exists no $X \in U^+ \setminus \mathbb{K}$ (resp. $Y \in U^- \setminus \mathbb{K}$) such that $[X, F_i] = 0$ $([E_i, Y] = 0)$ for all $i \in I$.

Note that $U^0 \subset U_0$. For $\lambda \in V_{\mathbb{Z}}$, let $U_\lambda^+ := U^+ \cap U_\lambda$ and $U_\lambda^- := U^- \cap U_\lambda$. Let $V_{\mathbb{Z}}^{\pi,+} := \oplus_{i \in I}\mathbb{Z}_{\geq 0}\pi(i)$. Then $U^+ = \oplus_{\lambda \in V_{\mathbb{Z}}^{\pi,+}} U_\lambda^+$ and $U^- = \oplus_{\lambda \in V_{\mathbb{Z}}^{\pi,+}} U_{-\lambda}^-$.

LEMMA 2.2. *Let $\grave{U}$ be a $\mathbb{K}$-algebra (with* 1*) generated by $\grave{K}_\lambda$, $\grave{L}_\lambda$, $\grave{E}_i$ and $\grave{F}_i$ satisfying conditions* (i)-(iv) *above (in Definition* 2.1*). Then there exists a $\mathbb{K}$-algebra epimorphism $\xi : \grave{U} \to U$ such that $\xi(\grave{K}_\lambda) = K_\lambda$, $\xi(\grave{L}_\lambda) = L_\lambda$ $(\lambda \in V_{\mathbb{Z}})$, $\xi(\grave{E}_i) = E_i$, $\xi(\grave{F}_i) = F_i$ $(i \in I)$.*

PROOF. Let $\lambda \in V_{\mathbb{Z}}^{\pi,+} \setminus \{0\}$. Assume that there exists $X \in \grave{U}_\lambda^+ \setminus \{0\}$ such that $[X, \grave{F}_i] = 0$ for all $i \in I$. Let $\mathcal{I}$ (resp. $\mathcal{I}^+$) be the two-sided ideal of $\grave{U}$ (resp. $\grave{U}^+$) generated by X. Then $\mathcal{I} = \mathrm{Span}_{\mathbb{K}}(\grave{U}^-\grave{U}^0\mathcal{I}^+)$. Let $\grave{U}'$ be the quotient $\mathbb{K}$-algebra

$\grave{U}/\mathcal{I}$. Then $\grave{U}'$ also satisfies the same conditions as Definition 2.1 (i)-(iv). Let $g : \grave{U} \to \grave{U}'$ be the canonical map. Then $g_{|\grave{U}^-} : \grave{U}^- \to (\grave{U}')^-$ and $g_{|\grave{U}^0} : \grave{U}^0 \to (\grave{U}')^0$ are the $\mathbb{K}$-algebra isomorphisms. We see that $\dim(\grave{U}')^+_{\mu_1} = \dim \grave{U}^+_{\mu_1}$ for $\mu_1 \in V_{\mathbb{Z}}^{\pi,+}$ with $\mu_1 - \lambda \notin V_{\mathbb{Z}}^{\pi,+}$, that $\dim(\grave{U}')^+_\lambda = \dim \grave{U}^+_\lambda - 1$, and that $\dim(\grave{U}')^+_{\mu_2} \leq \dim \grave{U}^+_{\mu_2}$ for $\mu_2 \in V_{\mathbb{Z}}^{\pi,+}$ with $\mu_2 - \lambda \in V_{\mathbb{Z}}^{\pi,+}$. We also have a similar property for $\grave{U}^-$. Then we can see the claim of this theorem by a standard argument using a direct limit. □

By Lemma 2.2, we have the $\mathbb{K}$-algebra automorphism $\Omega^{\chi,\pi} : U(\chi,\pi) \to U(\chi,\pi)$ with

$$\Omega^{\chi,\pi}(K_\lambda L_\mu) = K_{-\lambda} L_{-\mu} \quad (\lambda, \mu \in V_{\mathbb{Z}}),$$
$$\Omega^{\chi,\pi}(E_i) = F_i L_{-\pi(i)},\ \Omega^{\chi,\pi}(F_i) = K_{-\pi(i)} E_i \quad (i \in I).$$

Define the bicharacter $\chi^{\rm op} : V_{\mathbb{Z}} \times V_{\mathbb{Z}} \to \mathbb{K}^\times$ by $\chi^{\rm op}(x,y) := \chi(y,x)$ $(x,\, y \in V_{\mathbb{Z}})$. By Lemma 2.2, we have the $\mathbb{K}$-algebra isomorphism $\Upsilon^{\chi^{\rm op},\pi} : U(\chi^{\rm op},\pi) \to U(\chi,\pi)$ with

$$\Upsilon^{\chi^{\rm op},\pi}(K_\lambda L_\mu) = K_\mu L_\lambda,\ (\lambda, \mu \in V_{\mathbb{Z}}), \tag{2.1}$$
$$\Upsilon^{\chi^{\rm op},\pi}(E_i) = F_i,\ \Upsilon^{\chi^{\rm op},\pi}(F_i) = E_i\ (i \in I), \tag{2.2}$$

which will be referred as the *Chevalley involution*. We also have the $\mathbb{K}$-algebra anti-automorphism $\Gamma^{\chi^{\rm op},\pi} : U(\chi^{\rm op},\pi) \to U(\chi,\pi)$ with

$$\Gamma^{\chi^{\rm op},\pi}(K_\lambda L_\mu) = K_\mu L_\lambda,\ (\lambda,\mu \in V_{\mathbb{Z}}),\ \Gamma^{\chi^{\rm op},\pi}(E_i) = E_i,\ \Gamma^{\chi^{\rm op},\pi}(F_i) = F_i\ (i \in I).$$

For $U = U(\chi,\pi)$, we have the following equations.

$$\begin{aligned} E_i^k F_i^m &= \textstyle\sum_{r=0}^{\min\{k,m\}} (r)_{q_{ii}}! \binom{m}{r}_{q_{ii}} \binom{k}{r}_{q_{ii}} q_{ii}^{\frac{r(-2k+r+1)}{2}} \\ &\quad \cdot (\textstyle\prod_{s=0}^{r-1}(-K_{\pi(i)} + q_{ii}^{-m+k+s} L_{\pi(i)})) F_i^{m-r} E_i^{k-r}, \end{aligned}$$

where $\binom{m}{r}_{q_{ii}}$ means q-binomial coefficient. Hence

$$[E_i, F_i^m] = (m)_{q_{ii}}(-K_{\pi(i)} + q_{ii}^{-m+1} L_{\pi(i)}) F_i^{m-1},$$
$$[F_i, E_i^m] = (m)_{q_{ii}}(-L_{\pi(i)} + q_{ii}^{-m+1} K_{\pi(i)}) E_i^{m-1}.$$

For $m \in \mathbb{Z}_{\geq 0}$ and $i,\, j \in I$ with $i \neq j$, define $E_{m,i,j},\ E^\vee_{m,i,j} \in U^+_{\pi(j)+m\pi(i)},\ F_{m,i,j}$ $F_{m,i,j} \in U^-_{-\pi(j)-m\pi(i)}$ inductively by $E_{0,i,j} := E^\vee_{0,i,j} := E_j$, $F_{0,i,j} := E^\vee_{0,i,j} := F_j$, and

(2.3)
$$\begin{aligned} E_{m+1,i,j} &:= E_i E_{m,i,j} - q_{ii}^m q_{ij} E_{m,i,j} E_i,\ E^\vee_{m+1,i,j} := E^\vee_{m,i,j} E_i - q_{ii}^m q_{ji} E_i E^\vee_{m,i,j}, \\ F_{m+1,i,j} &:= F_i F_{m,i,j} - q_{ii}^m q_{ji} F_{m,i,j} F_i,\ F^\vee_{m+1,i,j} := F^\vee_{m,i,j} F_i - q_{ii}^m q_{ij} F_i F^\vee_{m,i,j}. \end{aligned}$$

We have

$$\begin{aligned} \Upsilon^{\chi^{\rm op},\pi}(E_{m,i,j}) &= F_{m,i,j}, \quad \Upsilon^{\chi^{\rm op},\pi}(F_{m,i,j}) = E_{m,i,j}, \\ \Upsilon^{\chi^{\rm op},\pi}(E^\vee_{m,i,j}) &= F^\vee_{m,i,j}, \quad \Upsilon^{\chi^{\rm op},\pi}(F^\vee_{m,i,j}) = E^\vee_{m,i,j}. \end{aligned} \tag{2.4}$$

We have

$$\begin{aligned} &[E_i, F_{m,i,j}] = -(m)_{q_{ii}}(m; q_{ii}, q_{ij}q_{ji}) K_{\pi(i)} F_{m-1,i,j}, \\ &[F_i, E_{m,i,j}] = -(m)_{q_{ii}}(m; q_{ii}, q_{ij}q_{ji}) L_{\pi(i)} E_{m-1,i,j}, \\ &[E_j, F_{m,i,j}] = (m; q_{ii}, q_{ij}q_{ji})! F_i^m L_{\pi(j)}, \\ &[F_j, E_{m,i,j}] = (m; q_{ii}, q_{ij}q_{ji})! E_i^m K_{\pi(j)}, \\ &[E_{m,i,j}, F_{m,i,j}] = (m)_{q_{ii}}!(m; q_{ii}, q_{ij}q_{ji})!(-K_{\pi(j)+m\pi(i)} + L_{\pi(j)+m\pi(i)}), \\ &[E_{m,i,j}, F_{m',i,k}] = 0 \quad (m' \in \mathbb{Z}_{\geq 0}, k \in I \setminus \{i,j\}). \end{aligned} \tag{2.5}$$

We have

$$\begin{aligned}
\Omega^{\chi,\pi}(E_{r,i,j}) &= q_{ii}^{-\frac{r(r-1)}{2}} q_{ji}^{-r} F_{r,i,j} L_{-\pi(j)-r\pi(i)}, \\
\Omega^{\chi,\pi}(E^{\vee}_{r,i,j}) &= q_{ii}^{-\frac{r(r-1)}{2}} q_{ij}^{-r} F^{\vee}_{r,i,j} L_{-\pi(j)-r\pi(i)}, \\
\Omega^{\chi,\pi}(F_{r,i,j}) &= q_{ii}^{\frac{r(r-1)}{2}} q_{ji}^{r} K_{-\pi(j)-r\pi(i)} E_{r,i,j}, \\
\Omega^{\chi,\pi}(F^{\vee}_{r,i,j}) &= q_{ii}^{\frac{r(r-1)}{2}} q_{ij}^{r} K_{-\pi(j)-r\pi(i)} E^{\vee}_{r,i,j}.
\end{aligned} \tag{2.6}$$

We have

$$\begin{aligned}
&\Gamma^{\chi^{\mathrm{op}},\pi}\Gamma^{\chi,\pi} = \mathrm{id}_{U(\chi,\pi)},\ \Gamma^{\chi,\pi}\Gamma^{\chi^{\mathrm{op}},\pi} = \mathrm{id}_{U(\chi^{\mathrm{op}},\pi)}, \\
&\Gamma^{\chi^{\mathrm{op}},\pi}(E_{r,i,j}) = E^{\vee}_{r,i,j},\ \Gamma^{\chi^{\mathrm{op}},\pi}(F_{r,i,j}) = F^{\vee}_{r,i,j}.
\end{aligned} \tag{2.7}$$

2.2. Kharchenko's Poincaré-Birkohoff-Witt theorem and Heckenberger's Lusztig isomorphisms. We recall the following theorem by Kharchenko. We also introduce some notations.

THEOREM 2.3. ([**16**]) *Let χ be a bicharacter and $U = U(\chi,\pi)$ be the generalized quantum group. Then there exists a unique pair $(R^{\pi,+}_{\chi}, \varphi^{\pi,+}_{\chi})$ of a subset $R^{\pi,+}_{\chi}$ of $V^{\pi,+}_{\mathbb{Z}}$ and a map $\varphi^{\pi,+}_{\chi} : R^{\pi,+}_{\chi} \to \mathbb{N}$ satisfying the following:*

Let $X := \{(\alpha,t) \in R^{\pi,+}_{\chi} \times \mathbb{N} | t \in \mathbb{J}_{1,\varphi^{\pi,+}_{\chi}(\alpha)}\}$. Define the map $z : X \to R^{\pi,+}_{\chi}$ by $z(\alpha,t) := \alpha$. Let Y be the set of maps $y : X \to \mathbb{Z}_{\geq 0}$ such that $|\{x \in X | y(x) \geq 1\}| < \infty$ and $(y(x))_{\chi(z(x),z(x))}! \neq 0$ for all $x \in X$. Then

(*) $\forall \lambda \in V^{\pi,+}_{\mathbb{Z}}$, $\dim U^{+}_{\lambda} = |\{y \in Y | \sum_{x\in X} y(x)z(x) = \lambda\}|$.

Moreover, letting $q_{i'j'} := \chi(\pi(i'),\pi(j'))$ $(i',j' \in I)$, for i, $j \in I$ with $i \neq j$, we have

$$\{t \in \mathbb{Z}_{\geq 0} | \pi(j) + t\pi(i) \in R^{\pi,+}_{\chi}\} = \{t \in \mathbb{Z}_{\geq 0} | (t)_{q_{ii}}!(t; q_{ii}, q_{ij}q_{ji})! \neq 0\}. \tag{2.8}$$

Let $R^{\pi}_{\chi} := R^{\pi,+}_{\chi} \cup (-R^{\pi,+}_{\chi})$. Define the map $\varphi^{\pi}_{\chi} : R^{\pi}_{\chi} \to \mathbb{N}$ by $\varphi^{\pi}_{\chi}(-\alpha) := \varphi^{\pi}_{\chi}(\alpha) := \varphi^{\pi,+}_{\chi}(\alpha)$ $(\alpha \in R^{\pi,+}_{\chi})$. We say that χ is (π,i)-*good* if it satisfies the condition that for every $j \in I \setminus \{i\}$, there exists $N^{\chi,\pi}_{ij} \in \mathbb{Z}_{\geq 0}$ such that $\pi(j) + N^{\chi,\pi}_{ij}\pi(i) \in R^{\pi,+}_{\chi}$ and $\pi(j) + (N^{\chi,\pi}_{ij} + 1)\pi(i) \notin R^{\pi,+}_{\chi}$. Let $N^{\chi,\pi}_{ii} := -2$.

Let $i \in I$ and χ be a (π,i)-good bicharacter. Define the map $\tau^{\chi}_{i}\pi : I \to V_{\mathbb{Z}}$ by $\tau^{\chi}_{i}\pi(j) := \pi(j) + N^{\chi,\pi}_{ij}\pi(i)$ $(j \in I)$. By (2.8), we see that $\tau^{\chi}_{i}\pi$ is a (π,i)-good bicharacter, and

$$N^{\chi,\tau^{\chi}_{i}\pi}_{ij} = N^{\chi,\pi}_{ij} \quad (j \in I), \tag{2.9}$$

whence $(\tau^{\chi}_{i})^{2}\pi = \pi$.

THEOREM 2.4. ([**5**]) *Let $a : I \to \mathbb{K}^{\times}$ be a map. Let $i \in I$. Let χ be a (π,i)-good bicharacter. Then there exists a $\mathbb{K}$-algebra isomorphism $T^{\chi,\tau^{\chi}_{i}\pi,a}_{i} : U(\chi,\tau^{\chi}_{i}\pi) \to U(\chi,\pi)$ such that*

$$\begin{aligned}
&T_i(K_{\lambda}) = K_{\lambda}, \quad T_i(L_{\lambda}) = L_{\lambda} \quad (\lambda \in V_{\mathbb{Z}}), \\
&T_i(E_i) = a(i)F_iL_{-\pi(i)}, \quad T_i(F_i) = \tfrac{1}{a(i)}K_{-\pi(i)}E_i, \\
&T_i(E_j) = a(j)E_{N^{\chi,\pi}_{ij},i,j} \quad (j \in I \setminus \{i\}) \\
&T_i(F_j) = \tfrac{1}{a(j)(N^{\chi,\pi}_{ij})_{q_{ii}}!(N^{\chi,\pi}_{ij};q_{ii},q_{ij}q_{ji})!}F_{N^{\chi,\pi}_{ij},i,j} \quad (j \in I \setminus \{i\}),
\end{aligned} \tag{2.10}$$

where $T_i := T^{\chi,\tau^{\chi}_{i}\pi,a}_{i}$ and $q_{i'j'} := \chi(\pi(i'),\pi(j'))$ $(i', j' \in I)$. Moreover we have

$$R^{\tau^{\chi}_{i}\pi}_{\chi} = R^{\pi}_{\chi},\ R^{\tau^{\chi}_{i}\pi,+}_{\chi} = (R^{\pi,+}_{\chi} \setminus \{\pi(i)\}) \cup \{-\pi(i)\},\ \varphi^{\pi}_{\chi} = \varphi^{\tau^{\chi}_{i}\pi}_{\chi}. \tag{2.11}$$

Let $i, j \in I$ be such that $i \neq j$. Let χ be a (π, i)-good and (π, j)-good bicharacters. We say that χ is *a* (π, i, j)*-good bicharacter* if $\tau^{\chi}_{i_1} \cdots \tau^{\chi}_{i_m} \pi$ can be defined for all $m \in \mathbb{N}$ and all $i_t \in \{i, j\}$ $(t \in \mathbb{J}_{1,m})$.

$$\text{Let } m^{\chi,\pi}_{ij} := |R^{\pi,+}_{\chi} \cap (\mathbb{Z}\pi(i) \oplus \mathbb{Z}\pi(j))| (\in \mathbb{J}_{2,\infty} \cup \{\infty\}). \tag{2.12}$$

By (2.11), using the same argument as that for [**6**, Lemma 4], we have the following result.

LEMMA 2.5. *Let χ be a (π, i, j)-good bicharacter. Let $X := \mathbb{Z}\pi(i) \oplus \mathbb{Z}\pi(j)$. Let $m := m^{\chi,\pi}_{ij}$. Let $i_{2y-1} := i$, $i_{2y} := j$ $(y \in \mathbb{N})$. Let $\pi_t := \tau^{\chi}_{i_t}\tau^{\chi}_{i_{t-1}} \cdots \tau^{\chi}_{i_1}\pi$ $(t \in \mathbb{N})$. Then for $k \in \mathbb{J}_{1,m}$, we have*

$$|R^{\pi,+}_{\chi} \cap (-R^{\pi_k,+}_{\chi})| = |R^{\pi,+}_{\chi} \cap (-R^{\pi_k,+}_{\chi}) \cap X| = k.$$

Moreover if $m < \infty$, we have

$$\pi_m = \tau^{\chi}_{i_{m+1}}\tau^{\chi}_{i_m} \cdots \tau^{\chi}_{i_2}\pi,$$

and

$$R^{\pi,+}_{\chi} \cap X = \{\pi(i_1) = i, \pi_1(i_2), \ldots, \pi_{m-2}(i_{m-1}), \pi_{m-1}(i_m) = j\}.$$

Furthermore $|\{\pi_t(i_{t+1}) | t \in \mathbb{N}\}| = \infty$ if $m = \infty$.

Let χ be a (π, i, j)-good bicharacter. We say that χ is *a* (π, i, j)*-good finite bicharacter* if m in Lemma 2.5 is finite.

THEOREM 2.6. ([**5**]) *Let $i, j \in I, i \neq j$ and χ be a (π, i, j)-good finite bicharacter with $m := m^{\chi,\pi}_{ij} < \infty$. For $t \in \mathbb{N}$, let $\pi'_t := \tau^{\chi}_{i_{t+1}}\tau^{\chi}_{i_t} \cdots \tau^{\chi}_{i_2}\pi$ ($i_k = i$ (resp. j) if k is odd (resp. even)), a_t, $a'_t : I \to \mathbb{K}^\times$ be maps, $\dot{T}_{(t)} := T^{\chi,\pi_1,a_1}_{i_1} \cdots T^{\chi,\pi_t,a_t}_{i_t}$ and $\ddot{T}_{(t)} := T^{\chi,\pi'_1,a'_1}_{i_2} \cdots T^{\chi,\pi'_t,a'_t}_{i_{t+1}}$. Then there exists a map $b : I \to \mathbb{K}^\times$ such that*

$$\dot{T}_{(m)}(E_k) = b(k)\ddot{T}_{(m)}(E_k),\ \dot{T}_{(m)}(F_k) = \frac{1}{b(k)}\ddot{T}_{(m)}(F_k) \quad (k \in I).$$

Moreover there exists $z \in \mathbb{K}^\times$ such that

$$\dot{T}_{(m-1)}(E_{i_m}) = zE_j,\ \dot{T}_{(m-1)}(F_{i_m}) = z^{-1}F_j.$$

2.3. Strict Heckenberger's Lusztig isomorphisms.

LEMMA 2.7. *Let $i, j \in I, i \neq j$ and χ be a (π, i, j)-good finite bicharacter with $m := m^{\chi,\pi}_{ij} < \infty$. Also $b : I \to \mathbb{K}^\times$ be the map and $z \in \mathbb{K}^\times$ in Theorem 2.6. Then the following statements hold.*

(1) *If $\Omega^{\chi,\pi}\dot{T}_{(m)} = \dot{T}_{(m)}\Omega^{\chi,\pi_m}$, then $b(k) \in \{1, -1\}$ for all $k \in I$.*

(2) *If $\Omega^{\chi,\pi}\dot{T}_{(m-1)} = \dot{T}_{(m-1)}\Omega^{\chi,\pi_{m-1}}$, then $z \in \{1, -1\}$.*

PROOF. (1) Let $k \in I$. We have

$$\begin{aligned} F_k &= b(k)\ddot{T}^{-1}_{(m)}\dot{T}_{(m)}(F_k) = b(k)\ddot{T}^{-1}_{(m)}\dot{T}_{(m)}\Omega^{\chi,\pi_m}(E_kL_{\pi(k)}) \\ &= b(k)\Omega^{\chi,\pi}\ddot{T}^{-1}_{(m)}\dot{T}_{(m)}(E_kL_{\pi(k)}) = b(k)^2\Omega^{\chi,\pi}(E_kL_{\pi(k)}) = b(k)^2F_k, \end{aligned} \tag{2.13}$$

whence $b(k)^2 = 1$, so $b(k) \in \{1, -1\}$.

(2) This can be proved similarly to (1). □

LEMMA 2.8. *Let $a : I \to \mathbb{K}^\times$ be a map. Let $i \in I$ and χ be a (π, i)-good bicharacter. Then*

$$\Omega^{\chi,\pi} T_i^{\chi,\tau_i^\chi\pi,a} = T_i^{\chi,\tau_i^\chi\pi,a}\Omega^{\chi,\tau_i^\chi\pi}$$

if and only if

$$a(i)^2 = 1 \quad \text{and} \quad a(j)^2 = \frac{q_{ii}^{\frac{N_{ij}^{\chi,\pi}(N_{ij}^{\chi,\pi}-1)}{2}} q_{ji}^{N_{ij}^{\chi,\pi}}}{(N_{ij}^{\chi,\pi})_{q_{ii}}!(N_{ij}^{\chi,\pi}; q_{ii}, q_{ij}q_{ji})!} \quad (j \in I \setminus \{i\}). \tag{2.14}$$

PROOF. The claim follows from Theorem 2.4 and (2.6) $\square$

LEMMA 2.9. *Let G be the $\mathbb{K}$-algebra (with 1) defined with the generators X, Y, Z and the relations $[Z, X] = 2X$, $[Z, Y] = -2Y$, $[X, Y] = Z$. (Namely G is isomorphic to the universal enveloping algebra of $\mathrm{sl}_2(\mathbb{K})$.) Let $k \in \mathbb{Z}_{\geq 0}$. Let Γ be the $(k+1)$-dimensional left G-module with the $\mathbb{K}$-basis $\{\gamma_r | r \in \mathbb{J}_{0,k}\}$ such that $Z\gamma_r := (k-2r)\gamma_r$, $Y\gamma_r := \gamma_{r+1}$, $X\gamma_r := r(k-r+1)\gamma_{r-1}$, where $\gamma_{-1} := \gamma_{k+1} := 0$. Let $a \in \mathbb{K}^\times$. Define $\xi \in \mathrm{End}_{\mathbb{K}}(\Gamma)$ by $\xi(v) := \exp(aX)\exp(-a^{-1}Y)\exp(aX) \cdot v$ $(v \in \Gamma)$. Then*

$$\xi(Y^r\gamma_0) = \frac{(-1)^{k-r}a^{2r-k}r!}{(k-r)!}Y^{k-r}\gamma_0, \quad \xi(X^r\gamma_k) = \frac{(-1)^r a^{k-2r}r!}{(k-r)!}X^{k-r}\gamma_k.$$

PROOF. Let $\bar{\Gamma}$, $\bar{\gamma}_y$ $(y \in \mathbb{J}_{0,1})$, $\bar{\xi}$ be Γ, γ_y, ξ respectively for $k = 1$. Then $\bar{\xi}(\bar{\gamma}_0) = (-a^{-1})\bar{\gamma}_1$ and $\bar{\xi}(\bar{\gamma}_1) = a\bar{\gamma}_0$. Regard $\bar{\Gamma}^{\otimes k}$ as the (k-fold tensor) G-module in a standard way. Let $g : M \to \bar{M}^{\otimes k}$ be the G-module monomorphism $g : M \to \bar{\Gamma}^{\otimes k}$ such that $g(\gamma_0) = \bar{\gamma}_0^{\otimes k}$. Then

$$g(\gamma_r) = g(Y^r\gamma_0) = Y^r \cdot g(\gamma_0) = r! \sum_{|\{x\in\mathbb{J}_{1,k}|i_x=1\}|=r} \bar{\gamma}_{i_1} \otimes \cdots \otimes \bar{\gamma}_{i_k}.$$

Note $g(\xi(v)) = \bar{\xi}^{\otimes k}(g(v))$ $(v \in \Gamma)$. Then

$$\begin{aligned} g(\xi(\gamma_r)) &= g(\xi(Y^r\gamma_0)) = \bar{\xi}^{\otimes k}(g(Y^r\gamma_0)) \\ &= \tfrac{(-a^{-1})^{k-r}a^r r!}{(k-r)!} g(Y^{k-r}\gamma_0) = g(\tfrac{(-1)^{k-r}a^{2r-k}r!}{(k-r)!}\gamma_{k-r}). \end{aligned}$$

Thus we can see the claim. $\square$

LEMMA 2.10. *Let $a_{11} := a_{22} := 2 (\in \mathbb{K})$. Let a_{12}, $a_{21} \in \mathbb{K}$ be such that $(a_{12}, a_{21}) \in \{(0,0), (-1,-1), (-2,-1), (-3,-1)\}$. Let $m \in \mathbb{Z}_{\geq 0}$ be* 2 (resp. 3, resp. 4, resp. 6) *if a_{12} is* 0 (resp. -1, resp. -2, resp. -3). *Let $i_{2x-1} := 1$, $i_{2x} := 2$ $(x \in \mathbb{N})$. Let G be a $\mathbb{K}$-algebra (with 1) satisfying the following conditions* (i) *and* (ii).

(i) *There exist X_i, Y_i, $Z_i \in G$ $(i \in \mathbb{J}_{1,2})$ such that*

$$\begin{aligned} &[Z_1, Z_2] = 0, [Z_i, X_j] = a_{ij}X_j, [Z_i, Y_j] = -a_{ij}Y_j, [X_i, Y_j] = \delta_{ij}Z_i, \\ &(\mathrm{ad}X_{i'})^{1-a_{i'j'}}(X_{j'}) = (\mathrm{ad}Y_{i'})^{1-a_{i'j'}}(Y_{j'}) = 0 \ (i' \neq j'). \end{aligned}$$

(ii) *For $\Re \in G$ and $\Im \in \{X_i, Y_i | i \in \mathbb{J}_{1,2}\}$, there exists $r \in \mathbb{N}$ such that $(\mathrm{ad}\Im)^r(\Re) = 0$.*

Let $b_1, b_2 \in \mathbb{K}^\times$. Assume $b_1 = b_2$ if $m = 3$. For $i \in \mathbb{J}_{1,2}$, define $\xi_i \in \mathrm{Aut}(G)$ by $\xi_i(\Re) := \exp(b_i\mathrm{ad}X_i)\exp(-b_i^{-1}\mathrm{ad}Y_i)\exp(b_i\mathrm{ad}X_i)$ $(\Re \in G)$. Then we have

(2.15)

$$\begin{aligned} &\xi_i(X_i) = -b_i^2 Y_i, \ \xi_i(Y_i) = -b_i^{-2}X_i, \\ &\xi_i(X_j) = \tfrac{b_i^{-a_{ij}}}{(-a_{ij})!}(\mathrm{ad}X_i)^{-a_{ij}}(X_j), \ \xi_i(Y_j) = \tfrac{(-b_i)^{a_{ij}}}{(-a_{ij})!}(\mathrm{ad}Y_i)^{-a_{ij}}(Y_j) \quad (i \neq j), \end{aligned}$$

$$\begin{aligned}&\xi_{i_t}\xi_{i_{t+1}}\cdots\xi_{i_{t+m-2}}(X_{i_{t+m-1}})=X_{i_{t-1}},\\ &\xi_{i_t}\xi_{i_{t+1}}\cdots\xi_{i_{t+m-2}}(Y_{i_{t+m-1}})=Y_{i_{t-1}}\\ &(t\in\mathbb{Z}),\end{aligned}\tag{2.16}$$

and

$$\xi_{i_1}\xi_{i_2}\cdots\xi_{i_m}=\xi_{i_2}\xi_{i_3}\cdots\xi_{i_{m+1}}.\tag{2.17}$$

PROOF. We can see (2.15) by Lemma 2.9. As for (2.16), for example, if $a_{12}=-3$, by Lemma 2.9, we have

$$\begin{aligned}&\xi_1\xi_2\xi_1\xi_2\xi_1(X_2)\\ &=\xi_1\xi_2\xi_1\xi_2(\tfrac{b_1^3}{3!}[X_1,[X_1,[X_1,X_2]]])\\ &=-\tfrac{b_1^3}{3!}\xi_1\xi_2\xi_1\xi_2([X_1,[X_1,[X_2,X_1]]])\\ &=-\tfrac{b_1^3}{3!}\xi_1\xi_2\xi_1([b_2[X_2,X_1],[b_2[X_2,X_1],(-1)b_2^{-1}X_1]])\\ &=-\tfrac{b_1^3b_2}{3!}\xi_1\xi_2([(-1)\tfrac{b_1}{2!}[X_1,[X_1,X_2]],\tfrac{2!}{b_1}[X_1,X_2]])\\ &=\tfrac{b_1^3b_2}{3!}\xi_1\xi_2([[X_1,[X_2,X_1]],[X_2,X_1]])\\ &=\tfrac{b_1^3b_2}{3!}\xi_1([[b_2[X_2,X_1],(-1)b_2^{-1}X_1],(-1)b_2^{-1}X_1])\\ &=-\tfrac{b_1^3}{3!}\xi_1([X_1,[X_1,[X_1,X_2]]])\\ &=(-1)\tfrac{b_1^3}{3!}\cdot(-1)\tfrac{3!}{b_1^3}X_2\\ &=X_2.\end{aligned}$$

Now we show (2.17). Let $\xi':=\xi_{i_1}\xi_{i_2}\cdots\xi_{i_{m-1}}$. By (2.16), for $\Re\in G$, we see

$$\begin{aligned}&\xi'\xi_{i_m}(\xi')^{-1}(\Re)\\ &=\exp(b_i\mathrm{ad}\xi'(X_{i_m}))\exp(-b_i^{-1}\mathrm{ad}\xi'(Y_{i_m}))\exp(b_i\mathrm{ad}\xi'(X_{i_m}))(\Re)\\ &=\xi_{i_2}(\Re).\end{aligned}$$

This completes the proof. □

DEFINITION 2.11. Let $\chi:V_{\mathbb{Z}}\times V_{\mathbb{Z}}\to\mathbb{K}^\times$ be a bi-character, $\pi:I\to V_{\mathbb{Z}}$ be a map such that $\pi(I)$ is a $\mathbb{Z}$-base of $V_{\mathbb{Z}}$ and $q_{ij}:=\chi(\pi(i),\pi(j))$ for all $i,j\in I$. Let $A=[a_{ij}]_{ij\in I}$ be a symmetrizable generalized Cartan matrix. Let $d_i\in\mathbb{N}$ be such that $d_ia_{ij}=d_ja_{ji}$. $(i,\ j\in I)$. Let $\dot q\in\mathbb{K}^\times$ be such that $\dot q^r\neq1$ for all $r\in\mathbb{N}$. Let $\sqrt{\chi}:V_{\mathbb{Z}}\times V_{\mathbb{Z}}\to\mathbb{K}^\times$ be a bicharacter and $\dot q_{ij}:=\sqrt{\chi}(\pi(i),\pi(j))$ $(i,\ j\in I)$. Assume that $\dot q_{ii}=\dot q^{2d_i}$ $(i\in I)$, $\dot q_{ij}\dot q_{ji}=\dot q^{2d_ia_{ij}}$ $(i,j\in I)$, $q_{ij}:=\dot q_{ij}^2$ for all $i,\ j\in I$. Also assume that for every $k\in I$, there exists $\Theta(q_{kk}-1)\in\mathbb{K}^\times$ such that $\Theta(q_{kk}-1)^2=q_{kk}-1$. Then we say that such χ is *a (π,A)-admissible bicharacter.* If A is the Cartan matrix of a finite-dimensional complex Lie algebra (i.e., A is a symmetrizable generalized Cartan matrix of finite-type), we call $U(\chi,\pi)$ *a finite-type multiparameter quantum group.*

For a (π,A)-admissible bicharacter χ, $U(\chi,\pi)$ is presented by the generators given by Definition 2.1 (i) and the relations composed of those of Definition 2.1 (ii) and $E_{1-a_{ij},i,j}=F_{1-a_{ij},i,j}=0$ $(i,j\in I,\ i\neq j)$, which is well-known and can be proved by a standard argument along with Theorem 3.1 below.

LEMMA 2.12. *Let χ be a $(\pi,A=[a_{ij}]_{ij\in I})$-admissible bicharacter.*

(1) *Then χ is (π,i)-good bi-character for every $i\in I$ and $N_{ij}^{\chi,\pi}=-a_{ij}$ for all $i,\ j\in I$.*

(2) *For $i\in I$ let $\ddot q_{jk}:=\sqrt{\chi}(\tau_i^\chi\pi(j),\tau_i^\chi\pi(k))$ $(j,k\in I)$. Then we have*

$$\ddot q_{jj}=\dot q_{jj}\ (j\in I),\ \ddot q_{jk}\ddot q_{kj}=\dot q_{jk}\dot q_{kj}\ (j,k\in I).$$

In particular, χ is $(\tau_i^\chi\pi, A)$-admissible.

(3) *For $i,\ j \in I$ with $i \neq j$, χ is (π,i,j)-finite if and only if $a_{ij}a_{ji} \in \mathbb{J}_{0,3}$. Moreover*

$$m_{ij}^{\chi,\pi} = \begin{cases} 2 & \text{if } a_{ij}a_{ji} = 0, \\ 3 & \text{if } a_{ij}a_{ji} = 1, \\ 4 & \text{if } a_{ij}a_{ji} = 2, \\ 6 & \text{if } a_{ij}a_{ji} = 3. \end{cases} \tag{2.18}$$

PROOF. Claim (1) follows from (2.8). Claim (2) can be proved directly. Claim (3) follows from Claims (1) and (2) and Lemma 2.5. □

Let χ be a $(\pi, A = [a_{ij}]_{ij\in I})$-admissible bicharacter. Let $\dot{q}_{jk}$ be as in Definition 2.11. Define the map $\varpi : I \to \mathbb{K}^\times$ by

$$\varpi(j) = \begin{cases} 1 & (j = i), \\ \frac{\dot{q}_{ij}^{a_{ij}}}{(-a_{ij})_{q_{ii}}!\Theta(q_{ii}-1)^{-a_{ij}}} & (j \in I \setminus \{i\}). \end{cases}$$

Let

$$T_i^{\chi,\tau_i^\chi\pi} := T_i^{\chi,\tau_i^\chi\pi,\varpi}.$$

We see directly that ϖ satisfies (2.14). As for (2.10), letting $T_i := T_i^{\chi,\tau_i^\chi\pi}$, we have

$$\begin{aligned} &T_i(K_\lambda L_\mu) = K_\lambda L_\mu\ (\lambda,\mu \in V_{\mathbb{Z}}),\ T_i(E_i) = F_i L_{-\pi(i)},\ T_i(F_i) = K_{-\pi(i)}E_i, \\ &T_i(E_{r,i,j}^\vee) = \frac{(r)_{q_{ii}}!\dot{q}_{ij}^{a_{ij}+2r}\Theta(q_{ii}-1)^{a_{ij}+2r}}{(-a_{ij}-r)_{q_{ii}}!}E_{-a_{ij}-r,i,j}, \\ &T_i(F_{r,i,j}^\vee) = \frac{(r)_{q_{ii}}!\dot{q}_{ii}^{(a_{ij}+2r)(a_{ij}-1)}\dot{q}_{ij}^{-a_{ij}-2r}\Theta(q_{ii}-1)^{a_{ij}+2r}}{(-a_{ij}-r)_{q_{ii}}!}F_{-a_{ij}-r,i,j} \\ &(j \in I \setminus \{i\},\ r \in \mathbb{J}_{0,-a_{ij}}). \end{aligned}$$

We also have

$$(T_i^{\chi,\pi})^{-1} = \Gamma^{\chi^{\rm op},\pi}T_i^{\chi^{\rm op},\tau_i^\chi\pi}\Gamma^{\chi,\tau_i^\chi\pi}. \tag{2.19}$$

For $i \in I$, define the $\mathbb{K}$-algebra automorphism $\zeta_i^{\chi,\pi} : U(\chi,\pi) \to U(\chi,\pi)$ by

$$\begin{aligned} &\zeta_i^{\chi,\pi}(K_\lambda L_\mu) := K_\lambda L_\mu\ (\lambda,\mu \in V_{\mathbb{Z}}), \\ &\zeta_i^{\chi,\pi}(E_j) := \tfrac{1}{\dot{q}_{ij}\dot{q}_{ji}}E_j,\ \zeta_i^{\chi,\pi}(F_j) := \dot{q}_{ij}\dot{q}_{ji}F_j\ (j \in I). \end{aligned}$$

Then we have

$$T_i^{\chi,\tau_i^\chi\pi}\Upsilon^{\chi^{\rm op},\tau_i^\chi\pi} = \zeta_i^{\chi,\pi}\Upsilon^{\chi^{\rm op},\pi}T_i^{\chi^{\rm op},\tau_i^\chi\pi} \quad (i \in I).$$

For $i \in I$, let

$$\bar{E}_i := \frac{E_i}{\Theta(q_{ii}-1)}, \quad \bar{F}_i := -\frac{F_i}{\Theta(q_{ii}-1)}, \quad \bar{H}_i := \frac{K_{\pi(i)} - L_{\pi(i)}}{q_{ii}-1}.$$

Note that

$$[\bar{E}_i, \bar{F}_i] = \delta_{ij}\bar{H}_i,$$

and

$$\bar{H}_i\bar{E}_j = q_{ij}\bar{E}_j\bar{H}_i + q_{ij}^{-1}\frac{q_{ij}q_{ji}-1}{q_{ii}-1}\bar{E}_j L_{\pi(i)}.$$

Let $\mathbb{O}$ be $\mathbb{Q}$-subalgebra of $\mathbb{K}$ generated by $\dot{q}_{ij}^{\pm 1}$, $\frac{1}{(-a_{ij})_{q_{ii}}!}$ for all $i, j \in I$. Let $U_{\mathbb{O}}$ (resp. $U_{\mathbb{O}}^0$, resp. $U_{\mathbb{O}}^+$, resp. $U_{\mathbb{O}}^-$) be the $\mathbb{O}$-subalgebra (with 1) of $U = U(\chi,\pi)$ (resp. U^0, resp. U^+, resp. U^-) generated by $K_{\pi(i)}^{\pm 1}$, $L_{\pi(i)}^{\pm 1}$, $\bar{H}_i$, $\bar{E}_i$, $\bar{F}_i$ (resp. $K_{\pi(i)}^{\pm 1}$, $L_{\pi(i)}^{\pm 1}$,

$\bar{H}_i$, resp. $\bar{E}_i$, resp. $\bar{F}_i$) for all $i \in I$. We can see $U_{\mathbb{O}} = U^-_{\mathbb{O}} \otimes_{\mathbb{O}} U^0_{\mathbb{O}} \otimes_{\mathbb{O}} U^+_{\mathbb{O}}$. We also see that the elements

$$\prod_{i \in I} K^{x(i)}_{\pi(i)} (K_{\pi(i)} L_{\pi(i)})^{y(i)} \bar{H}_i^{z(i)}$$

($x(i) \in \{0,1\}$, $z(i) \in \mathbb{Z}_{\geq 0}$, $y(i) \in \mathbb{Z}$) form $\mathbb{O}$-basis of $U^0_{\mathbb{O}}$.

THEOREM 2.13. *Let $i, j \in I, i \neq j$ and χ be a (π, i, j)-good finite bicharacter with $m := m^{\chi,\pi}_{ij} < \infty$. Assume that χ is (π, A)-admissible. Let $b : I \to \mathbb{K}^\times$ and $z \in \mathbb{K}^\times$ be as in Theorem 2.6. Assume that $T^{\chi,\pi_t,a_t}_{i_t} = T^{\chi,\pi_t}_{i_t}$ and $T^{\chi,\pi'_t,a'_t}_{i_{t+1}} = T^{\chi,\pi'_t}_{i_{t+1}}$ $(t \in \mathbb{N})$. Then we have*

$$b(k) = 1 \ (k \in I) \quad \textit{and} \quad z = 1. \tag{2.20}$$

PROOF. We divide the proof into steps.

Step 1. *Assume that $\dot{q}$ is transcendental over $\mathbb{Q}$ and that $\dot{q}_{j'i'} = 1$ for all i', $j' \in I$ with $i' < j'$.* Then $\mathbb{O}$ is the $\mathbb{Q}$-subalgebra of $\mathbb{K}$ generated by $\dot{q}^{\pm 1}$, $\frac{1}{(-a_{i'j'})_{q_{i'i'}}!}$ for all i', $j' \in I$ with $i' \neq j'$. So $\mathbb{O}$ is a principal integral domain.

Consider the $\mathbb{Q}$-algebra $(U_{\mathbb{O}})_1 := U_{\mathbb{O}}/(\dot{q}-1)U_{\mathbb{O}}$. Let $f : U_{\mathbb{O}} \to (U_{\mathbb{O}})_1$ be the canonical map. Note that for $k \in I$, we have $f(L_{\pi(k)}) = f(K_{\pi(k)})$, and we see that $f(K_{\pi(k)})$ is a central element of $(U_{\mathbb{O}})_1$. So we can consider the quotient $\mathbb{Q}$-algebra $(U_{\mathbb{O}})_1/(f(K_{\pi(k)}) - 1)$. By Lemmas 2.7, 2.8 and 2.10, we have (2.20).

Step 2. Let U' denote the U of Step 1, and let $\ddot{q}$ denote $\dot{q}$ for U'. Then $\ddot{q}$ is transcendental over $\mathbb{Q}$. Assume that that $\dot{q}_{j'i'} = 1$ for all i', $j' \in I$ with $i' < j'$. We use a specialization argument with $\ddot{q} \to \dot{q}$; this $\dot{q}$ is the one for U of this step. We can obtain (2.20) from Step 1 by considering the $\mathbb{O}$-subalgebra of U' generating by $K^{\pm 1}_{\pi(t)}$, $L^{\pm 1}_{\pi(t)}$, E_t, F_t $(t \in I)$ and using Lemma 2.2.

Step 3. *General cases.* Repeat the same arguments as in Step 2. □

2.4. Drinfeld pairing. Let $U = U(\chi, \pi)$ be the generalized quantum group. We regard $U = U(\chi, \pi)$ as a Hopf algebra $(U, \Delta, S, \varepsilon)$ with $\Delta(K_\lambda) = K_\lambda \otimes K_\lambda$, $\Delta(L_\lambda) = L_\lambda \otimes L_\lambda$, $\Delta(E_i) = E_i \otimes 1 + K_{\pi(i)} \otimes E_i$, $\Delta(F_i) = F_i \otimes L_{\pi(i)} + 1 \otimes F_i$, $S(K_\lambda) = K_{-\lambda}$, $S(L_\lambda) = L_{-\lambda}$, $S(E_i) = -K_{-\pi(i)}E_i$, $S(F_i) = -F_i L_{-\pi(i)}$, $\varepsilon(K_\lambda) = \varepsilon(L_\lambda) = 1$, and $\varepsilon(E_i) = \varepsilon(E_i) = 0$.

For i, $j \in I$ with $i \neq j$ and $r \in \mathbb{Z}_{\geq 0}$, if $E_{r,i,j} \neq 0$,we have

$$\Delta(E_{r,i,j}) = E_{r,i,j} \otimes 1 + \sum_{k=0}^{r} \frac{(r)_{q_{ii}}!(k; q_{ii}, q_{ii}^{r-k} q_{ij} q_{ji})!}{(k)_{q_{ii}}!(r-k)_{q_{ii}}!} E_i^k K_{(r-k)\pi(i)+\pi(j)} \otimes E_{r-k,i,j}.$$

Let $U^{+,\flat} = U^{+,\flat}(\chi, \pi) := \oplus_{\lambda \in V_{\mathbb{Z}}} U^+ K_\lambda$, and $U^{-,\flat} = U^{-,\flat}(\chi, \pi) := \oplus_{\lambda \in V_{\mathbb{Z}}} U^- L_\lambda$. Then $U = \mathrm{Span}_{\mathbb{K}}(U^{-,\flat} U^{+,\flat}) = \mathrm{Span}_{\mathbb{K}}(U^{+,\flat} U^{-,\flat})$ and in a standard way (see [**3**]), we have a bilinear form $\vartheta = \vartheta^{\chi,\pi} : U^{+,\flat} \times U^{-,\flat} \to \mathbb{K}$ with the following properties:

$$
\begin{aligned}
&\vartheta(K_\lambda, L_\mu) = \chi(\lambda,\mu), \vartheta(E_i,F_j) = \delta_{ij}, \vartheta(K_\lambda, F_j) = \vartheta(E_i, L_\lambda) = 0, \\
&\vartheta(X^+Y^+, X^-) = \textstyle\sum_{k^-} \vartheta(X^+, (X^-)^{(2)}_{k^-})\vartheta(Y^+, (X^-)^{(1)}_{k^-}), \\
&\vartheta(X^+, X^-Y^-) = \textstyle\sum_{k^+} \vartheta((X^+)^{(1)}_{k^+}, X^-)\vartheta((X^+)^{(2)}_{k^+}, Y^-), \\
&\vartheta(S(X^+), X^-) = \vartheta(X^+, S^{-1}(X^-)), \\
&\vartheta(X^+,1) = \varepsilon(X^+), \vartheta(1, X^-) = \varepsilon(X^-), \\
&X^-X^+ = \textstyle\sum_{r^+,r^-} \vartheta((X^+)^{\prime,(1)}_{r^+}, S((X^-)^{\prime,(1)}_{r^-}))\vartheta((X^+)^{\prime,(3)}_{r^+}, (X^-)^{\prime,(3)}_{r^-}) \\
&\qquad \cdot (X^+)^{\prime,(2)}_{r^+}(X^-)^{\prime,(2)}_{r^-}, \\
&X^+X^- = \textstyle\sum_{r^+,r^-} \vartheta((X^+)^{\prime,(3)}_{r^+}, S((X^-)^{\prime,(3)}_{r^-}))\vartheta((X^+)^{\prime,(1)}_{r^+}, (X^-)^{\prime,(1)}_{r^-}) \\
&\qquad \cdot (X^-)^{\prime,(2)}_{r^-}(X^+)^{\prime,(2)}_{r^+}
\end{aligned}
\tag{2.21}
$$

for λ, $\mu \in V_{\mathbb{Z}}$, i, $j \in I$, and X^+, $Y^+ \in U^{+,\flat}$, X^-, $Y^- \in U^{-,\flat}$, where $(X^+)^{(x)}_{k^+}$ and $(X^-)^{(x)}_{k^-}$ with $x \in \mathbb{J}_{1,2}$ (resp. $(X^+)^{\prime,(y)}_{r^+}$ and $(X^-)^{\prime,(y)}_{r^-}$ with $y \in \mathbb{J}_{1,3}$) are any elements of $U^{+,\flat}$ and $U^{-,\flat}$ respectively satisfying $\Delta(X^\pm) = \sum_{k^\pm}(X^\pm)^{(1)}_{k^\pm} \otimes (X^\pm)^{(2)}_{k^\pm}$, (resp. $((\mathrm{id}_U \otimes \Delta)\circ\Delta)(X^\pm) = \sum_{r^\pm}(X^\pm)^{\prime,(1)}_{r^\pm} \otimes (X^\pm)^{\prime,(2)}_{r^\pm} \otimes (X^\pm)^{\prime,(3)}_{r^\pm}$).

We have

$$
\begin{aligned}
&\vartheta^{\chi,\pi}(X^+K_\lambda, X^-L_\mu) = \vartheta^{\chi,\pi}(X^+, X^-)\chi(\lambda,\mu) \\
&(X^+ \in U^+,\ X^- \in U^-,\ \lambda,\mu \in V_{\mathbb{Z}}).
\end{aligned}
$$

It follows that $\vartheta^{\chi,\pi}_{|U^+\times U^-}$ is non-degenerate. We have $\vartheta^{\chi,\pi}(X^+, X^-) = 0$ for $\lambda, \mu \in V^{\pi,+}_{\mathbb{Z}}$ with $\lambda \neq \mu$ and $X^+ \in U^+_\lambda$, $X^- \in U^-_{-\mu}$.

3. Kostant-Lusztig $\mathbb{A}$-form

In this section we establish the Kostant-Lusztig $\mathbb{A}$-forms for the finite-type multiparameter quantum group $U = U(\chi,\pi)$ (see Definitions 2.1 and 2.11) and construct invariant bases for each factor of the triangular decomposition. Let $A = [a_{ij}]_{ij\in I}$ be a Cartan matrix associated with a finite dimensional complex simple Lie algebra, i.e., A is a symmetrizable generalized Cartan matrix of finite-type. We assume that χ is a (π, A)-admissible bicharacter (see Definition 2.11). Recall the symbols q_{ij}, $\dot{q}_{ij}$ and $\Theta(q_{ii}-1)$. Let W be the Weyl group associated to the Cartan matrix A and generated by the simple reflections s_i $(i \in I)$ with relations $s_i^2 = e$ $(i \in I)$ and $(s_is_j)^{m^{\chi,\pi}_{ij}} = e$, ($e$ is the identity element of W). Note that W is the finite Weyl group.

3.1. Some standard notations and results. Define the map $\ell : W \to \mathbb{Z}_{\geq 0}$ by $\ell(e) := 0$ and $\ell(w) := \min\{r \in \mathbb{N} | \exists i_t \in I\,(t \in \mathbb{J}_{1,r}), w = s_{i_1}\cdots s_{i_r}\}$. In fact ℓ is the length map of the Coxeter system $(W, \{s_i | i \in I\})$. It is well-known that

$$|\ell(ws_i) - \ell(w)| = 1 \text{ for } w \in W \text{ and } i \in I.$$

Let W act on $V_{\mathbb{Z}}$ by $s_i \cdot \pi(j) := \pi(j) - a_{ij}\pi(j)$ $(i, j \in I)$. Use the convention as follows

Let $s_{i_1}\cdots s_{i_t}$ (resp. $\tau^\chi_{i_{t-1}}\cdots\tau^\chi_{i_1}\pi$) mean e (resp. π) if $t = 0$.

We have

$$s_{i_1}\cdots s_{i_{t-1}} \cdot \pi(i_t) = \tau^\chi_{i_{t-1}}\cdots\tau^\chi_{i_1}\pi(i_t) \ (t \in \mathbb{N},\ i_x \in I\,(x \in \mathbb{J}_{1,t})).$$

For $w = s_{k_1} \ldots s_{k_r} \in W$, if $r = \ell(w)$, the expression $s_{k_1} \ldots s_{k_{\ell(w)}}$ is called *reduced*. For $w = s_{k_1} \ldots s_{k_{\ell(w)}} \in W$, let

$$\mathrm{id}^{\chi,\pi} T_w := T_{k_1}^{\chi,\pi_1} \cdots T_{k_{\ell(w)}}^{\chi,\pi_{\ell(w)}} \quad (\mathrm{id}^{\chi,\pi} T_e := \mathrm{id}_{U(\chi,\pi)}),$$

where $\pi_t := \tau_{k_t}^{\chi} \cdots \tau_{k_1}^{\chi} \pi$ ($t \in \mathbb{J}_{1,\ell(w)}$). By Theorem 2.13, $\mathrm{id}^{\chi,\pi} T_w$ is independent of the choice of reduced expressions for w.

It is well-known that there exists a unique $w_\circ \in W$ such that $\ell(w) \leq \ell(w_\circ)$ for all $w \in W$; $w_\circ$ is called *the longest element*. We also know that $\ell(w_\circ) = |R_\chi^{\pi,+}|$ and

$$\forall w \in W, \ \ell(w_\circ) = \ell(w) + \ell(w^{-1} w_\circ). \tag{3.1}$$

Let $n = (n_1, \ldots, n_{\ell(w_\circ)}) \in I^{\ell(w_\circ)}$ be such that $s_{n_1} \cdots s_{n_{\ell(w_\circ)}} = w_\circ$ (reduced expression of $w_\circ$). For $t \in \mathbb{J}_{1,\ell(w_\circ)}$, let $\beta_{n;t} := s_{n_1} \cdots s_{n_{t-1}} \cdot \pi(n_t)$. By (2.11) and Lemma 2.12, we have (see [**10**, 1.7])

$$R_\chi^{\pi,+} = \{\beta_{n;t} | t \in \mathbb{J}_{1,\ell(w_\circ)}\}.$$

Thus R_χ^{π} can be identified with the root system of W. For $t \in \mathbb{J}_{1,\ell(w_\circ)}$, let

$$\begin{aligned} E_{n;t} &:= \mathrm{id}^{\chi,\pi} T_{s_{n_1} \ldots s_{n_{t-1}}}(E_{n_t}), \ F_{n;t} := \mathrm{id}^{\chi,\pi} T_{s_{n_1} \ldots s_{n_{t-1}}}(F_{n_t}), \\ \bar{E}_{n;t} &:= \mathrm{id}^{\chi,\pi} T_{s_{n_1} \ldots s_{n_{t-1}}}(\bar{E}_{n_t}), \ \bar{F}_{n;t} := \mathrm{id}^{\chi,\pi} T_{s_{n_1} \ldots s_{n_{t-1}}}(\bar{F}_{n_t}), \end{aligned}$$

($E_{n;1} := E_{n_1}$, $F_{n;1} := F_{n_1}$, $\bar{E}_{n;1} := \bar{E}_{n_1}$, $\bar{F}_{n;1} := \bar{F}_{n_1}$), that is,

$$\bar{E}_{n;t} = \frac{E_{n;t}}{\Theta(q_{n_t n_t} - 1)} = \frac{E_{n;t}}{\Theta(\chi(\beta_{n;t}, \beta_{n;t}) - 1)}.$$

Let $n_0 \in I$ be such that $s_{n_0} w_\circ s_{n_{\ell(w_\circ)}} = w_\circ$. By Theorem 2.13, using a standard argument (see [**11**, Proposition 8.20]), we have

$$E_{n;\ell(w_\circ)} = E_{n_0}, \ F_{n;\ell(w_\circ)} = F_{n_0}, \ \bar{E}_{n;\ell(w_\circ)} = \bar{E}_{n_0}, \ \bar{F}_{n;\ell(w_\circ)} = \bar{F}_{n_0}. \tag{3.2}$$

The following result can be proved by a standard argument (see [**2**], [**9**] for example).

THEOREM 3.1. *Let $k := \ell(w_\circ)$, $J := \mathbb{J}_{1,k}$ and $\beta_t := s_{n_1} \cdots s_{n_{t-1}} \cdot \pi(n_t)$ ($t \in J$).*

(1) *Let $\sigma : J \to J$ be a bijection. Then the elements*

$$E_{n;\sigma(1)}^{x_1} \cdots E_{n;\sigma(k)}^{x_k} \quad (x_t \in \mathbb{Z}_{\geq 0} \, (t \in J)))$$

form a $\mathbb{K}$-basis of U^+.

(2) *Let y, $z \in J$ be such that $y < z$. Let X be a $\mathbb{K}$-subalgebra of U generated by the elements $E_{n;x}$ ($x \in \mathbb{J}_{y+1,z-1}$). Then*

$$E_{n;y} E_{n;z} - \chi(\beta_y, \beta_z) E_{n;z} E_{n;y} \in X. \tag{3.3}$$

(3) *We have*

$$\vartheta^{\chi,\pi}(E_{n;k}^{x_k} \cdots E_{n;1}^{x_1}, F_{n;k}^{y_k} \cdots F_{n;1}^{y_1}) = \prod_{t=1}^{k} \delta_{x_t, y_t} (x_t)_{\chi(\beta_t,\beta_t)}!,$$

where x_t, $y_t \in \mathbb{Z}_{\geq 0}$ ($t \in J$).

3.2. Type G_2**.** In this subsection we assume $\theta = 2$ (so $I = \mathbb{J}_{1,2}$), and $A = [a_{ij}]_{ij\in I}$ is the Cartan matrix of type G_2. So $a_{12} = -3$ and $a_{21} = -1$. Consider the generalized quantum group $U = U(\chi, \pi)$. Let q_{ij} be as in Definition 2.1 and χ be a (π, A)-admissible bicharacter. Let $q := q_{11}$ and $a := q_{12}$. Then $q_{22} = q^3$ and $q_{21} = q^{-3}a^{-1}$. Since $E_{4,1,2} = E_{2,2,1} = 0$, we have

$$\begin{aligned}
&E_1^4E_2 - (1+q)(1+q^2)aE_1^3E_2E_1 + q(1+q^2)(1+q+q^2)a^2E_1^2E_2E_1^2\\
&\quad -q^3(1+q)(1+q^2)a^3E_1^2E_2E_1^2 + q^6a^4E_2E_1^4 = 0,\\
&E_1^2E_2 - (1+q)(1-q+q^2)aE_2E_1E_2 + q^3a^2E_2^2E_1 = 0.
\end{aligned}$$

Let

$$\begin{aligned}
&K_1; = K_{\pi(1)},\ K_2; = K_{\pi(2)},\ E_{12} := E_{1,1,2},\ E_{112} := E_{2,1,2},\\
&E_{1112} := E_{3,1,2},\ E_{11212} := E_{112}E_{12} - aq^2E_{12}E_{112}.
\end{aligned}$$

Then we have

$$\begin{aligned}
&E_{12}E_2 = aq^3E_2E_{12},\ E_{1112}E_2 = a^3q^6E_2E_{1112} + a^2q^3(q^2-1)(q-1)E_{12}^3,\\
&E_{112}E_2 = a^2q^3E_2E_{112} + aq(q^2-1)E_{12}^2,\\
&E_{1112}E_2 = a^3q^3E_2E_{1112} + aq(q^2-q-1)E_{11212} + a^2q^2(q^3-1)E_{12}E_{112},\\
&E_1E_2 = aE_2E_1 + E_{12},\ E_{11212}E_{12} = aq^3E_{12}E_{11212},\\
&E_{112}E_{12} = aq^2E_{12}E_{112} + E_{11212},\ E_{1112}E_{12} = a^2q^3E_{12}E_{1112} + \tfrac{aq(q^3-1)}{q+1}E_{112}^2,\\
&E_1E_{12} = aqE_{12}E_1 + E_{112},\ E_{112}E_{11212} = aq^3E_{112}E_{11212},\\
&E_{1112}E_{11212} = a^3q^6E_{11212}E_{1112} + \tfrac{a^2q^3(q^3-1)(q-1)}{q+1}E_{112}^3,\\
&E_1E_{11212} = a^2q^3E_{11212}E_1 + \tfrac{aq(q^3-1)}{q+1}E_{112}^2,\ E_{1112}E_{112} = aq^3E_{112}E_{1112},\\
&E_1E_{112} = aq^2E_{112}E_1 + E_{1112},\ E_1E_{1112} = aq^3E_{1112}E_1.
\end{aligned}$$

We have

$$\begin{aligned}
\Delta(E_{12}) &= E_{12}\otimes 1 + (1-q^{-3})E_1K_2\otimes E_2 + K_1K_2\otimes E_{12},\\
\Delta(E_{112}) &= E_{112}\otimes 1 + (1-q^{-3})(1-q^{-2})E_1^3K_2\otimes E_2\\
&\quad +(1-q^{-2})(1+q)E_1K_1K_2\otimes E_{12} + K_1^2K_2\otimes E_{112},\\
\Delta(E_{1112}) &= E_{1112}\otimes 1 + (1-q^{-3})(1-q^{-2})(1-q^{-1})E_1^3K_2\otimes E_2\\
&\quad +(q^2-1)(1-q^{-3})E_1^2K_1K_2\otimes E_{12},\\
&\quad +q^{-1}(q^3-1)E_1K_1^2K_2\otimes E_{112} + K_1^3K_2\otimes E_{1112},
\end{aligned}$$

and

$$\begin{aligned}
\Delta(E_{11212}) = E_{11212}&\otimes 1 + \tfrac{(q^3-1)^2}{q^4}E_{112}E_1K_2\otimes E_2 + \tfrac{(q^3-1)(q^2-q-1)}{aq^5}E_{1112}K_2\otimes E_2\\
&+\tfrac{q^3-1}{q}E_{112}K_1K_2\otimes E_{12} + \tfrac{(q^3-1)^2(q^2-1)(q-1)}{aq^{12}}E_1^3K_1^2\otimes E_2^2\\
&+\tfrac{(q^3-1)^2(q^2-1)}{q^6}E_1^2K_1K_2\otimes E_2E_{12} + \tfrac{(q^3-1)(q^2-1)}{q^3}E_1K_1^2K_2^2\otimes E_{12}^2\\
&+K_1^3K_2^2\otimes E_{11212}.
\end{aligned}$$

Let

$$\begin{aligned}
&\hat{E}_1 := E_1,\ \hat{E}_2 := E_2,\ \hat{E}_{12} := \tfrac{q^3}{q^3-1}E_{12},\ \hat{E}_{112} := \tfrac{q^5}{(q^3-1)(q^2-1)}E_{112},\\
&\hat{E}_{1112} := \tfrac{q^6}{(q^3-1)(q^2-1)(q-1)}E_{1112},\ \hat{E}_{11212} := \tfrac{q^9}{(q^3-1)^2(q^2-1)(q-1)}E_{11212}.
\end{aligned}$$

Then we have

$$\begin{aligned}
&\hat{E}_{1112}\hat{E}_2 = a^3q^6\hat{E}_2\hat{E}_{1112} + a^2q^3(q^3-1)\hat{E}_{12}^3,\ \hat{E}_{112}\hat{E}_2 = a^2q^3\hat{E}_2\hat{E}_{112} + a(q^3-1)\hat{E}_{12}^2,\\
&\hat{E}_{1112}\hat{E}_2 = a^3q^3\hat{E}_2\hat{E}_{1112} + aq^{-2}(q^2-q-1)(q^3-1)\hat{E}_{11212}\\
&\qquad +a^2(q^3-1)(q^2+q+1)\hat{E}_{12}\hat{E}_{112},\\
&\hat{E}_1\hat{E}_2 = a\hat{E}_2\hat{E}_1 + q^{-3}(q^3-1)\hat{E}_{12},\ \hat{E}_{112}\hat{E}_{12} = aq^2\hat{E}_{12}\hat{E}_{112} + q^{-1}(q-1)\hat{E}_{11212},\\
&\hat{E}_{1112}E_{12} = a^2q^3\hat{E}_{12}\hat{E}_{1112} + a(q^3-1)\hat{E}_{112}^2,\ \hat{E}_1E_{12} = aq\hat{E}_{12}\hat{E}_1 + q^{-2}(q^2-1)\hat{E}_{112},\\
&\hat{E}_{1112}\hat{E}_{11212} = a^3q^6\hat{E}_{11212}\hat{E}_{1112} + a^2q^3(q^3-1)\hat{E}_{112}^3,\\
&\hat{E}_1\hat{E}_{11212} = a^2q^3\hat{E}_{11212}\hat{E}_1 + a(q^3-1)\hat{E}_{112}^2,\ \hat{E}_1\hat{E}_{112}{=}aq^2\hat{E}_{112}\hat{E}_1 + q^{-1}(q-1)\hat{E}_{1112}.
\end{aligned}$$

and

$$\begin{aligned}
&\Delta(\hat{E}_{12}) = \hat{E}_{12}\otimes 1 + \hat{E}_1K_2\otimes\hat{E}_2 + K_1K_2\otimes\hat{E}_{12},\\
&\Delta(\hat{E}_{112}) = \hat{E}_{112}\otimes 1 + \hat{E}_1^3K_2\otimes\hat{E}_2 + (q+1)\hat{E}_1K_1K_2\otimes\hat{E}_{12} + K_1^2K_2\otimes E_{112},\\
&\Delta(\hat{E}_{1112}) = \hat{E}_{1112}\otimes 1 + \hat{E}_1^3K_2\otimes\hat{E}_2 + (q^2+q+1)\hat{E}_1^2K_1K_2\otimes\hat{E}_{12},\\
&\qquad +(q^2+q+1)\hat{E}_1K_1^2K_2\otimes\hat{E}_{112} + K_1^3K_2\otimes\hat{E}_{1112},\\
&\Delta(\hat{E}_{11212}) = \hat{E}_{11212}\otimes 1 + (q^2+q+1)\hat{E}_{112}\hat{E}_1K_2\otimes\hat{E}_2\\
&\qquad +a^{-1}q^{-2}(q^2-q-1)\hat{E}_{1112}K_2\otimes\hat{E}_2\\
&\qquad +(q^2+q+1)\hat{E}_{112}K_1K_2\otimes\hat{E}_{12} + a^{-1}q^{-3}\hat{E}_1^3K_1^2\otimes\hat{E}_2^2\\
&\qquad +(q^2+q+1)\hat{E}_1^2K_1K_2\otimes\hat{E}_2\hat{E}_{12} + (q^2+q+1)\hat{E}_1K_1^2K_2^2\otimes\hat{E}_{12}^2\\
&\qquad +K_1^3K_2^2\otimes\hat{E}_{11212}.
\end{aligned}$$

For $\mathbf{a} = (a_1,\ldots,a_6)\in\mathbb{Z}_{\geq 0}^6$, let

$$\begin{aligned}
Q_1(\mathbf{a}) &:= \frac{E_2^{a_1}E_{12}^{a_2}((3)_q^{-1}!E_{11212})^{a_3}((2)_q^{-1}E_{112})^{a_4}((3)_q^{-1}!E_{1112})^{a_5}E_1^{a_6}}{(a_1)_{q^3}!(a_2)_q!(a_3)_{q^3}!(a_4)_q!(a_5)_{q^3}!(a_6)_q!},\\
Q_2(\mathbf{a}) &:= \hat{E}_2^{a_1}\hat{E}_{12}^{a_2}\hat{E}_{11212}^{a_3}\hat{E}_{112}^{a_4}\hat{E}_{1112}^{a_5}\hat{E}_1^{a_6}.
\end{aligned}$$

Then we have

$$\vartheta^{\chi,\pi}(Q_1(\mathbf{a}),\Upsilon^{\chi^{\mathrm{op}},\pi}(Q_2(\mathbf{b}))) = \delta_{\mathbf{a},\mathbf{b}}\quad (\mathbf{a},\,\mathbf{b}\in\mathbb{Z}_{\geq 0}^6). \tag{3.4}$$

By (3.4), $\{Q_1(\mathbf{a})|\mathbf{a}\in\mathbb{Z}_{\geq 0}^6\}$ and $\{Q_2(\mathbf{a})|\mathbf{a}\in\mathbb{Z}_{\geq 0}^6\}$ are $\mathbb{K}$-bases of $U(\chi,\pi)$. Let $\acute{\mathbb{A}}$ be the $\mathbb{Z}$-subalgebra of $\mathbb{K}$ generated by $q^{\pm 1}$ and $a^{\pm 1}$, i.e., $\acute{\mathbb{A}} = \mathbb{Z}[q^{\pm 1},a^{\pm 1}]$. For $t\in\mathbb{J}_{1,2}$, let $\mathfrak{g}_t$ be the $\acute{\mathbb{A}}$-submodules of $U^{+,\flat}$ with the $\acute{\mathbb{A}}$-bases $\{Q_t(\mathbf{a})K_\lambda|\mathbf{a}\in\mathbb{Z}_{\geq 0}^6,\ \lambda\in V_{\mathbb{Z}}\}$. Clearly $\mathfrak{g}_2$ is a $\acute{\mathbb{A}}$-subalgebra of U with $\Delta(\mathfrak{g}_2)\subset\mathfrak{g}_2\otimes_{\acute{\mathbb{A}}}\mathfrak{g}_2$. By (2.21) and (3.4), we see the following.

(3.5) As a $\acute{\mathbb{A}}$-subalgebra of U, $\mathfrak{g}_1$ is generated by $\frac{E_i^x}{(x)_{q_{ii}}!}$ $(i\in I(=\mathbb{J}_{1,2}),\ x\in\mathbb{Z}_{\geq 0})$ and K_λ $(\lambda\in V_{\mathbb{Z}})$. $\mathfrak{g}_1$ is a Hopf $\acute{\mathbb{A}}$-subalgebra of U.

Then $\mathfrak{g}_2$ is also a Hopf $\acute{\mathbb{A}}$-algebra.

Let $n := (1,2,1,2,1,2)\in I^{\ell(w_\circ)}$. Then we have

$$\begin{aligned}
&\bar{E}_{n;1} = \bar{E}_1, && \bar{F}_{n;1} = \bar{F}_1(=\Upsilon^{\chi^{\mathrm{op}},\pi}(\bar{E}_{n;1})),\\
&\bar{E}_{n;2} = \frac{\dot{q}_{12}^{-3}}{(3)_q!\Theta(q-1)^3\Theta(q^3-1)}E_{1112}, && \bar{F}_{n;2} = \dot{q}_{11}^3\Upsilon^{\chi^{\mathrm{op}},\pi}(\bar{E}_{n;2}),\\
&\bar{E}_{n;3} = \frac{\dot{q}_{12}^{-2}}{(2)_q!\Theta(q^3-1)\Theta(q-1)^2}E_{112}, && \bar{F}_{n;3} = \dot{q}_{11}^4\Upsilon^{\chi^{\mathrm{op}},\pi}(\bar{E}_{n;3}),\\
&\bar{E}_{n;4} = \frac{\dot{q}_{12}^{-4}}{(3)_q!\Theta(q^3-1)^2\Theta(q-1)^3}E_{11212}, && \bar{F}_{n;4} = \dot{q}_{11}^6\Upsilon^{\chi^{\mathrm{op}},\pi}(\bar{E}_{n;4}),\\
&\bar{E}_{n;5} = \frac{\dot{q}_{12}^{-1}}{\Theta(q-1)\Theta(q^3-1)}E_{12}, && \bar{F}_{n;5} = \dot{q}_{11}^3\Upsilon^{\chi^{\mathrm{op}},\pi}(\bar{E}_{n;5}),\\
&\bar{E}_{n;6} = \bar{E}_2, && \bar{F}_{n;6} = \bar{F}_2(=\Upsilon^{\chi^{\mathrm{op}},\pi}(\bar{E}_{n;6})),
\end{aligned} \tag{3.6}$$

where recall that $\dot{q}_{ij}^2 = a$. Let $n' := (2,1,2,1,2,1) \in I^{\ell(w_\circ)}$. By (2.19) and (3.2), we have

$$\bar{E}_{n';t} = \Gamma^{\chi^{\mathrm{op}},\pi}(\bar{E}_{n;7-t}),\ \bar{F}_{n';t} = \Gamma^{\chi^{\mathrm{op}},\pi}(\bar{F}_{n;7-t}) \quad (t \in \mathbb{J}_{1,6}). \tag{3.7}$$

Let $\bar{\mathfrak{g}}^+$ be the $\acute{\mathbb{A}}$-subalgebra of U generated by $\frac{\bar{E}_i^x}{(x)_{q_{ii}}!}$ $(i \in I(= \mathbb{J}_{1,2}),\ x \in \mathbb{Z}_{\geq 0})$ and K_λ $(\lambda \in V_{\mathbb{Z}})$. Let $y_{2t-1} := q$, $y_{2t} := q^3$ for $t \in \mathbb{J}_{1,3}$. Let $y'_t := y_{7-t}$ for $t \in \mathbb{J}_{1,6}$. By (3.3), (3.5), (3.6) and (3.7), we see the following.

(3.8) For a bijection $\sigma : \mathbb{J}_{1,6} \to \mathbb{J}_{1,6}$, the elements
$\frac{\bar{E}^{x_1}_{n;\sigma(1)}}{(x_1)_{y_{\sigma(1)}}!} \cdots \frac{\bar{E}^{x_6}_{n;\sigma(6)}}{(x_6)_{y_{\sigma(6)}}!} K_\lambda$ (resp. $\frac{\bar{E}^{x_1}_{n';\sigma(1)}}{(x_1)_{y'_{\sigma(1)}}!} \cdots \frac{\bar{E}^{x_6}_{n';\sigma(6)}}{(x_6)_{y'_{\sigma(6)}}!} K_\lambda$)
$(x_t \in \mathbb{Z}_{\geq 0}\ (t \in \mathbb{J}_{1,6}),\ \lambda \in V_{\mathbb{Z}})$ form a $\acute{\mathbb{A}}$-base of $\bar{\mathfrak{g}}^+$.
$\bar{\mathfrak{g}}^+$ is a Hopf $\acute{\mathbb{A}}$-subalgebra of U.

Let $\bar{\mathfrak{g}}^-$ be the $\acute{\mathbb{A}}$-subalgebra of U generated by $\frac{\bar{F}_i^x}{(x)_{q_{ii}}!}$ $(i \in I(= \mathbb{J}_{1,2}),\ x \in \mathbb{Z}_{\geq 0})$ and L_λ $(\lambda \in V_{\mathbb{Z}})$. By (3.6), (3.7) and (3.8), we see the following.

(3.9) For a bijection $\sigma : \mathbb{J}_{1,6} \to \mathbb{J}_{1,6}$, the elements
$\frac{\bar{F}^{x_1}_{n;\sigma(1)}}{(x_1)_{y_{\sigma(1)}}!} \cdots \frac{\bar{F}^{x_6}_{n;\sigma(6)}}{(x_6)_{y_{\sigma(6)}}!} L_\lambda$ (resp. $\frac{\bar{F}^{x_1}_{n';\sigma(1)}}{(x_1)_{y'_{\sigma(1)}}!} \cdots \frac{\bar{F}^{x_6}_{n';\sigma(6)}}{(x_6)_{y'_{\sigma(6)}}!} L_\lambda$)
$(x_t \in \mathbb{Z}_{\geq 0}\ (t \in \mathbb{J}_{1,6}),\ \lambda \in V_{\mathbb{Z}})$ form a $\acute{\mathbb{A}}$-base of $\bar{\mathfrak{g}}^-$.
$\bar{\mathfrak{g}}^-$ is a Hopf $\acute{\mathbb{A}}$-subalgebra of U.

3.3. General case. Let $\mathbb{A}$ be the $\mathbb{Z}$-subalgebra of $\mathbb{K}$ generated by $\dot{q}_{ij}^{\pm 1}$ for all $i,\ j \in I$. $n = (n_1, \ldots, n_{\ell(w_\circ)}) \in I^{\ell(w_\circ)}$ be as in Subsection 3.1. Let $\pi_0 := \pi$ and $\pi_t := \tau^\chi_{n_t} \cdots \tau^\chi_{n_1}\pi$ $(t \in \mathbb{J}_{1,\ell(w_\circ)})$. For $t \in \mathbb{J}_{1,\kappa}$ and $x \in \mathbb{Z}_{\geq 0}$, let

$$\bar{E}^{(x)}_{n;t} := \frac{\bar{E}^x_{n;t}}{(x)_{\chi(\pi_{t-1}(n_t),\pi_{t-1}(n_t))}!},\qquad \bar{F}^{(x)}_{n;t} := \frac{\bar{F}^x_{n;t}}{(x)_{\chi(\pi_{t-1}(n_t),\pi_{t-1}(n_t))}!}.$$

Let $U^+_{\mathbb{A}}$ (resp. $U^-_{\mathbb{A}}$) be the $\mathbb{A}$-subalgebra of U^+ (resp. U^-) generated by $\frac{\bar{E}_i^x}{(x)_{q_{ii}}!}$ (resp. $\frac{\bar{F}_i^x}{(x)_{q_{ii}}!}$) with $i \in I$ and $x \in \mathbb{Z}_{\geq 0}$.

THEOREM 3.2. *Let $\sigma : \mathbb{J}_{1,\ell(w_\circ)} \to \mathbb{J}_{1,\ell(w_\circ)}$ be a bijection.*
(1) *The elements*

$$\bar{E}^{(x_1)}_{n;\sigma(1)} \cdots \bar{E}^{(x_{\ell(w_\circ)})}_{n:\sigma(\ell(w_\circ))} \text{ with } x_t \in \mathbb{Z}_{\geq 0}\ (t \in \mathbb{J}_{1,\ell(w_\circ)}) \tag{3.10}$$

form an $\mathbb{A}$-base for $U^+_{\mathbb{A}}$.
(2) *The elements*

$$\bar{F}^{(x_1)}_{n;\sigma(1)} \cdots \bar{F}^{(x_{\ell(w_\circ)})}_{n:\sigma(\ell(w_\circ))} \text{ with } x_t \in \mathbb{Z}_{\geq 0}\ (t \in \mathbb{J}_{1,\ell(w_\circ)}) \tag{3.11}$$

form an $\mathbb{A}$-base for $U^-_{\mathbb{A}}$.

PROOF. We only prove (1), as (2) can be proved similarly or obtained from (1) via Chevalley isomorphism. If $|I| = 1$, the claim is clear.

If $|I| = 2$ and A is of type G_2, the result is obtained in (3.8) and (3.9). For other rank two types, the claim can be proved in a similar and in fact easier way.

Let us consider the higher rank cases. Let Z be the free $\mathbb{A}$-submodule of U^+ with the free $\mathbb{A}$-basis formed by the elements of (3.10) (for $\sigma := \mathrm{id}_{\mathbb{J}_{1,\ell(w_\circ)}}$, see Theorem 3.1 (1)). By an argument similar to [**11**, Proposition 8.20] and by the claim for rank-two cases, we see that $\bar{E}^{(x)}_{n;t} \in U^+_{\mathbb{A}}$, so $Z \subseteq U^+_{\mathbb{A}}$. Now by (3.1), we have $Z = U^+_{\mathbb{A}}$. Thus the claim follows from (3.3). □

For k, $r \in \mathbb{Z}_{\geq 0}$ and $x \in \mathbb{K}^\times$ with $(k+r)_x! \neq 0$, let $\binom{k}{r}_x := \frac{(k+r)_x!}{(k)_x!(r)_x!}$. For $x \in \mathbb{K}^\times$ with $x^b \neq 1$ for all $b \in \mathbb{N}$ and for X, $Y \in U^0$, $l \in \mathbb{Z}$ and $p \in \mathbb{Z}_{\geq 0}$, let

$$\begin{bmatrix} X, Y, l \\ p \end{bmatrix}_x := \prod_{t=1}^{p} \frac{x^{l-t+1}X - Y}{x^t - 1}.$$

Then we have

$$\begin{bmatrix} X, Y, l \\ p \end{bmatrix}_x - \begin{bmatrix} X, Y, l+1 \\ p \end{bmatrix}_x = -x^{l-p+1} \begin{bmatrix} X, Y, l \\ p-1 \end{bmatrix}_x X,$$
$$\begin{bmatrix} X, Y, 0 \\ l \end{bmatrix}_x \begin{bmatrix} X, Y, -l \\ p \end{bmatrix}_x = \binom{p+l}{p}_x \begin{bmatrix} X, Y, 0 \\ p+l \end{bmatrix}_x.$$

Let $U^0_{\mathbb{A}}$ be the $\mathbb{A}$-subalgebra generated by $K_\lambda L_\mu$ (λ, $\mu \in V_{\mathbb{Z}}$) and $\begin{bmatrix} K_{\pi(i)}, L_{\pi(i)}, l \\ p \end{bmatrix}_{q_{ii}}$ ($i \in I$, $l \in \mathbb{Z}$, $p \in \mathbb{Z}_{\geq 0}$). By a standard argument, we have the following two lemmas.

LEMMA 3.3. *The elements*

$$\prod_{i \in I} K^{x_i}_{\pi(i)} (K_{\pi(i)} L_{\pi(i)})^{y_i} \begin{bmatrix} K_{\pi(i)}, L_{\pi(i)}, 0 \\ z_i \end{bmatrix}_{q_{ii}} \qquad (x_i \in \mathbb{Z}_{0,1},\ y_i \in \mathbb{Z},\ z_i \in \mathbb{Z}_{\geq 0})$$

form an $\mathbb{A}$-base of $U^0_{\mathbb{A}}$.

Let $U_{\mathbb{A}}$ be the $\mathbb{A}$-subalgebra of U generated by $U^+_{\mathbb{A}}$ and $U^-_{\mathbb{A}}$.

LEMMA 3.4. *We have $U^0_{\mathbb{A}} \subset U_{\mathbb{A}}$, and we have the $\mathbb{A}$-module isomorphism*

$$\mathbf{m}_{\mathbb{A}} : U^-_{\mathbb{A}} \otimes_{\mathbb{A}} U^0_{\mathbb{A}} \otimes_{\mathbb{A}} U^+_{\mathbb{A}} \longrightarrow U_{\mathbb{A}}$$

given by $\mathbf{m}_{\mathbb{A}}(X \otimes Y \otimes Z) := XYZ$.

Acknowledgments

We thank the referee for careful reading and kind comments. The third author would like to express his heartfelt thanks to Professor Naihong Hu for valuable communication.

References

[1] N. Andruskiewitsch and H.-J. Schneider, *On the classification of finite-dimensional pointed Hopf algebras*, Ann. of Math. (2) **171** (2010), no. 1, 375–417. MR2630042

[2] I. Angiono and H. Yamane, *The R-matrix of quantum doubles of Nichols algebras of diagonal type*, J. Math. Phys. **56** (2015), no. 2, 021702, 19. MR3390862

[3] V. G. Drinfel′d, *Quantum groups*, Proceedings of the International Congress of Mathematicians, Vol. 1, 2 (Berkeley, Calif., 1986), Amer. Math. Soc., Providence, RI, 1987, pp. 798–820. MR934283

[4] Z. Fan and Y. Li, *Two-parameter quantum algebras, canonical bases, and categorifications*, Int. Math. Res. Not. IMRN **16** (2015), 7016–7062. MR3428954

[5] I. Heckenberger, *Lusztig isomorphisms for Drinfel′d doubles of bosonizations of Nichols algebras of diagonal type*, J. Algebra **323** (2010), no. 8, 2130–2182. MR2596372

[6] I. Heckenberger and H. Yamane, *A generalization of Coxeter groups, root systems, and Matsumoto's theorem*, Math. Z. **259** (2008), no. 2, 255–276. MR2390080

[7] I. Heckenberger, *Classification of arithmetic root systems*, Adv. Math. **220** (2009), no. 1, 59–124. MR2462836

[8] I. Heckenberger, *The Weyl groupoid of a Nichols algebra of diagonal type*, Invent. Math. **164** (2006), no. 1, 175–188. MR2207786

[9] I. Heckenberger and H. Yamane, *Drinfel'd doubles and Shapovalov determinants*, Rev. Un. Mat. Argentina **51** (2010), no. 2, 107–146. MR2840165

[10] J. E. Humphreys, *Reflection groups and Coxeter groups*, Cambridge Studies in Advanced Mathematics, vol. 29, Cambridge University Press, Cambridge, 1990. MR1066460

[11] J. C. Jantzen, *Lectures on quantum groups*, Graduate Studies in Mathematics, vol. 6, American Mathematical Society, Providence, RI, 1996. MR1359532

[12] J. C. Jantzen, *Representations of algebraic groups*, 2nd ed., Mathematical Surveys and Monographs, vol. 107, American Mathematical Society, Providence, RI, 2003. MR2015057

[13] M. Jimbo, *A q-difference analogue of* $U(\mathfrak{g})$ *and the Yang-Baxter equation*, Lett. Math. Phys. **10** (1985), no. 1, 63–69. MR797001

[14] K. Jeong, S.-J. Kang, and M. Kashiwara, *Crystal bases for quantum generalized Kac-Moody algebras*, Proc. London Math. Soc. (3) **90** (2005), no. 2, 395–438. MR2142133

[15] M. Kashiwara, *On crystal bases of the Q-analogue of universal enveloping algebras*, Duke Math. J. **63** (1991), no. 2, 465–516. MR1115118

[16] V. K. Kharchenko, *A quantum analogue of the Poincaré-Birkhoff-Witt theorem* (Russian, with Russian summary), Algebra Log. **38** (1999), no. 4, 476–507, 509; English transl., Algebra and Logic **38** (1999), no. 4, 259–276. MR1763385

[17] G. Lusztig, *Quantum groups at roots of* 1, Geom. Dedicata **35** (1990), no. 1-3, 89–113. MR1066560

[18] G. Lusztig, *Canonical bases arising from quantized enveloping algebras*, J. Amer. Math. Soc. **3** (1990), no. 2, 447–498. MR1035415

Department of Mathematics, North Carolina State University, Raleigh, North Carolina 27695-8205

Email address: jing@ncsu.edu

Department of Mathematics, North Carolina State University, Raleigh, North Carolina 27695-8205

Email address: misra@ncsu.edu

Department of Mathematics, Faculty of Science, University of Toyama, Gofuku, Toyama 930-8555, Japan

Email address: hiroyuki@sci.u-toyama.ac.jp

Contemporary Mathematics
Volume **713**, 2018
https://doi.org/10.1090/conm/713/14315

Catalan triangle numbers and binomial coefficients

Kyu-Hwan Lee and Se-jin Oh

Abstract. The binomial coefficients and Catalan triangle numbers appear as weight multiplicities of the finite-dimensional simple Lie algebras and affine Kac–Moody algebras. We prove that any binomial coefficient can be written as weighted sums along rows of the Catalan triangle. The coefficients in the sums form a triangular array, which we call the *alternating Jacobsthal triangle*. We study various subsequences of the entries of the alternating Jacobsthal triangle and show that they arise in a variety of combinatorial constructions. The generating functions of these sequences enable us to define their k-analogue of q-deformation. We show that this deformation also gives rise to interesting combinatorial sequences. The starting point of this work is certain identities in the study of Khovanov–Lauda–Rouquier algebras and fully commutative elements of a Coxeter group.

1. Introduction

It is widely accepted that Catalan numbers are the most frequently occurring combinatorial numbers after the binomial coefficients. As binomial coefficients can be defined inductively from the Pascal's triangle, so can Catalan numbers from a triangular array of numbers whose entry in the nth row and k^{th} column is denoted by $C(n,k)$ for $0 \le k \le n$. Set the first entry $C(0,0) = 1$, and then each subsequent entry is the sum of the entry above it and the entry to the left. All entries outside of the range $0 \le k \le n$ are considered to be 0. Then we obtain the array shown in (1.1) known as *Catalan triangle* introduced by L.W. Shapiro [**8**] in 1976. Notice that Catalan numbers C_n appear on the hypotenuse of the triangle, i.e. $C_n = C(n,n)$ for $n \ge 0$.

$$(1.1)\qquad \begin{array}{ccccccccc}
1 \\
1 & 1 \\
1 & 2 & 2 \\
1 & 3 & 5 & 5 \\
1 & 4 & 9 & 14 & 14 \\
1 & 5 & 14 & 28 & 42 & 42 \\
1 & 6 & 20 & 48 & 90 & 132 & 132 \\
1 & 7 & 27 & 75 & 165 & 297 & 429 & 429 \\
\vdots & \vdots & \vdots & \vdots & \vdots & \vdots & \vdots & \vdots & \ddots
\end{array}$$

2010 *Mathematics Subject Classification.* Primary 05A10; Secondary 05E10.
The first author was partially supported by a grant from the Simons Foundation (#318706).
The second author was supported by NRF Grant # 2016R1C1B2013135.

The first goal of this paper is to write each binomial coefficient as weighted sums along rows of the Catalan triangle. In the first case, we take the sums along the nth and $n+1^{\text{st}}$ rows of the Catalan triangle, respectively, and obtain 2-power weighted sums to express a binomial coefficient. More precisely, we prove:

THEOREM 1.1. *For integers $n \geq 1$ and $0 \leq k \leq n+1$, we have the identities*

$$\binom{n+k+1}{k} = \sum_{s=0}^{k} C(n,s)2^{k-s} = \sum_{s=0}^{k} C(n+1,s)2^{\max(k-1-s,0)} \tag{1.2}$$

and

$$\binom{n+1}{\lceil (n+1)/2 \rceil} = \sum_{k=0}^{\lceil n/2 \rceil} C(\lceil n/2 \rceil, k)2^{\max(\lfloor n/2 \rfloor - k, 0)}. \tag{1.3}$$

It is quite intriguing that the two most important families of combinatorial numbers are related in this way. By replacing 2-powers with x-powers in the identities, we define Catalan triangle polynomials and make a conjecture on stacked directed animals studied in [**1**] (see Section 2.3). It is also interesting that $\binom{n+1}{\lceil (n+1)/2 \rceil}$ appearing in (1.3) is exactly the number of fully commutative, involutive elements of the Coxeter group of type A_n. (See (4.2) in [**9**].) Actually, a clue to the identities (1.2) and (1.3) was found in the study of the homogeneous representations of Khovanov–Lauda–Rouquier algebras and the fully commutative elements of type D_n in the paper [**2**] of the first-named author and G. Feinberg, where they proved the following:

THEOREM 1.2. [**2**] *For $n \geq 1$, we have*

$$\frac{n+3}{2} C_n = \sum_{k=0}^{n-1} C(n,k)2^{|n-2-k|}, \tag{1.4}$$

where C_n is the nth Catalan number.

We note that $\frac{n+3}{2} C_n - 1$ is the number of the fully commutative elements of type D_n. (See [**9**]). The identity (1.4) is obtained by decomposing the set of fully commutative elements of type D_n into *packets*. Likewise, we expect interesting combinatorial interpretations and representation-theoretic applications of the identities (1.2) and (1.3). In particular, $C(n,k)$ appear as weight multiplicities of finite-dimensional simple Lie algebras and affine Kac–Moody algebras of types A and C [**6, 10, 11**].

To generalize Theorem 1.1, we use other rows of the Catalan triangle and there appears a natural sequence of numbers $A(m,t)$, defined by

$$A(m,0) = 1, \quad A(m,t) = A(m-1,t-1) - A(m-1,t),$$

to yield the following result:

THEOREM 1.3. *For any $n > k \geq m \geq t \geq 1$, we have*

$$\binom{n+k+1}{k} = \sum_{s=0}^{k-m} C(n+m,s)2^{k-m-s} + \sum_{t=1}^{m} A(m,t)C(n+m,k-m+t). \tag{1.5}$$

In particular, when $k = m$, we have

$$\binom{n+k+1}{k} = \sum_{t=0}^{k} A(k,t)C(n+k,t). \tag{1.6}$$

The sequence consisting of $A(m,t)$ is listed as A220074 in the On-line Encyclopedia of Integer Sequences (OEIS). However, the identity (1.6) does not seem to have been known.

The identity (1.6) clearly suggests that the triangle consisting of the numbers $A(m,t)$ be considered as a transition triangle from the Catalan triangle to the Pascal triangle. We call it the *alternating Jacobsthal triangle.* The triangle has (sums of) subsequences of the entries with interesting combinatorial interpretations. In particular, diagonal sums are related to the Fibonacci numbers and horizontal sums are related to the Jacobsthal numbers.

The second goal of this paper is to study a k-analogue of q-deformation of the Fibonacci and Jacobsthal numbers through a k-analogue of the alternating Jacobsthal triangle. This deformation is obtained by putting the parameters q and k into the generating functions of these numbers. Our constructions give rise to different polynomials than the Fibonacci and Jacobsthal polynomials which can be found in the literature (e.g. [**4**, **7**]).

For example, the k-analogue $J_{k,m}(q)$ of q-deformation of the Jacobsthal numbers is given by the generating function

$$\frac{x(1-qx)}{(1-kq^2x^2)(1-(q+1)x)} = \sum_{m=1}^{\infty} J_{k,m}(q)x^m.$$

When $q = 1$ and $k = 1$, we recover the usual generating function $\dfrac{x}{(1+x)(1-2x)}$ of the Jacobsthal numbers.

Interestingly enough, sequences given by special values of this deformation have various combinatorial interpretations. For example, the sequence

$$(J_{2,m}(1))_{m\geq 1} = (1, 1, 4, 6, 16, 28, 64, 120, \dots)$$

is listed as *A*007179 in OEIS and has the interpretation as the numbers of equal dual pairs of some integrals studied in [**3**]. (See Table 1 on p.365 in [**3**].) Similarly, many subsequences of a k-analogue of the alternating Jacobsthal triangle are found to have combinatorial meanings. See the triangle (5.2), for example.

An outline of this paper is as follows. In the next section, we prove Theorem 1.1 to obtain Catalan triangle expansions of binomial coefficients as 2-power weighted sums. We also introduce Catalan triangle polynomials and study some of their special values. In Section 3, we prove Theorem 1.3 and investigate the alternating Jacobsthal triangle to obtain generating functions and meaningful subsequences. The following section is concerned with q-deformation of the Fibonacci and Jacobsthal numbers. The last section is devoted to the study of a k-analogue of the q-deformation of the Fibonacci and Jacobsthal numbers using the k-analogue of the alternating Jacobsthal triangle.

2. Catalan expansion of binomial coefficients

In this section, we prove expressions of binomial coefficients as 2-power weighted sums along rows of the Catalan triangle. Catalan trapezoids are introduced for the proofs. In the last subsection, Catalan triangle polynomials are defined and some of their special values will be considered.

2.1. Catalan triangle. We begin with a formal definition of the Catalan triangle numbers.

DEFINITION 2.1. For $n \geq 0$ and $0 \leq k \leq n$, we define the (n, k)-*Catalan triangle number* $C(n, k)$ recursively by

$$C(n,k) = \begin{cases} 1 & \text{if } n = 0; \\ C(n,k-1) + C(n-1,k) & \text{if } 0 < k < n; \\ C(n-1,0) & \text{if } k = 0; \\ C(n,n-1) & \text{if } k = n, \end{cases} \tag{2.1}$$

and define the nth *Catalan number* C_n by

$$C_n = C(n,n) \qquad \text{for } n \geq 0.$$

The closed form formula for the Catalan triangle numbers is well known: for $n \geq 0$ and $0 \leq k \leq n$,

$$C(n,k) = \frac{(n+k)!(n-k+1)}{k!(n+1)!}.$$

In particular, we have

$$C_n = \frac{1}{n+1}\binom{2n}{n},$$

and it can be easily verified

$$C(n,k) = \binom{n+k}{k} - \binom{n+k}{k-1}. \tag{2.2}$$

THEOREM 2.2. [**2**] *For $n \geq 1$, we have*

$$\frac{n+3}{2} C_n = \sum_{k=0}^{n-1} C(n,k) 2^{|n-2-k|}. \tag{2.3}$$

As mentioned in the introduction, $\frac{n+3}{2}C_n - 1$ is the number of the fully commutative elements of type D_n ([**9**]) and the identity (2.3) is obtained by decomposing the set of fully commutative elements of type D_n into *packets*.

THEOREM 2.3. *For $n \in \mathbb{Z}_{\geq 0}$, we have*

$$\binom{n+1}{\lceil (n+1)/2 \rceil} = \sum_{s=0}^{\lceil n/2 \rceil} C(\lceil n/2 \rceil, s) 2^{\max(\lfloor n/2 \rfloor - s, 0)}. \tag{2.4}$$

PROOF. Set $\mathcal{Q}_n := \binom{n+1}{\lceil (n+1)/2 \rceil}$ for convenience. We will consider n even and n odd separately. First, assume $n = 2k$ for some $k \in \mathbb{Z}_{\geq 0}$. Then we have

$$\mathcal{Q}_{2k} = \binom{2k+1}{k+1}.$$

By (2.3) in Theorem 2.2, we have

$$\sum_{s=0}^{k-1} C(k,s) 2^{|k-2-s|} = \frac{k+3}{2} C_k = \frac{k+3}{2k+2}\binom{2k}{k}. \tag{2.5}$$

On the other hand,

$$\sum_{s=0}^{k-1} C(k,s)2^{|k-2-s|} = \sum_{s=0}^{k-2} C(k,s)2^{k-2-s} + 2C(k,k-1).$$

Note that $C(k,k-1) = C_k$. Multiplying (2.5) by 4, we have

$$\sum_{s=0}^{k} C(k,s)2^{k-s} + 5C(k,k-1) = \frac{2k+6}{k+1}\binom{2k}{k}.$$

Hence

$$\sum_{s=0}^{k} C(k,s)2^{k-s} = \frac{2k+6}{k+1}\binom{2k}{k} - 5C_k = \frac{2k+1}{k+1}\binom{2k}{k} = \binom{2k+1}{k+1} = \mathcal{Q}_{2k}.$$

Next, assume $n = 2k+1$ for some $k \in \mathbb{Z}_{\geq 0}$. Then we have

$$\mathcal{Q}_{2k+1} = \binom{2k+2}{k+1}.$$

Again from (2.3), we obtain

$$\sum_{s=0}^{k} C(k+1,s)2^{|k-1-s|} = \frac{k+4}{2}C_{k+1} = \frac{k+4}{2k+4}\binom{2k+2}{k+1}. \tag{2.6}$$

On the other hand,

$$\sum_{s=0}^{k} C(k+1,s)2^{|k-1-s|} = \sum_{s=0}^{k-1} C(k+1,s)2^{k-1-s} + 2C(k+1,k).$$

Note that $C(k+1,k) = C_{k+1}$. Multiplying (2.6) by 2, we have

$$\begin{aligned}\sum_{s=0}^{k-1} C(k+1,s)2^{k-s} + 4C(k+1,k) &= \sum_{s=0}^{k+1} C(k+1,s)2^{\max(k-s,0)} + 2C(k+1,k)\\ &= \frac{k+4}{k+2}\binom{2k+2}{k+1}.\end{aligned}$$

Since

$$\frac{k+4}{k+2}\binom{2k+2}{k+1} - 2C(k+1,k) = \binom{2k+2}{k+1} = \mathcal{Q}_{2k+1},$$

our assertion is true in this case as well. □

COROLLARY 2.4. *For $n \in \mathbb{Z}_{\geq 0}$, the dual form of (2.4) holds; i.e.,*

$$\binom{n+1}{\lfloor (n+1)/2 \rfloor} = \sum_{s=0}^{\lfloor n/2 \rfloor} C(\lfloor n/2 \rfloor, s)2^{\lceil n/2 \rceil - s}.$$

In particular, we have the following identity by replacing n with $2n-1$:

$$\binom{2n}{n} = \sum_{s=0}^{n-1} C(n-1,s)2^{n-s} = \sum_{s=0}^{n} \frac{(n+s-1)!(n-s)}{s!n!}2^{n-s} = \sum_{s=0}^{n}\binom{n}{s}^2.$$

PROOF. The assertion for $n = 2k$ follows from the fact that $\binom{2k+1}{k} = \binom{2k+1}{k+1}$ and for $n = 2k+1$ follows from the fact that $2 \times \binom{2k+1}{k} = \binom{2k+2}{k+1}$. □

From Theorem 2.3, we have, for $n = 2k-1$ ($k \in \mathbb{Z}_{\geq 1}$),

$$\binom{2k}{k} = \sum_{s=0}^{k} C(k,s) 2^{\max(k-1-s,0)}$$
$$= \sum_{s=0}^{k-1} C(k,s) 2^{k-1-s} + \frac{1}{k+1}\binom{2k}{k}.$$

Since $\binom{2k}{k} - \frac{1}{k+1}\binom{2k}{k} = \binom{2k}{k-1}$, we obtain a new identity:

$$\binom{2k}{k-1} = \sum_{s=0}^{k-1} C(k,s) 2^{k-1-s}. \tag{2.7}$$

More generally, we have the following identity which is an interesting expression of a binomial coefficient $\binom{n+k+1}{k}$ as a 2-power weighted sum of the Catalan triangle along the nth row.

THEOREM 2.5. *For $0 \leq k \leq n+1$, we have*

$$\binom{n+k+1}{k} = \sum_{s=0}^{\min(n,k)} C(n,s) 2^{k-s}. \tag{2.8}$$

PROOF. We will use an induction on $n+1-k$. The cases when $k = n+1, n, n-1$ are already proved. Indeed, the cases $k = n+1$ and $k = n$ follow from Corollary 2.4, and the case $k = n-1$ is nothing but (2.7). Assume that we have the identity (2.8) for any n and k such that $2 \leq n+1-k \leq m$. Consider the case $n+1-k = m+1$. By assumption, we have

$$\binom{n+k+2}{k+1} = \sum_{s=0}^{k+1} C(n,s) 2^{k+1-s}. \tag{2.9}$$

Since $C(n,k+1) = \dfrac{n-k}{n+k+2}\dbinom{n+k+2}{k+1}$, the identity (2.9) can be written as

$$\binom{n+k+2}{k+1} - \frac{n-k}{n+k+2}\binom{n+k+2}{k+1} = \sum_{s=0}^{k} C(n,s) 2^{k+1-s}.$$

Now, simplifying the left-hand side

$$\binom{n+k+2}{k+1} - \frac{n-k}{n+k+2}\binom{n+k+2}{k+1} = \frac{2k+2}{n+k+2}\binom{n+k+2}{k+1} = 2\binom{n+k+1}{k},$$

we obtain the desired identity

$$\binom{n+k+1}{k} = \sum_{s=0}^{k} C(n,s) 2^{k-s}.$$ □

EXAMPLE 2.6.

$$\binom{7}{3} = \sum_{s=0}^{3} C(3,s)2^{3-s} = 1 \times 8 + 3 \times 4 + 5 \times 2 + 5 = 35,$$

$$\binom{8}{3} = \sum_{s=0}^{3} C(4,s)2^{3-s} = 1 \times 8 + 4 \times 4 + 9 \times 2 + 14 = 56.$$

2.2. Catalan trapezoid. As a generalization of Catalan triangle, we define a *Catalan trapezoid* by considering a trapezoidal array of numbers with m complete columns ($m \geq 1$). Let the entry in the nth row and k^{th} column of the array be denoted by $C_m(n,k)$ for $0 \leq k \leq m+n-1$. Set the entries of the first row to be $C_m(0,0) = C_m(0,1) = \cdots = C_m(0,m-1) = 1$, and then each subsequent entry is the sum of the entry above it and the entry to the left as in the case of Catalan triangle. All entries outside of the range $0 \leq k \leq n+m-1$ are considered to be 0. Then, in particular, we obtain $C_m(n,0) = 1$ for all $n \geq 0$. For example, when $m = 3$, we obtain

$$\begin{array}{cccccccccc}
1 & 1 & 1 & & & & & & & \\
1 & 2 & 3 & 3 & & & & & & \\
1 & 3 & 6 & 9 & 9 & & & & & \\
1 & 4 & 10 & 19 & 28 & 28 & & & & \\
1 & 5 & 15 & 34 & 62 & 90 & 90 & & & \\
1 & 6 & 21 & 55 & 117 & 207 & 297 & 297 & & \\
1 & 7 & 28 & 83 & 200 & 407 & 704 & 1001 & 1001 & \\
1 & 8 & 36 & 119 & 319 & 726 & 1430 & 2431 & 3432 & 3432 \\
\vdots & \vdots & \vdots & \vdots & \vdots & \vdots & \vdots & \vdots & \vdots & \ddots
\end{array} \tag{2.10}$$

Alternatively, the numbers $C_m(n,k)$ can be defined in the following way.

DEFINITION 2.7. For an integer $m \geq 1$, set $C_1(n,k) = C(n,k)$ for $0 \leq k \leq n$ and $C_2(n,k) = C(n+1,k)$ for $0 \leq k \leq n+1$, and define inductively

(2.11)

$$C_m(n+1,k) = \begin{cases} \begin{pmatrix} n+1+k \\ k \end{pmatrix} & \text{if } 0 \leq k < m, \\ \begin{pmatrix} n+1+k \\ k \end{pmatrix} - \begin{pmatrix} n+m+1+k-m \\ k-m \end{pmatrix} & \text{if } m \leq k \leq n+m, \\ 0 & \text{if } n+m < k. \end{cases}$$

Using the numbers $C_m(n,k)$, we prove the following theorem.

THEOREM 2.8. *For any triple of integers (m,k,n) such that $1 \leq m \leq k \leq n+m$, we have*

(2.12)

$$\binom{n+k+1}{k} = \sum_{s=0}^{\min(n,k)} C(n,s)2^{k-s} = \sum_{s=0}^{k-m} C(n+m,s)2^{k-m-s} + \sum_{s=0}^{m-1} C(n+1+s,k-s).$$

PROOF. By Theorem 2.5, the second case in (2.11) can be re-written as follows:

$$C_m(n+1,k) = \sum_{s=0}^{\min(n,k)} C(n,s)2^{k-s} - \sum_{s=0}^{k-m} C(n+m,s)2^{k-m-s} \quad \text{if } m \le k \le n+m.$$

On the other hand, for $m \le k \le n+m$, we use (2.2) to get

$$\binom{n+1+k}{k} - \binom{n+m+1+k-m}{k-m} = \sum_{s=0}^{m-1} C(n+1+s, k-s).$$

Thus we obtain

$$\sum_{s=0}^{\min(n,k)} C(n,s)2^{k-s} - \sum_{s=0}^{k-m} C(n+m,s)2^{k-m-s} = \sum_{s=0}^{m-1} C(n+1+s,k-s)$$

for $m \le k \le n+m$. This completes the proof. □

By specializing (2.12) at $m = 1$, we obtain different expressions of a binomial coefficient $\binom{n+k+1}{k} = \binom{n+1+k}{k}$ as a 2-power weighted sum of the Catalan triangle along the $n+1^{\text{st}}$ row. (cf. (2.8))

COROLLARY 2.9. *We have the following identities: For $k \ge 1$,*

$$\begin{aligned} \binom{n+1+k}{k} &= \sum_{s=0}^{k} C(n,s)2^{k-s} = \sum_{s=0}^{k-1} C(n+1,s)2^{k-1-s} + C(n+1,k) \\ &= \sum_{s=0}^{k} C(n+1,s)2^{\max(k-1-s,0)}. \end{aligned} \tag{2.13}$$

Note that, combining Theorem 2.3 and Corollary 2.9, we have proven Theorem 1.1.

REMARK 2.10. The identities in Corollary 2.9 can be interpreted combinatorially, and one can prove them bijectively. As it reveals combinatorics behind the identities, we sketch a bijective proof of the first identity $\binom{n+1+k}{k} = \sum_{s=0}^{k} C(n,s)2^{k-s}$.

BIJECTIVE PROOF.[1] We interpret $\binom{n+1+k}{k}$ as the number of the lattice paths from $(0,0)$ to $(n+1+k, n+1-k)$ with up-steps $(1,1)$ and down-steps $(1,-1)$. For such a path λ, let $P_0, P_1, \dots, P_s$ be the intersections of λ with the x-axis, decomposing λ into $s+1$ parts $D_0, D_1, \dots, D_s$. Then the last part D_s of λ from P_s to the end point $(n+1+k, n+1-k)$ is a Dyck path, i.e. it stays at or above the x-axis.

We remove from each of D_i, $i = 0, \dots, s-1$, the first step and the last step and from D_s the first step, and flip them, if necessary, to obtain Dyck paths, and concatenate them, putting a step $(1,1)$, between them. The resulting path is a Dyck path from $(0,0)$ to $(n+k-s, n-k+s)$. This gives a 2^s-to-1 map from the

[1] This proof was communicated to us by Jang Soo Kim. We thank him for allowing us to use his proof.

set of the lattice paths from $(0,0)$ to $(n+1+k, n+1-k)$ to the set of the Dyck paths from $(0,0)$ to $(n+k-s, n-k+s)$. Since $C(n, k-s)$ is equal to the number of the Dyck paths from $(0,0)$ to $(n+k-s, n-k+s)$, we have proven

$$\binom{n+1+k}{k} = \sum_{s=0}^{k} C(n, k-s)2^s = \sum_{s=0}^{k} C(n,s)2^{k-s}. \qquad \square$$

2.3. Catalan triangle polynomials. The identities in the previous subsections naturally give rise to the following definition.

DEFINITION 2.11. For $0 \le k \le n$, we define the $(n,k)^{th}$ *Catalan triangle polynomial* $\mathfrak{F}_{n,k}(x)$ by

$$\mathfrak{F}_{n,k}(x) = \sum_{s=0}^{k} C(n,s)x^{k-s} = \sum_{s=0}^{k-1} C(n,s)x^{k-s} + C(n,k). \tag{2.14}$$

Note that the degree of $\mathfrak{F}_{n,k}$ is k.

Evaluations of $\mathfrak{F}_{n,k}(x)$ at the first three nonnegative integers are as follows:

- $\mathfrak{F}_{n,k}(0) = C(n,k)$
- $\mathfrak{F}_{n,k}(1) = C(n+1,k) = C_2(n,k)$
- $\mathfrak{F}_{n,k}(2) = \binom{n+k+1}{k} = C_n(n+1,k)$

Clearly,

$$\binom{n}{k} = \binom{n-1}{k-1} + \binom{n-1}{k}, \quad C(n,k) = C(n,k-1) + C(n-1,k),$$
$$\mathfrak{F}_{n,k}(d) = \mathfrak{F}_{n,k-1}(d) + \mathfrak{F}_{n-1,k}(d) \quad \text{for any } d \in \mathbb{Z}.$$

Let us recall the description of $C(n,k)$ in terms of binomial coefficients:

$$C(n,k) = 2\binom{n+k}{k} - (2-1)\binom{n+k+1}{k}. \tag{2.15}$$

THEOREM 2.12. *For any $d \in \mathbb{Z}_{\ge 1}$, we have*

$$C(n,k) = d\,\mathfrak{F}_{n-1,k}(d) - (d-1)\mathfrak{F}_{n,k}(d)$$

which recovers (2.15) *when $d = 2$.*

PROOF. Since $\mathfrak{F}_{n,k}(d) = \sum_{s=0}^{k-1} C(n,s)d^{k-s} + C(n,k)$, we have

$$\frac{\mathfrak{F}_{n,k}(d) - C(n,k)}{d} = \sum_{s=1}^{k-1} C(n,s-1)d^{k-s}.$$

Hence the equation (2.1) and the fact that $C(n,0) = C(n-1,0) = 1$ yield

$$\mathfrak{F}_{n,k}(d) - \frac{\mathfrak{F}_{n,k}(d) - C(n,k)}{d} = \mathfrak{F}_{n-1,k}(d). \tag{2.16}$$

By multiplying (2.16) by d, our assertion follows. $\square$

With the consideration of Corollary 2.9, we define a natural variation of $\mathfrak{F}_{n,k}(x)$.

DEFINITION 2.13. For $0 \leq k \leq n$, we define the *modified* $(n,k)^{th}$ *Catalan triangle polynomial* in the following way:

$$\widetilde{\mathfrak{F}}_{n,k}(x) = \sum_{s=0}^{k} C(n+1,s)\, x^{\max(k-1-s,0)}. \tag{2.17}$$

Note that the degree of $\widetilde{\mathfrak{F}}_{n,k}(x)$ is $k-1$.

Evaluations of $\widetilde{\mathfrak{F}}_{n,k}(x)$ at the first three nonnegative integers are as follows:

- $\widetilde{\mathfrak{F}}_{n,k}(0) = C(n+1,k-1) + C(n+1,k)$
- $\widetilde{\mathfrak{F}}_{n,k}(1) = C(n+2,k)$
- $\widetilde{\mathfrak{F}}_{n,k}(2) = \binom{n+k+1}{k}$

Similarly,

$$\widetilde{\mathfrak{F}}_{n,k}(d) = \widetilde{\mathfrak{F}}_{n,k-1}(d) + \widetilde{\mathfrak{F}}_{n-1,k}(d) \quad \text{for any } d \in \mathbb{Z}.$$

We have the same result in Theorem 2.12 for $\widetilde{\mathfrak{F}}_{n,k}(d)$ also:

THEOREM 2.14. *For any* $d \in \mathbb{Z}_{\geq 1}$, *we have*

$$C(n,k) = d\widetilde{\mathfrak{F}}_{n-1,k}(d) - (d-1)\widetilde{\mathfrak{F}}_{n,k}(d)$$

which recovers (2.15) *when* $d = 2$.

Let σ_n be the number of n-celled *stacked directed animals* in a square lattice. See [**1**] for definitions. The sequence $(\sigma_n)_{n\geq 0}$ is listed as *A059714* in OEIS. We conjecture

$$\sigma_n = \widetilde{\mathfrak{F}}_{n,n}(3) \quad \text{for } n \geq 0.$$

We expect that one can find a direct, combinatorial proof of this conjecture. A list of the numbers σ_n is below, and the conjecture is easily verified for the numbers in the list:

$$\sigma_0 = 1,\ \sigma_1 = 3,\ \sigma_2 = 11,\ \sigma_3 = 44,\ \sigma_4 = 184,\ \sigma_5 = 789,\ \sigma_6 = 3435,$$
$$\sigma_7 = 15100,\ \sigma_8 = 66806,\ \sigma_9 = 296870,\ \sigma_{10} = 1323318,\ \sigma_{11} = 5911972.$$

For example, when $n = 7$, we have the sequence $(C(8,k),\ 0 \leq k \leq 7)$ equal to

$$(1,\ 8,\ 35,\ 110,\ 275,\ 572,\ 1001,\ 1430),$$

and compute

$$\widetilde{\mathfrak{F}}_{7,7}(3) = 3^6 + 8\cdot 3^5 + 35\cdot 3^4 + 110\cdot 3^3 + 275\cdot 3^2 + 572\cdot 3 + 1001 + 1430 = 15100.$$

3. Alternating Jacobsthal triangle

In the previous section, the binomial coefficient $\binom{n+k+1}{k}$ is written as sums along the nth and the $n+1^{\text{st}}$ row of the Catalan triangle, respectively. In this section, we consider other rows of the Catalan triangle as well and obtain a more general result. In particular, the $n+k^{\text{th}}$ row will produce a canonical sequence of numbers, which form the alternating Jacobsthal triangle. We study some subsequences of the triangle and their generating functions in the subsections.

Define $A(m,t) \in \mathbb{Z}$ recursively for $m \geq t \geq 0$ by

$$A(m,0) = 1, \quad A(m,t) = A(m-1,t-1) - A(m-1,t). \tag{3.1}$$

Here we set $A(m,t) = 0$ when $t > m$. Then, by induction on m, one can see that

$$\sum_{t=1}^{m} A(m,t) = 1 \quad \text{and} \quad A(m,m) = 1.$$

Using the numbers $A(m,t)$, we prove the following theorem which is a generalization of the identities in Corollary 2.9:

THEOREM 3.1. *For any $n > k \geq m \geq t \geq 1$, we have*

$$\binom{n+k+1}{k} = \sum_{s=0}^{k-m} C(n+m,s)2^{k-m-s} + \sum_{t=1}^{m} A(m,t)C(n+m,k-m+t). \tag{3.2}$$

PROOF. We will use an induction on $m \in \mathbb{Z}_{\geq 1}$. For $m = 1$, we already proved the identity in Corollary 2.9. Assume that we have the identity (3.2) for some $m \in \mathbb{Z}_{\geq 1}$. By specializing (2.12) at $m+1$, we have

$$\begin{aligned}\binom{n+k+1}{k} &= \sum_{s=0}^{k-m-1} C(n+m+1,s)2^{k-m-1-s} + \sum_{s=0}^{m} C(n+1+s,k-s) \\ &= \sum_{s=0}^{k-m-1} C(n+m+1,s)2^{k-m-1-s} + \sum_{s=0}^{m-1} C(n+1+s,k-s) \\ &\quad + C(n+m+1,k-m).\end{aligned}$$

By the induction hypothesis applied to (2.12), we have

$$\sum_{s=0}^{m-1} C(n+1+s,k-s) = \sum_{t=1}^{m} A(m,t)C(n+m,k-m+t).$$

Then our assertion follows from the fact that

$$C(n+m,k-m+t) = C(n+m+1,k-m+t) - C(n+m+1,k-m+t-1). \quad \square$$

We obtain the triangle consisting of $A(m,t)$ $(m \geq t \geq 0)$:

$$\begin{array}{cccccccccc}
1 \\
1 & 1 \\
1 & 0 & 1 \\
1 & 1 & -1 & 1 \\
1 & 0 & 2 & -2 & 1 \\
1 & 1 & -2 & 4 & -3 & 1 \\
1 & 0 & 3 & -6 & 7 & -4 & 1 \\
1 & 1 & -3 & 9 & -13 & 11 & -5 & 1 \\
1 & 0 & 4 & -12 & 22 & -24 & 16 & -6 & 1 \\
\vdots & \vdots & \vdots & \vdots & \vdots & \vdots & \vdots & \vdots & \vdots & \ddots
\end{array} \tag{3.3}$$

The triangle in (3.3) will be called the *alternating Jacobsthal triangle*. The 0th column is colored in blue (online) to indicate the fact that some formulas do not take entries from this column.

EXAMPLE 3.2. For $m = 3$, we have

$$\binom{n+k+1}{k} = \sum_{s=0}^{k-3} C(n+3,s)2^{k-3-s} + C(n+3,k-2) - C(n+3,k-1) + C(n+3,k).$$

By specializing (3.2) at $m = k$, the kth row of alternating Jacobsthal triangle and $n+k^{\text{th}}$ row of Catalan triangle yield the binomial coefficient $\binom{n+k+1}{k}$:

COROLLARY 3.3. *For any $n > k$, we have*

$$\binom{n+k+1}{k} = \sum_{t=0}^{k} A(k,t)\, C(n+k,t). \tag{3.4}$$

EXAMPLE 3.4.

$$\begin{aligned}\binom{8}{3} &= A(3,0)C(7,0) + A(3,1)C(7,1) + A(3,2)C(7,2) + A(3,3)C(7,3)\\ &= 1 \times 1 + 1 \times 7 - 1 \times 27 + 1 \times 75 = 56.\end{aligned}$$

$$\begin{aligned}\binom{9}{3} &= A(3,0)C(8,0) + A(3,1)C(8,1) + A(3,2)C(8,2) + A(3,3)C(8,3)\\ &= 1 \times 1 + 1 \times 8 - 1 \times 35 + 1 \times 110 = 84.\end{aligned}$$

$$\begin{aligned}\binom{9}{4} &= A(4,0)C(8,0) + A(4,1)C(8,1) + A(4,2)C(8,2) + A(4,3)C(8,3) + A(4,3)C(8,3)\\ &= 1 \times 1 + 0 \times 8 + 2 \times 35 - 2 \times 110 + 1 \times 275 = 126.\end{aligned}$$

3.1. Generating function. The numbers $A(m,t)$ can be encoded into a generating function in a standard way. Indeed, from (3.1), we obtain

$$\begin{aligned} A(m,t) &= \sum_{k=t-1}^{m-1} (-1)^{m-1-k} A(k,t-1) \\ &= A(m-1,t-1) - A(m-2,t-1) - \cdots + (-1)^{m-t} A(t-1,t-1). \end{aligned} \tag{3.5}$$

LEMMA 3.5. *We have*

$$\frac{1}{(1-x)(1+x)^t} = \sum_{m=t}^{\infty} A(m,t) x^{m-t}. \tag{3.6}$$

PROOF. When $t = 0$, we have $\dfrac{1}{1-x} = 1 + x + x^2 + \cdots = \displaystyle\sum_{m=0}^{\infty} A(m,0)x^m$. Inductively, when $t > 0$, we have

$$\frac{1}{(1-x)(1+x)^t} = \frac{1}{(1+x)} \frac{1}{(1-x)(1+x)^{t-1}} = (1 - x + x^2 - \cdots) \sum_{m=t-1}^{\infty} A(m,t-1)x^{m-t+1}.$$

Then we obtain (3.6) from (3.5). □

3.2. Subsequences. The alternating Jacobsthal triangle has various subsequences with interesting combinatorial interpretations.

First, we write

$$\frac{1}{(1-x)(1+x)^t} = \sum_{m \geq 0} a_{m+1,t} x^{2m} - \sum_{m \geq 0} b_{m+1,t} x^{2m+1}$$

to define the subsequences $\{a_{m,t}\}$ and $\{b_{m,t}\}$ of $\{A(m,t)\}$. Then we have

$$a_{m,t} = A(t+2m-2,t) \quad \text{and} \quad b_{m,t} = -A(t+2m-1,t).$$

Clearly, $a_{m,t}, b_{m,t} \geq 0$. Using (3.5), we obtain

$$a_{m,t} = \sum_{k=1}^{m} a_{k,t-1} + \sum_{k=1}^{m-1} b_{k,t-1} \quad \text{and} \quad b_{m,t} = \sum_{k=1}^{m} b_{k,t-1} + \sum_{k=1}^{m} a_{k,t-1}.$$

It is easy to see that $a_{n,2} = n$ and $b_{n,2} = n$. Then we have

$$a_{n,3} = \sum_{k=1}^{n} a_{k,2} + \sum_{k=1}^{n-1} b_{k,2} = \frac{n(n+1)}{2} + \frac{n(n-1)}{2} = n^2.$$

Similarly,

$$b_{n,3} = \sum_{k=1}^{n} b_{k,2} + \sum_{k=1}^{n} a_{k,2} = \frac{n(n+1)}{2} + \frac{n(n+1)}{2} = n(n+1).$$

We compute more and obtain

$$a_{n,4} = \frac{n(n+1)(4n-1)}{6}, \qquad b_{n,4} = \frac{n(n+1)(4n+5)}{6},$$
$$a_{n,5} = \frac{n(n+1)(2n^2+2n-1)}{6}, \qquad b_{n,5} = \frac{n(n+1)^2(n+2)}{3}.$$

Note also that

$$\frac{1}{(1+x)(1-x)^t} = \sum_{m\geq 0} a_{m+1,t}x^{2m} + \sum_{m\geq 0} b_{m+1,t}x^{2m+1}.$$

Next, we define

$$B(m,t) = A(m,m-t) \qquad \text{for } m \geq t$$

to obtain the triangle

$$\begin{array}{ccccccccc}
1 & & & & & & & & \\
1 & 1 & & & & & & & \\
1 & 0 & 1 & & & & & & \\
1 & -1 & 1 & 1 & & & & & \\
1 & -2 & 2 & 0 & 1 & & & & \\
1 & -3 & 4 & -2 & 1 & 1 & & & \\
1 & -4 & 7 & -6 & 3 & 0 & 1 & & \\
1 & -5 & 11 & -13 & 9 & -3 & 1 & 1 & \\
1 & -6 & 16 & -24 & 22 & -12 & 4 & 0 & 1 \\
\vdots & \vdots & \vdots & \vdots & \vdots & \vdots & \vdots & \vdots & \vdots & \ddots
\end{array} \tag{3.7}$$

LEMMA 3.6. *For $m \geq t$, we have*

$$B(m,t) = 1 - \sum_{k=t}^{m-1} B(k,t-1).$$

PROOF. We use induction on m. When $m = t$, we have $B(m,t) = A(m,0) = 1$. Assume that the identity is true for some $m \geq t$. Since we have

$$\begin{aligned} B(m+1,t) &= A(m+1,m+1-t) = A(m,m-t) - A(m,m+1-t) \\ &= B(m,t) - B(m,t-1), \end{aligned}$$

we obtain

$$B(m+1,t)+\sum_{k=t}^{m} B(k,t-1) = B(m,t) - B(m,t-1) + \sum_{k=t}^{m} B(k,t-1)$$
$$= 1 - \sum_{k=t}^{m-1} B(k,t-1) - B(m,t-1) + \sum_{k=t}^{m} B(k,t-1) = 1$$

by the induction hypothesis. □

Using Lemma 3.6, one can derive the following formulas:

- $B(n,0) = A(n,n) = 1$ and $B(n,1) = A(n,n-1) = 2-n$,
- $B(n,2) = A(n,n-2) = 4 + \dfrac{n(n-5)}{2}$,
- $B(n,3) = A(n,n-3) = 8 - \dfrac{n(n^2-9n+32)}{6}$.

We consider the columns of the triangle (3.7) and let $c_{m,t} = (-1)^t B(m+t+1,t)$ for each t for convenience. Then the sequences $(c_{m,t})_{m\geq 1}$ for the first several t's appear in the OEIS. Specifically, we have:

- $(c_{m,2}) = (2,4,7,11,16,22,\dots)$ corresponds to *A*000124,
- $(c_{m,3}) = (2,6,13,24,40,62,\dots)$ corresponds to *A*003600,
- $(c_{m,4}) = (3,9,22,46,86,148,\dots)$ corresponds to *A*223718,
- $(c_{m,5}) = (3,12,34,80,166,314,\dots)$ corresponds to *A*257890,
- $(c_{m,6}) = (4,16,50,130,296,610,\dots)$ corresponds to *A*223659.

3.3. Diagonal sums. As we will see in this subsection, the sums along lines of slope 1 in the alternating Jacobsthal triangle are closely related to Fibonacci numbers. We begin with fixing a notation. For $s \geq 0$, define

$$B_s = \sum_{t+m-2=s,\, t>0} A(m,t).$$

Using the generating function (3.6), we have

$$F(x) := \sum_{t=1}^{\infty} \frac{x^{2t-2}}{(1-x)(1+x)^t} = \sum_{s=0}^{\infty} B_s x^s.$$

Then we obtain

$$(1-x)x^2F(x) = \sum_{t=1}^{\infty}\left(\frac{x^2}{1+x}\right)^t = \frac{x^2}{1+x-x^2}$$

and the formula

$$F(x) = \frac{1}{(1-x)(1+x-x^2)}. \tag{3.8}$$

It is known that the function $F(x)$ is the generating function of the sequence of the alternating sums of the Fibonacci numbers; precisely, we get

$$B_s = \sum_{k=1}^{s+1} (-1)^{k-1}\operatorname{Fib}(k) = 1 + (-1)^s \operatorname{Fib}(s) \quad (s \geq 0), \tag{3.9}$$

where $(\operatorname{Fib}(s))_{s\geq 0}$ is the Fibonacci sequence. (See A119282 in OEIS.) From the construction, the following is obvious:

$$B_{s+1} = -B_s + B_{s-1} + 1 \ (s \geq 1) \quad \text{and} \quad B_0 = 1,\ B_1 = 0.$$

4. q-deformation

In this section, we study a q-deformation of the Fibonacci and Jacobsthal numbers by putting the parameter q into the identities and generating functions we obtained in the previous section. We also obtain a family of generating functions of certain sequences by expanding the q-deformation of the generating function of the numbers $A(m,t)$ in terms of q.

4.1. q-Fibonacci numbers. For $s \geq 0$, define

$$B_s(q) = \sum_{t+m-2=s,\, t>0} A(m,t)q^{m-t} \in \mathbb{Z}[q].$$

Then we obtain

$$\begin{aligned} F(x,q) &:= \sum_{t=1}^{\infty} \frac{x^{2t-2}}{(1-qx)(1+qx)^t} \\ &= \sum_{s=0}^{\infty} B_s(q)x^s = \frac{1}{(1-qx)(1+qx-x^2)}. \end{aligned}$$

Note that

$$B_{s+1}(q) = -qB_s(q) + B_{s-1}(q) + q^s \ (s \geq 1) \quad \text{and} \quad B_0(q) = 1,\ B_1(q) = 0.$$

Motivated by (3.9), we define a q-analogue of Fibonacci number by

$$\widetilde{B}_s(q) := (-1)^s B_s(q) + (-1)^{s+1}q^s = \sum_{t+m-2=s,\, t>0} |A(m,t)|q^{m-t} + (-1)^{s+1}q^s.$$

In particular, we have

$$\widetilde{B}_1(q) = q,\ \widetilde{B}_2(q) = 1,\ \widetilde{B}_3(q) = q^3 + q,\ \widetilde{B}_4(q) = 2q^2 + 1,\ \widetilde{B}_5(q) = q^5 + 2q^3 + 2q,$$
$$\widetilde{B}_6(q) = 3q^4 + 4q^2 + 1,\ \widetilde{B}_7(q) = q^7 + 3q^5 + 6q^3 + 3q,\ \widetilde{B}_8(q) = 4q^6 + 9q^4 + 7q^2 + 1.$$

These polynomials can be readily read off from the alternating Jacobsthal triangle (3.3). We observe that $\widetilde{B}_{2s}(q) \in \mathbb{Z}_{\geq 0}[q^2]$ and $\widetilde{B}_{2s+1}(q) \in \mathbb{Z}_{\geq 0}[q^2]q$ and that $\widetilde{B}_s(q)$ is weakly unimodal.

Note that we have

$$F(x,q) - \frac{1}{1-qx} = \frac{1}{(1-qx)(1+qx-x^2)} - \frac{1}{1-qx} = \frac{x^2-qx}{(1-qx)(1+qx-x^2)}$$
$$= \sum_{s=0}^{\infty} B_s(q)x^s - \sum_{s=0}^{\infty}(qx)^s = \sum_{s=0}^{\infty} \widetilde{B}_s(q)(-x)^s.$$

Thus the generating function $CF(x,q)$ of $\widetilde{B}_s(q)$ is given by

$$CF(x,q) := \sum_{s=0}^{\infty} \widetilde{B}_s(q)x^s = \frac{x^2+qx}{(1+qx)(1-qx-x^2)}. \tag{4.1}$$

REMARK 4.1. The well-known Fibonacci polynomial $\mathcal{F}_s(q)$ can be considered as a different q-Fibonacci number whose generating function is given by

$$\sum_{t=1}^{\infty} \frac{x^{2t-2}}{(1-qx)^t} = \frac{1}{1-qx-x^2} = \sum_{s=0}^{\infty} \mathcal{F}_s(q)x^s.$$

Recall that the polynomial $\mathcal{F}_s(q)$ can be read off from the Pascal triangle. When $q = 2$, the number $\mathcal{F}_s(2)$ is nothing but the sth Pell number. On the other hand, it does not appear that the sequence

$$(\widetilde{B}_s(2))_{s\geq 1} = (2, 1, 10, 9, 52, 65, 278, 429, 1520, \dots)$$

has been studied in the literature.

4.2. q-Jacobsthal numbers. Recall that the Jacobsthal numbers J_m are defined recursively by $J_m = J_{m-1} + 2J_{m-2}$ with $J_1 = 1$ and $J_2 = 1$. Then the Jacobsthal sequence (J_m) is given by

$$(1, 1, 3, 5, 11, 21, 43, 85, 171, \dots).$$

Consider the function

$$Q(x,q) := \sum_{t=1}^{\infty} \frac{x^t}{(1-qx)(1+qx)^t}. \tag{4.2}$$

Define

$$H_m(q) := \sum_{t=1}^{m} A(m,t) q^{m-t} \qquad \text{for } m \geq 1.$$

For example, we can read off

$$H_5(q) = q^4 - 2q^3 + 4q^2 - 3q + 1$$

from the alternating Jacobsthal triangle (3.3). Using (3.6), we obtain

$$\begin{aligned} Q(x,q) &= \sum_{t=1}^{\infty}\sum_{m=t}^{\infty} A(m,t) q^{m-t} x^m \\ &= \sum_{m=1}^{\infty}\sum_{t=1}^{m} A(m,t) q^{m-t} x^m = \sum_{m=1}^{\infty} H_m(q) x^m. \end{aligned}$$

A standard computation also yields

$$Q(x,q) = \frac{x}{(1-qx)(1+(q-1)x)}. \tag{4.3}$$

By taking $q = 0$ or $q = 1$, the equation (4.3) becomes $\dfrac{x}{1-x}$. On the other hand, by taking $q = -1$, the equation (4.3) becomes

$$\frac{x}{(1+x)(1-2x)},$$

which is the generating function of the Jacobsthal numbers J_m. That is, we have

- $H_m(0) = H_m(1) = 1$ for all m,
- $H_m(-1)$ is the mth Jacobsthal number J_m for each m.

Since $J_m = H_m(-1) = \sum_{t=1}^{m} |A(m,t)|$, we see that an alternating sum of the entries along a row of the triangle (3.3) is equal to a Jacobsthal number.

Moreover, we have a natural q-deformation $J_m(q)$ of the Jacobsthal number J_m, which is defined by

$$J_m(q) := H_m(-q) = \sum_{t=1}^{m} |A(m,t)| q^{m-t}.$$

For example, we have

$$J_3(q) = q^2 + q + 1, \quad J_4(q) = 2q^2 + 2q + 1, \quad J_5(q) = q^4 + 2q^3 + 4q^2 + 3q + 1.$$

Note that $J_m(q)$ is weakly unimodal. We also obtain

$$Q(x,-q) = \frac{x}{(1+qx)(1-(q+1)x)} = \sum_{m=1}^{\infty} J_m(q)x^m.$$

The following identity is well-known ([**4, 5**]):

$$J_m = \sum_{r=0}^{\lfloor (m-1)/2 \rfloor} \binom{m-r-1}{r} 2^r.$$

Hence we have

$$J_m = \sum_{r=0}^{\lfloor (m-1)/2 \rfloor} \binom{m-r-1}{r} 2^r = \sum_{t=1}^{m} |A(m,t)| = H_m(-1) = J_m(1).$$

REMARK 4.2. In the literature, one can find different Jacobsthal polynomials. See [**7**], for example.

4.3. A family of generating functions. Now let us expand $Q(x,q)$ with respect to q to define the functions $L_\ell(x)$:

$$Q(x,q) = \sum_{t=1}^{\infty} \frac{x^t}{(1-qx)(1+qx)^t} = \sum_{\ell=0}^{\infty} L_\ell(x)q^\ell.$$

LEMMA 4.3. *For $\ell \geq 0$, we have*

$$L_{\ell+1}(x) = \frac{-x}{1-x}L_\ell(x) + \frac{x^{\ell+2}}{1-x}.$$

PROOF. Clearly, we have $L_0(x) = \sum_{n=1}^{\infty} x^n = \frac{x}{1-x}$. We see that

$$\frac{x}{1-x} + \sum_{\ell=0}^{\infty} L_{\ell+1}(x)q^{\ell+1} = Q(x,q) = \frac{x}{(1-qx)(1+(q-1)x)}.$$

On the other hand, we obtain

$$\begin{aligned}
&\frac{x}{1-x} + \sum_{\ell=0}^{\infty} \left\{ \frac{-x}{1-x}L_\ell(x)q^{\ell+1} + \frac{x}{1-x}(qx)^{\ell+1} \right\} \\
&= \frac{x}{1-x} - \frac{qx}{1-x} \cdot \frac{x}{(1-qx)(1+(q-1)x)} + \frac{x}{1-x} \cdot \frac{qx}{1-qx} \\
&= \frac{x}{(1-qx)(1+(q-1)x)} = Q(x,q).
\end{aligned}$$

This completes the proof. □

Using Lemma 4.3, we can compute first several $L_\ell(x)$:

- $L_0(x) = \sum_{n=1}^{\infty} x^n = \frac{x}{1-x}$,
- $L_1(x) = \frac{-x^2}{(1-x)^2} + \frac{x^2}{1-x} = \frac{-x^3}{(1-x)^2} = -\sum_{n=1}^{\infty} nx^{n+2}$,
- $L_2(x) = \frac{x^4}{(1-x)^3} + \frac{x^3}{1-x} = \frac{x^3(1-x+x^2)}{(1-x)^3}$.

- $L_3(x) = -\frac{x^4(1-x+x^2)}{(1-x)^4} + \frac{x^4}{1-x} = -\frac{x^5(2-2x+x^2)}{(1-x)^4}$.

One can check that $L_2(x)$ is the generating function of the sequence $A000124$ in OEIS and that $L_3(x)$ is the generating function of the sequence $A003600$. Note that the lowest degree of $L_\ell(x)$ in the power series expansion is larger than or equal to $\ell+1$. More precisely, the lowest degree of $L_\ell(x)$ is $\ell+1+\delta(\ell \equiv 1 (\text{mod } 2))$.

5. k-analogue of q-deformation

In this section, we consider k-analogues of the q-deformations we introduced in the previous section. This construction, in particular, leads to a k-analogue of the alternating Jacobsthal triangle for each $k \in \mathbb{Z} \setminus \{0\}$. Specializations of this construction at some values of k and q produce interesting combinatorial sequences.

Define $A_k(m,t)$ by

$$A_k(m,0) = k^{\lfloor m/2 \rfloor} \quad \text{and} \quad A_k(m,t) = A_k(m-1,t-1) - A_k(m-1,t).$$

Then we have

$$\frac{1}{(1-kx^2)(1+x)^{t-1}} = \sum_{m=t}^{\infty} A_k(m,t) x^{m-t}$$

in the same way as we obtained (3.6). As in Section 4.2, we also define

$$H_{k,m}(q) = \sum_{t=1}^{m} A_k(m,t) q^{m-k}.$$

We obtain the generating function $Q_k(x,q)$ of $H_{k,m}(q)$ by

$$Q_k(x,q) := \sum_{t=1}^{\infty} \frac{x^t}{(1-kq^2x^2)(1+qx)^{t-1}} = \frac{x(1+qx)}{(1-kq^2x^2)(1+(q-1)x)} = \sum_{m=1}^{\infty} H_{k,m}(q) x^m.$$

In particular, when $q=1$, we have

$$Q_k(x,1) = \sum_{t=1}^{\infty} \frac{x^t}{(1-kx^2)(1+x)^{t-1}} = \frac{x(1+x)}{1-kx^2} = \sum_{m=1}^{\infty} k^{m-1}\left(x^{2m-1} + x^{2m}\right).$$

Note that $H_{k,m}(1) = k^{\lfloor (m-1)/2 \rfloor}$.

Moreover, the triangle given by the numbers $A_k(m,t)$ can be considered as a k-analogue of the alternating Jacobsthal triangle (3.3). See the triangles (5.1) and (5.2). Thus we obtain infinitely many triangles as k varies in $\mathbb{Z} \setminus \{0\}$. Similarly, we define a k-analogue of the q-Jacobsthal number by

$$J_{k,m}(q) := H_{k,m}(-q),$$

and the number $J_{k,m}(1) = \sum_{t=1}^{m} |A_k(m,t)|$ can be considered as the k-analogue of the mth Jacobsthal number.

For example, if we take $k = 2$, the polynomial $H_{2,m}(q)$ can be read off from the following triangle consisting of $A_2(m,t)$:

$$
\begin{array}{ccccccccccccc}
1 \\
1 & 1 \\
2 & 0 & 1 \\
2 & 2 & -1 & 1 \\
4 & 0 & 3 & -2 & 1 \\
4 & 4 & -3 & 5 & -3 & 1 \\
8 & 0 & 7 & -8 & 8 & -4 & 1 \\
8 & 8 & -7 & 15 & -16 & 12 & -5 & 1 \\
16 & 0 & 15 & -22 & 31 & -28 & 17 & -6 & 1 \\
16 & 16 & -15 & 37 & -53 & 59 & -45 & 23 & -7 & 1 \\
32 & 0 & 31 & -52 & 90 & -112 & 104 & -68 & 30 & -8 & 1 \\
\vdots & \vdots & \vdots & \vdots & \vdots & \vdots & \vdots & \vdots & \vdots & \vdots & \vdots & \ddots
\end{array}
\tag{5.1}
$$

We have $J_{2,m}(1) = \sum_{t=1}^{m} |A_2(m,t)|$, and the sequence

$$(J_{2,m}(1))_{m\geq 1} = (1, 1, 4, 6, 16, 28, 64, 120, \dots)$$

appears as *A*007179 in OEIS. As mentioned in the introduction, this sequence has the interpretation as the numbers of equal dual pairs of some integrals studied in [**3**]. (See Table 1 on p.365 in [**3**].)

Define $B_k(m,t) = A_k(m, m-t)$. Then we obtain the following sequences from (5.1) which appear in OEIS:

- $(B_2(m,2))_{m\geq 3} = (2, 3, 5, 8, 12, 17, 23, 30, \dots) \leftrightarrow$ *A*002856, *A*152948,
- $(-B_2(m,3))_{m\geq 5} = (3, 8, 16, 28, 45, 68, \dots) \leftrightarrow$ *A*254875.

We also consider diagonal sums and find

- the positive diagonals

$$\left(\sum_{m+t=2s,\, t>0} A_2(m,t) \right)_{s\geq 1}$$

  corresponds to

$$(1, 3, 8, 21, 55, 144, 377, \dots) \leftrightarrow (\mathrm{Fib}(2s)),$$

  where $\mathrm{Fib}(s)$ is the Fibonacci number;
- the negative diagonals

$$\left(- \sum_{m+k=2s+1,\, t>0} A_2(m,k) \right)_{s\geq 1}$$

  corresponds to

$$(0, 1, 5, 18, 57, 169, \dots) \leftrightarrow A258109,$$

  whose s^{th} entry is the number of Dyck paths of length $2(s+1)$ and height 3.

Similarly, when $k=-1$, we obtain the following triangle consisting of $A_{-1}(m,t)$:

$$
\begin{array}{cccccccccccc}
1 \\
1 & 1 \\
-1 & 0 & 1 \\
-1 & -1 & -1 & 1 \\
1 & 0 & 0 & -2 & 1 \\
1 & 1 & 0 & 2 & -3 & 1 \\
-1 & 0 & 1 & -2 & 5 & -4 & 1 \\
-1 & -1 & -1 & 3 & -7 & 9 & -5 & 1 \\
1 & 0 & 0 & -4 & 10 & -16 & 14 & -6 & 1 \\
1 & 1 & 0 & 4 & -14 & 26 & -30 & 20 & -7 & 1 \\
-1 & 0 & 1 & -4 & 18 & -40 & 56 & -50 & 27 & -8 & 1 \\
\vdots & \vdots & \vdots & \vdots & \vdots & \vdots & \vdots & \vdots & \vdots & \vdots & \vdots & \ddots
\end{array}
\tag{5.2}
$$

We find some meaningful subsequences of this triangle and list them below.

- $(|A_{-1}(m,4)|)_{m\geq 4}=(1,3,5,7,10,14,18,22,\dots)=(\lfloor\binom{n}{2}/2\rfloor)_{n\geq 3}\leftrightarrow$ *A011848*,
- $(B_{-1}(m,2))_{m\geq 5}=(2,5,9,14,20,27,\dots)\leftrightarrow$ *A212342*,
- $(-B_{-1}(m,3))_{m\geq 6}=(2,7,16,30,50,77,\dots)\leftrightarrow$ *A005581*,
- $\left(\sum_{m+t=2s,\, t\geq 2} A_{-1}(m,t)\right)_{s\geq 2}=(1,1,4,9,25,64,169,441,\dots)=(\mathrm{Fib}(n)^2)_{n\geq 1}$,
- $\left(\sum_{t=2}|A_{-1}(m,t)|\right)_{m\geq 2}=(1,2,3,6,13,26,51,102,\dots)\leftrightarrow$ *A007910*.

Define $B_{k,s}(q)$ and $F_k(x,q)$ by $B_{k,s}(q):=\sum\limits_{t+m-2=s,\, t>0} A_k(m,t)q^{m-t}\in\mathbb{Z}[q]$ and

$$
\begin{aligned}
F_k(x,q) &:= \sum_{t=1}^{\infty}\frac{x^{2t-2}}{(1-kq^2x^2)(1+qx)^{t-1}}=\frac{1+qx}{(1-kq^2x^2)(1+qx-x^2)}\\
&=\sum_{s=1}^{\infty}B_{k,s}(q)x^s.
\end{aligned}
$$

Let us consider the following to define $\widetilde{B}_{k,s}(q)$:

$$
\frac{1+qx}{(1-kq^2x^2)(1+qx-x^2)}-\frac{1+qx}{1-kq^2x^2}=\frac{(1+qx)(-qx+x^2)}{(1-kq^2x^2)(1+qx-x^2)}=\sum_{s=0}^{\infty}\widetilde{B}_{k,s}(q)(-x)^s.
$$

Define a k-analogue $CF_k(x,q)$ of the function $CF(x,q)$ by

$$
CF_k(x,q):=\frac{(1-qx)(qx+x^2)}{(1-kq^2x^2)(1-qx-x^2)}=\sum_{s=0}^{\infty}\widetilde{B}_{k,s}(q)x^s.
$$

The polynomial $\widetilde{B}_{k,s}(q)$ can be considered as a k-analogue of the q-Fibonacci number $\widetilde{B}_s(q)$.

Finally, we define $L_{k,\ell+1}(x)$ by

$$
Q_k(x,q)=\sum_{t=1}^{\infty}\frac{x^t}{(1-kq^2x^2)(1+qx)^{t-1}}=\sum_{\ell=0}^{\infty}L_{k,\ell}(x)q^{\ell}.
$$

Then, using a similar argument as in the proof of Lemma 4.3, one can show that

$$
L_{k,\ell+1}(x)=\frac{-x}{1-x}L_{k,\ell}(x)+\frac{k^{\lfloor(\ell+1)/2\rfloor}x^{\ell+2}}{1-x}.
$$

REMARK 5.1. We can consider the Jacobsthal–Lucas numbers and the Jacobsthal–Lucas polynomials starting with the generating function

$$\frac{1+4x}{(1-x^2)(1-x)^{t-1}},$$

and study their (k-analogue of) q-deformation.

Acknowledgments

We would like to thank Jang Soo Kim for helpful comments. The second-named author would like to thank the faculty and staff of the Department of Mathematics at the University of Oregon, where part of this paper was completed. In particular, he is thankful to Benjamin Young for helpful discussions on topics related to this paper.

References

[1] M. Bousquet-Mélou and A. Rechnitzer, *Lattice animals and heaps of dimers*, Discrete Math. **258** (2002), no. 1-3, 235–274. MR2002764
[2] G. Feinberg and K.-H. Lee, *Fully commutative elements of type D and homogeneous representations of KLR-algebras*, J. Comb. **6** (2015), no. 4, 535–557. MR3382608
[3] J. Heading, *The equality of multiple integrals in the series development of a reflection coefficient for waves governed by a general linear differential equation*, J. Phys. A **14** (1981), no. 11, 2881–2895. MR633458
[4] V. E. Hoggatt Jr. and M. Bicknell-Johnson, *Convolution arrays for Jacobsthal and Fibonacci polynomials*, Fibonacci Quart. **16** (1978), no. 5, 385–402. MR515833
[5] A. F. Horadam, *Jacobsthal representation numbers*, Fibonacci Quart. **34** (1996), no. 1, 40–54. MR1371475
[6] J. S. Kim, K.-H. Lee and S.-j. Oh, *Weight multiplicities and Young tableaux through affine crystals*, arXiv:1703.10321.
[7] T. Koshy, *Fibonacci and Lucas numbers with applications*, Pure and Applied Mathematics (New York), Wiley-Interscience, New York, 2001. MR1855020
[8] L. W. Shapiro, *A Catalan triangle*, Discrete Math. **14** (1976), no. 1, 83–90. MR0387069
[9] J. R. Stembridge, *The enumeration of fully commutative elements of Coxeter groups*, J. Algebraic Combin. **7** (1998), no. 3, 291–320. MR1616016
[10] S. Tsuchioka, *Catalan numbers and level 2 weight structures of* $A^{(1)}_{p-1}$, New trends in combinatorial representation theory, RIMS Kôkyûroku Bessatsu, B11, Res. Inst. Math. Sci. (RIMS), Kyoto, 2009, pp. 145–154. MR2562491
[11] S. Tsuchioka and M. Watanabe, *Pattern avoidance seen in multiplicities of maximal weights of affine Lie algebra representations*, Proc. Amer. Math. Soc. **146** (2018), no. 1, 15–28. MR3723117

DEPARTMENT OF MATHEMATICS, UNIVERSITY OF CONNECTICUT, STORRS, CONNECTICUT 06269
Email address: khlee@math.uconn.edu

DEPARTMENT OF MATHEMATICS, EWHA WOMANS UNIVERSITY, SEOUL, 120-750, SOUTH KOREA
Email address: sejin092@gmail.com

Contemporary Mathematics
Volume **713**, 2018
https://doi.org/10.1090/conm/713/14317

Walk algebras, distinguished subexpressions, and point counting in Kac-Moody flag varieties

Dinakar Muthiah and Daniel Orr

ABSTRACT. We study walk algebras and Hecke algebras for Kac-Moody root systems. Each choice of orientation for the set of real roots gives rise to a corresponding "oriented" basis for each of these algebras. We show that the notion of distinguished subexpression naturally arises when studying the transition matrix between oriented bases. We then relate these notions to the geometry of Kac-Moody flag varieties and Bott-Samelson varieties. In particular, we show that the number of points over a finite field in certain intersections of these varieties is given by change of basis coefficients between oriented bases of the Hecke algebra. Using these results we give streamlined derivations of Deodhar's formula for R-polynomials and point-counting formulas for specializations of nonsymmetric Macdonald polynomials $E_\lambda(\mathsf{q}, t)$ at $\mathsf{q} = 0, \infty$.

1. Introduction

Alcove walk algebras were introduced in [**Ram**] to study the combinatorics of affine Hecke algebras and Hall-Littlewood polynomials. The purpose of this paper is to study walk algebras for general Kac-Moody groups. Roughly speaking, the walk algebra is obtained from the Hecke algebra by forgetting the braid relations and weakening the quadratic relations; thus the Hecke algebra is naturally a quotient of the walk algebra.

For each orientation of the set of real roots we define a basis of the walk algebra indexed by subexpressions. Our first aim is to investigate the combinatorics of these bases. We show that the change-of-basis matrix between oriented bases is controlled by a subset of distinguished subexpressions depending on the chosen pair of orientations (Theorem 4.6). The notion of distinguished subexpression (for certain choices of orientations) originated in the work of Deodhar [**Deo**] and has also played an important role in work of Billig and Dyer [**BD**]. It also appears in the works of Gaussent and Littelmann [**GL**] and Schwer [**Sch**] (as positively folded galleries); and in the works of Ram [**Ram**] and Parkinson, Ram, and Schwer [**PRS**] (as positively folded alcove walks). We demonstrate, generalizing [**Ram**], that the definition of distinguished subexpressions arises naturally as a consequence of the relations in the walk algebra.

2010 *Mathematics Subject Classification.* Primary 20C08.

D.M. was partially supported by a PIMS postdoctoral fellowship.
D.O. was partially supported by NSF grant DMS-1600653.

After establishing the basic result on change of oriented bases in walk algebras, we descend to the Hecke algebra and relate the change of basis coefficients to point counting in Kac-Moody flag varieties (Theorem 5.6). Up to a simple normalization, the change of basis coefficients from the standard orientation to another orientation are given by the number of points over a finite field in certain orbit intersections in the flag variety. This equality is already known as a consequence of [**BD**, Theorem 5], but our approach is different. We mention that oriented bases of Hecke algebras appeared in earlier works of Dyer [**Dy**, **Dy2**]. More recently, work of Gobet [**Go**] establishes the positivity of the expansion coefficients of the Kazhdan-Lusztig basis of the Hecke algebra into any fixed oriented basis.

Our primary goal in this paper is to provide a synthesized exposition of various results and ideas from [**BD**], [**GL**], [**Ram**], and [**PRS**]. In addition to our investigation of walk algebras, one novelty of our presentation is that we give a different, more geometric proof of the point-counting result described above. This proof, which is inspired by the methods of [**GL**], uses the Bott-Samelson resolution of Schubert varieties. We believe that for certain audiences this proof may be more approachable than existing proofs which involve direct manipulation of relations in the Kac-Moody group.

We use these results to give streamlined derivations of Deodhar's formula for R-polynomials [**Deo**] and point-counting formulas for nonsymmetric Macdonald polynomials $E_\lambda(\mathsf{q},t)$ at $\mathsf{q}=0,\infty$. In both of these situations we emphasize the role played by change of oriented bases. We discuss how the point-counting formula for nonsymmetric Macdonald polynomials at $\mathsf{q}=0$ relates to a similar result of [**Ion**].

In future work we aim to adapt the constructions of this paper to the setting of p-adic Kac-Moody groups and their Iwahori-Hecke algebras (see [**BKP**, **BGR**, **M**] for existing work on this topic). One promising direction is to incorporate the work of Gaussent and Rousseau ([**GR1**, **GR2**]) into the present exposition. Their work suggests that the notion of expression, which is a sequence of Weyl group elements indexed by integers, should be replaced with a certain type of continuous path in the coweight space indexed by the interval $[0,1]\subseteq\mathbb{R}$. For example, they work with a notion called "Hecke path" that appears to be a useful proxy for the notion of positively folded alcove path.

Another promising direction is to incorporate the combinatorial study of the object that replaces the Weyl group. This object is no longer a Coxeter group, but it carries a natural Bruhat order, introduced in [**BKP**], and a length function [**M**, **MO**] compatible with this order. Conjectures about point-counting in double-affine flag varieties were made using this length function in [**MO**, §8.2]. A full generalization of the results of this note would include answers to these conjectures. With such a goal in mind, one of our initial motivations for writing this note was to give an exposition that is likely to generalize to the p-adic Kac-Moody setting.

2. Notation and preliminaries

Let us consider a Kac-Moody root system with index set I, Weyl group W, simple roots $\Pi=\{\alpha_i\}_{i\in I}$, and simple reflections $S=\{s_i\}_{i\in I}$. For $s_i\in S$, we will abuse notation and sometimes write $\alpha_{s_i}=\alpha_i$.

Imaginary roots will not play a role in what we do, so when we say root we mean *real root*. Let us write Δ_+ (resp. Δ_-) for the set of positive (resp. negative) roots. Let $\ell: W\to\mathbb{N}$ be the length function.

For a positive integer r, we write $[r]=\{1,2,\ldots,r\}$.

2.1. Expressions and subexpressions. An *expression* of length r is a r-tuple $\sigma = (\sigma_1, \ldots, \sigma_r) \in S^r$. For any such σ there is a unique r-tuple $(i_1, \ldots, i_r) \in I^r$ such that $\sigma = (s_{i_1}, \ldots, s_{i_r})$.

A *subexpression* τ of σ is a r-tuple $\tau = (\tau_1, \ldots, \tau_r) \in S^r$ such that $\tau_j \in \{e, \sigma_j\}$. Note τ consists of the data of the r-tuple $(\tau_1, \ldots, \tau_r)$ *and* the expression σ, i.e., τ remembers of which expression it is a subexpression. Notationally, let us write $\tau \leq \sigma$ to indicate that τ is a subexpression of σ. Let us also write $\sigma \leq \sigma$ for the *full subexpression* that agrees with σ in every slot.

Given two subexpressions τ, ρ of σ, we say that ρ is a subexpression of τ, denoted $\rho \leq \tau \leq \sigma$, if $\tau_k = e$ implies $\rho_k = e$. Given a subexpression $\tau = (\tau_1, \ldots, \tau_r)$ let us define the *product* of τ by

$$\pi(\tau) = \tau_1 \cdots \tau_r \,. \tag{2.1}$$

2.2. Orientations. An *orientation* A is a decomposition of $\Delta = \Delta_+^A \cup \Delta_-^A$ such that $\Delta_+^A \cap \Delta_-^A = \varnothing$, $\Delta_+^A = -\Delta_-^A$, and Δ_+^A is convex:

$$\alpha, \beta \in \Delta_+^A \text{ and } a\alpha + b\beta \in \Delta \text{ for } a, b > 0 \text{ implies } a\alpha + b\beta \in \Delta_A^+. \tag{2.2}$$

Given a root β, we will write $\beta >_A 0$ if $\beta \in \Delta_+^A$ (and $\beta <_A 0$ if $\beta \in \Delta_-^A$). Let us write A_0 for the *standard orientation* where $\Delta_+^{A_0} = \Delta_+$.

EXAMPLE 2.3. The Weyl group acts on the set of orientations as follows. Given an orientation A and $x \in W$, we can form a new orientation $x(A)$ by defining $\beta >_{x(A)} 0$ if and only if $x^{-1}(\beta) >_A 0$.

EXAMPLE 2.4. For any orientation A, its negative $-A$ is the orientation with $\Delta_+^{-A} = -\Delta_+^A$.

Given an orientation and a root β, we define the sign of β with respect to A by

$$\operatorname{sgn}_A(\beta) = \begin{cases} +1 & \text{if } \beta >_A 0 \\ -1 & \text{if } \beta <_A 0\,. \end{cases} \tag{2.5}$$

EXAMPLE 2.6. Suppose that $\Delta = \{\alpha + n\delta \mid \alpha \in \Delta_\circ,\, n \in \mathbb{Z}\}$ is an irreducible root system of untwisted affine Kac-Moody type. Here $\Delta_\circ$ is a Kac-Moody root system of finite-type and the positive roots Δ_+ are those $\alpha + n\delta$ such that $n \geq 0$ and $n > 0$ if $\alpha \in (\Delta_\circ)_-$. Another important orientation is the *periodic orientation* A_∞, which is defined by $\alpha + n\delta >_{A_\infty} 0$ if and only if $\alpha \in (\Delta_\circ)_+$. We also define $A_{-\infty} = -A_\infty$. We note that the orientations A_∞ are not of the form $x(A_0)$ for any $x \in W$.

REMARK 2.7. In [**BD**, Introduction], one finds some remarks on the classification of orientations. Here we simply mention that in finite Kac-Moody type, all orientations are of the form $x(A_0)$ for $x \in W$, while in affine Kac-Moody type, there are finitely many orientations up to the action of the Weyl group.

REMARK 2.8. Let $\widehat{\Delta} \supset \Delta$ be the set of all real and imaginary roots of the given Kac-Moody root system. One may define an orientation of $\widehat{\Delta}$ as above, but with the convexity assumption relaxed by replacing $a\alpha + b\beta$ in (2.2) simply by $\alpha + \beta$. The intersection of such an orientation of $\widehat{\Delta}$ with Δ produces an orientation of Δ in the sense defined above (using (2.2)). In fact, it follows from the proof of [**BD**, Proposition 1] that the orientations of Δ obtained in this way are exactly those of the form $\Delta_+^A = \Delta \cap P$, where P is the cone of elements greater than or equal to 0 with respect to some lexicographic ordering of the $\mathbb{R}$-span of Δ. The latter are known as *geometric orientations* of Δ, which form in general a proper subset of all orientations of Δ. However, the applications considered in this paper involve only geometric orientations.

2.3. The length function. The following discussion of the length function will be used when we relate the change-of-basis coefficients for oriented bases to point-counting in Kac-Moody flag varieties.

Following Dyer [**Dy**], we define for any orientation A a length function $\ell_A : W \to \mathbb{Z}$ defined by

$$\ell_A(w) = \sum_{\alpha \in \mathrm{Inv}(w^{-1})} \mathrm{sgn}_A(\alpha) = \ell(w) - 2\#(\mathrm{Inv}(w^{-1}) \cap \Delta_-^A) \tag{2.9}$$

where $\mathrm{Inv}(w^{-1}) = \Delta_+ \cap w(\Delta_-)$. It satisfies the following basic properties.

LEMMA 2.10. *Let A be an orientation. Then:*

(1) *For $i \in I$, we have*

$$\ell_A(ws_i) - \ell_A(w) = \begin{cases} +1 & \text{if } w(\alpha_i) >_A 0 \\ -1 & \text{if } w(\alpha_i) <_A 0\,. \end{cases} \tag{2.11}$$

(2) $\ell_A(e) = 0$.
(3) *For any $w \in W$, $\ell_{-A}(w) = -\ell_A(w)$.*
(4) *For any $x, w \in W$, $\ell_{x(A)}(w) = \ell_A(x^{-1}w) - \ell_A(x^{-1})$.*

For an orientation A and an expression $\tau = (\tau_1, \ldots, \tau_r)$, we define

$$\zeta_\tau^+ = \#\{k \in [r] \mid \tau_k = \sigma_k \text{ and } -\tau_1 \cdots \tau_k(\alpha_{\sigma_k}) >_A 0\} \tag{2.12}$$

$$\zeta_\tau^- = \#\{k \in [r] \mid \tau_k = \sigma_k \text{ and } -\tau_1 \cdots \tau_k(\alpha_{\sigma_k}) <_A 0\} \tag{2.13}$$

$$\kappa_\tau = \#\{k \in [r] \mid \tau_k = e\}\,. \tag{2.14}$$

LEMMA 2.15. *Let A be an orientation and let σ be a reduced expression. Then for any subexpression $\tau \leq \sigma$, we have*

$$\kappa_\tau + 2\zeta_\tau^+ = \ell(\pi(\sigma)) + \ell_A(\pi(\tau))\,. \tag{2.16}$$

PROOF. Applying (2.11) successively to the Weyl group elements in the sequence $e, \tau_1, \tau_1\tau_2, \ldots$, we deduce that $\ell_A(\pi(\tau)) = \zeta_\tau^+ - \zeta_\tau^-$. We add this to $\ell(\pi(\sigma)) = \zeta_\tau^+ + \zeta_\tau^- + \kappa_\tau$ to obtain (2.16). □

REMARK 2.17. When $A = -A_0$, formula (2.16) reduces to $\kappa_\tau + 2\zeta_\tau^+ = \ell(\pi(\sigma)) - \ell(\pi(\tau))$. This is [**Deo**, Lemma 5.1].

2.3.1. *Length for the periodic orientation in affine type.* Suppose that $W = Q_\circ^\vee \rtimes W_\circ$ is the Weyl group of an irreducible Kac-Moody root system of untwisted affine type. Here $W_\circ$ and $Q_\circ^\vee$ are the Weyl group and coroot lattice of a finite type Kac-Moody algebra $\mathfrak{g}_\circ$. Let 2ρ be the sum of all positive roots of $\mathfrak{g}_\circ$.

Let us denote an arbitrary element $x = (\lambda, u) \in W$ by $\varpi^\lambda u$. For such x we write $\mathrm{wt}(x) = \lambda \in Q_\circ^\vee$ and $\mathrm{dir}(x) = u \in W_\circ$. The elements $\varpi^\lambda = \varpi^\lambda e \in W$ for $\lambda \in Q_\circ^\vee$ pairwise commute and are called translation elements.

LEMMA 2.18. *For $W = Q_\circ^\vee \rtimes W_\circ$, we have*

$$\ell_{A_\infty}(\varpi^\lambda u) = -\langle \lambda, 2\rho \rangle + \ell(u) \tag{2.19}$$

$$\ell_{A_{-\infty}}(\varpi^\lambda u) = \langle \lambda, 2\rho \rangle - \ell(u)\,. \tag{2.20}$$

PROOF. The two formulas are equivalent, so let us prove the second formula. Let $\Delta = \{\alpha + r\delta\}$ be the set of affine roots as in Example 2.6. We have the following formula for the action of $u^{-1}\varpi^{-\lambda}$ on affine roots

$$u^{-1}\varpi^{-\lambda}(\alpha + r\delta) = u^{-1}(\alpha) + (r + \langle\lambda, \alpha\rangle)\delta\,. \tag{2.21}$$

Thus $\alpha + r\delta \in \mathrm{Inv}(u^{-1}\varpi^{-\lambda})$ if and only if one of the following holds:

$$\alpha > 0,\ u^{-1}(\alpha) > 0, \text{ and } 0 \le r < -\langle\lambda, \alpha\rangle \tag{2.22}$$

$$\alpha > 0,\ u^{-1}(\alpha) < 0, \text{ and } 0 \le r \le -\langle\lambda, \alpha\rangle \tag{2.23}$$

$$\alpha < 0,\ u^{-1}(\alpha) > 0, \text{ and } 0 < r < -\langle\lambda, \alpha\rangle \tag{2.24}$$

$$\alpha < 0,\ u^{-1}(\alpha) < 0, \text{ and } 0 < r \le -\langle\lambda, \alpha\rangle\,. \tag{2.25}$$

Therefore,

(2.26)

$$\begin{aligned}\ell_{A_{-\infty}}(\varpi^\lambda u) = &\sum_{\substack{\alpha\in\Delta_\circ \\ -\langle\lambda,\alpha\rangle>0}} -\operatorname{sgn}(\alpha)\left(-\langle\lambda,\alpha\rangle + \begin{cases}1 & \text{if } \alpha>0 \text{ and } u^{-1}(\alpha)<0\\ -1 & \text{if } \alpha<0 \text{ and } u^{-1}(\alpha)>0\end{cases}\right)\\ &- \#\{\alpha\in\Delta_\circ : \alpha>0,\ u^{-1}(\alpha)<0, \text{ and } \langle\lambda,\alpha\rangle = 0\}\\ = &\ \langle\lambda, 2\rho\rangle - \ell(u)\,.\end{aligned}$$

□

3. Hecke algebras and walk algebras

3.1. Walk algebras. Following [**Ram**], we define the *walk algebra* $\mathcal{A}$ to be the unital $\mathbb{Z}$-algebra generated by symbols $\{c_i^+, c_i^-, f_i^+, f_i^-\}_{i\in I}$ subject to the following relations for each $i \in I$

$$c_i^+ = c_i^- + f_i^+ \tag{3.1}$$

$$c_i^- = c_i^+ + f_i^-\,. \tag{3.2}$$

We note that for all $i \in I$

$$f_i^- = -f_i^+\,. \tag{3.3}$$

REMARK 3.4. In affine Kac-Moody type, [**Ram**] interprets the algebra $\mathcal{A}$ in terms of alcove walks. The $c_i^\pm$ (resp. the $f_i^\pm$) generators correspond to crossing (resp. folding along) the i-wall of an alcove. For this reason, in this case $\mathcal{A}$ is called the *alcove* walk algebra in [**Ram**].

3.2. Hecke algebras. Let $\mathcal{H}$ be the Hecke algebra corresponding to W. Explicitly, $\mathcal{H}$ is the unital $\mathbb{Z}[v, v^{-1}]$-algebra generated by symbols $\{T_i\}_{i\in I}$ subject to the braid relations and the following quadratic relations

$$(T_i - v)(T_i + v^{-1}) = 0, \quad i \in I\,. \tag{3.5}$$

We immediately have $T_i^{-1} = T_i - (v - v^{-1})$. For each $w \in W$, we define

$$T_w = T_{i_1} \cdots T_{i_r} \tag{3.6}$$

where $\sigma = (s_{i_1}, \ldots, s_{i_r})$ is a reduced expression for w. Following [**Ram**], we have the following.

PROPOSITION 3.7. *There is a surjective $\mathbb{Z}$-algebra map*

$$\Phi : \mathcal{A} \to \mathcal{H} \tag{3.8}$$

given by

$$c_i^{\pm} \mapsto T_i^{\pm 1} \tag{3.9}$$

$$f_i^{\pm} \mapsto \pm(v - v^{-1}) \tag{3.10}$$

for all $i \in I$.

Ram explicitly describes the kernel of this map in the affine case, but since it is not necessary for our purposes we will not reproduce it here.

3.3. Oriented bases. Fix an orientation A. Let σ be an expression of length r, and let τ be a subexpression. We define

$$L(\tau \leq \sigma, A) = x_1 \cdots x_r \in \mathcal{A} \tag{3.11}$$

where $x_k = c_{\sigma_k}^{\epsilon_k}$ if $\tau_k = \sigma_k$ and $x_k = f_{\sigma_k}^{\epsilon_k}$ if $\tau_k = e$; the sign ϵ_k is defined by

$$\epsilon_k = \operatorname{sgn}_A(-\tau_1 \cdots \tau_k(\alpha_{\sigma_k}))\,. \tag{3.12}$$

We also define

$$T(\tau \leq \sigma, A) = \Phi\,(L(\tau \leq \sigma, A)) \in \mathcal{H}\,. \tag{3.13}$$

To simplify notation, we will write $L(\sigma \leq \sigma, A) = L(\sigma, A)$ and $T(\sigma \leq \sigma, A) = T(\sigma, A)$.

REMARK 3.14. In affine Kac-Moody type, expressions σ exactly correspond to "unfolded alcove walks", and subexpressions $\tau \leq \sigma$ correspond to foldings of the alcove walk σ. Ram considers the orientation $A_{-\infty}$, and in Ram's terminology, $L(\tau \leq \sigma, A_{-\infty})$ is a word in $\mathcal{A}$ corresponding to the "alcove walk" $\tau \leq \sigma$. Our definition is the natural generalization of his to arbitrary orientations.

Immediately from the definition of $L(\tau \leq \sigma, A)$, we have the following recurrence relation for these elements:

PROPOSITION 3.15. *Fix an orientation A. Let $\sigma = (\sigma_1, \ldots, \sigma_r)$ be an expression of length r, and let $\tau = (\tau_1, \ldots, \tau_r)$ be a subexpression of σ. Then we have:*

$$L(\tau \leq \sigma, A) = L(\tau_1 \leq \sigma_1, A) L((\tau_2, \ldots, \tau_r) \leq (\sigma_2, \ldots, \sigma_r), \tau_1^{-1}(A))\,. \tag{3.16}$$

The following is [**Ram**, Lemma 3.1] for the periodic orientation (see also [**Gö**, Lemma 2.4.2]), and it holds in general by the same argument.

PROPOSITION 3.17. *The set of $\{L(\tau \leq \sigma, A)\}$ form a basis of $\mathcal{A}$ as we vary over all expressions σ and subexpressions $\tau \leq \sigma$.*

We will call a basis of $\mathcal{A}$ of this form an *oriented basis.*

REMARK 3.18. In [**Gö**, Theorem 3.1.1], it is shown that $T(\sigma, A)$ depends only on $\pi(\sigma)$ for any expression σ. For any $w \in W$ we may therefore unambiguously define $T(w, A) \in \mathcal{H}$ to be $T(\sigma, A)$ for any expression σ that $\pi(\sigma) = w$. Correspondingly, for any fixed A, the set $\{T(w, A) \mid w \in W\}$ is a basis for $\mathcal{H}$, which we call an oriented basis. These oriented bases of the Hecke algebra $\mathcal{H}$ appeared (with slightly different conventions) in earlier work of Dyer [**Dy**] (see also [**Dy2**, §9]).

4. Change of basis formula

In this section we study the change of basis between oriented bases of $\mathcal{A}$. This gives rise to the notion of distinguished subexpressions.

DEFINITION 4.1. Let A, A' be orientations. Let σ be an expression of length r, let τ be a subexpression of σ, and let ρ be a subexpression of τ. Let $K = \{k \in [r] \mid \rho_k = e, \tau_k = \sigma_k\}$. Then we say ρ is a *distinguished subexpression of* τ *(with respect to* A *and* A'*)* if for all $k \in K$, we have

$$\operatorname{sgn}_A(-\tau_1 \cdots \tau_k(\alpha_{\sigma_k})) = \operatorname{sgn}_{A'}(-\rho_1 \cdots \rho_k(\alpha_{\sigma_k})) . \tag{4.2}$$

Let us write $\mathscr{D}_A^{A'}(\tau \le \sigma)$ for the set of ρ satisfying the above conditions.

REMARK 4.3. In the setting of Remark 3.14, distinguished subexpressions for $A = A_0$, $A' = A_{-\infty}$, and $\tau = \sigma$, correspond to "positively folded alcove walks".

DEFINITION 4.4. Let A, A' be orientations. Let σ be an expression of length r, let τ be a subexpression of σ, and let ρ be a subexpression of τ. We define

(4.5)
$$\begin{aligned} &\mathcal{K}_A^{A'}(\rho \le \tau \le \sigma) \\ &= \{k \in [r] \mid \rho_k = e, \tau_k = e, \operatorname{sgn}_A(-\tau_1 \cdots \tau_k(\alpha_{\sigma_k})) \ne \operatorname{sgn}_{A'}(-\rho_1 \cdots \rho_k(\alpha_{\sigma_k}))\} . \end{aligned}$$

THEOREM 4.6. *Let* A *and* A' *be any orientations, and let* τ *be a subexpression of an expression* σ*. Then we have*

$$L(\tau \le \sigma, A) = \sum_{\rho \in \mathscr{D}_A^{A'}(\tau \le \sigma)} (-1)^{\#\mathcal{K}_A^{A'}(\rho \le \tau \le \sigma)} L(\rho \le \sigma, A') . \tag{4.7}$$

This is best understood in terms of repeated application of (3.1), (3.2), and (3.3) to "straighten" each factor in $L(\tau \le \sigma, A)$ from left to right. Let us demonstrate this with an example before proving Theorem 4.6.

EXAMPLE 4.8. For $W = \langle s_1, s_2 \rangle$, the Weyl group of SL_3, consider $A = s_1(A_0)$, $A' = s_2 s_1(A_0)$, and $\tau = (s_1, s_2, e) \le (s_1, s_2, s_1) = \sigma$. We have

$$\begin{aligned} &L(\tau \le \sigma, A) \\ &= c_1^- c_2^+ f_1^- \\ &= (c_1^+ + f_1^-) c_2^+ f_1^- \\ &= c_1^+ c_2^+ f_1^- + f_1^- c_2^+ f_1^- \\ &= c_1^+ (c_2^- + f_2^+) f_1^- + f_1^- (c_2^- + f_2^+) f_1^- \\ &= -c_1^+ c_2^- f_1^+ - c_1^+ f_2^+ f_1^+ - f_1^- c_2^- f_1^+ + f_1^- f_2^+ f_1^- \\ &= -L((s_1, s_2, e), A') - L((s_1, e, e), A') - L((e, s_2, e), A') + L((e, e, e), A') . \end{aligned}$$

In this example all subexpressions of τ are distinguished.

The proof of Theorem 4.6 is based on the following, which is straightforward.

PROPOSITION 4.9. *Let* A *and* A' *be any orientations, and let* τ *be a subexpression of an expression* σ *of length* r*. Let* $\widetilde{\sigma} = (\sigma_2, \ldots, \sigma_r)$ *and* $\widetilde{\tau} = (\tau_2, \ldots, \tau_r)$*.*

Consider three cases for this data:

Case A: $\mathrm{sgn}_A(\tau_1(\alpha_{\sigma_1})) = \mathrm{sgn}_{A'}(\tau_1(\alpha_{\sigma_1}))$

Case B: $\mathrm{sgn}_A(\tau_1(\alpha_{\sigma_1})) \neq \mathrm{sgn}_{A'}(\tau_1(\alpha_{\sigma_1}))$ and $\tau_1 = e$

Case C: $\mathrm{sgn}_A(\tau_1(\alpha_{\sigma_1})) \neq \mathrm{sgn}_{A'}(\tau_1(\alpha_{\sigma_1}))$ and $\tau_1 = \sigma_1$

We have

$$L(\tau_1 \leq \sigma_1, A) = \begin{cases} L(\tau_1 \leq \sigma_1, A') & \textit{if } \text{Case A} \\ -L(\tau_1 \leq \sigma_1, A') & \textit{if } \text{Case B} \\ L(\tau_1 \leq \sigma_1, A') + L(e \leq \sigma_1, A') & \textit{if } \text{Case C} \end{cases} \tag{4.10}$$

and $\mathscr{D}_A^{A'}(\tau \leq \sigma)$ *is equal to*

(4.11)

$$\begin{cases} \{(\tau_1, \widetilde{\rho}) \mid \widetilde{\rho} \in \mathscr{D}_{\tau_1^{-1}(A)}^{\tau_1^{-1}(A')}(\widetilde{\tau} \leq \widetilde{\sigma})\} & \textit{if } \text{Case A} \\ \{(\tau_1, \widetilde{\rho}) \mid \widetilde{\rho} \in \mathscr{D}_{\tau_1^{-1}(A)}^{\tau_1^{-1}(A')}(\widetilde{\tau} \leq \widetilde{\sigma})\} & \textit{if } \text{Case B} \\ \{(\tau_1, \widetilde{\rho}) \mid \widetilde{\rho} \in \mathscr{D}_{\tau_1^{-1}(A)}^{\tau_1^{-1}(A')}(\widetilde{\tau} \leq \widetilde{\sigma})\} \bigsqcup \{(e, \widetilde{\rho}) \mid \widetilde{\rho} \in \mathscr{D}_A^{\tau_1^{-1}(A')}(\widetilde{\tau} \leq \widetilde{\sigma})\} & \textit{if } \text{Case C} \end{cases}$$

where we write $\widetilde{\rho} = (\rho_2, \ldots, \rho_r)$.

Proof of Theorem 4.6. Let us write $\sigma = (\sigma_1, \ldots, \sigma_r)$ and $\tau = (\tau_1, \ldots, \tau_r)$. We will proceed by induction on r.

There are three cases according to the three cases in Proposition 4.9. We will argue for the first case, i.e., let us assume $\mathrm{sgn}_A(\tau_1(\alpha_{\sigma_1})) = \mathrm{sgn}_{A'}(\tau_1(\alpha_{\sigma_1}))$. The other cases are similar and we omit them to avoid repetition.

By (4.10) and Proposition 3.15 we have

$$L(\tau \leq \sigma, A) = L(\tau_1 \leq \sigma_1, A')L(\widetilde{\tau} \leq \widetilde{\sigma}, \tau_1^{-1}(A)). \tag{4.12}$$

By induction we have

$$\begin{aligned} &L(\widetilde{\tau} \leq \widetilde{\sigma}, \tau_1^{-1}(A)) \\ &= \sum_{\widetilde{\rho}=(\rho_2,\ldots,\rho_r)\in\mathscr{D}_{\tau_1^{-1}(A)}^{\tau_1^{-1}(A')}(\widetilde{\tau}\leq\widetilde{\sigma})} (-1)^{\#\mathcal{K}_{\tau_1^{-1}(A)}^{\tau_1^{-1}(A')}(\widetilde{\rho}\leq\widetilde{\tau}\leq\widetilde{\sigma})} L(\widetilde{\rho} \leq \widetilde{\tau}, \tau_1^{-1}(A')). \end{aligned} \tag{4.13}$$

Thus (4.12) equals

$$\sum_{\widetilde{\rho}=(\rho_2,\ldots,\rho_r)\in\mathscr{D}_{\tau_1^{-1}(A)}^{\tau_1^{-1}(A')}(\widetilde{\tau}\leq\widetilde{\sigma})} (-1)^{\#\mathcal{K}_{\tau_1^{-1}(A)}^{\tau_1^{-1}(A')}(\widetilde{\rho}\leq\widetilde{\tau}\leq\widetilde{\sigma})} L((\tau_1, \rho_2, \ldots, \rho_r) \leq \tau, A') \tag{4.14}$$

by Proposition 3.15. Finally, by (4.11), this is equal to

$$\sum_{\rho=(\rho_1,\ldots,\rho_r)\in\mathscr{D}_A^{A'}(\tau\leq\sigma)} (-1)^{\#\mathcal{K}_{\tau_1^{-1}(A)}^{\tau_1^{-1}(A')}(\widetilde{\rho}\leq\widetilde{\tau}\leq\widetilde{\sigma})} L(\rho \leq \tau, A'). \tag{4.15}$$

Finally, by our assumption that $\mathrm{sgn}_A(\tau_1(\alpha_{\sigma_1})) = \mathrm{sgn}_{A'}(\tau_1(\alpha_{\sigma_1}))$, we have

$$\#\mathcal{K}_A^{A'}(\rho \leq \tau \leq \sigma) = \#\mathcal{K}_{\tau_1^{-1}(A)}^{\tau_1^{-1}(A')}(\widetilde{\rho} \leq \widetilde{\tau} \leq \widetilde{\sigma}) \tag{4.16}$$

and the change of basis formula follows. □

5. Counting points in Kac-Moody flag varieties

Let G be the Kac-Moody group associated with the root system of §2, let $B \subset G$ be the positive Borel subgroup, and let $T \subset B$ be the maximal torus. (The reader may refer to [**Kum**] for a thorough treatment of these notions.) We will be concerned with the Kac-Moody flag variety G/B, which is an ind-scheme. The fixed points for the T-action on G/B are precisely indexed by $w \in W$. Let us abuse notation and write $w \in G/B$ for the corresponding fixed point and also $w \in G$ for fixed lifts of these points.

We call a cocharacter $\eta : \mathbb{G}_m \to T$ regular if its pairing with any real root is nonzero. For any regular cocharacter η, we define an orientation A_η by $\alpha >_{A_\eta} 0$ if any only if $\langle \eta, \alpha \rangle > 0$. We will consider the change of basis between oriented bases corresponding to A_0 and A_η.

If $\sigma = (\sigma_1, \ldots, \sigma_r)$ is a reduced expression, we have by Theorem 4.6 that

$$L(\sigma, A_0) = \sum_{\tau \in \mathscr{D}^\eta(\sigma)} L(\tau \leq \sigma, A_\eta) \tag{5.1}$$

where, to simplify notation, we write $\mathscr{D}^\eta(\sigma) = \mathscr{D}_{A_0}^{A_\eta}(\sigma \leq \sigma)$. Mapping to the Hecke algebra gives

$$T(\sigma, A_0) = \sum_{\tau \in \mathscr{D}^\eta(\sigma)} m^\eta_{\tau \leq \sigma} T(\pi(\tau), A_\eta) \tag{5.2}$$

where

$$m^\eta_{\tau \leq \sigma} = (v - v^{-1})^{\kappa_\tau}\,. \tag{5.3}$$

(Note that we do not get any minus signs here because σ is reduced and $A = A_0$.) Let us write

$$M^{\eta,u}_\sigma = \sum_{\tau \in \mathscr{D}^\eta(\sigma) : \pi(\tau) = u} v^{\kappa_\tau + 2\zeta^+_\tau} m^\eta_{\tau \leq \sigma} = \sum_{\tau \in \mathscr{D}^\eta(\sigma) : \pi(\tau) = u} v^{2\zeta^+_\tau} (v^2 - 1)^{\kappa_\tau}\,. \tag{5.4}$$

Now let us define $X^{\eta,u}_w \subset G/B$ by

$$X^{\eta,u}_w = \{x \in Bw \cdot B \mid \lim_{z \to 0} \eta(z) \cdot x = u\}\,. \tag{5.5}$$

In the next section we will prove the following:

THEOREM 5.6. *Let $w \in W$, and let σ be a reduced expression for w. Let η be a regular cocharacter. Then for every $u \in W$, we have:*

$$M^{\eta,u}_\sigma|_{v^2 = q} = \# X^{\eta,u}_w(\mathbb{F}_q)\,. \tag{5.7}$$

REMARK 5.8. For any orientation A, let $Q_A \subset G$ be the subgroup generated by the one-parameter subgroups $e_\alpha : \mathbb{G}_a \to G$ corresponding to $\alpha \in \Delta^A_+$. Then [**BD**, Theorem 1] asserts that the Bruhat-type decomposition $G = \sqcup_{w \in W} Q_A w B$ holds. In particular, it follows that for $A = A_\eta$ the limit $\lim_{z \to 0} \eta(z) \cdot x$ exists and is equal to an element of W for any $x \in G/B$. Moreover, we see that $X^{\eta,u}_w = Q_A u \cdot B \cap Bw \cdot B$. We note that Theorem 5.6 is a consequence of [**BD**, Theorem 5], but we will give a different proof below.

REMARK 5.9. By Lemma 2.15, we have

$$M^{\eta,u}_\sigma = v^{\ell(w) + \ell_{A_\eta}(u)} \sum_{\tau \in \mathscr{D}^\eta(\sigma) : \pi(\tau) = u} m^\eta_{\tau \leq \sigma}\,. \tag{5.10}$$

In particular, up to an explicit constant depending only on w, u, and η, the point-count in Theorem 5.6 is given exactly by summing coefficients $m^{\eta}_{\tau\leq\sigma}$.

5.1. Bott-Samelson varieties. For each $i \in I$, let us write $P_i \supset B$ for the corresponding rank-one parabolic subgroup; we will also write $P_{s_i} = P_i$. Let us also write $e_{\pm\alpha_i} = e_{\pm\alpha_{s_i}} : \mathbb{G}_a \to G$ for the corresponding one-parameter subgroups.

Fix an expression $\sigma = (\sigma_1, \ldots, \sigma_r)$. Then we can form the Bott-Samelson variety:

$$\Sigma(\sigma) = P_{\sigma_1} \times^B P_{\sigma_2} \times^B \cdots \times^B P_{\sigma_r}/B \tag{5.11}$$

which is a smooth projective variety. Let us denote the points of $\Sigma(\sigma)$ by tuples $[g_1, \ldots, g_r]$ where $g_j \in P_{\sigma_j}$ and we consider such tuples up to the following equivalence

$$[g_1, \ldots, g_r] = [g_1 b_1, b_1^{-1} g_2 b_2, \ldots, b_{r-1}^{-1} g_r b_r] \tag{5.12}$$

for all $(b_1, \ldots, b_r) \in B^r$. There is a natural B-equivariant map

$$\pi : \Sigma(\sigma) \to G/B \tag{5.13}$$

given by

$$\pi([g_1, \ldots, g_r]) = g_1 \cdots g_r B \tag{5.14}$$

and the B-action on $\Sigma(\sigma)$ is given by left multiplication on the first component. For any subexpression $\tau = (\tau_1, \ldots, \tau_r)$, we can form the corresponding point $[\tau_1, \ldots, \tau_r] \in \Sigma(\sigma)$. These points are precisely the T-fixed points of $\Sigma(\sigma)$.

For each subexpression $\tau \leq \sigma$ let us define

$$\mathcal{U}_\tau = \{[\tau_1 e_{-\alpha_{\sigma_1}}(z_1), \ldots, \tau_r e_{-\alpha_{\sigma_r}}(z_r)] \mid z_i \in \mathbb{G}_a\}\,. \tag{5.15}$$

Then $\mathcal{U}_\tau$ is an open subset of $\Sigma(\sigma)$ isomorphic to the affine space $\mathbb{A}^r$. The coordinates $(z_1, \ldots, z_r)$ allow us to view $\mathcal{U}_\tau$ as the tangent space to the point $\tau \in \Sigma(\sigma)$.

Let $\eta : \mathbb{G}_m \to T$ be a regular cocharacter. Then we can restrict the T-action on $\Sigma(\sigma)$ to a $\mathbb{G}_m$-action via η. Because η is regular, the fixed points of η are exactly equal to the T-fixed points. Parallel to (5.5), we can therefore form the Białynicki-Birula cell

$$C^\eta_\tau = \{x \in \Sigma(\sigma) \mid \lim_{z\to 0} \eta(z)\cdot x = \tau\}\,. \tag{5.16}$$

Because $\mathcal{U}_\tau$ is a T-invariant open subset containing τ, we must have $C^\eta_\tau \subset \mathcal{U}_\tau$. Furthermore, the coordinates (5.15) on $\mathcal{U}_\tau$ allow us to explicitly compute the equations defining C^η_τ. It is given by the vanishing of the coordinates z_k that have negative weight under the action of $\eta(z)$. Under the action of $\eta(z)$, the coordinate z_k is sent to $z^{\langle \tau_k \cdots \tau_1(\eta), -\alpha_{\sigma_k}\rangle} z_k$. Therefore, we must set $z_k = 0$ if $\langle \eta, -\tau_1 \cdots \tau_k(\alpha_{\sigma_k})\rangle < 0$. To summarize:

(5.17)
$$C^\eta_\tau = \{[\tau_1 e_{-\alpha_{\sigma_1}}(z_1), \ldots, \tau_r e_{-\alpha_{\sigma_r}}(z_r)] \in \mathcal{U}_\tau \mid z_k = 0 \text{ if } \langle \eta, -\tau_1 \cdots \tau_k(\alpha_{\sigma_k})\rangle < 0\}\,.$$

THEOREM 5.18. *Let σ be a reduced expression. If $\tau \notin \mathscr{D}^\eta(\sigma)$, then*

$$\pi^{-1}(B\pi(\sigma)\cdot B) \cap C^\eta_\tau = \varnothing\,. \tag{5.19}$$

If $\tau \in \mathscr{D}^\eta(\sigma)$, then:

$$\pi^{-1}(B\pi(\sigma)\cdot B) \cap C^\eta_\tau = \{[\tau_1 e_{-\alpha_{\sigma_1}}(z_1), \cdots, \tau_r e_{-\alpha_{\sigma_r}}(z_r)] \in \mathcal{U}_\tau \mid z_k = 0 \textit{ if } \langle \eta, -\tau_1 \cdots \tau_k(\alpha_{\sigma_k})\rangle < 0 \text{ and } z_k \neq 0 \textit{ if } \tau_k = e\}\,.$$

In the proof of Theorem 5.18 we will use the following three standard facts.

LEMMA 5.20. *Suppose $w \in W$ and $s \in S$ with $\ell(ws) = \ell(w)+1$. Then, we have*

$$BwBsB = BwsB\,. \tag{5.21}$$

LEMMA 5.22. *Let $s \in S$. Then we have*

$$Be_{-\alpha_s}(z)B \subseteq BsB\,. \tag{5.23}$$

for $z \in \mathbb{G}_m \subset \mathbb{G}_a$.

LEMMA 5.24. *Let $\sigma^{(k)} = (\sigma_1, \dots, \widehat{\sigma_k}, \dots, \sigma_r)$. Then we have a closed embedding*

$$i^{(k)} : \Sigma(\sigma^{(k)}) \hookrightarrow \Sigma(\sigma) \tag{5.25}$$

given by

$$[g_1, \dots, g_{k-1}, g_{k+1}, \dots, g_r] \mapsto [g_1, \dots, g_{k-1}, e, g_{k+1}, \dots, g_r]\,. \tag{5.26}$$

Furthermore, the map $i^{(k)}$ intertwines the two maps $\pi : \Sigma(\sigma^{(k)}) \to G/B$ and $\pi : \Sigma(\sigma) \to G/B$.

The following Lemma is a simple application of the Borel fixed point theorem (see [**Bri**, Corollary 2.2.2] for a similar argument).

LEMMA 5.27. *Suppose $v \in W$ is in the image of $\pi : \Sigma(\sigma) \to G/B$. Then there exists $\tau \leq \sigma$ such that $\pi(\tau) = v$.*

From this lemma we can deduce the following.

LEMMA 5.28. *Suppose $\sigma = (\sigma_1, \dots, \sigma_r)$ is a reduced expression. Let $\sigma^{(k)} = (\sigma_1, \dots, \widehat{\sigma_k}, \dots, \sigma_r)$. Then $\pi(\Sigma(\sigma^{(k)})) \cap B\pi(\sigma) \cdot B = \varnothing$ for all $k \in [r]$.*

PROOF OF THEOREM 5.18. Define

$$\begin{aligned} S^\eta_\tau = \{ & [\tau_1 e_{-\alpha_{\sigma_1}}(z_1), \dots, \tau_r e_{-\alpha_{\sigma_r}}(z_r)] \in \mathcal{U}_\tau \mid \\ & z_k = 0 \text{ if } \langle \eta, -\tau_1 \cdots \tau_k(\alpha_{\sigma_k}) \rangle < 0 \text{ and } z_k \neq 0 \text{ if } \tau_k = e \} \end{aligned}$$

and consider a point $[\tau_1 e_{-\alpha_{\sigma_1}}(z_1), \dots, \tau_r e_{-\alpha_{\sigma_r}}(z_r)] \in C^\eta_\tau \backslash S^\eta_\tau$. We must have $\tau_k = e$ and $z_k = 0$ for some k. Hence, the point $[\tau_1 e_{-\alpha_{\sigma_1}}(z_1), \dots, \tau_r e_{-\alpha_{\sigma_r}}(z_r)]$ belongs to $i^{(k)}(\Sigma((\sigma_1, \dots, \widehat{\sigma_k}, \dots, \sigma_r))$. Therefore, $\pi([\tau_1 e_{-\alpha_{\sigma_1}}(z_1), \dots, \tau_r e_{-\alpha_{\sigma_r}}(z_r)])$ does not belong to $B\pi(\sigma) \cdot B$ by Lemmas 5.24 and 5.28. We conclude that

$$\pi^{-1}(B\pi(\sigma) \cdot B) \cap C^\eta_\tau \subseteq S^\eta_\tau\,.$$

Conversely suppose we have $[\tau_1 e_{-\alpha_{\sigma_1}}(z_1), \dots, \tau_r e_{-\alpha_{\sigma_r}}(z_r)] \in S^\eta_\tau$. We need to show

$$\tau_1 e_{-\alpha_{\sigma_1}}(z_1) \cdots \tau_r e_{-\alpha_{\sigma_r}}(z_r) \in B\pi(\sigma)B\,. \tag{5.29}$$

We proceed by induction on r. By induction, we have

$$\tau_1 e_{-\alpha_{\sigma_1}}(z_1) \cdots \tau_{r-1} e_{-\alpha_{\sigma_{r-1}}}(z_{r-1}) \in B\sigma_1 \cdots \sigma_{r-1} B\,. \tag{5.30}$$

We also have

$$\tau_r e_{-\alpha_{\sigma_r}}(z_r) \in B\sigma_r B\,. \tag{5.31}$$

Indeed, if $\tau_r = e$, then $z_r \in \mathbb{G}_m$ and (5.31) holds by Lemma 5.22. If $\tau_r = \sigma_r$, then we have $\sigma_r e_{-\alpha_{\sigma_r}}(z_r) = e_{\alpha_{\sigma_r}}(z_r)\sigma_r$, hence we have (5.31). Then by (5.30) and Lemma 5.20 we have (5.29). Therefore:

$$\pi^{-1}(B\pi(\sigma) \cdot B) \cap C^\eta_\tau = S^\eta_\tau .$$

Finally, if $\tau \notin \mathscr{D}^{A_\eta}_{A_0}(\sigma)$, then we must have $\tau_k = e$ and $\langle \eta, -\tau_1 \cdots \tau_k(\alpha_{\sigma_k})\rangle < 0$ for some k. Then we must have $z_k = 0$ by (5.17), which contradicts the conditions on $[\tau_1 e_{-\alpha_{\sigma_1}}(z_1), \dots, \tau_r e_{-\alpha_{\sigma_r}}(z_r)]$ in the definition of S^η_τ. Hence $S^\eta_\tau = \varnothing$ in this case. □

Proof of Theorem 5.6. Let $w \in W$, and let σ be a reduced expression for w. Because the map $\pi : \Sigma(\sigma) \to G/B$ is a bijection over $Bw \cdot B \subset G/B$, it induces a bijection of sets:

$$\bigsqcup_{\tau \in \mathscr{D}^\eta(\sigma) : \pi(\tau) = u} \pi^{-1}(Bw \cdot B) \cap C^\eta_\tau \cong X^{\eta,u}_w . \tag{5.32}$$

By Theorem 5.18 we conclude that

$$\begin{aligned} \#X^{\eta,u}_w(\mathbb{F}_q) &= \#_{\mathbb{F}_q}\left(\bigsqcup_{\tau \in \mathscr{D}^\eta(\sigma) : \pi(\tau) = u} \pi^{-1}(Bw \cdot B) \cap C^\eta_\tau \right) \\ &= \sum_{\tau \in \mathscr{D}^\eta(\sigma) : \pi(\tau) = u} v^{2\zeta^+_\tau}(v^2 - 1)^{\kappa_\tau}|_{v^2 = q} \end{aligned} \tag{5.33}$$

where ζ^+_τ and κ_τ are defined by (2.12) and (2.14). □

6. Some connections to other work

6.1. Deodhar's formula for R-polynomials. The Kazhdan-Lusztig involution [**KL**] is the $\mathbb{Z}$-linear algebra involution $\overline{\cdot} : \mathcal{H} \to \mathcal{H}$ on the Hecke algebra given on generators by $\overline{T}_i = T_i^{-1}$ for all $i \in I$ and $\overline{v} = v^{-1}$. The R-polynomials are essentially the matrix of the Kazhdan-Lusztig involution. More precisely, for any pair $u, w \in W$, there is a polynomial $R_{u,w} \in \mathbb{Z}[v, v^{-1}]$ defined by

$$T_w = \sum_{u \in W} v^{\ell(u) - \ell(w)} R_{u,w} \overline{T}_u . \tag{6.1}$$

It is clear that $\overline{T}_u = T(u, -A_0)$, so applying (5.2), we have the following formula for the $R_{u,w}$.

Theorem 6.2. [**Deo**, Theorem 1.3] *Let σ be a reduced expression for w. Then we have*

$$R_{u,w} = \sum_{\tau \in \mathscr{D}^{-A_0}_{A_0}(\sigma) : \pi(\tau) = u} v^{\ell(w) - \ell(u)} m^{-A_0}_{\tau \leq \sigma} . \tag{6.3}$$

Theorem 5.6 translates in this special case to the fact that $R_{u,w}|_{v^2 = q}$ is equal to $\#_{\mathbb{F}_q}(B^- u \cdot B \cap Bw \cdot B)$, where B^- is the Borel subgroup opposite to B.

6.2. Affine type. For the rest of the paper, let us consider the case when W is the Weyl group of an affine Kac-Moody group G. For simplicity we assume that G is of untwisted affine type. Thus $W = Q_\circ^\vee \rtimes W_\circ$, where $W_\circ$ and $Q_\circ^\vee$ are the Weyl group and coroot lattice of a finite type Kac-Moody group $G_\circ$. Write $I = I_\circ \sqcup \{0\}$, where $I_\circ$ is the index set for the simple roots of $G_\circ$.

The Hecke algebra $\mathcal{H}$ of W, the affine Hecke algebra, has another presentation known as the Bernstein presentation, which is analogous to the semidirect product decomposition of W. This is expressed as a linear isomorphism $\mathcal{H} \cong \mathbb{Z}[v^{\pm 1}][Q_\circ^\vee] \otimes_{\mathbb{Z}[v^{\pm 1}]} \mathcal{H}_\circ$ where $\mathcal{H}_\circ$ is the Hecke algebra of $W_\circ$ and $\mathbb{Z}[v^{\pm 1}][Q_\circ^\vee]$ is the group algebra of $Q_\circ^\vee$, with basis elements X^λ for $\lambda \in Q_\circ^\vee$. These elements satisfy $X^\lambda X^\mu = X^{\lambda+\mu}$, and the tensor factors in the isomorphism above each form subalgebras of $\mathcal{H}$. The relations between these subalgebras are given by

$$T_i X^\lambda = X^{s_i(\lambda)} T_i + (v - v^{-1}) \frac{X^\lambda - X^{s_i(\lambda)}}{1 - X^{\alpha_i}} \tag{6.4}$$

for $i \in I_\circ$ and $\lambda \in Q_\circ^\vee$. For more details on this presentation of $\mathcal{H}$, we refer the reader to [**Lus**].

Let θ be the highest (long) root of $G_\circ$. The element T_0 from $\mathcal{H}$ is given in the Bernstein presentation as $T_0 = T_{s_\theta}^{-1} X^{-\theta^\vee}$.

We have a basis of $\mathcal{H}$ given by $\{X^\lambda T_u \mid \lambda \in Q_\circ^\vee,\, u \in W_\circ\}$ in the Bernstein presentation. It turns out that this is the oriented basis for the periodic orientation. More precisely, for any element $x = \varpi^\lambda u \in W$, one has $T(x, A_\infty) = X^{\mathrm{wt}(x)} T_{\mathrm{dir}(x)} = X^\lambda T_u$ (see, e.g., [**Mac**, (3.2.10) and (3.5.1)]).

Let $\mathbf{I} \subset G$ be the Iwahori subgroup (i.e., the standard Borel subgroup of G) and let $\mathbf{U}^\pm$ be the subgroups of G corresponding to $A_{\pm\infty}$ as in Remark 5.8. By Theorem 5.6, the coefficient of the Bernstein basis element $T(x, A_\infty)$ in $T(w, A_0)$ for $x, w \in W$ in the expansion (5.2) is proportional to $\#_{\mathbb{F}_q}(\mathbf{U}^+ x \cdot \mathbf{I} \cap \mathbf{I} w \cdot \mathbf{I})$, with the coefficient of proportionality given by $v^{\ell(w) + \ell_{A_\infty}(x)} = v^{\ell(w) + \ell(u) - \langle \lambda, 2\rho \rangle}$ if $x = \varpi^\lambda u$. A similar interpretation holds for $T(x, A_{-\infty})$, with the $\mathbf{U}^-$-orbit of x instead of the $\mathbf{U}^+$-orbit; in this special case Theorem 5.6 was proved in [**PRS**, Theorem 7.1].

REMARK 6.5. The groups $\mathbf{U}^\pm$ also have the following concrete realization. As G is untwisted affine, it can be realized as a central extension of a loop group of $G_\circ$. For us the relevant point is that there is a map $G \to G_\circ$ given by evaluating a loop at 1. Let $U_\circ^\pm$ be the positive and negative unipotent subgroups of $G_\circ$ corresponding to the set of positive and negative roots. Modulo the center of G, $\mathbf{U}^\pm$ is precisely the preimage of $U_\circ^\pm$ under the map $G \to G_\circ$. Because the central extension splits over $U_\circ^\pm$, we can use the splitting to explicitly see that $\mathbf{U}^\pm$ is a copy of the loop group of $U_\circ^\pm$ embedded as a subgroup of G.

6.3. Coefficients of nonsymmetric Macdonald polynomials. Continuing with the notation from the previous subsection, let us consider the polynomial representation of the affine Hecke algebra $\mathcal{H}$. This can be described abstractly as the induced representation $\mathrm{Ind}_{\mathcal{H}_\circ}^{\mathcal{H}}(\mathrm{triv})$, where triv is the one-dimensional representation in which T_i for $i \in I_\circ$ acts by the scalar v. Using the Bernstein presentation and in particular (6.4), one sees that the polynomial representation is linearly isomorphic to $\mathbb{Z}[v^{\pm 1}][Q_\circ^\vee]$, with the action of X^λ given by multiplication in the group algebra and the action of T_i for $i \in I_\circ$ given by the Demazure-Lusztig operator

$$T_i = vs_i + \frac{v - v^{-1}}{1 - X^{\alpha_i}}(1 - s_i)\,. \tag{6.6}$$

The action of the element T_0 from $\mathcal{H}$ in the polynomial representation is given via the Bernstein presentation by the formula $T_0 = T_{s_\theta}^{-1} X^{-\theta^\vee}$.

The nonsymmetric Macdonald polynomials $E_\lambda(\mathsf{q}, v)$ for $\lambda \in Q_\circ^\vee$ form a distinguished basis in the polynomial representation of $\mathcal{H}$ (with coefficients extended to $\mathbb{Q}(\mathsf{q}, v)$ — we use the letter q to distinguish this parameter from the q in $\mathbb{F}_q$). We refer the interested reader to [**Ch2**, §3.3] for an introduction to these remarkable polynomials and a discussion of their origins. Our present aim is to relate the specializations of $E_\lambda(\mathsf{q}, v)$ at $\mathsf{q} = 0, \infty$ to the change of basis and point-counting problems considered in this paper.

To this end, we make use of Cherednik's "intertwiner" construction of the $E_\lambda(\mathsf{q}, v)$, which is based on the operators in the polynomial representation (see [**Ch1**, Theorem 5.1], or [**RY**, (3.3)] for our conventions). Fix a reduced expression $\sigma_\lambda = (s_{i_1}, \dots, s_{i_r})$ such that $\pi(\sigma_\lambda) = m_\lambda$, where $m_\lambda = \varpi^\lambda u_\lambda^{-1}$ is the minimum coset representative for ϖ^λ in $W/W_\circ$. Then the intertwiner construction is as follows:

(6.7)

$$\begin{aligned} &E_\lambda(\mathsf{q}, v) \\ &= v^{-\ell(u_\lambda)} \left(T_{i_1} + \frac{v - v^{-1}}{\mathsf{q}^{a_1} v^{b_1} - 1} \right) \left(T_{i_2} + \frac{v - v^{-1}}{\mathsf{q}^{a_2} v^{b_2} - 1} \right) \cdots \left(T_{i_r} + \frac{v - v^{-1}}{\mathsf{q}^{a_r} v^{b_r} - 1} \right) \cdot 1 \end{aligned}$$

where $a_k \in \mathbb{Z}_{>0}$ and $b_k \in \mathbb{Z}$ are explicit integers.

At the specializations $\mathsf{q} = 0, \infty$ we obtain

$$\begin{aligned} E_\lambda(\infty, v) &= v^{-\ell(u_\lambda)} T_{i_1} T_{i_2} \cdots T_{i_r} \cdot 1 \\ &= v^{-\ell(u_\lambda)} T(\sigma_\lambda, A_0) \cdot 1 \qquad (6.8) \\ E_\lambda(0, v) &= v^{-\ell(u_\lambda)} \left(T_{i_1} - (v - v^{-1})\right) \left(T_{i_2} - (v - v^{-1})\right) \cdots \left(T_{i_r} - (v - v^{-1})\right) \cdot 1 \\ &= v^{-\ell(u_\lambda)} T_{i_1}^{-1} T_{i_2}^{-1} \cdots T_{i_r}^{-1} \cdot 1 \\ &= v^{-\ell(u_\lambda)} T(\sigma_\lambda, -A_0) \cdot 1\,. \qquad (6.9) \end{aligned}$$

In order to compute more effectively in the polynomial representation, we consider the change of basis to the Bernstein basis (the oriented basis corresponding to A_∞):

$$T(\sigma_\lambda, A_0) = \sum_{\tau \in \mathscr{D}_{A_0}^{A_\infty}(\sigma_\lambda)} m_{\tau \leq \sigma}^{A_\infty} T(\pi(\tau), A_\infty) \tag{6.10}$$

$$\begin{aligned} T(\sigma_\lambda, -A_0) &= \overline{T(\sigma_\lambda, A_0)} \qquad (6.11) \\ &= \sum_{\tau \in \mathscr{D}_{A_0}^{-A_\infty}(\sigma_\lambda)} \overline{m_{\tau \leq \sigma}^{-A_\infty} T(\pi(\tau), -A_\infty)} \\ &= \sum_{\tau \in \mathscr{D}_{A_0}^{-A_\infty}(\sigma_\lambda)} \overline{m_{\tau \leq \sigma}^{-A_\infty}} T(\pi(\tau), A_\infty)\,. \end{aligned}$$

Applying these to 1 in the polynomial representation, we obtain

$$E_\lambda(\infty, v) = v^{-\ell(u_\lambda)} T(\sigma_\lambda, A_0) \cdot 1 \tag{6.12}$$
$$= v^{-\ell(u_\lambda)} \sum_{\tau \in \mathscr{D}_{A_0}^{A_\infty}(\sigma_\lambda)} m_{\tau \leq \sigma}^{A_\infty} T(\pi(\tau), A_\infty) \cdot 1$$
$$= v^{-\ell(u_\lambda)} \sum_{\tau \in \mathscr{D}_{A_0}^{A_\infty}(\sigma_\lambda)} m_{\tau \leq \sigma}^{A_\infty} X^{\mathrm{wt}(\pi(\tau))} v^{\ell(\mathrm{dir}(\pi(\tau)))}$$

$$E_\lambda(0, v) = v^{-\ell(u_\lambda)} T(\sigma_\lambda, -A_0) \cdot 1 \tag{6.13}$$
$$= v^{-\ell(u_\lambda)} \sum_{\tau \in \mathscr{D}_{A_0}^{-A_\infty}(\sigma_\lambda)} \overline{m_{\tau \leq \sigma}^{-A_\infty}} T(\pi(\tau), A_\infty) \cdot 1$$
$$= v^{-\ell(u_\lambda)} \sum_{\tau \in \mathscr{D}_{A_0}^{-A_\infty}(\sigma_\lambda)} \overline{m_{\tau \leq \sigma}^{-A_\infty}} X^{\mathrm{wt}(\pi(\tau))} v^{\ell(\mathrm{dir}(\pi(\tau)))}.$$

Let $[X^\nu]f$ denote the coefficient of the monomial X^ν in $f \in \mathbb{Z}[v^{\pm 1}][Q_\circ^\vee]$. Using Theorem 5.6 and Lemma 2.18, we deduce the following:

THEOREM 6.14. *For any $\lambda, \nu \in Q_\circ^\vee$, we have*

$$[X^\nu]E_\lambda(\infty, v)\big|_{v^2 = q} = q^{\langle \lambda_- + \nu, \rho \rangle} \#_{\mathbb{F}_q}\left(\sqcup_{u \in W_\circ} \mathbf{U}^+ \varpi^\nu u \cdot \mathbf{I} \cap \mathbf{I} m_\lambda \cdot \mathbf{I}\right) \tag{6.15}$$

$$[X^\nu]E_\lambda(0, v)\big|_{v^{-2} = q} = q^{\langle \lambda_- - \nu, \rho \rangle + \ell(u_\lambda)} \#_{\mathbb{F}_q}\left(\sqcup_{u \in W_\circ} \mathbf{U}^- \varpi^\nu u \cdot \mathbf{I} \cap \mathbf{I} m_\lambda \cdot \mathbf{I}\right) \tag{6.16}$$

where λ_- is the unique antidominant weight in the $W_\circ$-orbit of λ.

REMARK 6.17. Formulas (6.12) and (6.13) are special cases of the Ram-Yip formula [**RY**, Theorem 3.1], which gives a combinatorial expression for $E_\lambda(\mathsf{q}, v)$ with both parameters. Our change of basis computations leading to these formulas are entirely parallel to the proof of the Ram-Yip formula, but they are simpler at $\mathsf{q} = 0, \infty$.

6.3.1. *Comparison with Ion's formula.* Ion has proved a slightly different version of (6.16) in [**Ion**, Theorem 4.2], namely

$$[X^\nu]E_\lambda(0, v)\big|_{v^{-2} = q} = q^{\langle \lambda_- + \nu, \rho \rangle + \ell(u_\lambda) - \ell(w_0)} \#_{\mathbb{F}_q}(\mathbf{U}^- \varpi^{-\nu} \cdot \mathbf{I} \cap \mathbf{K} \varpi^{-\lambda} \cdot \mathbf{I}) \tag{6.18}$$

where $\mathbf{K} = \sqcup_{u \in W_\circ} \mathbf{I} w \mathbf{I}$ and w_0 is the long element of $W_\circ$. One can establish the equality of the two formulas as follows.[1]

Let $G \xrightarrow{r} G/\mathbf{I} \xrightarrow{p} G/\mathbf{K}$ be the canonical projections. Observe that $p(X \cap Y) = p(X) \cap p(Y)$ for any $X, Y \subset G/\mathbf{I}$ such that $r^{-1}(X)$ is right $\mathbf{K}$-stable. Both of the sets $\sqcup_{u \in W_\circ} \mathbf{U}^- \varpi^\nu u \mathbf{I}$ and $\sqcup_{u \in W_\circ} \mathbf{I} m_\lambda u \mathbf{I}$ are right $\mathbf{K}$-stable (for the former, see, e.g., [**BD**, Corollary 1]).

Since m_λ is a minimal coset representative, the restriction $p_{|_{\mathbf{I} m_\lambda \cdot \mathbf{I}}} : \mathbf{I} m_\lambda \cdot \mathbf{I} \to \mathbf{I} \varpi^\lambda \cdot \mathbf{K}$ is an isomorphism. Hence it restricts further to a bijection between $(\sqcup_{u \in W_\circ} \mathbf{U}^- \varpi^\nu u \cdot \mathbf{I}) \cap \mathbf{I} m_\lambda \cdot \mathbf{I}$ and its image under p. We obtain

$$\#_{\mathbb{F}_q}(\sqcup_{u \in W_\circ} \mathbf{U}^- \varpi^\nu u \cdot \mathbf{I} \cap \mathbf{I} m_\lambda \cdot \mathbf{I}) = \#_{\mathbb{F}_q}(\mathbf{U}^- \varpi^\nu \cdot \mathbf{K} \cap \mathbf{I} \varpi^\lambda \cdot \mathbf{K}). \tag{6.19}$$

---

[1] It may help the reader to be aware that [**Ion**] uses different conventions for the Macdonald polynomial parameters—namely our $\mathsf{q} = 0$ and v arbitrary corresponds to his $\mathsf{q} = \infty$ and $t = v^{-2}$.

On the other hand, one can verify each fiber of the map $p_{|_{\mathbf{U}^-\varpi^\nu\cdot\mathbf{I}}} : \mathbf{U}^-\varpi^\nu \cdot \mathbf{I} \to \mathbf{U}^-\varpi^\nu \cdot \mathbf{K}$ contains exactly $q^{\ell(w_\circ)}$ points, and that the same is true for the further restriction of p to $\mathbf{U}^-\varpi^\nu \cdot \mathbf{I} \cap (\sqcup_{u\in W_\circ} \mathbf{I} m_\lambda u \cdot \mathbf{I})$. Therefore

$$\#_{\mathbb{F}_q}(\mathbf{U}^-\varpi^\nu \cdot \mathbf{K} \cap \mathbf{I}\varpi^\lambda \cdot \mathbf{K}) = q^{-\ell(w_\circ)} \#_{\mathbb{F}_q}(\mathbf{U}^-\varpi^\nu \cdot \mathbf{I} \cap (\sqcup_{u\in W_\circ} \mathbf{I} m_\lambda u \cdot \mathbf{I}))\,. \tag{6.20}$$

Finally, we claim that for any $x \in W$ and $\nu \in Q_\circ^\vee$ one has

$$\#_{\mathbb{F}_q}(\mathbf{U}^-\varpi^\nu \cdot \mathbf{I} \cap \mathbf{I}x \cdot \mathbf{I}) = q^{\langle \nu, 2\rho\rangle} \#_{\mathbb{F}_q}(\mathbf{U}^-\varpi^{-\nu}\mathbf{I} \cap \mathbf{I}x^{-1} \cdot \mathbf{I})\,. \tag{6.21}$$

This follows from Theorem 5.6 and the following lemma, which gives a termwise equality of the respective sums from (5.7), up to the factor $\ell_{A_{-\infty}}(\varpi^\nu) = \langle \nu, 2\rho\rangle$.

LEMMA 6.22. *Suppose $\nu \in Q_\circ^\vee$ and σ is a reduced expression of length r. Then the map $\tau = (\tau_1, \dots, \tau_r) \mapsto (\tau_r, \dots, \tau_1) =: \tau'$ reversing a subexpression gives a bijection*

$$\{\tau \in \mathscr{D}_{A_0}^{A_{-\infty}}(\sigma) \mid \pi(\tau) = \varpi^\nu\} \xrightarrow{\sim} \{\tau \in \mathscr{D}_{A_0}^{A_{-\infty}}(\sigma') \mid \pi(\tau) = \varpi^{-\nu}\} \tag{6.23}$$

such that $\kappa_\tau = \kappa_{\tau'}$ and $\zeta_\tau^\pm = \zeta_{\tau'}^\mp$.

Therefore

$$\begin{aligned} \#_{\mathbb{F}_q}(\mathbf{U}^-\varpi^\nu \cdot \mathbf{I} \cap (\sqcup_{u\in W_\circ} \mathbf{I} m_\lambda u \cdot \mathbf{I})) &= q^{\langle \nu, 2\rho\rangle} \#_{\mathbb{F}_q}(\mathbf{U}^-\varpi^{-\nu} \cdot \mathbf{I} \cap (\sqcup_{u\in W_\circ} \mathbf{I} u^{-1} m_\lambda^{-1} \cdot \mathbf{I})) \\ &= q^{\langle \nu, 2\rho\rangle} \#_{\mathbb{F}_q}(\mathbf{U}^-\varpi^{-\nu} \cdot \mathbf{I} \cap \mathbf{K}\varpi^{-\lambda} \cdot \mathbf{I})\,. \end{aligned} \tag{6.24}$$

By stringing together (6.19),(6.20), and (6.24), our formula (6.16) is converted to Ion's formula (6.18).

Acknowledgments

D.M. thanks Abid Ali, Gerald Cliff, Manish Patnaik, and Anna Puskás for their talks and discussion in the learning seminar on the topic of [**GL**] which took place in Fall 2015 at the University of Alberta. He especially thanks Manish Patnaik and Anna Puskás for their work in organizing the seminar. D.O. thanks Elizabeth Milićević for helpful discussions about point counting in affine flag varieties. We also wish to thank an anonymous referee whose comments led to several improvements in our exposition.

References

[BGR] N. Bardy-Panse, S. Gaussent, and G. Rousseau, *Iwahori-Hecke algebras for Kac-Moody groups over local fields*, Pacific J. Math. **285** (2016), no. 1, 1–61. MR3554242

[BD] Y. Billig and M. Dyer, *Decompositions of Bruhat type for the Kac-Moody groups*, Nova J. Algebra Geom. **3** (1994), no. 1, 11–39. MR1332937

[BKP] A. Braverman, D. Kazhdan, and M. M. Patnaik, *Iwahori-Hecke algebras for p-adic loop groups*, Invent. Math. **204** (2016), no. 2, 347–442. MR3489701

[Bri] M. Brion, *Lectures on the geometry of flag varieties*, Topics in cohomological studies of algebraic varieties, Trends Math., Birkhäuser, Basel, 2005, pp. 33–85. MR2143072

[Ch1] I. Cherednik, *Intertwining operators of double affine Hecke algebras*, Selecta Math. (N.S.) **3** (1997), no. 4, 459–495. MR1613515

[Ch2] I. Cherednik, *Double affine Hecke algebras*, London Mathematical Society Lecture Note Series, vol. 319, Cambridge University Press, Cambridge, 2005. MR2133033

[Deo] V. V. Deodhar, *On some geometric aspects of Bruhat orderings. I. A finer decomposition of Bruhat cells*, Invent. Math. **79** (1985), no. 3, 499–511. MR782232

[Dy] M. J. Dyer, *Hecke algebras and shellings of Bruhat intervals. II. Twisted Bruhat orders*, Kazhdan-Lusztig theory and related topics (Chicago, IL, 1989), Contemp. Math., vol. 139, Amer. Math. Soc., Providence, RI, 1992, pp. 141–165. MR1197833

[Dy2] M. J. Dyer. "Modules for the dual nil Hecke ring." http://www3.nd.edu/ dyer/papers/nilhecke.pdf

[GL] S. Gaussent and P. Littelmann, *LS galleries, the path model, and MV cycles*, Duke Math. J. **127** (2005), no. 1, 35–88. MR2126496

[GR1] S. Gaussent and G. Rousseau, *Kac-Moody groups, hovels and Littelmann paths* (English, with English and French summaries), Ann. Inst. Fourier (Grenoble) **58** (2008), no. 7, 2605–2657. MR2498360

[GR2] S. Gaussent and G. Rousseau, *Spherical Hecke algebras for Kac-Moody groups over local fields*, Ann. of Math. (2) **180** (2014), no. 3, 1051–1087. MR3245012

[Go] T. Gobet, *Twisted filtrations of Soergel bimodules and linear Rouquier complexes*, J. Algebra **484** (2017), 275–309. MR3656723

[Gö] U. Görtz, *Alcove walks and nearby cycles on affine flag manifolds*, J. Algebraic Combin. **26** (2007), no. 4, 415–430. MR2341858

[Ion] B. Ion, *A weight multiplicity formula for Demazure modules*, Int. Math. Res. Not. **5** (2005), 311–323. MR2130799

[KL] D. Kazhdan and G. Lusztig, *Representations of Coxeter groups and Hecke algebras*, Invent. Math. **53** (1979), no. 2, 165–184. MR560412

[Kum] S. Kumar, *Kac-Moody groups, their flag varieties and representation theory*, Progress in Mathematics, vol. 204, Birkhäuser Boston, Inc., Boston, MA, 2002. MR1923198

[Lus] G. Lusztig, *Affine Hecke algebras and their graded version*, J. Amer. Math. Soc. **2** (1989), no. 3, 599–635. MR991016

[Mac] I. G. Macdonald, *Affine Hecke algebras and orthogonal polynomials*, Cambridge Tracts in Mathematics, vol. 157, Cambridge University Press, Cambridge, 2003. MR1976581

[M] D. Muthiah, *On Iwahori-Hecke algebras for p-adic loop groups: double coset basis and Bruhat order*, Amer. J. Math. **140** (2018), no. 1, 221–244, DOI 10.1353/ajm.2018.0004. MR3749194

[MO] D. Muthiah, and D. Orr. "On the double-affine Bruhat order: the $\varepsilon = 1$ conjecture and classification of covers in ADE type." arXiv:1609.03653

[PRS] J. Parkinson, A. Ram, and C. Schwer, *Combinatorics in affine flag varieties*, J. Algebra **321** (2009), no. 11, 3469–3493. MR2510057

[Ram] A. Ram, *Alcove walks, Hecke algebras, spherical functions, crystals and column strict tableaux*, Pure Appl. Math. Q. **2** (2006), no. 4, Special Issue: In honor of Robert D. MacPherson., 963–1013. MR2282411

[RY] A. Ram and M. Yip, *A combinatorial formula for Macdonald polynomials*, Adv. Math. **226** (2011), no. 1, 309–331. MR2735761

[Sch] C. Schwer, *Galleries, Hall-Littlewood polynomials, and structure constants of the spherical Hecke algebra*, Int. Math. Res. Not. (2006), Art. ID 75395, 31. MR2264725

Department of Mathematics and Statistics, Lederle Graduate Research Tower, 1623D, University of Massachusetts Amherst, 710 North Pleasant Street, Amherst, Massachusetts 01003-9305

Email address: muthiah@math.umass.edu

Department of Mathematics (MC 0123), 460 McBryde Hall, Virginia Tech, 225 Stanger Street, Blacksburg, Virginia 24061

Email address: dorr@vt.edu

Contemporary Mathematics
Volume **713**, 2018
https://doi.org/10.1090/conm/713/14318

Critical points of master functions and mKdV hierarchy of type $A_{2n}^{(2)}$

Alexander Varchenko and Tyler Woodruff

Abstract. We consider the population of critical points generated from the critical point of the master function with no variables, which is associated with the trivial representation of the twisted affine Lie algebra $A_{2n}^{(2)}$. The population is naturally partitioned into an infinite collection of complex cells $\mathbb{C}^m$, where m are some positive integers. For each cell we define an injective rational map $\mathbb{C}^m \to \mathcal{M}(A_{2n}^{(2)})$ of the cell to the space $\mathcal{M}(A_{2n}^{(2)})$ of Miura opers of type $A_{2n}^{(2)}$. We show that the image of the map is invariant with respect to all mKdV flows on $\mathcal{M}(A_{2n}^{(2)})$ and the image is point-wise fixed by all mKdV flows $\frac{\partial}{\partial t_r}$ with index r greater than $4m$.

1. Introduction

Let $\mathfrak{g}$ be a Kac-Moody algebra with invariant scalar product $(\, , \,)$, $\mathfrak{h} \subset \mathfrak{g}$ Cartan subalgebra, $\alpha_0, \dots, \alpha_n$ simple roots. Let $\Lambda_1, \dots, \Lambda_N$ be dominant integral weights, $k_0, \dots, k_n$ nonnegative integers, $k = k_0 + \dots + k_n$.

Consider $\mathbb{C}^N$ with coordinates $z = (z_1, \dots, z_N)$. Consider $\mathbb{C}^k$ with coordinates u collected into $n+1$ groups, the j-th group consisting of k_j variables,

$$u = (u^{(0)}, \dots, u^{(n)}), \qquad u^{(j)} = (u_1^{(j)}, \dots, u_{k_j}^{(j)}).$$

The *master function* is the multivalued function on $\mathbb{C}^k \times \mathbb{C}^N$ defined by the formula

$$\begin{aligned}(1.1) \qquad \Phi(u,z) = & \sum_{a<b} (\Lambda_a, \Lambda_b) \ln(z_a - z_b) - \sum_{a,i,j} (\alpha_j, \Lambda_a) \ln(u_i^{(j)} - z_a) + \\ & + \sum_{j<j'} \sum_{i,i'} (\alpha_j, \alpha_{j'}) \ln(u_i^{(j)} - u_{i'}^{(j')}) + \sum_j \sum_{i<i'} (\alpha_j, \alpha_j) \ln(u_i^{(j)} - u_{i'}^{(j)}),\end{aligned}$$

with singularities at the places where the arguments of the logarithms are equal to zero.

A point in $\mathbb{C}^k \times \mathbb{C}^N$ can be interpreted as a collection of particles in $\mathbb{C}$: $z_a, u_i^{(j)}$. A particle z_a has weight Λ_a, a particle $u_i^{(j)}$ has weight $-\alpha_j$. The particles interact pairwise. The interaction of two particles is determined by the scalar product of their weights. The master function is the "total energy" of the collection of particles.

2010 *Mathematics Subject Classification.* Primary 17B80; Secondary 17B67, 37K10, 81R12.

A.Varchenko was supported in part by NSF grants DMS-1362924, DMS-1665239, and Simons Foundation grant 336826.

Notice that all scalar products are integers. So the master function is the logarithm of a rational function. From a "physical" point of view, all interactions are integer multiples of a certain unit of measurement. This is important for what will follow.

The variables u are the *true* variables, variables z are *parameters*. We may think that the positions of z-particles are fixed and the u-particles can move.

There are "global" characteristics of this situation,

$$I(z,\kappa) = \int e^{\Phi(u,z)/\kappa} A(u,z)du,$$

where $A(u,z)$ is a suitable density function, κ a parameter, and there are "local" characteristics – critical points of the master function with respect to the u-variables,

$$d_u\Phi(u,z) = 0.$$

A critical point is an equilibrium position of the u-particles for fixed positions of the z-particles. In this paper we are interested in the equilibrium positions of the u-particles.

Examples of master functions associated with $\mathfrak{g} = \mathfrak{sl}_2$ were considered by Stieltjes and Heine in 19th century, see for example [**Sz**]. Master functions we introduced in [**SV**] to construct integral representations for solutions of the KZ equations, see also [**V1**, **V2**].

The critical points of master functions with respect to the u-variables were used to find eigenvectors in the associated Gaudin models by the Bethe ansatz method, see [**BF**, **RV**, **V3**]. In important cases the algebra of functions on the critical set of a master function is closely related to Schubert calculus, see [**MTV09**].

In [**ScV**, **MV1**] it was observed that the critical points of master functions with respect to the u-variables can be deformed and form families. Having one critical point, one can construct a family of new critical points. The family is called a population of critical points. A point of the population is a critical point of the same master function or of another master function associated with the same $\mathfrak{g}, \Lambda_1, \dots, \Lambda_N$ but with different integer parameters $k_0, \dots, k_n$. The population is a variety isomorphic to the flag variety of the Kac-Moody algebra $\mathfrak{g}^t$ Langlands dual to $\mathfrak{g}$, see [**MV1**, **MV2**, **F**].

In [**VW**], it was discovered that the population originated from the critical point of the master function associated with the affine Lie algebra $\widehat{\mathfrak{sl}}_{n+1}$ and the parameters $N = 0, k_0 = \cdots = k_n = 0$ is connected with the mKdV integrable hierarchy associated with $\widehat{\mathfrak{sl}}_{n+1}$. Namely, that population can be naturally embedded into the space of $\widehat{\mathfrak{sl}}_{n+1}$ Miura opers so that the image of the embedding is invariant with respect to all mKdV flows on the space of Miura opers. For $n = 1$, that result follows from the classical paper by M. Adler and J. Moser [**AM**], which served as a motivation for [**VW**].

In this paper we prove the analogous statement for the twisted affine Lie algebra $A^{(2)}_{2n}$. The special case $A^{(2)}_2$ was considered in [**VWW14**].

In Sections 2 - 4 we follow the paper [**DS**] by V. Drinfled and V. Sokolov. We review the affine Lie algebras $A^{(1)}_{2n}$ and $A^{(2)}_{2n}$, the associated mKdV and KdV hierarchies, Miura maps. For example, the $A^{(2)}_{2n}$ mKdV hierarchy is a sequence of commuting flows on the infinite-dimensional space $\mathcal{M}(A^{(2)}_{2n})$ of the $A^{(2)}_{2n}$ Miura opers.

In Section 5 we study tangent maps to Miura maps. In Section 6 formula (6.1), we introduce our master functions,

$$
\begin{aligned}
(1.2)\ \Phi(u^{(0)}_1,\dots,u^{(0)}_{k_0},\dots,u^{(n)}_1,\dots,u^{(n)}_{k_n})
&= -4\sum_{i,i'}\ln(u^{(n-1)}_i-u^{(n)}_{i'})-2\sum_{j=0}^{n-2}\sum_{i,i'}\ln(u^{(j)}_i-u^{(j+1)}_{i'})\\
&+8\sum_{i<i'}\ln(u^{(n)}_i-u^{(n)}_{i'})+4\sum_{j=1}^{n-1}\sum_{i<i'}\ln(u^{(j)}_i-u^{(j)}_{i'})+2\sum_{i<i'}\ln(u^{(0)}_i-u^{(0)}_{i'}).
\end{aligned}
$$

This master function is the special case of the master function in (1.1). The master function in (1.2) is defined by formula (1.1) if $\mathfrak{g}$ is the Langlands dual to $A^{(2)}_{2n}$ and $N=0$, see a remark in Section 6.1.

Following [**MV1**, **MV2**, **VW**], we describe the generation procedure of new critical points starting from a given one. We define the population of critical points generated from the critical point of the function with no variables, namely, the function corresponding to the parameters $k_0=\dots=k_n=0$. That population is partitioned into complex cells $\mathbb{C}^m$ labeled by degree increasing sequences $J=(j_1,\dots,i_m)$, see the definition in Section 6.5.

In Theorem 6.5 we deduce from [**MV3**] that every critical point of the master function in (1.2) with arbitrary parameters $k_0,\dots,k_n$ belongs a cell of our population. Moreover, a function in (1.2) with some parameters $k_0,\dots,k_n$ either does not have critical points at all or its critical points form a cell $\mathbb{C}^m$ corresponding to a degree increasing sequence.

In Section 7, to every degree increasing sequence J we assign a rational injective map $\mu^J:\mathbb{C}^m\to\mathcal{M}(A^{(2)}_{2n})$ of the cell corresponding to J to the space $\mathcal{M}(A^{(2)}_{2n})$ of Miura opers of type $A^{(2)}_{2n}$. We describe properties of that map.

In Section 8, we formulate and prove our main result. Theorem 8.1 says that for any degree increasing sequence, the variety $\mu^J(\mathbb{C}^m)$ is invariant with respect to all mKdV flows on $\mathcal{M}(A^{(2)}_{2n})$ and that variety is point-wise fixed by all flows $\frac{\partial}{\partial t_r}$ with index r greater than $4m$.

Our result shows that there is a deep interrelation between the critical set of the master functions of the form (1.2) and rational finite-dimensional submanifolds of the space $\mathcal{M}(A^{(2)}_{2n})$, invariant with respect to all flows of the $A^{(2)}_{2n}$ mKdV hierarchy.

Initially the critical points of the master functions were related to quantum integrable systems of the Gaudin type through the Bethe ansatz, [**SV**, **BF**, **RV**, **V3**]. Our result shows that the critical points are also related to the classical integrable systems, namely, the mKdV hierarchies.

In the next papers we plan to extend this result to other affine Lie algebras.

2. Kac-Moody algebra of type $A_{2n}^{(1)}$

In this section we follow [**DS**, Section 5].

2.1. Definition. For $n \geqslant 2$, consider the $(2n+1)\times(2n+1)$ Cartan matrix of type $A_{2n}^{(1)}$,

$$A_{2n}^{(1)} = \begin{pmatrix} a_{0,0} & a_{0,1} & \dots & a_{0,2n} \\ \dots & \dots & \dots & \dots \\ \dots & \dots & \dots & \dots \\ a_{2n,0} & a_{2n,1} & \dots & a_{2n,2n} \end{pmatrix} = \begin{pmatrix} 2 & -1 & 0 & \dots & 0 & -1 \\ -1 & 2 & -1 & \dots & 0 & 0 \\ 0 & -1 & 2 & \dots & \dots & \dots \\ \dots & \dots & \dots & \dots & \dots & \dots \\ 0 & 0 & 0 & \dots & 2 & -1 \\ -1 & 0 & 0 & \dots & -1 & 2 \end{pmatrix}.$$

The Kac-Moody algebra $\mathfrak{g}(A_{2n}^{(1)})$ *of type* $A_{2n}^{(1)}$ is the Lie algebra with *canonical generators* $E_i, H_i, F_i \in \mathfrak{g}(A_{2n}^{(1)}), i = 0, \dots, 2n$, subject to the relations:

$$[E_i, F_j] = \delta_{i,j} H_i,$$
$$[H_i, E_j] = a_{i,j} E_j, \qquad [H_i, F_j] = -a_{i,j} F_j, \qquad (\operatorname{ad} E_i)^{1-a_{i,j}} E_j = 0,$$
$$(\operatorname{ad} F_i)^{1-a_{i,j}} F_j = 0, \qquad [H_i, H_j] = 0, \qquad \textstyle\sum_{i=0}^{2n} H_i = 0,$$

see [**DS**, Section 5]. The Lie algebra $\mathfrak{g}(A_{2n}^{(1)})$ is graded with respect to the *standard grading*, $\deg E_i = 1, \deg F_i = -1$, $i = 0, \dots, 2n$. Let $\mathfrak{g}(A_{2n}^{(1)})^j = \{x \in \mathfrak{g}(A_{2n}^{(1)}) \mid \deg x = j\}$, then $\mathfrak{g}(A_{2n}^{(1)}) = \oplus_{j\in\mathbb{Z}}\, \mathfrak{g}(A_{2n}^{(1)})^j$.

Notice that $\mathfrak{g}(A_{2n}^{(1)})^0$ is the $2n$-dimensional space generated by the H_i. Denote $\mathfrak{h} = \mathfrak{g}(A_{2n}^{(1)})^0$. Introduce elements α_j of the dual space $\mathfrak{h}^*$ by the conditions $\langle \alpha_j, H_i\rangle = a_{i,j}$ for $i, j = 0, \dots, 2n$. For $j = 0, 1, \dots 2n$, we denote by $\mathfrak{n}_j^- \subset \mathfrak{g}(A_{2n}^{(1)})$ the Lie subalgebra generated by F_i, $i \in \{0, 1, \dots, 2n\}$, $i \neq j$. For example, $\mathfrak{n}_0^-$ is generated by $F_1, F_2, \dots, F_{2n}$.

2.2. Realizations of $\mathfrak{g}(A_{2n}^{(1)})$. Consider the complex Lie algebra $\mathfrak{sl}_{2n+1}$ with standard basis $e_{i,j}$, $i, j = 1, \dots, 2n+1$.

Let $w = e^{2\pi i/(2n+1)}$. Define the *Coxeter automorphism* $C : \mathfrak{sl}_{2n+1} \to \mathfrak{sl}_{2n+1}$ of order $2n+1$ by the formula

$$C(X) = SXS^{-1}, \quad S = \operatorname{diag}(1, w, \dots, w^{2n}).$$

Denote $(\mathfrak{sl}_{2n+1})_j = \{x \in \mathfrak{sl}_{2n+1} \mid Cx = w^j x\}$. The twisted Lie subalgebra $L(\mathfrak{sl}_{2n+1}, C) \subset \mathfrak{sl}_{2n+1}[\xi, \xi^{-1}]$ is the subalgebra

$$L(\mathfrak{sl}_{2n+1}, C) = \oplus_{j\in\mathbb{Z}}\, \xi^j \otimes (\mathfrak{sl}_{2n+1})_{j \bmod 2n+1}.$$

The isomorphism $\tau_C : \mathfrak{g}(A_{2n}^{(1)}) \to L(\mathfrak{sl}_{2n+1}, C)$ is defined by the formula, for $i = 1, \dots, 2n$

$$E_0 \mapsto \xi \otimes e_{1,2n+1}, \qquad E_i \mapsto \xi \otimes e_{i+1,i},$$
$$F_0 \mapsto \xi^{-1} \otimes e_{2n+1,1}, \qquad F_i \mapsto \xi^{-1} \otimes e_{i,i+1},$$
$$H_0 \mapsto 1 \otimes (e_{1,1} - e_{2n+1,2n+1}), \qquad H_i \mapsto 1 \otimes (-e_{i,i} + e_{i+1,i+1}).$$

Under this isomorphism we have $\mathfrak{g}(A_{2n}^{(1)})^j = \xi^j \otimes (\mathfrak{sl}_{2n+1})_j$.

The *standard automorphism* $\sigma_0 : \mathfrak{sl}_{2n+1} \to \mathfrak{sl}_{2n+1}$ is the identity: $\sigma_0(X) = X$. The isomorphism $\tau_0 : \mathfrak{g}(A_{2n}^{(1)}) \to L(\mathfrak{sl}_{2n+1}, \sigma_0)$ is defined by the formula, for $i = 1, \dots, 2n$

$$E_0 \mapsto \lambda \otimes e_{1,2n+1}, \qquad E_i \mapsto 1 \otimes e_{i+1,i},$$
$$F_0 \mapsto \lambda^{-1} \otimes e_{2n+1,1}, \qquad F_i \mapsto 1 \otimes e_{i,i+1},$$
$$H_0 \mapsto 1 \otimes (e_{1,1} - e_{2n+1,2n+1}), \qquad H_i \mapsto 1 \otimes (-e_{i,i} + e_{i+1,i+1}).$$

2.3. Element $\Lambda^{(1)}$. Denote by $\Lambda^{(1)}$ the element $\sum_{j=0}^{2n} E_j \in \mathfrak{g}(A_{2n}^{(1)})$. Then $\mathfrak{z}(A_{2n}^{(1)}) = \{x \in \mathfrak{g}(A_{2n}^{(1)}) \mid [\Lambda^{(1)}, x] = 0\}$ is an Abelian Lie subalgebra of $\mathfrak{g}(A_{2n}^{(1)})$. Denote $\mathfrak{z}(A_{2n}^{(1)})^j = \mathfrak{z}(A_{2n}^{(1)}) \cap \mathfrak{g}(A_{2n}^{(1)})^j$, then $\mathfrak{z}(A_{2n}^{(1)}) = \oplus_{j\in\mathbb{Z}}\, \mathfrak{z}(A_{2n}^{(1)})^j$. We have $\dim \mathfrak{z}(A_{2n}^{(1)})^j = 1$ if $j \neq 0 \bmod 2n+1$ and $\dim \mathfrak{z}(A_{2n}^{(1)})^j = 0$ otherwise.

Let $\mathfrak{g}(A_{2n}^{(1)})$ be realized as $L(\mathfrak{sl}_{2n+1}, C)$ and written out as $(2n+1) \times (2n+1)$-matrices. For $m \in \mathbb{Z}$ and $1 \leqslant j < 2n+1$ introduce the element

$$A_{(2n+1)m+j} = \xi^{(2n+1)m+j} \otimes \begin{pmatrix} 0 & I_j \\ I_{2n+1-j} & 0 \end{pmatrix} \quad \in \quad L(\mathfrak{sl}_{2n+1}, C),$$

where I_j is the $j \times j$ identity matrix. We have $A_{(2n+1)m+j} = (A_1)^{(2n+1)m+j}$.

If $\mathfrak{g}(A_{2n}^{(1)})$ is realized as $L(\mathfrak{sl}_{2n+1}, \sigma_0)$, introduce the element

$$B_{(2n+1)m+j} = \begin{pmatrix} 0 & \lambda^{m+1} \otimes I_j \\ \lambda^m \otimes I_{2n+1-j} & 0 \end{pmatrix} \quad \in \quad L(\mathfrak{sl}_{2n+1}, \sigma_0).$$

We have $B_{(2n+1)m+j} = (B_1)^{(2n+1)m+j}$.

LEMMA 2.1. *For any $m \in \mathbb{Z}$, $1 \leqslant j < 2n+1$, the elements $(\tau_C)^{-1}(A_{(2n+1)m+j})$, $(\tau_0)^{-1}(B_{(2n+1)m+j})$ of $\mathfrak{z}(A_{2n}^{(1)})^{(2n+1)m+j}$ are equal.* □

Denote by $\Lambda^{(1)}_{(2n+1)m+j}$ the elements $(\tau_C)^{-1}(A_{(2n+1)m+j})$ and $(\tau_0)^{-1}(B_{(2n+1)m+j})$ of $\mathfrak{g}(A_{2n}^{(1)})$. Notice that $\Lambda_1^{(1)} = \sum_{i=0}^{2n} E_i = \Lambda^{(1)}$. For any $m \in \mathbb{Z}, 1 \leqslant j < 2n+1$, the element $\Lambda^{(1)}_{(2n+1)m+j}$ generates $\mathfrak{z}(A_{2n}^{(1)})^{(2n+1)m+j}$.

Let $T = \sum_{j=-\infty}^{m} T_j$ be a formal series with $T_j \in \mathfrak{g}(A_{2n}^{(1)})^j$. Denote $T^+ = \sum_{j=0}^{n} T_j$, $T^- = \sum_{j<0} T_j$. Let $\mathfrak{g}(A_{2n}^{(1)})$ be realized as $\mathfrak{sl}_{2n+1}[\lambda, \lambda^{-1}]$. Consider $\Lambda^{(1)} = B_1$ as a $(2n+1) \times (2n+1)$ matrix depending on the parameter λ. By [**DS**, Lemma 3.4], we may represent T uniquely in the form $T = \sum_{j=-\infty}^{k} b_j\, (\Lambda^{(1)})^j$, $b_j \in \mathrm{Diag}$, where $\mathrm{Diag} \subset \mathfrak{gl}_{2n+1}$ is the space of diagonal $(2n+1) \times (2n+1)$ matrices. Denote $(T)^+_{\Lambda^{(1)}} = \sum_{j=0}^{k} b_j\, (\Lambda^{(1)})^j$, $(T)^-_{\Lambda^{(1)}} = \sum_{j<0} b_j\, (\Lambda^{(1)})^j$.

LEMMA 2.2. *We have $(T)^+_{\Lambda^{(1)}} = T^+$, $(T)^-_{\Lambda^{(1)}} = T^-$, $b_0 = T^0$.*

PROOF. The isomorphism $\iota : \mathfrak{sl}_{2n+1}[\lambda, \lambda^{-1}] \to L(\mathfrak{sl}_{2n+1}, C^{(1)})$ is given by the formula $\lambda^m \otimes e_{k,l} \mapsto \xi^{(2n+1)m+k-l} \otimes e_{k,l}$. We have $\iota(b_0) = \iota(1 \otimes (b_0^1 e_{1,1} + \cdots + b_0^{2n+1} e_{2n+1,2n+1})) = 1 \otimes (b_0^1 e_{1,1} + \cdots + b_0^{2n+1} e_{2n+1,2n+1}) \in \mathfrak{g}(A_{2n}^{(1)})^0$, $\iota(b_1 \Lambda^{(1)}) = \iota((b_1^1 e_{1,1} + b_1^2 e_{2,2} + \cdots + b_1^{2n+1} e_{2n+1,2n+1})(e_{2,1} + \cdots + e_{2n+1,2n} + \lambda e_{1,2n+1}))$ $= \iota(b_1^1 \lambda e_{1,2n+1} + b_1^2 e_{2,1} + \cdots + b_1^{2n+1} e_{2n+1,2n}) = \xi \otimes (b_1^1 e_{1,2n+1} + b_1^2 e_{2,1} + \cdots + b_1^{2n+1} e_{2n+1,2n}) \in \mathfrak{g}(A_{2n}^{(1)})^1$, $\iota(b_{-1}(\Lambda^{(1)})^{-1}) = \iota((b_{-1}^1 e_{1,1} + b_{-1}^2 e_{2,2} + \dots$ $+ b_{-1}^{2n+1} e_{2n+1,2n+1})(e_{1,2} + \cdots + e_{2n,2n+1} + \lambda^{-1} e_{2n+1,1})) = \iota(b_{-1}^1 e_{1,2} + \cdots + b_{-1}^{2n} e_{2n,2n+1} + b_{-1}^{2n+1} \lambda^{-1} e_{2n+1,1}) = \xi^{-1} \otimes (b_{-1}^1 e_{1,2} + \cdots + b_{-1}^{2n} e_{2n,2n+1} + b_{-1}^{2n+1} e_{2n+1,1}) \in \mathfrak{g}(A_{2n}^{(1)})^{-1}$. Similarly one checks that $\iota(b_j\, (\Lambda^{(1)})^j) \in \mathfrak{g}(A_{2n}^{(1)})^j$ for any j. □

We have $(\Lambda^{(1)})^{-1} = \sum_{i=1}^{2n} e_{i,i+1} + \lambda^{-1} e_{2n+1,1}$,

$$E_0 = \Lambda^{(1)} e_{2n+1,2n+1}, \qquad E_i = \Lambda^{(1)} e_{i,i},$$
$$F_0 = e_{2n+1,2n+1} (\Lambda^{(1)})^{-1}, \qquad F_i = e_{i,i} (\Lambda^{(1)})^{-1},$$

for $i = 1, \dots, 2n$.

LEMMA 2.3. *Consider the elements $F_0, F_i + F_{2n+1-i}, 2(F_n + F_{n+1})$ for $i = 1, \dots, n-1$ as $(2n+1) \times (2n+1)$ matrices. Let $g \in \mathbb{C}$. Then*

$$\begin{aligned} e^{gF_0} &= 1 + g e_{2n+1,2n+1} (\Lambda^{(1)})^{-1}, \\ e^{g(F_i + F_{2n+1-i})} &= 1 + g(e_{i,i} + e_{2n+1-i,2n+1-i}) (\Lambda^{(1)})^{-1}, \\ e^{g2(F_n+F_{n+1})} &= 1 + 2g(e_{n,n} + e_{n+1,n+1})(\Lambda^{(1)})^{-1} + 4g^2 e_{n,n} (\Lambda^{(1)})^{-2}. \end{aligned} \tag{2.1}$$

□

LEMMA 2.4. *We have*

$$e_{i+1,i+1} \Lambda^{(1)} = \Lambda^{(1)} e_{i,i}, \qquad e_{i,i} (\Lambda^{(1)})^{-1} = (\Lambda^{(1)})^{-1} e_{i+1,i+1}, \tag{2.2}$$

for all i, where we set $e_{2n+2,2n+2} = e_{1,1}$. □

3. Kac-Moody algebra of type $A_{2n}^{(2)}$

In this section we follow [**DS**, Section 5].

3.1. Definition. For $n \geqslant 2$, consider the $(n+1) \times (n+1)$ Cartan matrix of type $A_{2n}^{(2)}$,

$$A_{2n}^{(2)} = \begin{pmatrix} a_{0,0} & a_{0,1} & \dots & a_{0,n} \\ \dots & \dots & \dots & \dots \\ \dots & \dots & \dots & \dots \\ a_{n,0} & a_{n,1} & \dots & a_{n,n} \end{pmatrix} = \begin{pmatrix} 2 & -1 & 0 & \dots & \dots & \dots & 0 \\ -2 & 2 & -1 & 0 & \dots & \dots & \dots \\ 0 & -1 & 2 & -1 & \dots & \dots & \dots \\ \dots & 0 & -1 & \dots & \dots & \dots & \dots \\ \dots & \dots & \dots & \dots & 2 & -1 & 0 \\ \dots & \dots & \dots & \dots & -1 & 2 & -1 \\ 0 & \dots & \dots & \dots & 0 & -2 & 2 \end{pmatrix}.$$

The Kac-Moody algebra $\mathfrak{g}(A_{2n}^{(2)})$ *of type* $A_{2n}^{(2)}$ is the Lie algebra with *canonical generators* $e_i, h_i, f_i \in \mathfrak{g}(A_{2n}^{(2)})$, $i = 0, \dots, n$, subject to the relations

$$\begin{aligned} &[e_i, f_j] = \delta_{i,j} h_i, \qquad [h_i, e_j] = a_{i,j} e_j, \qquad [h_i, f_j] = -a_{i,j} f_j, \\ &(\operatorname{ad} e_i)^{1-a_{i,j}} e_j = 0, \qquad (\operatorname{ad} f_i)^{1-a_{i,j}} f_j = 0, \qquad [h_i, h_j] = 0, \\ &2(h_0 + h_1 + \dots + h_{n-1}) + h_n = 0, \end{aligned}$$

see [**DS**, Section 5].

The Lie algebra $\mathfrak{g}(A_{2n}^{(2)})$ is graded with respect to the *standard grading*, $\deg e_i = 1, \deg f_i = -1$, $i = 0, \dots, n$. Let $\mathfrak{g}(A_{2n}^{(2)})^j = \{x \in \mathfrak{g}(A_{2n}^{(2)}) \mid \deg x = j\}$, then $\mathfrak{g}(A_{2n}^{(2)}) = \oplus_{j \in \mathbb{Z}} \mathfrak{g}(A_{2n}^{(2)})^j$.

Notice that $\mathfrak{g}(A_{2n}^{(2)})^0$ is the n-dimensional space generated by the h_i. Denote $\mathfrak{h} = \mathfrak{g}(A_{2n}^{(2)})^0$. Introduce elements α_j of the dual space $\mathfrak{h}^*$ by the conditions $\langle \alpha_j, h_i \rangle = a_{i,j}$ for $i, j = 0, \dots, n$.

3.2. Realizations of $\mathfrak{g}(A_{2n}^{(2)})$. Consider the complex Lie algebra $\mathfrak{sl}_{2n+1}$ with standard basis $e_{i,j}$, $i,j=1,\dots,2n+1$.

Let $w=e^{2\pi i/(4n+2)}$. Define the *Coxeter automorphism* $C:\mathfrak{sl}_{2n+1}\to\mathfrak{sl}_{2n+1}$ of order $4n+2$ by the formula

$$C(X)=-SX^TS^{-1},\quad S=\operatorname{diag}(1,-w,w^2,\dots,w^{2n-2},-w^{2n-1},w^{2n}),$$

where the $\{\}^T$ denotes transposition across the antidiagonal. Denote $(\mathfrak{sl}_{2n+1})_j=\{x\in\mathfrak{sl}_{2n+1}\mid Cx=w^jx\}$. The twisted Lie subalgebra $L(\mathfrak{sl}_{2n+1},C)\subset\mathfrak{sl}_{2n+1}[\xi,\xi^{-1}]$ is the subalgebra

$$L(\mathfrak{sl}_{2n+1},C)=\oplus_{j\in\mathbb{Z}}\,\xi^j\otimes(\mathfrak{sl}_{2n+1})_{j \bmod 4n+2}.$$

The isomorphism $\tau_C:\mathfrak{g}(A_{2n}^{(2)})\to L(\mathfrak{sl}_{2n+1},C)$ is defined by the formula

$$\begin{aligned}
&e_0\mapsto\xi\otimes e_{1,2n+1}, \qquad e_n\mapsto\xi\otimes(e_{n+1,n}+e_{n+2,n+1})\\
&f_0\mapsto\xi^{-1}\otimes e_{2n+1,1}, \qquad f_n\mapsto\xi^{-1}\otimes 2(e_{n,n+1}+e_{n+1,n+2})\\
&e_i\mapsto\xi\otimes(e_{i+1,i}+e_{2n+2-i,2n+1-i}), \qquad f_i\mapsto\xi^{-1}\otimes(e_{i,i+1}+e_{2n+1-i,2n+2-i}),\\
&h_0\mapsto 1\otimes(e_{1,1}-e_{2n+1,2n+1}), \qquad h_n\mapsto 1\otimes 2(-e_{n,n}+e_{n+2,n+2}),\\
&h_i\mapsto 1\otimes(-e_{i,i}+e_{i+1,i+1}-e_{2n+1-i,2n+1-i}+e_{2n+2-i,2n+2-i}).
\end{aligned}$$

Under this isomorphism we have $\mathfrak{g}(A_{2n}^{(2)})^j=\xi^j\otimes(\mathfrak{sl}_{2n+1})_j$. Define the *standard automorphism* $\sigma_0:\mathfrak{sl}_{2n+1}\to\mathfrak{sl}_{2n+1}$ of order 2 by the formula

$$\sigma_0(X)=-QX^TQ^{-1}=-X^T,\quad Q=\operatorname{diag}(1,-1,\dots,-1,1).$$

Where the $\{\}^T$ is again transposition across the antidiagonal. Let $(\mathfrak{sl}_{2n+1})_{0,j}=\{x\in\mathfrak{sl}_{2n+1}\mid\sigma_0x=(-1)^jx\}$. Then the twisted Lie subalgebra $L(\mathfrak{sl}_{2n+1},\sigma_0)\subset\mathfrak{sl}_{2n+1}[\lambda,\lambda^{-1}]$ is the subalgebra

$$L(\mathfrak{sl}_{2n+1},\sigma_0)=\oplus_{j\in\mathbb{Z}}\,\lambda^j\otimes(\mathfrak{sl}_{2n+1})_{0,\,j \bmod 2}.$$

The isomorphism $\tau_0:\mathfrak{g}(A_{2n}^{(2)})\to L(\mathfrak{sl}_{2n+1},\sigma_0)$ is defined by the formula

$$\begin{aligned}
&e_0\mapsto\lambda\otimes e_{1,2n+1}, \qquad e_i\mapsto 1\otimes(e_{i+1,i}+e_{2n+2-i,2n+1-i}),\\
&f_0\mapsto\lambda^{-1}\otimes e_{2n+1,1}, \qquad f_i\mapsto 1\otimes(e_{i,i+1}+e_{2n+1-i,2n+2-i}),\\
&f_n\mapsto 1\otimes 2(e_{n,n+1}+e_{n+1,n+2}),\\
&h_0\mapsto 1\otimes(e_{1,1}-e_{2n+1,2n+1}), \qquad h_n\mapsto 1\otimes 2(-e_{n,n}+e_{n+2,n+2}),\\
&h_i\mapsto 1\otimes(-e_{i,i}+e_{i+1,i+1}-e_{2n+1-i,2n+1-i}+e_{2n+2-i,2n+2-i}).
\end{aligned}$$

3.3. Element $\Lambda^{(2)}$. Denote by $\Lambda^{(2)}$ the element $\sum_{i=0}^n e_i\in\mathfrak{g}(A_{2n}^{(2)})$. Then $\mathfrak{z}(A_{2n}^{(2)})=\{x\in\mathfrak{g}(A_{2n}^{(2)})\mid[\Lambda^{(2)},x]=0\}$ is an Abelian Lie subalgebra of $\mathfrak{g}(A_{2n}^{(2)})$. Denote $\mathfrak{z}^j(A_{2n}^{(2)})=\mathfrak{z}(A_{2n}^{(2)})\cap\mathfrak{g}(A_{2n}^{(2)})^j$, then $\mathfrak{z}(A_{2n}^{(2)})=\oplus_{j\in\mathbb{Z}}\mathfrak{z}(A_{2n}^{(2)})^j$. We have $\dim\mathfrak{z}(A_{2n}^{(2)})^j=0$ if j is even or $j=2n+1 \mod 4n+2$, and $\dim\mathfrak{z}(A_{2n}^{(2)})^j=1$ otherwise.

For example, $\dim\mathfrak{z}^j(A_2^{(2)})=1$ if $j=1,5 \mod 6$ and the dimension equals zero otherwise; $\dim\mathfrak{z}^j(A_4^{(2)})=1$ if $j=1,3,7,9 \mod 10$ and the dimension equals zero otherwise.

If $\mathfrak{g}(A_{2n}^{(2)})$ is realized as $L(\mathfrak{sl}_{2n+1}, C)$ and written out as $(2n+1)\times(2n+1)$ matrices, then for odd j, $1 \leqslant j < 2n+1$, introduce the element:

$$A_{(4n+2)m+j} = \xi^{(4n+2)m+j} \otimes \begin{pmatrix} 0 & I_j \\ I_{2n+1-j} & 0 \end{pmatrix},$$

$$A_{(4n+2)m-j} = \xi^{(4n+2)m-j} \otimes \begin{pmatrix} 0 & I_{2n+1-j} \\ I_j & 0 \end{pmatrix},$$

where I_j is the $j \times j$ identity matrix. We have $A_{(4n+2)m\pm j} = A_1^{(4n+2)m\pm j}$.

If $\mathfrak{g}(A_{2n}^{(2)})$ is realized as $L(\mathfrak{sl}_{2n+1}, \sigma_0)$ and written out as $(2n+1)\times(2n+1)$ matrices, then for odd j, $1 \leqslant j < 2n+1$, introduce the elements:

$$B_{(4n+2)m+j} = \begin{pmatrix} 0 & \lambda^{2m+1} \otimes I_j \\ \lambda^{2m} \otimes I_{2n+1-j} & 0 \end{pmatrix},$$

$$B_{(4n+2)m-j} = \begin{pmatrix} 0 & \lambda^{2m} \otimes I_{2n+1-j} \\ \lambda^{2m-1} \otimes I_j & 0 \end{pmatrix}.$$

We have $B_{(4n+2)m\pm j} = B_1^{(4n+2)m\pm j}$.

LEMMA 3.1. *For any $m \in \mathbb{Z}$, odd j, $1 \leqslant j < 2n+1$, the elements*

$$(\tau_C)^{-1}(A_{(4n+2)m\pm j}), \qquad (\tau_0)^{-1}(B_{(4n+2)m\pm j}),$$

of $\mathfrak{z}(A_{2n}^{(2)})^{(4n+2)m\pm j}$ are equal. □

Denote the elements $(\tau_C)^{-1}(A_{(4n+2)m+j})$, and $(\tau_0)^{-1}(A_{(4n+2)m-j})$ of $\mathfrak{g}(A_{2n}^{(2)})$ by $\Lambda^{(2)}_{(4n+2)m+j}$ and $\Lambda^{(2)}_{(4n+2)m-j}$, respectively. Notice that $\Lambda^{(2)}_1 = \sum_{i=0}^n e_i = \Lambda^{(2)}$. We set $\Lambda^{(2)}_j = 0$ if j is even or $j = 2n+1 \mod 4n+2$. The element $\Lambda^{(2)}_{(4n+2)m\pm j}$ generates $\mathfrak{z}(A_{2n}^{(2)})^{(4n+2)m\pm j}$.

3.4. Lie algebra $\mathfrak{g}(A_{2n}^{(2)})$ as a subalgebra of $\mathfrak{g}(A_{2n}^{(1)})$. The map $\rho : \mathfrak{g}(A_{2n}^{(2)}) \to \mathfrak{g}(A_{2n}^{(1)})$,

$$\begin{array}{lll} e_0 \mapsto E_0, & e_i \mapsto E_i + E_{2n+1-i}, & e_n \mapsto E_n + E_{n+1}, \\ f_0 \mapsto F_0, & f_i \mapsto F_i + F_{2n+1-i}, & f_n \mapsto 2(F_n + F_{n+1}), \\ h_0 \mapsto H_0, & h_i \mapsto H_i + H_{2n+1-i}, & h_n \mapsto 2(H_n + H_{n+1}), \end{array}$$

where $i = 1, \dots, n-1$, realizes the Lie algebra $\mathfrak{g}(A_{2n}^{(2)})$ as a subalgebra of $\mathfrak{g}(A_{2n}^{(1)})$. This embedding preserves the standard grading and $\rho(\Lambda^{(2)}) = \Lambda^{(1)}$. We have $\rho(\mathfrak{z}(A_{2n}^{(2)})^j) \subset \mathfrak{z}(A_{2n}^{(1)})^j$.

4. mKdV equations

In this section we follow [**DS**].

4.1. mKdV equations of type $A_{2n}^{(1)}$. Denote by $\mathcal{B}$ the space of complex-valued functions of one variable x. Given a finite dimensional vector space W, denote by $\mathcal{B}(W)$ the space of W-valued functions of x. Denote by ∂ the differential operator $\frac{d}{dx}$.

Consider the Lie algebra $\tilde{\mathfrak{g}}(A_{2n}^{(1)})$ of the formal differential operators of the form $c\partial + \sum_{i=-\infty}^{k} p_i$, $c \in \mathbb{C}, p_i \in \mathcal{B}(\mathfrak{g}(A_{2n}^{(1)})^i)$. Let $U = \sum_{i<0} U_i$, $U_i \in \mathcal{B}(\mathfrak{g}(A_{2n}^{(1)})^i)$. If $\mathcal{L} \in \tilde{\mathfrak{g}}(A_{2n}^{(1)})$, define

$$e^{\mathrm{ad}U}(\mathcal{L}) = \mathcal{L} + [U, \mathcal{L}] + \frac{1}{2!}[U, [U, \mathcal{L}]] + \dots .$$

The operator $e^{\mathrm{ad}U}(\mathcal{L})$ belongs to $\tilde{\mathfrak{g}}(A_{2n}^{(1)})$. The map $e^{\mathrm{ad}U}$ is an automorphism of the Lie algebra $\tilde{\mathfrak{g}}(A_{2n}^{(1)})$. The automorphisms of this type form a group. If elements of $\mathfrak{g}(A_{2n}^{(1)})$ are realized as matrices depending on a parameter as in Section 2.2, then $e^{\mathrm{ad}U}(\mathcal{L}) = e^U \mathcal{L} e^{-U}$.

A *Miura oper* of type $A_{2n}^{(1)}$ is a differential operator of the form

$$\mathcal{L} = \partial + \Lambda^{(1)} + V, \tag{4.1}$$

where $\Lambda^{(1)} = \sum_{i=0}^{2n} E_i \in \mathfrak{g}(A_{2n}^{(1)})$ and $V \in \mathcal{B}(\mathfrak{g}(A_{2n}^{(1)})^0)$. Any Miura oper of type $A_{2n}^{(1)}$ is an element of $\tilde{\mathfrak{g}}(A_{2n}^{(1)})$. Denote by $\mathcal{M}(A_{2n}^{(1)})$ the space of all Miura opers of type $A_{2n}^{(1)}$.

PROPOSITION 4.1 ([**DS**, Proposition 6.2]). *For any Miura oper $\mathcal{L}$ of type $A_{2n}^{(1)}$ there exists an element $U = \sum_{i<0} U_i$, $U_i \in \mathcal{B}(\mathfrak{g}(A_{2n}^{(1)})^i)$, such that the operator $\mathcal{L}_0 = e^{\mathrm{ad}U}(\mathcal{L})$ has the form*

$$\mathcal{L}_0 = \partial + \Lambda^{(1)} + H,$$

where $H = \sum_{j<0} H_j, H_j \in \mathcal{B}(\mathfrak{z}(A_{2n}^{(1)})^j)$. If $U, \tilde{U}$ are two such elements, then $e^{\mathrm{ad}U} e^{-\mathrm{ad}\tilde{U}} = e^{\mathrm{ad}T}$, where $T = \sum_{j<0} T_j$, $T_j \in \mathfrak{z}(A_{2n}^{(1)})^j$. □

Let $\mathcal{L}, U$ be as in Proposition 4.1. Let $r \neq 0 \bmod 2n+1$. The element $\phi(\Lambda_r^{(1)}) = e^{-\mathrm{ad}U}(\Lambda_r^{(1)})$ does not depend on the choice of U in Proposition 4.1.

The element $\phi(\Lambda_r^{(1)})$ is of the form $\sum_{i=-\infty}^{k} \phi(\Lambda_r^{(1)})^i$, $\phi(\Lambda_r^{(1)})^i \in \mathcal{B}(\mathfrak{g}(A_{2n}^{(1)})^i)$. We set $\phi(\Lambda_r^{(1)})^+ = \sum_{i=0}^{k} \phi(\Lambda_r^{(1)})^i$, $\phi(\Lambda_r^{(1)})^- = \sum_{i<0} \phi(\Lambda_r^{(1)})^i$.

Let $r \in \mathbb{Z}_{>0}$ and $r \neq 0 \bmod 2n+1$. The differential equation

$$\frac{\partial \mathcal{L}}{\partial t_r} = [\phi(\Lambda_r^{(1)})^+, \mathcal{L}] \tag{4.2}$$

is called the *r-th mKdV equation* of type $A_{2n}^{(1)}$.

Equation (4.2) defines vector fields $\frac{\partial}{\partial t_r}$ on the space $\mathcal{M}(A_{2n}^{(1)})$ of Miura opers of type $A_{2n}^{(1)}$. For all r, s, the vector fields $\frac{\partial}{\partial t_r}, \frac{\partial}{\partial t_s}$ commute, see [**DS**, Section 6].

LEMMA 4.2 ([**DS**]). *We have*

$$\frac{\partial \mathcal{L}}{\partial t_r} = -\frac{d}{dx}\phi(\Lambda_r^{(1)})^0. \tag{4.3}$$

□

4.2. mKdV equations of type $A_{2n}^{(2)}$. A *Miura oper* of type $A_{2n}^{(2)}$ is a differential operator of the form

$$\mathcal{L} = \partial + \Lambda^{(2)} + V, \tag{4.4}$$

where $\Lambda^{(2)} = \sum_{i=0}^{n} e_i \in \mathfrak{g}(A_{2n}^{(2)})$ and $V \in \mathcal{B}(\mathfrak{g}(A_{2n}^{(2)})^0)$. Denote by $\mathcal{M}(A_{2n}^{(2)})$ the space of all Miura opers of type $A_{2n}^{(2)}$.

PROPOSITION 4.3 ([**DS**, Proposition 6.2]). *For any Miura oper $\mathcal{L}$ of type $A_{2n}^{(2)}$ there exists an element $U = \sum_{i<0} U_i$, $U_i \in \mathcal{B}(\mathfrak{g}(A_{2n}^{(2)})^i)$, such that the operator $\mathcal{L}_0 = e^{\operatorname{ad}U}(\mathcal{L})$ has the form*

$$\mathcal{L}_0 = \partial + \Lambda^{(2)} + H,$$

where $H = \sum_{j<0} H_j, H_j \in \mathcal{B}(\mathfrak{z}(A_{2n}^{(2)})^j)$. If $U, \tilde{U}$ are two such elements, then $e^{\operatorname{ad}U}e^{-\operatorname{ad}\tilde{U}} = e^{\operatorname{ad}T}$, where $T = \sum_{j<0} T_j$, $T_j \in \mathfrak{z}(A_{2n}^{(2)})^j$. □

Let $\mathcal{L}, U$ be as in Proposition 4.3. Let r be odd, $r \neq 2n+1 \mod 4n+2$. The element $\phi(\Lambda_r^{(2)}) = e^{-\operatorname{ad}U}(\Lambda_r^{(2)})$ does not depend on the choice of U in Proposition 4.3.

The element $\phi(\Lambda_r^{(2)})$ is of the form $\sum_{i=-\infty}^{k} \phi(\Lambda_r^{(2)})^i$, $\phi(\Lambda_r^{(2)})^i \in \mathcal{B}(\mathfrak{g}(A_{2n}^{(2)})^i)$. We set $\phi(\Lambda_r^{(2)})^+ = \sum_{i=0}^{k} \phi(\Lambda_r^{(2)})^i$, $\phi(\mathcal{L}^{(2)}a_r)^- = \sum_{i<0} \phi(\Lambda_r^{(2)})^i$.

Let $r \in \mathbb{Z}_{>0}$, r odd and $r \neq 2n+1 \mod 4n+2$. The differential equation

$$\frac{\partial \mathcal{L}}{\partial t_r} = [\phi(\Lambda_r^{(2)})^+, \mathcal{L}] \tag{4.5}$$

is called the *r-th mKdV equation* of type $A_{2n}^{(2)}$.

Equation (4.5) defines vector fields $\frac{\partial}{\partial t_r}$ on the space $\mathcal{M}(A_2^{(2)})$ of Miura opers. For all r, s, the vector fields $\frac{\partial}{\partial t_r}$, $\frac{\partial}{\partial t_s}$ commute, see [**DS**, Section 6].

LEMMA 4.4 ([**DS**]). *We have*

$$\frac{\partial \mathcal{L}}{\partial t_r} = -\frac{d}{dx}\phi(\Lambda_r^{(2)})^0. \tag{4.6}$$

□

4.3. Comparison of mKdV equations of types $A_{2n}^{(2)}$ and $A_{2n}^{(1)}$. Consider $\mathfrak{g}(A_{2n}^{(2)})$ as a Lie subalgebra of $\mathfrak{g}(A_{2n}^{(1)})$, see Section 3.4. If $\mathcal{L}$ is a Miura oper of type $A_{2n}^{(2)}$, then it is also a Miura oper of type $A_{2n}^{(1)}$. We have $\mathcal{M}(A_{2n}^{(2)}) \subset \mathcal{M}(A_{2n}^{(1)})$,

$$\mathcal{M}(A_{2n}^{(1)}) = \{\mathcal{L} = \partial + \Lambda^{(1)} + \sum_{i=1}^{2n+1} v_i e_{i,i} \mid \sum_{i=1}^{2n+1} v_i = 0\}, \tag{4.7}$$

$$\mathcal{M}(A_{2n}^{(2)}) = \{\mathcal{L} = \partial + \Lambda^{(1)} + \sum_{i=1}^{2n+1} v_i e_{i,i} \mid \sum_{i=1}^{2n+1} v_i = 0,$$
$$v_j + v_{2n+2-j} = 0, j = 1, \ldots, 2n+1\}.$$

LEMMA 4.5. *Let r be odd, $r \neq 2n+1$ mod $4n+2$, $r > 0$. Let $\mathcal{L}^{A_{2n}^{(2)}}(t_r)$ be the solution of the r-th mKdV equation of type $A_{2n}^{(2)}$ with initial condition $\mathcal{L}^{A_{2n}^{(2)}}(0) = \mathcal{L}$. Let $\mathcal{L}^{A_{2n}^{(1)}}(t_r)$ be the solution of the r-th mKdV equation of type $A_{2n}^{(1)}$ with initial condition $\mathcal{L}^{A_{2n}^{(1)}}(0) = \mathcal{L}$. Then $\mathcal{L}^{A_{2n}^{(2)}}(t_r) = \mathcal{L}^{A_{2n}^{(1)}}(t_r)$ for all values of t_r.* □

PROOF. The element U in Proposition 4.3 which is used to construct the mKdV equation of type $A_{2n}^{(2)}$ can be used also to construct the mKdV equation of type $A_{2n}^{(1)}$. □

4.4. KdV equations of type $A_{2n}^{(1)}$. Let $\mathcal{B}((\partial^{-1}))$ be the algebra of formal pseudodifferential operators of the form $a = \sum_{i\in\mathbb{Z}} a_i\partial^i$, with $a_i \in \mathcal{B}$ and finitely many terms with $i > 0$. The relations in this algebra are

$$\partial^k u - u\partial^k = \sum_{i=1}^{\infty} k(k-1)\dots(k-i+1)\frac{d^i u}{dx^i}\partial^{k-i}$$

for any $k \in \mathbb{Z}$ and $u \in \mathcal{B}$. For $a = \sum_{i\in\mathbb{Z}} a_i\partial^i \in \mathcal{B}((\partial^{-1}))$, define $a^+ = \sum_{i\geqslant 0} a_i\partial^i$.

Denote $\mathcal{B}[\partial] \subset \mathcal{B}((\partial^{-1}))$ the subalgebra of differential operators $a = \sum_{i=0}^{m} a_i\partial^i$ with $m \in \mathbb{Z}_{\geqslant 0}$. Denote $\mathcal{D} \subset \mathcal{B}[\partial]$ the affine subspace of differential operators of the form

$L = \partial^{2n+1} + \sum\limits_{i=0}^{2n-1} u_i\partial^i.$

For $L \in \mathcal{D}$, there exists a unique $L^{\frac{1}{2n+1}} = \partial + \sum_{i\leqslant 0} a_i\partial^i \in \mathcal{B}((\partial^{-1}))$ such that $(L^{\frac{1}{2n+1}})^{2n+1} = L$. For $r \in \mathbb{N}$, we have $L^{\frac{r}{2n+1}} = \partial^r + \sum_{i=-\infty}^{r-1} b_i\partial^i$, $b_i \in \mathcal{B}$. We set $(L^{\frac{r}{2n+1}})^+ = \partial^r + \sum_{i=0}^{r-1} b_i\partial^i$.

For $r \in \mathbb{N}$, the differential equation

$$\frac{\partial L}{\partial t_r} = [L, (L^{\frac{r}{2n+1}})^+] \tag{4.8}$$

is called the *r-th KdV equation* of type $A_{2n}^{(1)}$.

Equation (4.8) defines flows $\frac{\partial}{\partial t_r}$ on the space $\mathcal{D}$. For all $r, s \in \mathbb{N}$ the flows $\frac{\partial}{\partial t_r}$ and $\frac{\partial}{\partial t_s}$ commute, see [**DS**].

4.5. Miura maps. Let $\mathcal{L} = \partial + \Lambda^{(1)} + V$ be a Miura oper of type $A_{2n}^{(1)}$ with $V = \sum_{k=1}^{2n+1} v_k e_{k,k}$, $\sum_{k=1}^{2n+1} v_k = 0$. For $i = 0, \dots, 2n+1$, define the scalar differential operator $L_i = \partial^{2n+1} + \sum_{j=0}^{2n-1} u_{j,i}\partial^j \in \mathcal{D}$ by the formula:

$$\begin{aligned} L_0 = L_{2n+1} &= (\partial - v_{2n+1})(\partial - v_{2n})\dots(\partial - v_2)(\partial - v_1), \\ L_i &= (\partial - v_i)(\partial - v_{i-1})\dots(\partial - v_1)(\partial - v_{2n+1})\dots(\partial - v_{i+2})(\partial - v_{i+1}), \end{aligned} \tag{4.9}$$

for $i = 1, \dots, 2n$.

THEOREM 4.6 ([**DS**, Proposition 3.18]). *Let a Miura oper $\mathcal{L}$ satisfy the mKdV equation* (4.2) *for some r. Then for every $i = 0, \dots, 2n$ the differential operator L_i satisfies the KdV equation* (4.8).

For $i = 0, \dots, 2n+1$, we define the *i-th Miura map* by the formula

$$\mathfrak{m}_i \,:\, \mathcal{M}(A_{2n}^{(1)}) \,\to\, \mathcal{D}, \quad \mathcal{L} \,\mapsto\, L_i,$$

see (4.9).

For $i = 0, 1, \dots, 2n$, an *i-oper* is a differential operator of the form

$$\mathcal{L} = \partial + \Lambda^{(1)} + V + W,$$

with $V \in \mathcal{B}(\mathfrak{g}(A_{2n}^{(1)})^0)$ and $W \in \mathcal{B}(\mathfrak{n}_i^-)$. For $w \in \mathcal{B}(\mathfrak{n}_i^-)$ and an i-oper $\mathcal{L}$, the differential operator $e^{\operatorname{ad} w}(\mathcal{L})$ is an i-oper. The i-opers $\mathcal{L}$ and $e^{\operatorname{ad} w}(\mathcal{L})$ are called *i-gauge equivalent.* A Miura oper is an i-oper for any i.

THEOREM 4.7 ([**DS**, Proposition 3.10]). *If Miura opers $\mathcal{L}$ and $\tilde{\mathcal{L}}$ are i-gauge equivalent, then $\mathfrak{m}_i(\mathcal{L}) = \mathfrak{m}_i(\tilde{\mathcal{L}})$.* □

5. Tangent maps to Miura maps

5.1. Tangent spaces. Consider the spaces of Miura opers $\mathcal{M}(A_{2n}^{(2)}) \subset \mathcal{M}(A_{2n}^{(1)})$. The tangent space to $\mathcal{M}(A_{2n}^{(2)})$ at a point $\mathcal{L}$ is

$$T_{\mathcal{L}}\mathcal{M}(A_{2n}^{(2)}) = \{X = \sum_{i=1}^{2n+1} X_i e_{i,i} \mid \sum_{i=1}^{2n+1} X_i = 0, \quad X_j + X_{2n+2-j} = 0, \ j = 1, \dots, 2n+1\}, \tag{5.1}$$

where X_i are functions of variable x. Recall $\mathcal{D} = \{L = \partial^{2n+1} + \sum_{i=0}^{2n-1} u_i \partial^i\}$. The tangent space to $\mathcal{D}$ at a point D is $T_D\mathcal{D} = \{Z = \sum_{i=0}^{2n-1} Z_i \partial^i\}$, where Z_i are functions of x.

Consider the restrictions of Miura maps to $\mathcal{M}(A_{2n}^{(2)})$ and the corresponding tangent maps

$$d\mathfrak{m}_i : T_{\mathcal{L}}\mathcal{M}(A_{2n}^{(2)}) \to T_{\mathfrak{m}_i(\mathcal{L})}\mathcal{D}, \quad i = 1, \dots, 2n+1. \tag{5.2}$$

By definition, if $\mathcal{L} = \partial + \Lambda^{(1)} + \sum_{i=1}^{2n+1} v_i e_{i,i} \in \mathcal{M}(A_{2n}^{(2)})$, $X = \sum_{i=1}^{2n+1} X_i e_{i,i} \in T_{\mathcal{L}}\mathcal{M}(A_{2n}^{(2)})$, $d\mathfrak{m}_i(X) = Z^i = \sum_{j=0}^{2n-1} Z_j^i \partial^j$, then

$$\begin{aligned} Z^i &= (-X_i)(\partial - v_{i-1}) \dots (\partial - v_1)(\partial - v_{2n+1}) \dots (\partial - v_{i+1}) \\ &+ (\partial - v_i)(-X_{i-1}) \dots (\partial - v_1)(\partial - v_{2n+1}) \dots (\partial - v_{i+1}) + \dots \\ &+ (\partial - v_i)(\partial - v_{i-1}) \dots (-X_1)(\partial - v_{2n+1}) \dots (\partial - v_{i+1}) \\ &+ (\partial - v_i)(\partial - v_{i-1}) \dots (\partial - v_1)(-X_{2n+1}) \dots (\partial - v_{i+1}) + \dots \\ &+ (\partial - v_i)(\partial - v_{i-1}) \dots (\partial - v_1)(\partial - v_{2n+1}) \dots (-X_{i+1}). \end{aligned} \tag{5.3}$$

In what follows we study the intersection of kernels of these tangent maps when i runs through certain subsets of $\{1, \dots, 2n+1\}$.

5.2. Formula for the first coefficient.

PROPOSITION 5.1. *Let $\mathcal{L} = \partial + \Lambda^{(1)} + \sum_{i=1}^{2n+1} v_i e_{i,i} \in \mathcal{M}(A_{2n}^{(1)})$, $X = \sum_{i=1}^{2n+1} X_i e_{i,i} \in T_{\mathcal{L}}\mathcal{M}(A_{2n}^{(2)})$, $d\mathfrak{m}_i(X) = Z^i = \sum_{j=0}^{2n-1} Z_j^i \partial^j$. Then*

$$Z_{2n-1}^i = -\left(\sum_{k=1}^{2n+1} v_k X_k + \sum_{k=1}^{i} (i-k) X_k' + \sum_{k=i+1}^{2n+1} (i + 2n + 1 - k) X_k' \right). \tag{5.4}$$

PROOF. The proof uses only the identity $\sum_{j=1}^{2n+1} v_j = 0$ and is straightforward. □

5.3. Intersection of kernels of $d\mathfrak{m}_i$.

LEMMA 5.2. *Let $\mathcal{L} = \partial + \Lambda^{(1)} + \sum_{k=1}^{2n+1} v_k e_{k,k} \in \mathcal{M}(A_{2n}^{(2)})$, $X = \sum_{k=1}^{2n+1} X_k e_{k,k} \in T_{\mathcal{L}}\mathcal{M}(A_{2n}^{(2)})$, $d\mathfrak{m}_i(X) = Z^i = \sum_{j=0}^{2n-1} Z_j^i \partial^j$. Assume that $Z_{2n-1}^i = 0$ for $i = 1, \dots, 2n$, then*

$$X_1' - 2v_1 X_1 = \sum_{k=2}^{2n} v_k X_k, \qquad X_i' = 0, \quad i = 2, \dots, 2n. \tag{5.5}$$

PROOF. By assumption we have the system of equations

$$(5.6)\qquad \begin{aligned} &X'_{2n-1}+2X'_{2n-2}+\cdots+(2n-1)X'_1+2nX'_{2n+1}+\sum_{k=1}^{2n+1}v_kX_k=0,\\ &X'_{2n-2}+2X'_{2n-3}+\cdots+(2n-1)X'_{2n+1}+2nX'_{2n}+\sum_{k=1}^{2n+1}v_kX_k=0,\\ &X'_{2n-3}+2X'_{2n-4}+\cdots+(2n-1)X'_{2n}+2nX'_{2n-1}+\sum_{k=1}^{2n+1}v_kX_k=0,\\ &\dots\\ &X'_1+2X'_{2n+1}+\cdots+(2n-1)X'_4+2nX'_3+\sum_{k=1}^{2n+1}v_kX_k=0,\\ &X'_{2n+1}+\cdots+(2n-1)X'_3+2nX'_2+\sum_{k=1}^{2n+1}v_kX_k=0. \end{aligned}$$

By subtracting the first equation from the second we get $2nX'_{2n}-X'_{2n-1}-X'_{2n-2}-\cdots-X'_1-X'_{2n+1}=0$, equivalently $(2n+1)X'_{2n}-\sum_{k=1}^{2n+1}X'_k=0$. Since $\sum_{k=1}^{2n+1}X_k=0$, we get $X'_{2n}=0$. By subtracting the second from the third we get $X'_{2n-1}=0$. Similarly we obtain

$$(5.7)\qquad X'_i=0,\quad i=2,\dots,2n.$$

Applying (5.7) to the last equation in (5.6) yields $X'_{2n+1}+\sum_{k=1}^{2n+1}v_kX_k=0$. By pulling out the terms for $k=1,2n+1$ we obtain $X'_{2n+1}+v_1X_1+v_{2n+1}X_{2n+1}=-(X'_1-2v_1X_1)=-\sum_{k=2}^{2n}v_kX_k$. □

LEMMA 5.3. *Let $j\in\{1,\dots,n-1\}$. Let $\mathcal{L}=\partial+\Lambda^{(1)}+\sum_{k=1}^{2n+1}v_ke_{k,k}\in\mathcal{M}(A_{2n}^{(2)})$, $X=\sum_{k=1}^{2n+1}X_ke_{k,k}\in T_{\mathcal{L}}\mathcal{M}(A_{2n}^{(2)})$, $d\mathfrak{m}_i(X)=Z^i=\sum_{j=0}^{2n-1}Z^i_j\partial^j$. Assume that $Z^i_{2n-1}=0$ for all $i\notin\{j,2n+1-j\}$, then*

$$X'_j+v_jX_j+v_{j+1}X_{j+1}=-\sum_{k=1,k\neq j,j+1}^{n}v_kX_k,\qquad X'_j+X'_{j+1}=0,\qquad X'_i=0$$

for $i\notin\{j,j+1,2n+1-j,2n+2-j\}$.

PROOF. By assumption we have the system of equations

$$
\begin{aligned}
&X'_{2n} + 2X'_{2n-1} + \cdots + (2n-1)X'_2 + 2nX'_1 + \sum_{k=1}^{2n+1} v_k X_k = 0,\\
&X'_{2n-1} + 2X'_{2n-2} + \cdots + (2n-1)X'_1 + 2nX'_{2n+1} + \sum_{k=1}^{2n+1} v_k X_k = 0,\\
&\dots\\
&X'_{2n+1-j} + \cdots + (2n+1-j)X'_1 + (2n+2-j)X'_{2n+1} + \dots\\
&\qquad\qquad \cdots + 2nX'_{2n+3-j} + \sum_{k=1}^{2n+1} v_k X_k = 0,\\
&X'_{2n-1-j} + \cdots + (2n-1-j)X'_1 + (2n-j)X'_{2n+1} + \cdots + 2nX'_{2n+1-j}\\
&\qquad\qquad + \sum_{k=1}^{2n+1} v_k X_k = 0,\\
&\dots\\
&X'_j + \cdots + jX'_1 + (j+1)X'_{2n+1} + \cdots + 2nX'_{j+2} + \sum_{k=1}^{2n+1} v_k X_k = 0,\\
&\dots
\end{aligned}
$$

Subtracting the second line from the first gives $X'_{2n+1} = 0$, cf. the proof of Lemma 5.2. Similarly, for $i \notin \{j, j+1, 2n+1-j, 2n+2-j\}$ considering the difference $Z^{i-1}_{2n-1} - Z^i_{2n-1} = 0$ we obtain $X'_i = 0$.

Considering the difference $Z^{2n+2-j}_{2n-1} - Z^{2n-j}_{2n-1} = 0$ we obtain

$$
\begin{aligned}
&X'_{2n+1-j} + \cdots + (2n+1-j)X'_1 + (2n+2-j)X'_{2n+1} + \cdots + 2nX'_{2n+3-j}\\
&\qquad\qquad + \sum_{k=1}^{2n+1} v_k X_k\\
&-\Big(X'_{2n-1-j} + \cdots + (2n-1-j)X'_1 + (2n-j)X'_{2n+1} + \cdots + 2nX'_{2n+1-j}\\
&\qquad\qquad + \sum_{k=1}^{2n+1} v_k X_k\Big)\\
&= -(2n+1)(X'_{2n+1-j} + X'_{2n+2-j}) + 2\sum_{k=1}^{2n+1} X'_k = 0.
\end{aligned}
$$

Hence $X'_{2n+1-j} + X'_{2n+2-j} = 0$ and $X'_j + X'_{j+1} = 0$. Now we can rewrite equation $Z^{2n+1}_{2n-1} = 0$ as

$$
(j-1)X'_{2n+2-j} + jX'_{2n+1-j} + (2n-j)X'_{j+1} + (2n+1-j)X'_j + \sum_{k=1}^{2n+1} v_k X_k = 0.
$$

Or equivalently

$$2X_j' + \sum_{k=1}^{2n+1} v_k X_k = 2X_j' + 2\sum_{k=1}^{n} v_k X_k = 2(X_j' + v_j X_j + v_{j+1}X_{j+1} + \sum_{k=1,k\neq j,j+1}^{n} v_k X_k) = 0.$$

□

LEMMA 5.4. *Let* $\mathcal{L} = \partial + \Lambda^{(1)} + \sum_{k=1}^{2n+1} v_k e_{k,k} \in \mathcal{M}(A_{2n}^{(2)})$, $X = \sum_{k=1}^{2n+1} X_k e_{k,k} \in T_{\mathcal{L}}\mathcal{M}(A_{2n}^{(2)})$, $d\mathfrak{m}_i(X) = Z^i = \sum_{j=0}^{2n-1} Z_j^i \partial^j$. *Assume that* $Z_{2n-1}^i = 0$ *for all* $i \notin \{n, n+1\}$, *then*

$$X_n' + v_n X_n = -\sum_{k=1}^{n-1} v_k X_k, \qquad X_i' = 0, \quad i \notin \{n, n+2\}.$$

PROOF. By assumption we have the system of equations

$$X_{2n}' + 2X_{2n-1}' + \cdots + (2n-1)X_2' + (2n)X_1' + \sum_{k=1}^{2n+1} v_k X_k = 0,$$

$$X_{2n-1}' + 2X_{2n-2}' + \cdots + (2n-1)X_1' + (2n)X_{2n+1}' + \sum_{k=1}^{2n+1} v_k X_k = 0,$$

$$\ldots$$

$$X_{n+1}' + \cdots + (n+1)X_1' + (n+2)X_{2n+1}' + \cdots + 2nX_{n+3}' + \sum_{k=1}^{2n+1} v_k X_k = 0,$$

$$X_{n-2}' + \cdots + (n-2)X_1' + (n-1)X_{2n+1}' + \cdots + 2nX_n' + \sum_{k=1}^{2n+1} v_k X_k = 0,$$

$$\ldots$$

$$X_{2n+1}' + 2X_{2n}' + \cdots + 2nX_2' + \sum_{k=1}^{2n+1} v_k X_k = 0.$$

Subtracting the second line from the first gives $X_{2n+1}' = 0$, cf. the proof of Lemma 5.2. Similarly, for $i \notin \{n, n+1, n+2\}$ considering the difference $Z_{2n-1}^{i-1} - Z_{2n-1}^i = 0$ we obtain $X_i' = 0$. Notice that $X_{n+1} = 0$ by assumption.

Now we can rewrite equation $Z_{2n-1}^{2n+1} = 0$ as

$$(n-1)X_{n+2}' + (n+1)X_n' + \sum_{k=1}^{2n+1} v_k X_k = 2(X_n' + v_n X_n + \sum_{k=1}^{n-1} v_k X_k) = 0.$$

□

6. Critical points of master functions and generation of tuples of polynomials

In this section we follow [**MV1**]. For functions $f(x), g(x)$, we denote

$$\mathrm{Wr}(f,g) = f(x)g'(x) - f'(x)g(x)$$

the Wronskian determinant, and $f'(x) := \frac{df}{dx}(x)$.

6.1. Master function. Choose nonnegative integers $k=(k_0,k_1,\dots,k_n)$. Consider variables $u=(u_i^{(j)})$, where $j=0,1,\dots,n$ and $i=1,\dots,k_j$. The *master function* $\Phi(u;k)$ is defined by the formula:

$$\begin{aligned}(6.1)\quad \Phi(u,k) &= -4\textstyle\sum_{i,i'}\ln(u_i^{(n-1)}-u_{i'}^{(n)}) - 2\sum_{j=0}^{n-2}\sum_{i,i'}\ln(u_i^{(j)}-u_{i'}^{(j+1)})\\ &+8\textstyle\sum_{i<i'}\ln(u_i^{(n)}-u_{i'}^{(n)}) + 4\sum_{j=1}^{n-1}\sum_{i<i'}\ln(u_i^{(j)}-u_{i'}^{(j)}) + 2\sum_{i<i'}\ln(u_i^{(0)}-u_{i'}^{(0)}).\end{aligned}$$

The product of symmetric groups $\Sigma_{\boldsymbol{k}}=\Sigma_{k_0}\times\Sigma_{k_1}\times\dots\times\Sigma_{k_n}$ acts on the set of variables by permuting the coordinates with the same upper index. The function Φ is symmetric with respect to the $\Sigma_{\boldsymbol{k}}$-action. A point u is a *critical point* if $d\Phi=0$ at u. In other words, u is a critical point if and only if the following expressions equal zero:

$$(6.2)\qquad \sum_{l=1}^{k_1}\frac{-2}{u_j^{(0)}-u_l^{(1)}}+\sum_{s\neq j}\frac{2}{u_j^{(0)}-u_s^{(0)}},\qquad j=1,\dots,k_0,$$

$$\sum_{l=1}^{k_{i-1}}\frac{-2}{u_j^{(i)}-u_l^{(i-1)}}+\sum_{l=1}^{k_{i+1}}\frac{-2}{u_j^{(i)}-u_l^{(i+1)}}+\sum_{s\neq j}\frac{4}{u_j^{(i)}-u_s^{(i)}},$$
$$i=1,\dots,n-2,\ j=1,\dots,k_i,$$

$$\sum_{l=1}^{k_{n-2}}\frac{-2}{u_j^{(n-1)}-u_l^{(n-2)}}+\sum_{l=1}^{k_n}\frac{-4}{u_j^{(n-1)}-u_l^{(n)}}+\sum_{s\neq j}\frac{4}{u_j^{(n-1)}-u_s^{(n-1)}},$$
$$j=1,\dots,k_{n-1},$$

$$\sum_{l=1}^{k_{n-1}}\frac{-4}{u_j^{(n)}-u_l^{(n-1)}}+\sum_{s\neq j}\frac{8}{u_j^{(n)}-u_s^{(n)}},\qquad j=1,\dots,k_n.$$

All the orbits have the same cardinality $\prod_{i=0}^n k_i!$. We do not make distinction between critical points in the same orbit.

REMARK 6.1. The definition of master functions can be found in [**SV**], see also [**MV1,MV2**]. The master functions $\Phi(u,k)$ in (6.1) are associated with the Kac-Moody algebra with Cartan matrix of type

$$(6.3)\qquad A=(a_{i,j})=\begin{pmatrix} 2 & -2 & 0 & 0 & \dots & \dots & 0\\ -1 & 2 & -1 & 0 & \dots & \dots & \dots\\ 0 & -1 & 2 & -1 & \dots & \dots & \dots\\ 0 & 0 & -1 & \dots & \dots & \dots & \dots\\ \dots & \dots & \dots & \dots & 2 & -1 & 0\\ \dots & \dots & \dots & \dots & -1 & 2 & -2\\ 0 & \dots & \dots & \dots & 0 & -1 & 2\end{pmatrix},$$

which is dual to the Cartan matrix $A_{2n}^{(2)}$, see this Langlands duality in [**MV1,MV2, VWW14**].

6.2. Polynomials representing critical points. Let $u=(u_i^{(j)})$ be a critical point of the master function Φ. Introduce the $(n+1)$-tuple of polynomials $y=(y_0(x),\dots,y_n(x))$,

$$(6.4)\qquad y_j(x) = \prod_{i=1}^{k_j}(x-u_i^{(j)}).$$

This tuple of polynomials defines a point in the direct product $(\mathbb{C}[x])^{n+1}$. We say that the tuple *represents the critical point.*

Each polynomial of the tuple will be considered up to multiplication by a nonzero number.

It is convenient to think that the $(n+1)$-tuple $y^{\emptyset} = (1, \dots, 1)$ of constant polynomials represents in $(\mathbb{C}[x])^{n+1}$, the critical point of the master function with no variables. This corresponds to the case $k = (0, \dots, 0)$.

We say that a given tuple $y \in (\mathbb{C}[x])^{n+1}$ is *generic* if each polynomial $y_i(x)$ has no multiple roots and for $i = 0, \dots, n-1$ the polynomials $y_i(x)$ and $y_{i+1}(x)$ have no common roots. If a tuple represents a critical point, then it is generic, see (6.2). For example, the tuple $y^{\emptyset}$ is generic.

6.3. Elementary generation. An $(n+1)$-tuple is called *fertile* if there exist polynomials $\tilde{y}_0, \dots, \tilde{y}_n \in (\mathbb{C}[x])^{n+1}$ such that

$$\operatorname{Wr}(\tilde{y}_j, y_j) = \prod_{i \neq j} y_i^{-a_{i,j}}, \qquad j = 0, 1, \dots, n, \tag{6.5}$$

where $a_{i,j}$ are the entries of the Cartan matrix of type $A_{2n}^{(2)}$, that is,

$$\operatorname{Wr}(\tilde{y}_0, y_0) = y_1^2, \qquad \operatorname{Wr}(\tilde{y}_i, y_i) = y_{i-1} y_{i+1}, \qquad i = 1, \dots, n-2, \tag{6.6}$$
$$\operatorname{Wr}(\tilde{y}_{n-1}, y_{n-1}) = y_{n-2} y_n^2, \qquad \operatorname{Wr}(\tilde{y}_n, y_n) = y_{n-1}.$$

For example, $y^{\emptyset}$ is fertile and $\tilde{y}_j = x + c_j$, where the c_j are arbitrary numbers.

Assume that an $(n+1)$-tuple of polynomials $y = (y_0, \dots, y_n)$ is fertile. Equations (6.5) give us first order inhomogeneous differential equations with respect to $\tilde{y}_i$. The solutions are

$$\tilde{y}_0 = y_0 \int \tfrac{y_1^2}{y_0^2} dx + c_0 y_0, \tag{6.7}$$

$$\tilde{y}_i = y_i \int \tfrac{y_{i-1} y_{i+1}}{y_i^2} dx + c_i y_i, \qquad i = 1, \dots, n-2, \tag{6.8}$$

$$\tilde{y}_{n-1} = y_{n-1} \int \tfrac{y_{n-2} y_n^2}{y_{n-1}^2} dx + c_{n-1} y_{n-1}, \tag{6.9}$$

$$\tilde{y}_n = y_n \int \tfrac{y_{n-1}}{y_n^2} dx + c_n y_n, \tag{6.10}$$

where $c_0, \dots, c_n$ are arbitrary numbers. For each $i = 0, \dots, n$, the tuple

$$y^{(i)}(x, c_i) = (y_0(x), \dots, y_{i-1}(x), \tilde{y}_i(x, c_i), y_{i+1}(x), \dots, y_n(x)) \in (\mathbb{C}[x])^{n+1} \tag{6.11}$$

forms a one-parameter family. This family is called the *generation of tuples from y in the i-th direction.* A tuple of this family is called an immediate descendant of y in the i-th direction.

THEOREM 6.1 ([**MV1**]).

(i) *A generic tuple $y = (y_0, \dots, y_n)$, $\deg y_i = k_i$, represents a critical point of the master function $\Phi(u; k)$ if and only if y is fertile.*

(ii) *If y represents a critical point, then for any $c \in \mathbb{C}$ the tuple $y^{(j)}(x, c)$, $j = 0, \dots, n$ is fertile.*

(iii) *If y is generic and fertile, then for almost all values of the parameter $c \in \mathbb{C}$ the tuple $y^{(j)}(x, c)$ is generic. The exceptions form a finite set in $\mathbb{C}$.*

(iv) *Assume that a sequence $y_{[\ell]}$, $\ell = 1, 2, \dots$, of fertile tuples has a limit $y_{[\infty]}$ in $(\mathbb{C}[x])^{n+1}$ as ℓ tends to infinity.*

  (a) *Then the limiting tuple $y_{[\infty]}$ is fertile.*

(b) *For $j = 0, \dots, n$, let $y^{(j)}_{[\infty]}$ be an immediate descendant of $y_{[\infty]}$. Then for all j there exist immediate descendants $y^{(j)}_{[\ell]}$ of $y_{[\ell]}$ such that $y^{(j)}_{[\infty]}$ is the limit of $y^{(j)}_{[\ell]}$ as ℓ tends to infinity.*

□

6.4. Degree increasing generation. Let $y = (y_0, \dots, y_n)$ be a generic fertile $(n+1)$-tuple of polynomials. Define $k_j = \deg y_j$ for $j = 0, \dots, n$.

The polynomial $\tilde{y}_0$ in (6.7) is of degree k_0 or $\tilde{k}_0 = 2k_1 + 1 - k_0$. We say that the generation $(y_0, \dots, y_n) \to (\tilde{y}_0, \dots, y_n)$ is *degree increasing* in the 0-th direction if $\tilde{k}_0 > k_0$. In that case $\deg \tilde{y}_0 = \tilde{k}_0$ for all c.

For $i = 1, \dots, n-2$, the polynomial $\tilde{y}_i$ in (6.8) is of degree k_i or $\tilde{k}_i = k_{i-1} + k_{i+1} + 1 - k_i$. We say that the generation $(y_0, \dots, y_i, \dots, y_n) \to (y_0, \dots, \tilde{y}_i, \dots, y_n)$ is *degree increasing* in the i-th direction if $\tilde{k}_i > k_i$. In that case $\deg \tilde{y}_i = \tilde{k}_i$ for all c.

The polynomial $\tilde{y}_{n-1}$ in (6.9) is of degree k_{n-1} or $\tilde{k}_{n-1} = k_{n-2} + 2k_n + 1 - k_{n-1}$. We say that the generation $(y_0, \dots, y_{n-1}, y_n) \to (y_0, \dots, \tilde{y}_{n-1}, y_n)$ is *degree increasing* in the $n-1$-st direction if $\tilde{k}_{n-1} > k_{n-1}$. In that case $\deg \tilde{y}_{n-1} = \tilde{k}_{n-1}$ for all c.

The polynomial $\tilde{y}_n$ in (6.10) is of degree k_n or $\tilde{k}_n = k_{n-1} + 1 - k_n$. We say that the generation $(y_0, \dots, y_{n-1}, y_n) \to (y_0, \dots, y_{n-1}, \tilde{y}_n)$ is *degree increasing* in the n-th direction if $\tilde{k}_n > k_n$. In that case $\deg \tilde{y}_n = \tilde{k}_n$ for all c.

For $i = 0, \dots, n$, if the generation is degree increasing in the i-th direction we normalize family (6.11) and construct a map $Y_{y,i} : \mathbb{C} \to (\mathbb{C}[x])^{n+1}$ as follows. First we multiply the polynomials $y_0, \dots, y_n$ by numbers to make them monic. Then we choose a monic polynomial $y_{i,0}$ satisfying the equation $\mathrm{Wr}(y_{i,0}, y_i) = \epsilon \prod_{j \neq i} y_j^{-a_{j,i}}$, for some nonzero integer ϵ, and such that the coefficient of x^{k_i} in $y_{i,0}$ equals zero. Set

$$\tilde{y}_i(x, c) = y_{i,0}(x) + c y_i(x), \tag{6.12}$$

and define

$$Y_{y,i} \, : \, \mathbb{C} \, \to \, (\mathbb{C}[x])^{n+1}, \qquad c \mapsto \, y^{(i)}(x, c) = (y_0(x), \dots, \tilde{y}_i(x, c), \dots, y_n(x)). \tag{6.13}$$

The polynomials of this $(n+1)$-tuple are monic.

6.5. Degree-transformations and generation of vectors of integers. The degree-transformations

$$\begin{aligned}
(6.14) \quad & k := (k_0, \dots, k_n) \;\mapsto\; k^{(0)} = (2k_1 + 1 - k_0, \dots, k_n), \\
& k := (k_0, \dots, k_n) \;\mapsto\; k^{(i)} = (k_0, \dots, k_{i-1} + k_{i+1} + 1 - k_i, \dots, k_n), \\
& \qquad\qquad\qquad\qquad\qquad\qquad\qquad i = 1, \dots, n-2, \\
& k := (k_0, \dots, k_{n-1}, k_n) \;\mapsto\; k^{(n-1)} = (k_0, \dots, k_{n-2} + 2k_n + 1 - k_{n-1}, k_n), \\
& k := (k_0, \dots, k_n) \;\mapsto\; k^{(n)} = (k_0, \dots, k_{n-1} + 1 - k_n),
\end{aligned}$$

correspond to the shifted action of reflections $w_0, \dots, w_n \in W$, where W is the Weyl group associated with the Cartan matrix A in (6.3) and $w_0, \dots, w_n$ are the standard generators, see [**MV1**, Lemma 3.11] for more detail.

We take formula (6.14) as the definition of *degree-transformations*:

$$\begin{aligned}
(6.15)\qquad & w_0 \ : k \ \mapsto \ k^{(0)} = (2k_1+1-k_0,\dots,k_n),\\
& w_i \ : k \ \mapsto \ k^{(i)} = (k_0,\dots,k_{i-1}+k_{i+1}+1-k_i,\dots,k_n), \quad i=1,\dots,n-2,\\
& w_{n-1} \ : k \ \mapsto \ k^{(n-1)} = (k_0,\dots,k_{n-2}+2k_n+1-k_{n-1},k_n),\\
& w_n \ : k \ \mapsto \ k^{(n)} = (k_0,\dots,k_{n-1}+1-k_n),
\end{aligned}$$

acting on arbitrary vectors $k=(k_0,\dots,k_n)$.

We start with the vector $k^{\emptyset}=(0,\dots,0)$ and a sequence $J=(j_1,j_2,\dots,j_m)$ of integers such that $j_i\in\{0,\dots,n\}$ for all i. We apply the corresponding degree transformations to $k^{\emptyset}$ and obtain a sequence of vectors $k^{\emptyset}$, $k^{(j_1)}=w_{j_1}k^{\emptyset}$, $k^{(j_1,j_2)}=w_{j_2}w_{j_1}k^{\emptyset}$,$\dots$,

$$(6.16)\qquad k^J = w_{j_m}\dots w_{j_2}w_{j_1}k^{\emptyset}.$$

We say that the *vector k^J is generated from $(0,\dots,0)$ in the direction of J.*

We call a sequence J *degree increasing* if for every i the transformation w_{j_i} applied to $w_{j_{i-1}}\dots w_{j_1}k^{\emptyset}$ increases the j_i-th coordinate.

6.6. Multistep generation. Let $J=(j_1,\dots,j_m)$ be a degree increasing sequence. Starting from $y^{\emptyset}=(1,\dots,1)$ and J, we construct a map

$$Y^J:\mathbb{C}^m\to(\mathbb{C}[x])^{n+1}$$

by induction on m. If $J=\emptyset$, the map $Y^{\emptyset}$ is the map $\mathbb{C}^0=(pt)\ \mapsto y^{\emptyset}$. If $m=1$ and $J=(j_1)$, the map $Y^{(j_1)}:\mathbb{C}\to(\mathbb{C}[x])^{n+1}$ is given by formula (6.13) for $y=y^{\emptyset}$ and $j=j_1$. More precisely, equation (6.5) takes the form $\mathrm{Wr}(\tilde y_{j_1},1)=1$. Then $\tilde y_{j_1,0}=x$ and

$$Y^{(j_1)}\ :\ \mathbb{C}\mapsto(\mathbb{C}[x])^{n+1},\qquad c\mapsto(1,\dots,x+c,\dots,1).$$

By Theorem 6.1 all tuples in the image are fertile and almost all tuples are generic (in this example all tuples are generic). Assume that for $\tilde J=(j_1,\dots,j_{m-1})$, the map $Y^{\tilde J}$ is constructed. To obtain Y^J we apply the generation procedure in the j_m-th direction to every tuple of the image of $Y^{\tilde J}$. More precisely, if

$$(6.17)\qquad Y^{\tilde J}\ :\ \tilde c=(c_1,\dots,c_{m-1})\ \mapsto\ (y_0(x,\tilde c),\dots,y_n(x,\tilde c)),$$

then

$$(6.18)\qquad Y^J:(\tilde c,c_m)\mapsto(y_0(x,\tilde c),\dots,y_{j_m,0}(x,\tilde c)+c_my_{j_m}(x,\tilde c),\dots,y_n(x,\tilde c)).$$

The map Y^J is called the *generation of tuples from $y^{\emptyset}$ in the J-th direction.*

LEMMA 6.2. *All tuples in the image of Y^J are fertile and almost all tuples are generic. For any $c\in\mathbb{C}^m$ the $(n+1)$-tuple $Y^J(c)$ consists of monic polynomials. The degree vector of this tuple equals k^J.* □

LEMMA 6.3. *The map Y^J sends distinct points of $\mathbb{C}^m$ to distinct points of $(\mathbb{C}[x])^{n+1}$.*

PROOF. The lemma is easily proved by induction on m. □

6.7. Critical points and the population generated from $y^\emptyset$. The set of all tuples $(y_0,\dots,y_n)\in(\mathbb{C}[x])^{n+1}$ obtained from $y^\emptyset=(1,\dots,1)$ by generations in all directions $J=(j_1,\dots,j_m)$, $m\geqslant 0$, (not necessarily degree increasing) is called the *population of tuples* generated from $y^\emptyset$, see [**MV1, MV2**].

THEOREM 6.4 ([**MV3**]). *If a tuple of polynomials $(y_0,\dots,y_n)$ represents a critical point of the master function $\Phi(u,k)$ defined in* (6.1) *for some parameters $k=(k_0,\dots,k_n)$, then $(y_0,\dots,y_n)$ is a point of the population generated from $y^\emptyset$ by a degree increasing generation, that is, there exist a degree increasing sequence $J=(j_1,\dots,j_m)$ and a point $c\in\mathbb{C}^m$ such that $(y_0(x),\dots,y_n(x))=Y^J(x,c)$. Moreover, for any other critical point of that function $\Phi(u,k)$ there is a point $c'\in\mathbb{C}^m$ such that the tuple $Y^J(x,c')$ represents that other critical point.*

By Theorem 6.4 a function $\Phi(u,k)$ either does not have critical points at all or all of its critical points form one cell $\mathbb{C}^m$.

PROOF. Theorem 3.8 in [**MV2**] says that $(y_0,\dots,y_n)$ is a point of the population generated from $y^\emptyset$. The fact that $(y_0,\dots,y_n)$ can be generated from $y^\emptyset$ by a degree increasing generation is a corollary of Lemmas 3.5 and 3.7 in [**MV2**]. The same lemmas show that any other critical point of the master function $\Phi(u,k)$ is represented by the tuple $Y^J(x,c')$ for a suitable $c'\in\mathbb{C}^m$. $\square$

7. Critical points of master functions and Miura opers

7.1. Miura oper associated with a tuple of polynomials, [MV2]. We say that a Miura oper of type $A^{(2)}_{2n}$, $\mathcal{L}=\partial+\Lambda^{(2)}+V$, is *associated to the* $(n+1)$-*tuple of polynomials* y if $V=-\sum_{i=0}^n \ln'(y_i)\,h_i$, where $\ln'(f(x))=\frac{f'(x)}{f(x)}$. If $\mathcal{L}$ is associated to y and $V=\sum_{i=1}^{2n+1} v_ie_{i,i}$, then

$$v_i=\ln'\frac{y_i}{y_{i-1}}=-v_{2n+2-i},\quad i=1,\dots,n-1,\tag{7.1}$$

$$v_n=\ln'\frac{y_n^2}{y_{n-1}}=-v_{n+2},\quad v_{n+1}=0.$$

We also have

$$\langle\alpha_j,V\rangle=\ln'\big(\prod_{i=0}^n y_i^{-a_{i,j}}\big),\tag{7.2}$$

where $a_{i,j}$ are entries of the Cartan matrix of type $A^{(2)}_{2n}$. More precisely,

$$\langle\alpha_0,V\rangle=\ln'\big(\frac{y_1^2}{y_0^2}\big),\qquad \langle\alpha_i,V\rangle=\ln'\big(\frac{y_{i-1}y_{i+1}}{y_i^2}\big),\quad i=1,\dots,n-2,\tag{7.3}$$

$$\langle\alpha_{n-1},V\rangle=\ln'\big(\frac{y_{n-2}y_n^2}{y_{n-1}^2}\big),\qquad \langle\alpha_n,V\rangle=\ln'\big(\frac{y_{n-1}}{y_n^2}\big).$$

For example,

$$\mathcal{L}^\emptyset:=\partial+\Lambda^{(2)}\tag{7.4}$$

is associated to the tuple $y^\emptyset=(1,\dots,1)$.

Define the map

$$\mu:(\mathbb{C}[x]\setminus\{0\})^{n+1}\to\mathcal{M}(A^{(2)}_{2n}),$$

which sends a tuple $y = (y_0, \dots, y_n)$ to the Miura oper $\mathcal{L} = \partial + \Lambda^{(2)} + V$ associated to y.

7.2. Deformations of Miura opers of type $A_{2n}^{(2)}$, [MV2].

LEMMA 7.1 ([**MV2**]). *Let $\mathcal{L} = \partial + \Lambda^{(2)} + V$ be a Miura oper of type $A_{2n}^{(2)}$. Let $g \in \mathcal{B}$ and $j \in \{0, \dots, n\}$. Then*

$$e^{\operatorname{ad} gf_j}\mathcal{L} = \partial + \Lambda^{(2)} + V - gh_j - (g' - \langle \alpha_j, V\rangle g + g^2)f_j. \tag{7.5}$$

□

COROLLARY 7.2 ([**MV2**]). *Let $\mathcal{L} = \partial + \Lambda^{(2)} + V$ be a Miura oper of type $A_{2n}^{(2)}$. Then $e^{\operatorname{ad} gf_j}\mathcal{L}$ is a Miura oper if and only if the scalar function g satisfies the Riccati equation*

$$g' - \langle \alpha_j, V\rangle g + g^2 = 0\ . \tag{7.6}$$

□

Let $\mathcal{L} = \partial + \Lambda^{(2)} + V$ be a Miura oper. For $j \in \{0, \dots, n\}$, we say that $\mathcal{L}$ is *deformable in the j-th direction* if equation (7.6) has a nonzero solution g, which is a rational function.

THEOREM 7.3 ([**MV2**]). *Let $\mathcal{L} = \partial + \Lambda^{(2)} + V$ be the Miura oper associated to the tuple of polynomials $y = (y_0, \dots, y_n)$. Let $j \in \{0, \dots, n\}$. Then $\mathcal{L}$ is deformable in the j-th direction if and only if there exists a polynomial $\tilde{y}_j$ satisfying equation (6.5). Moreover, in that case any nonzero rational solution g of the Riccati equation (7.6) has the form $g = \ln'(\tilde{y}_j/y_j)$ where $\tilde{y}_j$ is a solution of equation (6.5). If $g = \ln'(\tilde{y}_j/y_j)$, then the Miura oper*

$$e^{\operatorname{ad} gf_j}\mathcal{L} = \partial + \Lambda^{(2)} + V - gh_j \tag{7.7}$$

is associated to the tuple $y^{(j)}$, which is obtained from the tuple y by replacing y_j with $\tilde{y}_j$.

7.3. Miura opers associated with the generation procedure. Let $J = (j_1, \dots, j_m)$ be a degree increasing sequence, see Section 6.5. Let $Y^J : \mathbb{C}^m \to (\mathbb{C}[x])^{n+1}$ be the generation of tuples from $y^\emptyset$ in the J-th direction. We define the associated family of Miura opers by the formula:

$$\mu^J\ :\ \mathbb{C}^m\ \to\ \mathcal{M}(A_{2n}^{(2)}), \qquad c\ \mapsto\ \mu(Y^J(c)).$$

The map μ^J is called the *generation of Miura opers from $\mathcal{L}^\emptyset$ in the J-th direction,* see $\mathcal{L}^\emptyset$ in (7.4).

For $\ell = 1, \dots, m$, denote $J_\ell = (j_1, \dots, j_\ell)$ the beginning ℓ-interval of the sequence J. Consider the associated map $Y^{J_\ell} : \mathbb{C}^\ell \to (\mathbb{C}[x])^{n+1}$. Denote

$$Y^{J_\ell}(c_1, \dots, c_\ell) = (y_0(x, c_1, \dots, c_\ell; \ell), \dots, y_n(x, c_1, \dots, c_\ell; \ell)).$$

Introduce

$$\begin{aligned} g_1(x, c_1, \dots, c_m) &= \ln'(y_{j_1}(x, c_1; 1)), \\ g_\ell(x, c_1, \dots, c_m) &= \ln'(y_{j_\ell}(x, c_1 \dots, c_\ell; \ell)) - \ln'(y_{j_\ell}(x, c_1, \dots, c_{\ell-1}; \ell - 1)), \end{aligned} \tag{7.8}$$

for $\ell = 2, \dots, m$. For $c \in \mathbb{C}^m$, define $U^J(c) = \sum_{i<0}(U^J(c))_i$, $(U^J(c))_i \in \mathcal{B}(\mathfrak{g}(A_{2n}^{(2)})^i)$, depending on $c \in \mathbb{C}^m$, by the formula

$$e^{-\operatorname{ad} U^J(c)} = e^{\operatorname{ad} g_m(x,c)f_{j_m}} \cdots e^{\operatorname{ad} g_1(x,c)f_{j_1}}. \tag{7.9}$$

LEMMA 7.4. *For $c \in \mathbb{C}^m$, we have*

$$\mu^J(c) = e^{-\operatorname{ad} U^J(c)}(\mathcal{L}^{\emptyset}), \tag{7.10}$$

$$\mu^J(c) = \partial + \Lambda^{(2)} - \sum_{\ell=1}^{m} g_\ell(x,c) h_{j_\ell}. \tag{7.11}$$

PROOF. The lemma follows from Theorem 7.3. □

COROLLARY 7.5. *Let $r > 0$, odd and $r \neq n+1 \bmod 4n+2$. Let $c \in \mathbb{C}^m$. Let $\frac{\partial}{\partial t_r}\big|_{\mu^J(c)}$ be the value at $\mu^J(c)$ of the vector field of the r-th mKdV flow on the space $\mathcal{M}(A_{2n}^{(2)})$, see (4.5). Then*

$$\frac{\partial}{\partial t_r}\Big|_{\mu^J(c)} = -\frac{\partial}{\partial x}\left(e^{-\operatorname{ad} U^J(c)}(\Lambda^{(2)})^r\right)^0. \tag{7.12}$$

PROOF. The corollary follows from (4.6) and (7.10). □

We have the natural embedding $\mathcal{M}(A_{2n}^{(2)}) \hookrightarrow \mathcal{M}(A_{2n}^{(1)})$, see Section 3.4. Let $J = (j_1, j_2, \dots, j_m)$. Denote $\tilde{J} = (j_1, \dots, j_{m-1})$. Consider the associated family $\mu^{\tilde{J}} : \mathbb{C}^{m-1} \to \mathcal{M}(A_{2n}^{(2)})$. Denote $\tilde{c} = (c_1, \dots, c_{m-1})$.

PROPOSITION 7.6. *For any $r > 0$ the difference $\frac{\partial}{\partial t_r}\big|_{\mu^J(c)} - \frac{\partial}{\partial t_r}\big|_{\mu^{\tilde{J}}(\tilde{c})}$ has the following form for some scalar functions $u_1(x,c)$, $u_2(x,c)$, $u_3(x,c)$:*

(i) *if $j_m \in \{1, 2, \dots, n-1\}$, then*

$$\begin{aligned}\frac{\partial}{\partial t_r}\Big|_{\mu^J(c)} - \frac{\partial}{\partial t_r}\Big|_{\mu^{\tilde{J}}(\tilde{c})} &= u_1(x,c)(e_{j_m+1,j_m+1} - e_{j_m,j_m}) \\ &+ u_2(x,c)(e_{2n+2-j_m,2n+2-j_m} - e_{2n+1-j_m,2n+1-j_m}),\end{aligned} \tag{7.13}$$

(ii) *if $j_m = 0$, then*

$$\frac{\partial}{\partial t_r}\Big|_{\mu^J(c)} - \frac{\partial}{\partial t_r}\Big|_{\mu^{\tilde{J}}(\tilde{c})} = u_1(x,c)(e_{2n+1,2n+1} - e_{1,1}), \tag{7.14}$$

(iii) *if $j_m = n$, then*

$$\frac{\partial}{\partial t_r}\Big|_{\mu^J(c)} - \frac{\partial}{\partial t_r}\Big|_{\mu^{\tilde{J}}(\tilde{c})} = u_1(x,c)e_{n,n} + u_2(x,c)e_{n+1,n+1} + u_3(x,c)e_{n+2,n+2}. \tag{7.15}$$

PROOF. We will write Λ for $\Lambda^{(2)} = \Lambda^{(1)}$. Denote

$$A_r = e^{g_{m-1} f_{j_{m-1}}} \dots e^{g_1 f_{j_1}} \Lambda^r e^{-g_1 f_{j_1}} \dots e^{-g_{m-1} f_{j_{m-1}}}.$$

Expand $A_r = \sum_i A_r^i \Lambda^i$ where $A_r^i = \sum_{l=1}^{2n+1} A_r^{i,l} e_{l,l}$ with scalar coefficients $A_r^{i,l}$. Then $\frac{\partial}{\partial t_r}\big|_{\mu^{\tilde{J}}(\tilde{c})} = -\frac{\partial}{\partial x} A_r^0$. Assume that $j_m \in \{1, \ldots, n-1\}$. Then

$$\begin{aligned}\frac{\partial}{\partial t_r}\Big|_{\mu^J(c)} &= -\frac{\partial}{\partial x}\big[(1 + g_m(e_{j_m,j_m} + e_{2n+1-j_m,2n+1-j_m})\Lambda^{-1})A_r \\ &\times(1 - g_m(e_{j_m,j_m} + e_{2n+1-j_m,2n+1-j_m})\Lambda^{-1})\big]^0 = -\frac{\partial}{\partial x} A_r^0 \\ &-\frac{\partial}{\partial x}\big[g_m(e_{j_m,j_m} + e_{2n+1-j_m,2n+1-j_m})\Lambda^{-1}A_r\big]^0 \\ &+\frac{\partial}{\partial x}\big[A_r g_m(e_{j_m,j_m} + e_{2n+1-j_m,2n+1-j_m})\Lambda^{-1}\big]^0 \\ &+\frac{\partial}{\partial x}\big[g_m(e_{j_m,j_m} + e_{2n+1-j_m,2n+1-j_m})\Lambda^{-1}A_r g_m \\ &\qquad\times(e_{j_m,j_m} + e_{2n+1-j_m,2n+1-j_m})\Lambda^{-1}\big]^0.\end{aligned}$$

The last term is zero since

$$\begin{aligned}&\big[g_m(e_{j_m,j_m} + e_{2n+1-j_m,2n+1-j_m})\Lambda^{-1}A_r g_m(e_{j_m,j_m} + e_{2n+1-j_m,2n+1-j_m})\Lambda^{-1}\big]^0 \\ &= g_m^2(e_{j_m,j_m} + e_{2n+1-j_m,2n+1-j_m})[\Lambda^{-1}A_r\Lambda^{-1}]^0 \\ &\qquad\times(e_{j_m+1,j_m+1} + e_{2n+2-j_m,2n+2-j_m}) = 0,\end{aligned}$$

and we get

$$\begin{aligned}\frac{\partial}{\partial t_r}\Big|_{\mu^J(c)} - \frac{\partial}{\partial t_r}\Big|_{\mu^{\tilde{J}}(\tilde{c})} &= g_m A_r^{1,j_m+1}(e_{j_m+1,j_m+1} - e_{j_m,j_m}) \\ &+ g_m A_r^{1,2n+2-j_m}(e_{2n+2-j_m,2n+2-j_m} - e_{2n+1-j_m,2n+1-j_m}).\end{aligned}$$

The cases $j_m = 0, n$ are proved similarly. □

Let $\mathfrak{m}_i : \mathcal{M}(A_{2n}^{(1)}) \to \mathcal{D}$, $\mathcal{L} \mapsto L_i$, be the Miura maps defined in Section 4.5 for $i = 0, \ldots, 2n$. Below we consider the composition of the embedding $\mathcal{M}(A_{2n}^{(2)}) \hookrightarrow \mathcal{M}(A_{2n}^{(1)})$ and a Miura map.

LEMMA 7.7. *If $j_m = 0$, we have $\mathfrak{m}_i \circ \mu^J(\tilde{c}, c_m) = \mathfrak{m}_i \circ \mu^{\tilde{J}}(\tilde{c})$ for all $i \neq 0$. If $j_m = 1, \ldots, n$, we have $\mathfrak{m}_i \circ \mu^J(\tilde{c}, c_m) = \mathfrak{m}_i \circ \mu^{\tilde{J}}(\tilde{c})$ for all $i \neq j_m, 2n+1-j_m$.*

PROOF. The lemma follows from formula (7.10) and Theorem 4.7. □

LEMMA 7.8. *If $j_m = 0$, then*

$$\frac{\partial \mu^J}{\partial c_m}(\tilde{c}, c_m) = -a\frac{y_1(x,\tilde{c},m-1)^2}{y_0(x,\tilde{c},c_m,m)^2} h_0 \tag{7.16}$$

for some positive integer a. If $j_m = 1, \ldots, n-2$, then

$$\frac{\partial \mu^J}{\partial c_m}(\tilde{c}, c_m) = -a\frac{y_{j_m-1}(x,\tilde{c},m-1)y_{j_m+1}(x,\tilde{c},m-1)}{y_{j_m}(x,\tilde{c},c_m,m)^2} h_{j_m} \tag{7.17}$$

for some positive integer a. If $j_m = n-1$, then

$$\frac{\partial \mu^J}{\partial c_m}(\tilde{c}, c_m) = -a\frac{y_{n-2}(x,\tilde{c},m-1)y_n^2(x,\tilde{c},m-1)}{y_{n-1}(x,\tilde{c},c_m,m)^2} h_{n-1} \tag{7.18}$$

for some positive integer a. If $j_m = n$, then

$$\frac{\partial \mu^J}{\partial c_m}(\tilde{c}, c_m) = -a\frac{y_{n-1}(x,\tilde{c},m-1)}{y_n(x,\tilde{c},c_m,m)^2} h_n \tag{7.19}$$

for some positive integer a.

Notice that the right-hand side of these formulas can be written as

$$-a\prod_{i=0}^{n} y_i(x,c,m)^{-a_{i,j}}h_j. \tag{7.20}$$

PROOF. Let $j_m = 0$. Then $y_0(x,\tilde c, c_m, m) = y_{0,0}(x,\tilde c) + c_m y_0(x,\tilde c, m-1)$, where $y_{0,0}(x,\tilde c)$ is such that

$$\mathrm{Wr}(y_{0,0}(x,\tilde c), y_0(x,\tilde c, m-1)) \;=\; a\; y_1(x,\tilde c, m-1)^2,$$

for some positive integer a, see (6.12). We have $g_m = \ln'(y_0(x,\tilde c, c_m, m)) - \ln'(y_0(x,\tilde c, m-1))$.

By formula (7.11), we have

$$\begin{aligned}\frac{\partial\mu^J}{\partial c_m}(\tilde c, c_m) &= -\frac{\partial g_m}{\partial c_m}(\tilde c, c_m)h_0 = -\frac{\partial}{\partial c_m}\left(\frac{y'_{0,0}(x,\tilde c) + c_m y'_0(x,\tilde c, m-1)}{y_{0,0}(x,\tilde c) + c_m y_0(x,\tilde c, m-1)}\right)h_0\\ &= -\frac{\mathrm{Wr}(y_{0,0}(x,\tilde c), y_0(x,\tilde c, m-1))}{(y_{0,0}(x,\tilde c) + c_m y_0(x,\tilde c, m-1))^2}h_0 = -a\,\frac{y_1(x,\tilde c, m-1)^2}{y_0(x,\tilde c, c_m, m)^2}h_0\,.\end{aligned}$$

This proves formula (7.16). The other formulas are proved similarly. □

7.4. Intersection of kernels of $d\mathfrak{m}_i$. Let $J = (j_1,\dots,j_m)$ be a degree increasing sequence and $\mu^J : \mathbb{C}^m \to \mathcal{M}(A^{(2)}_{2n})$ the generation of Miura opers from $\mathcal{L}^\emptyset$ in the J-th direction. We have $\mu^J(c) = \partial + \Lambda^{(1)} + \sum_{k=1}^{2n+1} v_k(x,c)e_{k,k}$, where

$$\sum_{k=1}^{2n+1} v_k(x,c) = 0, \quad v_i(x,c) + v_{2n+2-i}(x,c) = 0,\ i = 1,\dots,2n+1.$$

Let $X(c) = \sum_{k=1}^{2n+1} X_k(x,c)e_{k,k} \in T_{\mu^J(c)}\mathcal{M}(A^{(2)}_{2n})$ be a field of tangent vectors to $\mathcal{M}(A^{(2)}_{2n})$ at the points of the image of μ^J,

$$\sum_{k=1}^{2n+1} X_k(x,c) = 0, \quad X_i(x,c) + X_{2n+2-i}(x,c) = 0,\ i = 1,\dots,2n+1.$$

Our goal is to show that under certain conditions we have

$$X(c) = A(c)\frac{\partial\mu^J}{\partial c_m}(c) \tag{7.21}$$

for some scalar function $A(c)$ on $\mathbb{C}^m$.

PROPOSITION 7.9. *Let $j_m = 0$ and $X(c) \in T_{\mu^J(c)}\mathcal{M}(A^{(2)}_{2n})$. Assume that $d\mathfrak{m}_i\big|_{\mu^J(c)}(X(c)) = 0$ for all $i = 1,\dots,2n$ and all $c \in \mathbb{C}^m$. Assume that $X(c)$ has the form indicated in the right-hand side of formula (7.14). Then equation (7.21) holds.*

PROOF. Since $X_k(x,c) = 0$ for $k = 2,\dots,2n$, equation (5.5) takes the form $X_1' - 2v_1X_1 = 0$, or more precisely, $X_1' = 2\ln'\big(\frac{y_1(x,\tilde c, m-1)}{y_0(x,\tilde c, c_m, m)}\big)X_1$. Hence $X_1(x,c) = -X_{2n+1} = A(c)\frac{y_1(x,\tilde c, m-1)^2}{y_0(x,\tilde c, c_m, m)^2}$ for some scalar $A(c)$. Lemma 7.8 implies equation (7.21). □

PROPOSITION 7.10. *Let $j_m \in \{1, \dots, n-1\}$ and $X(c) \in T_{\mu^J(c)}\mathcal{M}(A_{2n}^{(2)})$. Assume that $d\mathfrak{m}_i\big|_{\mu^J(c)}(X(c)) = 0$ for all $i \notin \{j_m, 2n+1-j_m\}$ and all $c \in \mathbb{C}^m$. Assume that $X(c)$ has the form indicated in the right-hand side of formula* (7.13). *Then equation* (7.21) *holds.*

PROOF. By Lemma 5.3 we have $X'_{j_m} + (v_{j_m} - v_{j_m+1})X_{j_m} = 0$. Then for $j_m = 1, \dots, n-2$, we have

$$\begin{aligned} X_{j_m} &= -X_{j_m+1} = X_{2n+1-j_m} = -X_{2n+2-j_m} \\ &= A(c)\, \frac{y_{j_m-1}(x, \tilde{c}, m-1) y_{j_m+1}(x, \tilde{c}, m-1)}{y_{j_m}(x, \tilde{c}, c_m, m)^2} \end{aligned}$$

and for $j_m = n-1$, we have

$$X_{n-1} = -X_n = X_{n+2} = -X_{n+3} = A(c)\, \frac{y_{n-2}(x, \tilde{c}, m-1) y_n^2(x, \tilde{c}, m-1)}{y_{n-1}(x, \tilde{c}, c_m, m)^2}.$$

Lemma 7.8 yields equation (7.21). □

PROPOSITION 7.11. *Let $j_m = n$ and $X(c) \in T_{\mu^J(c)}\mathcal{M}(A_{2n}^{(2)})$. Assume that $d\mathfrak{m}_i\big|_{\mu^J(c)}(X(c)) = 0$ for all $i \neq n, n+1$, and $c \in \mathbb{C}^m$. Assume that $X(c)$ has the form indicated in the right-hand side of formula* (7.15). *Then equation* (7.21) *holds.*

PROOF. By assumptions we have $X_{n+1} = 0$ as well as $X_i = 0$ for $i \neq n, n+1, n+2$. By Lemma 5.4 we have $X'_n + v_n X_n = 0$ where $v_n = \ln' \frac{y_n^2}{y_{n-1}}$. Hence $X_n = -X_{n+2} = A(c) \frac{y_{n-1}(x, \tilde{c}, m-1)}{y_n(x, \tilde{c}, c_m, m)^2}$ for some scalar function $A(c)$. Lemma 7.8 yields equation (7.21). □

8. Vector fields

8.1. Statement. Let $r > 0$ be odd, and $r \neq 2n+1 \bmod 4n+2$. Recall that we denote by $\frac{\partial}{\partial t_r}$ the r-th mKdV vector field on the space $\mathcal{M}(A_{2n}^{(2)})$ of Miura opers of type $A_{2n}^{(2)}$. We also denote by $\frac{\partial}{\partial t_r}$ the r-th mKdV vector field of type $A_{2n}^{(1)}$ on the space $\mathcal{M}(A_{2n}^{(1)})$ of Miura opers of type $A_{2n}^{(1)}$. We have a natural embedding $\mathcal{M}(A_{2n}^{(2)}) \hookrightarrow \mathcal{M}(A_{2n}^{(1)})$. Under this embedding the vector $\frac{\partial}{\partial t_r}$ on $\mathcal{M}(A_{2n}^{(2)})$ equals the vector filed $\frac{\partial}{\partial t_r}$ on $\mathcal{M}(A_{2n}^{(1)})$ restricted to $\mathcal{M}(A_{2n}^{(2)})$, see Section 4.3. We also denote by $\frac{\partial}{\partial t_r}$ the r-th KdV vector field on the space $\mathcal{D}$, see Section 4.4.

For a Miura map $\mathfrak{m}_i : \mathcal{M} \to \mathcal{D}$, $\mathcal{L} \mapsto L_i$, denote by $d\mathfrak{m}_i$ the associated derivative map $T\mathcal{M}(A_{2n}^{(1)}) \to T\mathcal{D}$ of tangent spaces. By Theorem 4.6 we have $d\mathfrak{m}_i : \frac{\partial}{\partial t_r}\big|_{\mathcal{L}} \mapsto \frac{\partial}{\partial t_r}\big|_{L_i}$.

Fix a degree increasing sequence $J = (j_1, \dots, j_m)$. Consider the associated family $\mu^J : \mathbb{C}^m \to \mathcal{M}(A_{2n}^{(2)})$ of Miura opers. For a vector field Γ on $\mathbb{C}^m$, we denote by $\mathfrak{L}_\Gamma \mu^J$ the derivative of μ^J along the vector field. The derivative is well-defined since $\mathcal{M}(A_{2n}^{(2)})$ is an affine space.

THEOREM 8.1. *Let $r > 0$ be odd, $r \neq 2n+1$ mod $4n+2$. Then there exists a polynomial vector field Γ_r on $\mathbb{C}^m$ such that*

$$\frac{\partial}{\partial t_r}\bigg|_{\mu^J(c)} = \mathfrak{L}_{\Gamma_r} \mu^J(c) \tag{8.1}$$

for all $c \in \mathbb{C}^m$. If $r > 4m$, then $\frac{\partial}{\partial t_r}\big|_{\mu^J(c)} = 0$.

COROLLARY 8.2. *The family μ^J of Miura opers is invariant with respect to all mKdV flows of type $A_{2n}^{(2)}$ and is point-wise fixed by flows with $r > 4m$.*

In other words, every mKdV flow corresponds to a flow on the space of integration parameters $c \in \mathbb{C}^m$. Informally speaking, we may say, that the integration parameters $c = (c_1, \dots, c_m)$ are times of the mKdV flows.

8.2. Proof of Theorem 8.1 for $m = 1$. Let $J = (j_1)$. Then $\mu^J(c_1) = e^{g_1 f_{j_1}} \mathcal{L}^\emptyset e^{-g_1 f_{j_1}} = \partial + \Lambda - g_1 h_{j_1}$, where $g_1 = \frac{1}{x+c_1}$, see formula (7.9). We have

$$\frac{\partial}{\partial t_r}\Big|_{\mu^J(c_1)} = -\frac{\partial}{\partial x}\left[e^{g_1 f_{j_1}} \Lambda^r e^{-g_1 f_{j_1}}\right]^0. \tag{8.2}$$

Assume $j_1 \in \{1, \dots, n-1\}$. Then $e^{g_1 f_{j_1}} = 1 + g_1(e_{j_1,j_1} + e_{2n+1-j_1,2n+1-j_1})\Lambda^{-1}$. If r is odd and $r > 1$, then the right-hand side of (8.2) is zero, hence $\frac{\partial}{\partial t_r}|_{\mu^J(c_1)} = \Gamma_r = 0$. Let $r = 1$, then

$$\frac{\partial}{\partial t_1}\Big|_{\mu^J(c_1)} = -\frac{\partial}{\partial x}\left[e^{g_1 f_{j_1}} \Lambda e^{-g_1 f_{j_1}}\right]^0 = \frac{\partial}{\partial x} g_1 h_{j_1} = -\frac{1}{(x+c_1)^2} h_{j_1} = -\frac{\partial \mu^J}{\partial c_1}(c_1).$$

Hence $\Gamma_1 = -\frac{\partial}{\partial c_1}$. Assume $j_1 = n$, then by formula (7.12),

$$\begin{aligned}\frac{\partial}{\partial t_r}\Big|_{\mu^J(c_1)} &= -\frac{\partial}{\partial x}\Big[(1 + 2g_1(e_{n,n} + e_{n+1,n+1})\Lambda^{-1} + 4g_1^2 e_{n,n}\Lambda^{-2}) \\ &\times \ \Lambda^r(1 - 2g_1(e_{n,n} + e_{n+1,n+1})\Lambda^{-1} + 4g_1^2 e_{n,n}\Lambda^{-2}\Big]^0.\end{aligned} \tag{8.3}$$

It follows from (8.3) and Lemma 2.4 that $\frac{\partial}{\partial t_r}\Big|_{\mu^J(c_1)} = 0$ for r odd and $r > 1$, hence $\Gamma_r = 0$. For $r = 1$ we have $\frac{\partial}{\partial t_r}\Big|_{\mu^J(c_1)} = -2\frac{dg_1}{dx}(e_{n,n} - e_{n+2,n+2}) = \frac{dg_1}{dx} h_n = -\frac{1}{(x+c_1)^2} h_n = -\frac{\partial \mu^J}{\partial c_1}(c_1)$. Hence $\Gamma_1 = -\frac{\partial}{\partial c_1}$. Similarly, if $j_1 = 0$, then $\Gamma_1 = -\frac{\partial}{\partial c_1}$ and $\Gamma_r = 0$ for $r > 1$. Theorem 8.1 is proved for $m = 1$.

8.3. Beginning of proof of Theorem 8.1 for $m > 1$. We prove the first statement of Theorem 8.1 by induction on m. Let $J = (j_1, \dots, j_m)$. Assume that the statement is proved for $\tilde{J} = (j_1, \dots, j_{m-1})$. Let

$$Y^{\tilde{J}} \ : \ \mathbb{C}^{m-1} \to (\mathbb{C}[x])^{n+1}, \quad \tilde{c} = (c_1, \dots, c_{m-1}) \ \mapsto \ (y_0(x, \tilde{c}), \dots, y_n(x, \tilde{c}))$$

be the generation of tuples in the $\tilde{J}$-th direction. Then the generation of tuples in the J-th direction is

$$Y^J : \mathbb{C}^m \mapsto (\mathbb{C}[x])^{n+1}, \ (\tilde{c}, c_m) \mapsto (y_0(x, \tilde{c}), \dots, y_{j_m,0}(x, \tilde{c}) + c_m y_{j_m}(x, \tilde{c}), \dots, y_n(x, \tilde{c}),$$

see (6.17) and (6.18). We have $g_m = \ln'(y_{j_m,0}(x, \tilde{c}) + c_m y_{j_m}(x, \tilde{c})) - \ln'(y_{j_m}(x, \tilde{c}))$, see (7.8).

By the induction assumption, there exists a polynomial vector field $\Gamma_{r,\tilde{J}} = \sum_{i=1}^{m-1} \gamma_i(\tilde{c})\frac{\partial}{\partial c_i}$ on $\mathbb{C}^{m-1}$ such that for all $\tilde{c} \in \mathbb{C}^{m-1}$ we have

$$\frac{\partial}{\partial t_r}\Big|_{\mu^{\tilde{J}}(\tilde{c})} = \mathfrak{L}_{\Gamma_{r,\tilde{J}}} \mu^{\tilde{J}}(\tilde{c}). \tag{8.4}$$

PROPOSITION 8.3. *There exists a scalar polynomial $\gamma_m(\tilde{c}, c_m)$ on $\mathbb{C}^m$ such that the vector field $\Gamma_r = \Gamma_{r,\tilde{J}} + \gamma_m(\tilde{c}, c_m)\frac{\partial}{\partial c_m}$ satisfies (8.1) for all $(\tilde{c}, c_m) \in \mathbb{C}^m$.*

8.4. Proof of Proposition 8.3.

LEMMA 8.4. *We have*

$$dm_i\Big|_{\mu^J(\tilde{c},c_m)}\left(\frac{\partial}{\partial t_r}\Big|_{\mu^J(\tilde{c},c_m)} - \mathfrak{L}_{\Gamma_{r,\bar{J}}}\mu^J(\tilde{c},c_m)\right) = 0. \tag{8.5}$$

for all $i \notin \{j_m, 2n+1-j_m\}$.

PROOF. The proof is the same as the proof of Lemma 5.5 in [**VW**]. Namely, by Theorem 4.7 we have $\mathfrak{m}_i \circ \mu^J(\tilde{c},c_m) = \mathfrak{m}_i \circ \mu^{\bar{J}}(\tilde{c})$ for all $i \notin \{j_m, 2n+1-j_m\}$. Hence,

$$dm_i\big|_{\mu^J(\tilde{c},c_m)}\left(\mathfrak{L}_{\Gamma_{r,\bar{J}}}\mu^J(\tilde{c},c_m)\right) = \mathfrak{L}_{\Gamma_{r,\bar{J}}}(\mathfrak{m}_i\circ\mu^J)(\tilde{c},c_m) = \mathfrak{L}_{\Gamma_{r,\bar{J}}}(\mathfrak{m}_i\circ\mu^{\bar{J}})(\tilde{c}). \tag{8.6}$$

By Theorems 4.6 and 4.7, we have

$$dm_i\Big|_{\mu^J(\tilde{c},c_m)}\left(\frac{\partial}{\partial t_r}\Big|_{\mu^J(\tilde{c},c_m)}\right) = \frac{\partial}{\partial t_r}\Big|_{\mathfrak{m}_i\circ\mu^J(\tilde{c},c_m)} = \frac{\partial}{\partial t_r}\Big|_{\mathfrak{m}_i\circ\mu^{\bar{J}}(\tilde{c})}. \tag{8.7}$$

By the induction assumption, we have

$$\frac{\partial}{\partial t_r}\Big|_{\mathfrak{m}_i\circ\mu^{\bar{J}}(\tilde{c})} = \mathfrak{L}_{\Gamma_{r,\bar{J}}}(\mathfrak{m}_i\circ\mu^{\bar{J}})(\tilde{c}). \tag{8.8}$$

These three formulas prove the lemma. □

LEMMA 8.5. *The difference* $\frac{\partial}{\partial t_r}\big|_{\mu^J(c)} - \mathfrak{L}_{\Gamma_{r,\bar{J}}}\mu^J(\tilde{c},c_m)$ *has the form indicated in the right-hand side of formula* (7.13) *if* $j_m \in \{1,\dots,n-1\}$, *has the form indicated in the right-hand side of formula* (7.14) *if* $j_m = 0$, *has the form indicated in the right-hand side of formula* (7.15) *if* $j_m = n$.

PROOF. We have

$$\begin{aligned}
&\frac{\partial}{\partial t_r}\Big|_{\mu^J(c)} - \mathfrak{L}_{\Gamma_{r,\bar{J}}}\mu^J(\tilde{c},c_m)\\
&\quad = \frac{\partial}{\partial t_r}\Big|_{\mu^J(c)} - \frac{\partial}{\partial t_r}\Big|_{\mu^{\bar{J}}(\tilde{c})} + \frac{\partial}{\partial t_r}\Big|_{\mu^{\bar{J}}(\tilde{c})} - \mathfrak{L}_{\Gamma_{r,\bar{J}}}\mu^J(\tilde{c},c_m)\\
&\quad = \frac{\partial}{\partial t_r}\Big|_{\mu^J(c)} - \frac{\partial}{\partial t_r}\Big|_{\mu^{\bar{J}}(\tilde{c})} + \mathfrak{L}_{\Gamma_{r,\bar{J}}}\mu^{\bar{J}}(\tilde{c}) - \mathfrak{L}_{\Gamma_{r,\bar{J}}}\mu^J(\tilde{c},c_m)\\
&\quad = \frac{\partial}{\partial t_r}\Big|_{\mu^J(c)} - \frac{\partial}{\partial t_r}\Big|_{\mu^{\bar{J}}(\tilde{c})} + \mathfrak{L}_{\Gamma_{r,\bar{J}}}g_m(x,\tilde{c},c_m)\,h_{j_m},
\end{aligned}$$

see formula (7.11). If $j_m \in \{1,\dots,n-1\}$, then $\frac{\partial}{\partial t_r}\Big|_{\mu^J(c)} - \frac{\partial}{\partial t_r}\Big|_{\mu^{\bar{J}}(\tilde{c})}$ has the form indicated in the right-hand side of formula (7.13) by Proposition 7.6 and $\mathfrak{L}_{\Gamma_{r,\bar{J}}}g_m(x,\tilde{c},c_m)\,h_{j_m}$ has that form since

$$h_{j_m} = -e_{j_m,j_m} + e_{j_m+1,j_m+1} - e_{2n+1-j_m,2n+1-j_m} + e_{2n+2-j_m,2n+2-j_m}.$$

This proves the lemma for $j_m \in \{1,\dots,n-1\}$. The other two cases of the lemma are proved similarly. □

Let us finish the proof of Proposition 8.3. By Lemmas 8.4 and 8.5 the difference $\frac{\partial}{\partial t_r}\big|_{\mu^J(c)} - \mathfrak{L}_{\Gamma_{r,\bar{J}}}\mu^J(\tilde{c},c_m)$ has the form indicated in the right-hand side of one of the formulas (7.13)-(7.15) and lies in the kernels of the differentials of Miura maps $\mathfrak{m}_i$ for all $i \notin \{j_m, 2n+1-j_m\}$. By Propositions 7.9, 7.10, 7.11 we conclude that the difference has the form $\gamma_m(\tilde{c},c_m)\frac{\partial\mu^J}{\partial c_m}$ for some scalar function $\gamma_m(\tilde{c},c_m)$

on $\mathbb{C}^m$. Therefore, $\frac{\partial}{\partial t_r}\big|_{\mu^J(\tilde c, c_m)} = \mathfrak{L}_{\Gamma_{r,\tilde J}}\mu^J(\tilde c, c_m) + \gamma_m(\tilde c, c_m)\frac{\partial \mu^J}{\partial c_m}(\tilde c, c_m)$. If we set $\Gamma_r = \Gamma_{r,\tilde J} + \gamma_m(\tilde c, c_m)\frac{\partial}{\partial c_m}$, then the vector field Γ_r will satisfy formula (8.1).

We need to prove that $\gamma_m(\tilde c, c_m)$ is a polynomial. The proof of that fact is the same as the proof of [**VW**, Proposition 5.9]. Proposition 8.3 is proved.

8.5. End of proof Theorem 8.1 for $m > 1$. Proposition 8.3 implies the first statement of Theorem 8.1. The second statement says that if $r > 4m$, then $\frac{\partial}{\partial t_r}\big|_{\mu^J(c)} = 0$. But that follows from Corollary 7.5 and Lemma 2.3.

Acknowledgments

The first author thanks the Max Plank Institute for Mathematics in Bonn for hospitality in 2015-2016 and the Hausdorff Institute for Mathematics in Bonn for hospitality in 2017.

References

[AM] M. Adler and J. Moser, *On a class of polynomials connected with the Korteweg-de Vries equation*, Comm. Math. Phys. **61** (1978), no. 1, 1–30. MR0501106

[BF] H. M. Babujian and R. Flume, *Off-shell Bethe ansatz equation for Gaudin magnets and solutions of Knizhnik-Zamolodchikov equations*, Modern Phys. Lett. A **9** (1994), no. 22, 2029–2039. MR1290286

[DS] V. G. Drinfel′d and V. V. Sokolov, *Lie algebras and equations of Korteweg-de Vries type* (Russian), Current problems in mathematics, Vol. 24, Itogi Nauki i Tekhniki, Akad. Nauk SSSR, Vsesoyuz. Inst. Nauchn. i Tekhn. Inform., Moscow, 1984, pp. 81–180. MR760998

[F] E. Frenkel, *Opers on the projective line, flag manifolds and Bethe ansatz* (English, with English and Russian summaries), Mosc. Math. J. **4** (2004), no. 3, 655–705, 783. MR2119144

[MTV09] E. Mukhin, V. Tarasov, and A. Varchenko, *Schubert calculus and representations of the general linear group*, J. Amer. Math. Soc. **22** (2009), no. 4, 909–940. MR2525775

[MV1] E. Mukhin and A. Varchenko, *Critical points of master functions and flag varieties*, Commun. Contemp. Math. **6** (2004), no. 1, 111–163. MR2048778

[MV2] E. Mukhin and A. Varchenko, *Miura opers and critical points of master functions*, Cent. Eur. J. Math. **3** (2005), no. 2, 155–182. MR2129922

[MV3] E. Mukhin and A. Varchenko, *On the number of populations of critical points of master functions*, J. Singul. **8** (2014), 31–38. MR3193226

[RV] N. Reshetikhin and A. Varchenko, *Quasiclassical asymptotics of solutions to the KZ equations*, Geometry, topology, & physics, Conf. Proc. Lecture Notes Geom. Topology, IV, Int. Press, Cambridge, MA, 1995, pp. 293–322. MR1358621

[SV] V. V. Schechtman and A. N. Varchenko, *Arrangements of hyperplanes and Lie algebra homology*, Invent. Math. **106** (1991), no. 1, 139–194. MR1123378

[ScV] I. Scherbak and A. Varchenko, *Critical points of functions, $\mathfrak{sl}_2$ representations, and Fuchsian differential equations with only univalued solutions* (English, with English and Russian summaries), Mosc. Math. J. **3** (2003), no. 2, 621–645, 745. Dedicated to Vladimir I. Arnold on the occasion of his 65th birthday. MR2025276

[Sz] G. Szegö, *Orthogonal Polynomials*, American Mathematical Society, New York, 1939. American Mathematical Society Colloquium Publications, v. 23. MR0000077

[V1] A. Varchenko, *Multidimensional hypergeometric functions and representation theory of Lie algebras and quantum groups*, Advanced Series in Mathematical Physics, vol. 21, World Scientific Publishing Co., Inc., River Edge, NJ, 1995. MR1384760

[V2] A. Varchenko, *Special functions, KZ type equations, and representation theory*, CBMS Regional Conference Series in Mathematics, vol. 98, Published for the Conference Board of the Mathematical Sciences, Washington, DC; by the American Mathematical Society, Providence, RI, 2003. MR2016311

[V3] A. Varchenko, *Quantum integrable model of an arrangement of hyperplanes*, SIGMA Symmetry Integrability Geom. Methods Appl. **7** (2011), Paper 032, 55. MR2804564

[VW] A. Varchenko and D. Wright, *Critical points of master functions and integrable hierarchies*, Adv. Math. **263** (2014), 178–229. MR3239138

[VWW14] A. Varchenko, T. Woodruff, and D. Wright, *Critical points of master functions and the mKdV hierarchy of type* $A_2^{(2)}$, Bridging algebra, geometry, and topology, Springer Proc. Math. Stat., vol. 96, Springer, Cham, 2014, pp. 167–195. MR3297115

.

Department of Mathematics, University of North Carolina at Chapel Hill Chapel Hill, North Carolina 27599-3250

Email address: anv@email.unc.edu

Department of Mathematics, University of North Carolina at Chapel Hill Chapel Hill, North Carolina 27599-3250

Email address: tykwood@gmail.com

SELECTED PUBLISHED TITLES IN THIS SERIES

713 **Naihuan Jing and Kailash C. Misra, Editors,** Representations of Lie Algebras, Quantum Groups and Related Topics, 2018

710 **Raphaël Danchin, Reinhard Farwig, Jiří Neustupa, and Patrick Penel, Editors,** Mathematical Analysis in Fluid Mechanics, 2018

709 **Fernando Galaz-García, Juan Carlos Pardo Millán, and Pedro Solórzano, Editors,** Contributions of Mexican Mathematicians Abroad in Pure and Applied Mathematics, 2018

708 **Christian Ausoni, Kathryn Hess, Brenda Johnson, Ieke Moerdijk, and Jérôme Scherer, Editors,** An Alpine Bouquet of Algebraic Topology, 2018

707 **Nitya Kitchloo, Mona Merling, Jack Morava, Emily Riehl, and W. Stephen Wilson, Editors,** New Directions in Homotopy Theory, 2018

706 **Yeonhyang Kim, Sivaram K. Narayan, Gabriel Picioroaga, and Eric S. Weber, Editors,** Frames and Harmonic Analysis, 2018

705 **Graham J. Leuschke, Frauke Bleher, Ralf Schiffler, and Dan Zacharia, Editors,** Representations of Algebras, 2018

704 **Alain Escassut, Cristina Perez-Garcia, and Khodr Shamseddine, Editors,** Advances in Ultrametric Analysis, 2018

703 **Andreas Malmendier and Tony Shaska, Editors,** Higher Genus Curves in Mathematical Physics and Arithmetic Geometry, 2018

702 **Mark Grant, Gregory Lupton, and Lucile Vandembroucq, Editors,** Topological Complexity and Related Topics, 2018

701 **Joan-Carles Lario and V. Kumar Murty, Editors,** Number Theory Related to Modular Curves, 2018

700 **Alexandre Girouard, Dmitry Jakobson, Michael Levitin, Nilima Nigam, Iosif Polterovich, and Frédéric Rochon, Editors,** Geometric and Computational Spectral Theory, 2017

699 **Mark L. Agranovsky, Matania Ben-Artzi, Catherine Bénéteau, Lavi Karp, Dmitry Khavinson, Simeon Reich, David Shoikhet, Gilbert Weinstein, and Lawrence Zalcman, Editors,** Complex Analysis and Dynamical Systems VII, 2017

698 **Alexander M. Blokh, Leonid A. Bunimovich, Paul H. Jung, Lex G. Oversteegen, and Yakov G. Sinai, Editors,** Dynamical Systems, Ergodic Theory, and Probability: in Memory of Kolya Chernov, 2017

697 **Fabrizio Broglia, Françoise Delon, Max Dickmann, Danielle Gondard-Cozette, and Victoria Ann Powers, Editors,** Ordered Algebraic Structures and Related Topics, 2017

696 **Ara S. Basmajian, Yair N. Minsky, and Alan W. Reid, Editors,** In the Tradition of Ahlfors–Bers, VII, 2017

695 **Katrina Barron, Elizabeth Jurisich, Antun Milas, and Kailash Misra, Editors,** Lie Algebras, Vertex Operator Algebras, and Related Topics, 2017

694 **Manjul Bhargava, Robert Guralnick, Gerhard Hiss, Klaus Lux, and Pham Huu Tiep, Editors,** Finite Simple Groups: Thirty Years of the Atlas and Beyond, 2017

693 **Michael Cwikel and Mario Milman, Editors,** Functional Analysis, Harmonic Analysis, and Image Processing, 2017

692 **Anatole Katok, Yakov Pesin, and Federico Rodriguez Hertz, Editors,** Modern Theory of Dynamical Systems, 2017

691 **Farrell Brumley, Maria Paula Gómez Aparicio, and Alberto Mínguez, Editors,** Around Langlands Correspondences, 2017

690 **Andrés Eduardo Caicedo, James Cummings, Peter Koellner, and Paul B. Larson, Editors,** Foundations of Mathematics, 2017

For a complete list of titles in this series, visit the
AMS Bookstore at **www.ams.org/bookstore/conmseries/**.